Intermediates for

ORGANIC SYNTHESIS

Intermediates for
ORGANIC SYNTHESIS

V. K. AHLUWALIA
Visiting Professor
Dr. B.R. Ambedkar Centre for Biomedical Research
University of Delhi

POOJA BHAGAT
Reader
Acharya Narendra Dev College
University of Delhi

RENU AGGARWAL
Reader
Ctargi College
University of Delhi

RAMESH CHANDRA
Director
Dr. B. R. Ambedkar Centre for Biomedical Research
Univerisity of Delhi

TECHSAR™

NEW DELHI

Published and Printed by
TechSar Pvt. Ltd.
4435-36/7, Ansari Road, Daryaganj,
New Delhi 110 002 (India)
E-mail: tech_info@techsarworld.com
Website: www.techsarworld.com

ISBN 978-81-88237-33-3

Reprint 2025

Preface

The present book is designed to make available procedures for the synthesis of various intermediates, which are generally required by organic chemists working in the universities, industries and by the research scholars at different levels. It is observed that no single publication is available describing the intermediates required for organic synthesis. One has to refer to literature for most of the intermediates. It is felt that at times the publications are not easily available: this hinders the progress of the work.

The intermediates described in this book include different types of phenols, aldehydes, carboxylic acids and ketones (acetophenones, ω-substituted acetophenones, propiophenones, butyrophenones, benzophenones, phenyl ketones and some miscellaneous ketones). The preparation of heterocyclic compounds (O-containing, S-containing, N-containing, N & S-containing) is also described. The synthesis of certain miscellaneous compounds of the type benzyl cyanides, β-ketoesters, chalcones, naphthaquinones, benzoquinones, stilbene and certain catalysts and reagents required for organic synthesis are also described.

Attempts have been made to describe the best possible procedures with ample experimental details keeping in mind the maximum yield. However, it is not claimed that the methods are completely perfect. All the procedures described have been verified by the authors and their associates.

Authors

Contents

Preface *v*

1. PHENOLS **1**

1.1 o-Methoxyphenol (Catechol Monomethyl Ether) (Guaiacol) 5
1.2 1,2–Dimethoxybenzene (Catechol Dimethyl Ether) (Veratrole) 6
1.3 3-methoxyphenol (Resorcinol Monomethyl Ether) 6
1.4 1,3-dimethoxybenzene (Resorcinol Dimethyl Ether) 7
1.5 4-methoxyphenol (Hydroquinone Monomethyl Ether) 7
1.6 1,4-dimethoxybenzene (Hydroquinone Dimethyl Ether) 8
1.7 2,3 – Dihydroxyanisole (Pyrogallol Monomethyl Ether) 9
1.8 2,3- Dimethoxyphenol (Pyragallol-2,3–Dimethyl Ether) 10
1.9 1,2,3 – Trimethoxybenzene (Pyrogallol Trimethyl Ether) 11
1.10 2,4–Dihydroxyanisole (4- Methoxy Resorcinol) 11
1.11 3,4- Dimethoxyphenol 12
1.12 1,3-dihydroxy-5-methoxybenzene (Phloroglucinol Monomethyl Ether) 13
1.13 3,5-dimethoxyphenol (Phloroglucinol Dimethyl Ether) 14
1.14 1,3,5- Trimethoxybenzene (Phloroglucinol Trimethyl Ether) 15
1.15 3-methoxy-5-methylphenol (Orcinol Monomethyl Ether) 15
1.16 3,5–Dimethoxytoluene (Orcinol Dimethyl Ether) 17

1.17	2,6-dihydroxytoluene (2- Methyl Resorcinol)	17
1.18	2,4–Dihydroxytoluene (4–Methyl Resorcinol)	18
1.19	2,4- Dihydroxy-1-ethylbenzene	19
1.20	P- Benzoquinone	19
1.21	1,2,4-triacetoxybenzene (Hydroxyhydroquinone Triacetate)	20
1.22	1,2,4- Trimethoxybenzene	20
1.23	1,2,4-trihydroxybenzene (Hydroxyquinol)	21
1.24	2-methoxy-1,4-dihydroxybenzene (o-methoxyhydroquinone)	21
1.25	2-methoxy-p-benzoquinone	22
1.26	2,3- Dimethoxy-p-benzoquinone	22
1.27	2,5- Dimethoxy-1,4-benzoquinone	23
1.28	3,4,5-trihydroxytoluene (5-Methyl Pyrogallol)	23
1.29	2-benzyl-5-methoxy-3-methylphenol	23
1.30	2-benzyl-3,5-dimethoxyphenol	24
1.31	2,4,6-trihydroxytoluene (C-methyl Phloroglucinol)	25
1.32	2,6-dimethoxy-4-hydroxytoluene	25
1.33	2,4,6-trimethoxytoluene (c-methyl Phloroglucinol Trimethyl Ether)	26
1.34	1,3-dihydroxy-2,5-dimethoxybenzene	26
1.35	1,2-dihydroxy-3,5-dimethoxybenzene	27
1.36	1,2,3,4-tetramethoxybenzene	27
1.37	1,2,4-triacetoxynaphthalene	28
1.38	3,4- Methylenedioxyphenol	29
1.39	6-allylguaiacol (Eugenol)	29
1.40	Methoxyphenols: General Method of Preparation	30
	References	31

2. ALDEHYDES **34**

2.1	o-methoxybenzaldehyde	35
2.2	p-methoxybenzaldehyde (Anisaldehyde)	36
2.3	m-hydroxybenzaldehyde	36
2.4	m-methoxybenzaldehyde	38
2.5	2,4-dihydroyxbenzaldehyde (β-resorcylic Aldehyde)	38

2.6 2-hydroxy-4-methoxybenzaldehyde 39
2.7 2,4-dimethoxybenzaldehyde 39
2.8 2,5 Dihydroxybenzaldehyde (Gentisic Aldehyde) 40
2.9 2,5-dimethoxybenzaldehyde 41
2.10 2,6 Dihydroxybenzaldehyde (γ-resorcylic Aldehyde) 42
2.11 3,4-dimethoxybenzaldehyde (Veratraldehyde) 43
2.12 3,4 Dihydroxybenzaldehyde (Protocatechualdehyde) 43
2.13 4,5-dihydroxy-3-methoxybenzaldehyde 44
2.14 2,4,6-trihydroxybenzaldehyde (Phloroglucinaldehyde) 45
2.15 4-hydroxy-2,6-dimethoxybenzaldehyde and 45
2.16 2-hydroxy-4,6-dimethoxybenzaldehyde 45
2.17 2,4-dihydroxy-6-methoxybenzaldehyde 46
2.18 2,4,6-trimethoxybenzaldehyde 47
2.19 2,4,6-trimethylbenzaldehyde (Mesitaldehyde) 47
2.20 2,3,4-trihydroxybenzaldehyde (Pyrogallol Aldehyde) 48
2.21 2,3,4-trimethoxybenzaldehyde 49
2.22 2-hydroxy-3,4-dimethoxybenzaldehyde 49
2.23 2,4,5-trihydroxybenzaldehyde 50
2.24 2-hydroxy-4,5-dimethoxybenzaldehyde 51
2.25 2,4,5-trimethoxybenzaldehyde (Asarlaldehyde) 51
2.26 3,4,5-trimethoxybenzaldehyde (Tri-o-methylgallaldehyde) 52
2.27 2,4-dihydroxy-3,6-dimethoxybenzaldehyde 53
2.28 2,4-dihydroxy-6-methylbenzaldehyde (β-orcinaldehyde) 54
2.29 2-hydroxy-4-methoxy-6-methylbenzaldehyde (Everninaldehyde) 54
2.30 2,4-dihydroxy-5-methylbenzaldehyde 55
2.31 2,4-dihydroxy-3-methylbenzaldehyde 55
2.32 2-hydroxy-4-methoxy-3-methylbenzaldehyde 56
2.33 2,5-dihydroxy-3-methoxybenzaldehyde 57
2.34 2,4,5-trihydroxy-6-methylbenzaldehyde 57
2.35 2,5-dihydroxy-4,6-dimethoxybenzaldehyde 58
2.36 2,5-dihydroxy-4-methoxy-6-methylbenzaldehyde 59

2.37	2-methoxy-4,5-methylenedioxybenzaldehyde	59
2.38	2,3,45- Tetramethoxybenzaldehyde	61
2.39	2-hydroxy-3,4,5,6-tetramethoxybenzaldehyde	62
2.40	2-methoxy-3-nitrobenzaldehyde	63
2.41	2-methoxy-5-nitrobenzaldehyde	63
2.42	3,5-dinitrosalicylaldehyde	64
2.43	2,4-dihydroxy-5-nitrobenzaldehyde (5-nitroresorcylic Aldehyde)	65
2.44	2-hydroxy-3-methoxy-5-nitrobenzaldehyde (5-nitro-o-vanillin)	65
2.45	2-hydroxy-3-methoxy-4-nitrobenzaldehyde (4-nitro-o-vanillin)	66
2.46	3,4-dimethoxy-6-nitrobenzaldehyde (6-nitroveratraldehyde)	66
	References	66

3. CARBOXYLIC ACIDS **70**

3.1	o-methoxybenzoic Acid	74
3.2	p-hydroxybenzoic Acid	75
3.3	2-hydroxy-3-methoxybenzoic Acid (o-vanillic Acid)	75
3.4	2,3-dihydroxybenzoic Acid	76
3.5	2,3-dimethoxybenzoic Acid	76
3.6	2,4-dihydroxybenzoic Acid (β-resorcyclic Acid)	77
3.7	Methyl 2,4-dihydroxybenzoate	77
3.8	2-hydroxy-4-methoxybenzoic Acid	78
3.9	2,5-dihydroxybenzoic Acid (Gentisic Acid)	78
3.10	2,5-dimethoxybenzoic Acid	79
3.11	2-hydroxy-5-methoxybenzoic Acid	79
3.12	2-hydroxy-6-methoxybenzoic Acid	80
3.13	3-methoxy-4-hydroxybenzoic Acid (Vanillic Acid)	80
3.14	3,4-dihydroxybenzoic Acid (Protocatechuic Acid)	81
3.15	Methyl 3,4-dihydroxybenzoate (Methyl Protocatechuate)	82
3.16	3-hydroxy-4-methoxybenzoic Acid (Isovanillic Acid)	82
3.17	3,4-dimethoxybenzoic Acid (Veratric Acid)	83

3.18 3,4-dimethoxy-6-nitrobenzoic Acid (6-nitro Veratric Acid) 83
3.19 3,4-methylenedioxybenzoic Acid (Piperonylic Acid) 84
3.20 3,5-dimethoxybenzoic Acid 84
3.21 3,5-dihydroxybenzoic Acid (α-resorcylic Acid) 85
3.22 Methyl 3,5-dimethoxybenzoate 86
3.23 2,3,4-trihydroxybenzoic Acid (Pyrogallol Carboxylic Acid) 86
3.24 Methyl 2,3,4-trihydroxy Benzoate 87
3.25 2-hydroxy-3,4-dimethoxybenzoic Acid 87
3.26 3,4,5-trimethoxybenzoic Acid (Tri-o-methyl Gallic Acid) 88
3.27 Methyl 3,4,5-trihydroxybenzoate (Methyl Gallate) 88
3.28 Methyl 3,4,5-trimethoxybenzoate 89
3.29 Methyl 3,4-dimethoxy-5-hydroxybenzoate[30] (Methyl 3,4 Di-o-methylgallate) 89
3.30 Methyl 3-methoxy-4,5-dihydroxybenzoate (Methyl 3-o-methyl Gallate) 90
3.31 3,5-dimethoxy-4-hydroxybenzoic Acid (Syringic Acid) 90
3.32 3,4-dimethoxy-5-hydroxybenzoic Acid (3,4-di-o-methyl Gallic Acid 91
3.33 2,4,6-trihydroxybenzoic Acid (Phloroglucinol Carboxylic Acid) 92
3.34 2-hydroxy-4,6-dimethoxybenzoic Acid 92
3.35 2,5-dihydroxy-4-methoxybenzoic Acid 93
3.36 2,4,5-trimethoxybenzoic Acid 93
3.37 2,4,5-trihydroxybenzoic Acid 94
3.38 2,6-dihydroxy-4-methylbenzoic Acid (p-orcellinic Acid) 95
3.39 2,4-dihydroxy-3-methylbenzoic Acid 95
3.40 2,4-dihydroxy-5-methoxybenzoic Acid 96
3.41 2,5-dihydroxy-4-methoxy-3-methylbenzoic Acid 98
3.42 2-hydroxy-3,4,5-trimethoxybenzoic Acid[48] 99

3.43 p-methoxophenylglyoxalic Acid ($p\text{-}OCH_3C_6H_4COCOOH$) 99

3.44 β-benzoylpropionic Acid ($C_6H_5COCH_2CH_2COOH$) 100

3.45 4-phenylbutyric Acid ($C_6H_5CH_2CH_2CH_2COOH$) 100

3.46 p-methoxyphenylacetic Acid 101

3.47 p-nitrophenylacetic Acid 101

3.48 2,4,Dinitrophenylacetic Acid 102

3.49 Cinnamic Acid 102

3.50 3-phenylpropionic Acid (Hydrocinnamic Acid) 103

3.51 Methyl Cinnamate 103

3.52 P-hydroxycinnamic Acid 104

References 104

4. KETONES 108

4.A.1 o-hydroxyacetophenone 112

4.A.2 o-methoxyacetophenone 113

4.A.3 p-hydroxyacetophenone 113

4.A.4 p-methoxyacetophenone 113

4.A.5 m-hydroxyacetophenone 114

4.A.6 2,4-dihydroxyacetophenone (β-resacetophenone) 115

4.A.7 2-hydroxy-4-methoxyacetophenone (Peanol) 116

4.A.8 2,4-dimethoxyacetophenone 116

4.A.9 2,4-dihydroxy-5-nitroacetophenone (5-nitro-β-resacetophenone) 117

4.A.10 2,5-dihydroxyacetophenone (Quinacetophenone) 117

4.A.11 2-hydroxy-5-methoxyacetophenone (Quinacetophenone Monomethyl Ether) 118

4.A.12 2,5-dimethoxyacetophenone (Quinacetophenone Dimethyl Ether) 119

4.A.13 2,3-dimethoxyacetophenone 119

4.A.13A 5-acetyl-2,4-dihydroxyacetophenone (1,5-diacetyl-2,4-dihydroxybenzene) 120

4.A.14 2-hydroxy-3-methoxyacetophenone 120

4.A.15 2,3-dihydroxyacetophenone 121

4.A.16 2,6-dihydroxyacetophenone (Resacetophenone) 121

4.A.17 2-hydroxy-6-methoxyacetophenone (γ-resacetophenone Monomethyl Ether) 123

4.A.18 2,6-dimethoxyacetophenone (γ-resacetophenone Dimethyl Ether) 123

4.A.19 3,4-dimethoxyacetophenone 124

4.A.20 3-hydroxy-4-methoxyacetophenone and 125

4.A.21 4-hydroxy-3-methoxyacetophenone 125

4.A.22 3,4-dihydroxyacetophenone 126

4.A.23 2,5-dichloroacetophenone 126

4.A.24 2,5-dibromoacetophenone 127

4.A.25 2,4-dihydroxy-3-methylacetophenone 127

4.A.26 2-hydroxy-4-methoxy-3-methylacetophenone 128

4.A.27 2,4-dihydroxy-5-methylacetophenone 128

4.A.28 2,4-dihydroxy-6-methylacetophenone (β-orcacetophenone) 129

4.A.29 2-hydroxy-4-methoxy-6-ethylacetophenone (β-orcacetophenone 4-0-methyl Ether) 129

4.A.30 2,4-dimethoxy-6-methylacetophenone (β-orcacetophenone Dimethyl Ether) 130

4.A.31 2,6-dihydroxy-4-methylacetophenone (γ-orcacetophenone) 130

4.A.32 2-hydroxy-6-methoxyacetophenone (γ-orcacetophenone-6-0-methyl Ether) 131

4.A.33 2,6-dimethoxy-4-methylacetophenone (γ-orcacetophenone Dimethyl Ether) 132

4.A.34 5-acetyl-2,4-dihydroxy-3-methylacetophenone (1,5-diacetyl-2,4- Dihydroxy-3-methylbenzene) 133

4.A.35 2,3,4-trithydroxyacetophenone (Gallacetophenone) 133

4.A.36 2,3-dihydroxy-4-methoxyacetophenone (Gallacetophenone-4-0-methyl Ether) 134

4.A.37 2-hydroxy-3,4-dimethoxyacetophenone (Gallacetophenone Dimethyl Ether) 134

4.A.38 2,3,4-trimethoxyacetophenone (Gallacetophenone Trimethyl Ether) 136

4.A.39 2,3,6-trimethoxyacetophenone 136

4.A.40 2-hydroxy-3,6-dimethoxyacetophenone 137
4.A.41 2-hydroxy-4,5-dimethoxyacetophenone 137
4.A.42 2,4,5-trimethoxyacetophenone 138
4.A.43 2,5-dihydroxy-4-methoxyacetophenone 138
4.A.44 2,4,6-trihydroxyacetophenone (Phloroacetophenone) 139
4.A.45 2-hydroxy-4,6-dimethoxyacetophenone (Phloroacetophenone Dimethyl Ether) 140
4.A.46 2,4,6-trimethoxyacetophenone (Phloroacetophenone Trimethyl Ether) 141
4.A.47 2,4-dihydroxy-6-methoxyacetophenone (Phloroacetophenone-6-0-methyl Ether) 141
4.A.48 2,6-dihydroxy-4-methoxyacetophenone (Phloroacetophenone-4-0- Methyl Ether) 142
4.A.49 2,5-dihydroxy-6-methoxyacetophenone 143
4.A.50 2-hydroxy-5,6-dimethoxyacetophenone 144
4.A.51 2,5-dihydroxy-3,4-dimethoxyacetophenone 144
4.A.52 2-hydroxy-3,4,5-tetrimethoxyacetophenone 145
4.A.53 2,3,4,5-tetramethoxyacetophenone 145
4.A.54 2-hydroxy-3,4,6-trimethoxyacetophenone 146
4.A.55 2,3,4,6-tetramethoxyacetophenone 148
4.A.56 2,4-dihydroxy-3,6-dimethoxyacetophenone 148
4.A.57 2,5-dihydroxy-4,6-dimethoxyacetophenone 149
4.A.58 2,5-dihydroxy-3,4,6-trimethoxyacetophenone 149
4.A.59 2-hydroxy-4,5,6-trimethoxyacetophenone 150
4.A.60 2-hydroxy-3,4,5,6-tetramethoxyacetophenone 152
4.A.61 2,3,4,5,6-pentamethoxyacetophenone 152
4.A.62 2-hydroxy-3-methyl-4,6-dimethoxyacetophenone (3-c-methyl-2-hydroxy-4,6-dimethoxy Acetophnone) (C-methylphloroacetophenone Dimethyl Ether) 153
4.A.63 2-hydroxy-3,4-dimethoxy-6-methylacetophenone 155
4.A.64 2,4,6-trihydroxy-3-methylacetophenone (3-c-methylphloroacetophenone) 156
4.B.65 ω-methoxyacetophenone ($C_6H_5COCH_2OCH_3$) 157
4.B.66 2,4-dihydroxy-ω-methoxyacetophenone (ω-methoxyresacetophenone) 158

4.B.67 2-hydroxy-ω,4-dimethoxyacetophenone (ω-methoxyresacetophenone-4-0-Methyl Ether) 159
4.B.68 ω,2,4-trihydroxyacetophenone (Fisetol) 160
4.B.69 ω,2-dihydroxy-4-methoxyacetophenone 160
4.B.70 2,4,6-trihydroxy-ω-methoxyacetophenone (ω-methoxyphloroacetophenone) 161
4.B.71 2-hydroxy-ω,4,6-trimethoxyacetophenone (ω-methoxyphloroacetophenone-4,6-dimethyl Ether) 161
4.B.72 2,4-dihydroxy-ω,6-dimethoxyacetophenone[88] (ω-methoxyphloroacetophenone-6-0-Methyl Ether) 162
4.B.73 ω,2-dihydroxy-4,6-dimethoxyacetophenone 163
4.B.74 ω-hydroxy-2,4,6-trimethoxyacetophenone 163
4.B.75 2,4-dihydroxy-6-methyl-ω-methoxyacetophenone[89] (ω-methoxy-orcacetophenone) 164
4.B.76 2,5-dihydroxy-ω,4-dimethoxyacetophenone 164
4.B.77 2-hydroxy-ω,4,5-trimethoxyacetophenone 165
4.B.78 2,5-dihydroxy-ω,4,6-trimethoxyacetophenone 165
4.B.79 2-hydroxy-ω,4,5,6-tetramethoxyacetophenone 166
4.B.80 2,4- Dihydroxy-ω,3,6-trimethoxyacetophenone 167
4.B.81 2-hydroxy-ω,3,4,6-tetramethoxyacetophenone 169
4.B.82 2,5-dihydroxy-ω,3,4,6-tetramethoxyacetophenone 169
4.B.83 2-hydroxy-ω-3,4,5,6-pentamethoxyacetophenone 170
4.B.84 2,4,6-trihydroxy-ω-ethoxyacetophenone (ω-ethoxyphloro-acetophenone) 171
4.B.85 2,4-dihydroxy-ω-benzoyloxyacetophenone (ω-benzoyloxy Resactophenone) 172
4.B.86 2,4,6-trihydroxy-ω-benzoyloxyacetophenone (ω-benzoyloxyphloroacetophenone) 173
4.B.87 2,4-dihydroxy-ω-acetoxyacetophenone (ω-acetoxyresacetophenone) 174
4.B.88 2-hydroxy-4-methoxy-ω-nitroacetophenone 175
4.B.89 2,4-dihydroxy-ω-chloroacetophenone (ω-chlororesacetophenone) 175
4.B.90 2-hydroxy-4-methoxy-ω-chloroacetophenone (ω-chlororesacetophenone-4-o-Methyl Ether) 176

4.B.91	2,4,6-trihydroxy-ω-chloroacetophenone (ω-chlorophloroacetophenone)	176
4.B.92	2-hydroxy-4,6-dimethoxy-ω-chloroacetophenone (ω-chloro-phloroacetophenone Dimethyl Ether)	177
4.C.93	2-4-dihydroxypropiophenone (Respropiophenone)	177
4.C.94	2-hydroxy-4-methoxypropiophenone	178
4.C.95	2-hydroxy-3-dimethoxypropiophenone[110]	178
4.C.96	2,4,6-trihydroxypropiophenone (Phloropropiophenone)	179
4.C.97	2-hydroxy-4,6-dimethoxypropiophenone	180
4.C.98	2,4-dihydroxy-6-methylpropiophenone	180
4.C.99	2-hydroxy-4-methoxy-6-methylpropiophenone[114]	181
4.C.100	2,4-dimethoxy-6-methylpropiophenone	181
4.D.101	2-hydroxy-4-methoxybutyrophenone[110]	182
4.D.102	2-hydroxy-3,4-dimethoxybutyrophenone	182
4.D.103	2-hydroxy-4,6-dimethoxybutyrophenone[110]	183
4.D.104	2,4-dihydroxy-ω-phenylbutyrophenone	184
4.E.105	p-hydroxybenzophenone	185
4.E.106	2,4-dihydroxybenzophenone (Resbenzophenone)	186
4.E.107	2-hydroxy-4-methoxybenzophenone	186
4.E.108	2,6 Dihydroxybenzophenone	187
4.E.109	2,5-dihydroxy-4-methoxybenzophenone	188
4.E.110	2-hydroxy-4,5-dimethoxybenzophenone	189
4.E.111	2,4-dihydroxy-3-methylbenzophenone	190
4.E.112	2,4,6-trihydroxybenzophenone (Phlorobenzophenone)	191
4.E.113	2-hydroxy-4,6-dimethoxybenzophenone	191
4.E.114	2,4-dihydroxy-6-methoxybenzophenone[133]	193
4.E.115	2,4-dihydroxy-6-methylbenzophenone[132]	193
4.E.116	2-hydroxy-4-methoxy-6-methylbenzophenone[132]	194
4.F.117	2,4-dihydroxyphenylbenzyl Ketone (2,4-dihydroxydeoxybenzoin)	194
4.F.118	2-hydroxy-4-methoxyphenylbenzyl Ketone	196
4.F.119	2,4,6-trihydroxyphenyl Benzyl Ketone	196

4.F.120 2-hydroxy-4,6-dimethoxyphenyl Benzyl Ketone 197
4.F.121 2,4-dihydroxy-6-methoxyphenyl Benzyl Ketone 197
4.G.122 2,4,6-trihydroxyisobutyrophenone (Phloroisobutyrophenone) 198
4.G.123 2,4,6-trihydroxyisovalerophenone (Phloroisovalerophenone) 199
4.G.124 Bis(2,4-dihydroxy-3-acetophenyl) Methane R-CH_2-R (R=2,4-dihydroxy-3-acetylphenol) 199
4.G.125 5,5-dimethyl-1,3-cyclohexanedione (Dimedone) 200
References 200

5. HETEROCYCLIC COMPOUNDS 207

5.A.1 7-hydroxycoumarin (Umbelliferone) 211
5.A.2 6-hydroxycoumarin 212
5.A.3 5,7-dihydroxycoumarin 212
5.A.4 5,7-dimethoxycoumarin 213
5.A.5 6,7-dihydroxycoumarin (Aesculetin) 213
5.A.6 7,8-dihydroxycoumarin (Daphnetin) 214
5.A.7 3-hydroxycoumarin 214
5.A.8 3-hydroxy-7-methoxycoumarin 215
5.A.9 4-hyroxycoumarin 216
5.A.10 4-hydroxy-7-methoxycoumarin 217
5.A.11 4,7-dihydroxycoumarin 218
5.A.12 7-hydroxy-4-methoxycoumarin[11] 219
5.A.13 4,5,7-trihydroxycoumarin 219
5.A.14 4-hydroxy-5,7-dimethoxycoumarin 220
5.A.15 7-hydroxy-3-methylcoumarin (3-methylumbelliferone) 221
5.A.16 7-hydroxy-4-methylcoumarin (4-methylumbelliferone) 222
5.A.17 5-methoxy-4-methylcoumarin15 (Wittig Reaction) 223
5.A.18 7-hydroxy-4-phenylcoumarin (4-phenylumbelliferone) 223
5.A.19 7-methoxy-4-phenylcoumarin 224
5.A.20 5,7-dihydroxy-4-phenylcoumarin 224

5.A.21	5,7-dimethoxy-4-phenylcoumarin	225
5.A.22	7-hydroxy-3-phenylcoumarin (3-phenylumbelliferone)	225
5.A.23	4,7-dihydroxy-3-phenylcoumarin	226
5.A.24	4-hydroxy-7-methoxy-3-phenylcoumarin	227
5.A.25	3,7-dimethoxy-4-hydroxycoumarin[22]	227
5.A.26	3-methoxy-4,7-dihydroxycoumarin[22]	228
5.A.27	7-hydroxy-3,4-dimethoxycoumarin[22]	228
5.A.28	7-hydroxy-3,4-dimethylcoumarin	229
5.A.29	7-methoxy-3,4-diphenylcoumarin[24]	230
5.A.30	Benzofuran (Coumarone)	230
5.A.31	6-hydroxycoumaran-3-one	231
5.A.32	4,6-dimethoxycoumaran-3-one	232
5.A.33	2-methyl Chromones 7,8-dihydroxy-2-methylchromone	232
5.A.34	7-hydroxy-2-methylchromone (K.R. Acetylation)	234
5.A.35	5,7-dihydroxy-2-methylchromone	234
5.A.36	2-methyl-3-methoxy-5,7-dihydroxychromone	235
5.A.37	Flavone	236
5.A.38	7-hydroxy Flavone	237
5.A.39	6-hydroxyflavone	239
5.A.40	5,7-dihydroxyflavone	240
5.A.41	3-methoxy-7-hydroxyflavone	241
5.A.42	7-methoxy-3-hydroxyflavone (7-methoxyflavonol)	241
5.A.43	5,7-dimethoxyflavanone	242
5.A.44	7-hydroxyisoflavone (Farkas Method)	243
5.A.45	7,4'-dihydroxyisoflavone (Diadzin)	244
5.A.46	5,7-dihydroxyisoflavone	245
5.A.47	2-methyl-7-hydroxyisoflavone	247
5.A.48	1-hydroxyxanthone	248
5.A.49	1,3-dihydroxyxanthone	248
5.A.50	1-hydroxy-3-methoxyxanthone	249
5.A.51	1,3-dihydroxy-5-methoxyxanthone	249
5.A.52	1,3-dihydroxy-6-methoxyxanthone	250

5.A.53	1,3-dihydroxy-7-methoxyxanthone	250
5.A.54	2,5-dimethylthiophen	251
5.C.55	2-aminopyridine	252
5.C.56	Pyridine-2-carboxylic Acid (Picolinic Acid)	252
5.C.57	Pyridine-3-carboxylic Acid (Nicotinic Acid)	253
5.C.58	Pyridine-4-carboxylic Acid (Isonicotinic Acid)	254
5.C.59	6-hydroxynicotinic Acid	254
5.C.60	Quinoline	256
5.C.61	2-methylquinoline (Quinaldine)	256
5.C.62	8-hydroxyquinoline (Oxine)	257
5.C.63	8-nitroquinoline	258
5.C.64	5,8-dimethoxyquinoline	258
5.C.65	5,8-dimethoxy-2-methylquinoline	260
5.C.66	2,4-dimethyl-5,8-dimethoxyquinoline	260
5.C.67	2,4,6-trimethylquinoline	261
5.C.68	2-hydroxy-4-methylquinoline (4-methylcarbostyril)	262
5.C.69	2-phenylquinoline-4-carboxylic Acid (Atophan)	262
5.C.70	Quinoline-2-carboxylic Acid (Quinaldinic Acid)	263
5.C.71	Quinoline-5-carboxylic Acid	264
5.C.72	5,8-dimethoxyisoquinoline	264
5.C.72A	5,8-dimethoxy-1-methylisoquinoline	266
5.C.73	2,3-diphenylquinoxaline	266
5.C.74	Piperazine-2,5-dione (Diketopiperazine)	267
5.C.75	2,5-dimethylpyrrole	267
5.C.76	3,5-dimethylpyrrole	268
5.C.77	Pyrazolones (Pyrazolin-5-ones)	269
5.C.78	4-ethoxymethylene-3-methyl-1-phenyl-2-pyrazolin-5-one	270
5.C.79	5-chloro-3-methyl-1-phenyl-1h-Pyrazole-4-carboxaldehyde	270
5.C.80	2,3-dimethyl-1-phenyl-3-pyrazolin-5-One (Antipyrin)	270
5.C.81	Benzimidazole	271
5.C.82	2-methyl Benzimidazole	271

5.C.83	Indole	272
5.C.84	2-phenylindole	272
5.C.85	Benzotriazole	273
5.D.86	2,4-dimethylthiazole	273
5.D.87	2-aminothiazole	274
5.D.88	2-amino-4-methylthiazole	275
5.D.89	4-phenyl Subsituted-2-aminothiazoles	275
5.D.90	2-aminobenzothiazole	277
5.D.91	4-phenylsubstituted 2-mercaptothiazoles	279
5.D.92	2-hydroxy Pyrimidine	280
5.D.93	Barbituricacid	280
5.D.94	Aminobarbituric Acid (Uramil)	281
5.D.95	1,3-diarylthiobarbituric Acids (1,3-diaryl-1,2,3,4,5,6-hexahydro-4,6-dioxo-2-thioxopyrimidine	282
5.D.96	1,2,3,4,7-pentahydro-5,7,7-trimethyl-4-oxo-1,3-diphenyl-2-thioxopyrano [2,3-d] Pyrimidine123	285
5.D.97	1,3-bis(4-chlorophenyl)-6-chloro-5-formyl-2-thiobarbituric Acid18 (1,3-bis (4-chlorophenyl)-5-carboxaldehyde-6-chloro-1,3-dihydro-4-oxo-2-thioxopyrimidine	286
5.D.98	5-ethoxymethylene-1,3-bis (2-methoxyphenyl)-2-thiobabituric Acid	286
	References	287

6. MISCELLANEOUS COMPOUNDS **292**

6.A.1	Aluminium Isopropoxide	292
6.A.2	Aluminium Tert-butoxide	292
6.A.3	2-aminobenzenethiols	293
6.A.4	P-anisyl Alcohol	294
6.A.5	Benzal Chloride5 ($C_6H_5CHCL_2$) (Benzylidene Chloride)	294
6.A.6	Benzoyl Peroxide	295
6.A.7	1-bromo-3-methylbut-2-ene (Prenyl Bromide)	295
6.A.7A	3-bromo-3-methyl-1-butyne	296
6.A.8	2-bromocyclohexanone	297

6.A.9 N-bromosuccinimide 297
6.A.9A n-butyl Lithium 297
6.A.10 2-carbethoxycyclopentanone 298
6.A.11 1-carbethoxyethylidenetriphenyl Phosphorane 298
6.A.12 Chloroacetonitrile 299
6.A.13 3-chloro-3-methylbut-1-yne 300
6.A.14 M-chloroperbenzoic Acid 300
6.A.15 P-chlorophenylmercuryl Chloride[18] 301
6.A.16 Cinnamyl Bromide 301
6.A.17 Diazomethane 302
6.A.18 2,3-dichloro-5,6-dicyanobenzoquinone (DDQ) 303
6.A.18A o,o-diethylchlorothiophosphonate 304
6.A.18B 2,4-dimethoxybenzonitrile 305
6.A.19 Diphenyl Carbinol 306
6.A.20 Ethoxycarbonylmethylenetriphenyl-phosphorane 306
6.A.21 Ethoxyacetyl Chloride 307
6.A.22 Ethoxymethyl Chloride 308
6.A.23 Ethyl Cyanoacetate 308
6.A.24 Triethyl Orthoformate, $CH(OC_2H_5)_3$ 309
6.A.25 o- And p-hydroxyphenylmercuryl Chloride 309
6.A.25A Jones Reagent[33A] 310
6.A.25B Lead Tetraacetate 310
6.A.26 Active Manganese Dioxide 311
6.A.27 Methyl Iodide 311
6.A.28 Methyl Thioglycollate 311
6.A.29 Palladium Catalysts 312
6.A.30 Peracids 314
6.A.31 Phenylhydrazine 315
6.A.32 Subsituted Phenylhydrazine 316
6.A.33 Phenyl Mercuryl Chloride (PHHGCL) 320
6.A.34 Pyridinium Bromide[49] 320
6.A.34A Pyridinium Bromide Perbromide 320
6.A.34B Pyridinium Chloro Chromate[51a] 321

6.A.34C	Pyridinium Chlorochromate Adsorbed on Alumina[51b]	321
6.A.35	Raney Nickel	321
6.A.35A	Sodamide	322
6.A.35B	Sodium Amalgam (Contains 3% of Sodium)	322
6.A.35C	Tetrabutylammonium Chromate[53a]	323
6.A.36	p-toluenesulfonylhydrazide ($p-CH_3C_6H_4SO_2NHNH_2$)	323
6.A.36A	Tri-n-butyltin Hydride[54a]	324
6.B.37	p-methoxybenzyl Cyanide (p-methoxyphenylacetonitriles)	324
6.B.38	2,4-dimethoxy Benzyl Cyanide	325
6.B.39	3,4- Methylenedioxy Benzyl Cyanide	328
6.B.40	2,4,5-trimethoxy Benzyl Cyanide	329
6.C.41	Ethyl Benzoyl Acetate	331
6.C.42	Ethyl O-methoxybenzoyl Acetate (o-methoxybenzoyl Acetic Ester) (General Method of Preparation)	333
6.C.43	Ethyl Propionyl Acetate[68]	334
6.D.44	Chalcone (Benzylideneacetophenone) $C_6H_5CH=CHCOC_6H_5$	335
6.D.45	Benzylidene Acetone (4-phenylbut-3-en-2-one) $C_6H_5CH=CHCOCH_3$ (Benzal Acetone)	336
6.D.46	4'-methoxybenzylidene Acetone [4-(4'-methoxyphenyl)But-3-en-2-one)] $p-OCH_3C_6H_4CH=CHCOCH_3$	336
6.E.47	1,2-naphthoquinone	337
6.E.48	1,4-naphthoquinone	338
6.E.49	2-hydroxy-1,4-naphthoquinone	339
6.E.50	5- Hydroxy-1,4-naphthoquinone (Junglone)	340
6.E.51	3-hydroxymethyl-5-hydroxy-1,4-naphthoquinone	340
6.E.52	2-methyl-1,4-naphthoquinone	341
6.E.53	2-bromo-1,4-naphthoquinone	341
6.E.54	2-chloro-1,4-naphthoquinone	342

6.E.55 2-bromo-5-hydroxy-1, 4-napthoquinone [2-bromojuglone] 342
6.E.56 3-chloro-5-hydroxy-1, 4-naphthoquinone (3-chlorojuglone) 343
6.E.57 2-methylbenzoquinone 343
6.F.58 Trans Stilbene (E) 343
6.F.59 Cis-stilbene (Z) 344
6.F.60 4-hydroxy Stilbene[89] 345
6.F.61 3,4-dimethoxy Stilbene 346
6.F.62 o-carboxystilbene (2-COOH $C_6H_4CH=CHC_6H_5$) 347
References 348

1

Phenols

Phenols and phenyl ethers are obtained by the following methods.

DAKIN'S OXIDATION

The phenolic aldehydes having phenolic group in ortho or para position to the aldehydic group on oxidation with alkaline hydrogen peroxide give the corresponding hydroxy compounds, i.e., the –CHO group is replaced by the –OH group

Mechanism

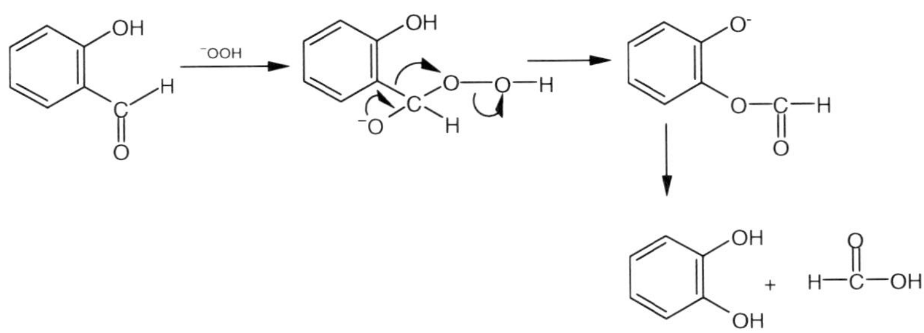

The above mechanism is confirmed by taking an aldehyde in which oxygen of the CHO group is labelled by ^{18}O; the formic acid formed has the ^{18}O labelling.

Dakin's oxidation has been used to prepare a number of phenols, viz. 1.7, 8, 11, 24 and 35.

Methoxyphenols (partial methyl ethers of phenols) have been obtained by oxidation of methoxybenzaldehydes with hydrogen peroxide at room temperature in presence of p-toluenesulphonic acid in methanol. Using this method a number of methoxyphenols (1.40) have been obtained.

Performic Acid Oxidation

Aromatic aldehydes can be converted into phenolic compounds by oxidation with performic acid. This reaction is similar to Dakin's oxidation. Thus the aldehydes on treatment with peracids (viz. Peracetic acid or perbenzoic acid) give the corresponding formate esters, which on hydrolysis with alkali give phenols.

Mechanism

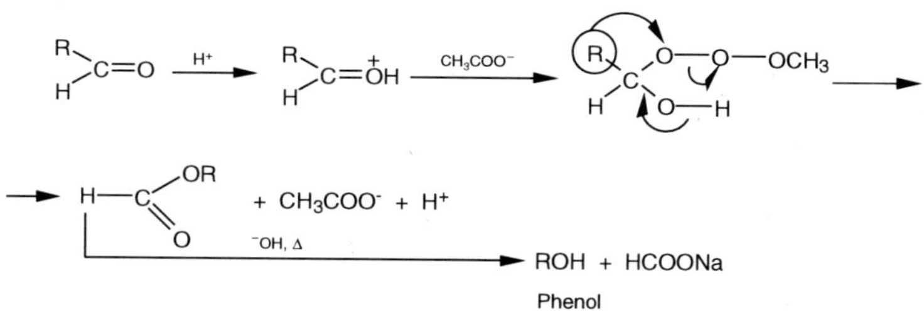

Using this method following phenols have been prepared: 1.10, 11, 36, 38.

Clemmensen Reduction

Certain alkyl (viz. methyl, ethyl, benzyl etc.) substituted phenols can be prepared by reduction of the corresponding carbonyl compounds with zinc amalgam and hydrochloric acid.

Mechanism

The mechanism of Clemmensen reduction is uncertain. Most likely, the reaction pathway involves transfer of electrons from the metal surface to the carbonyl carbon atom.

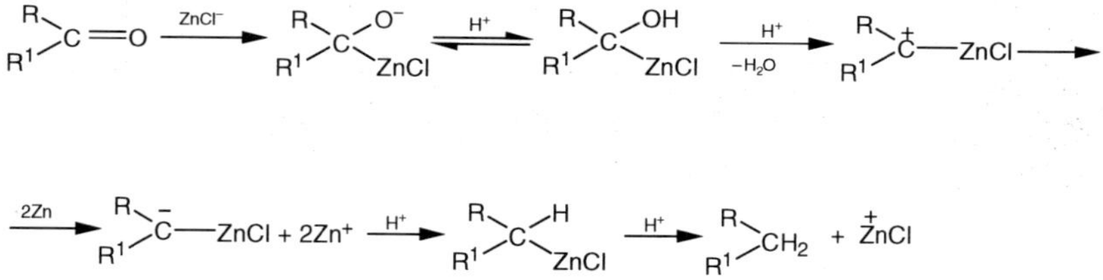

Using this method some alkyl substituted phenols, viz. 1.18, 19, 29, 30, 32 and 33 have been prepared.

Thiele Acetylation

A hydroxyl group can be introduced into p-benzoquinones by acetylation with acetic anhydride in presence of concentrated sulphuric acid. The triacetate so obtained is deacetylated to give trihydroxy compound. Thus

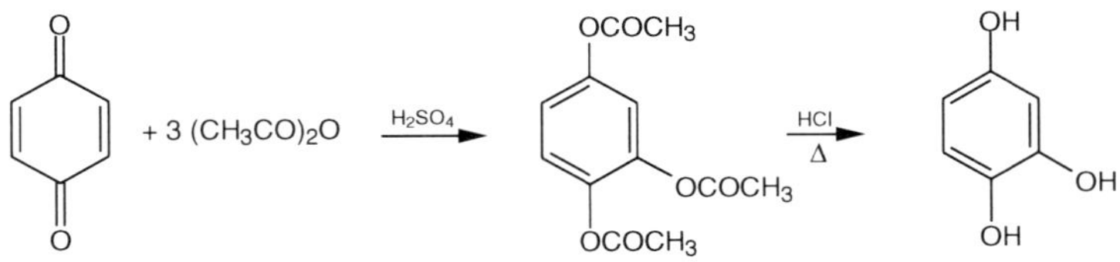

Mechanism

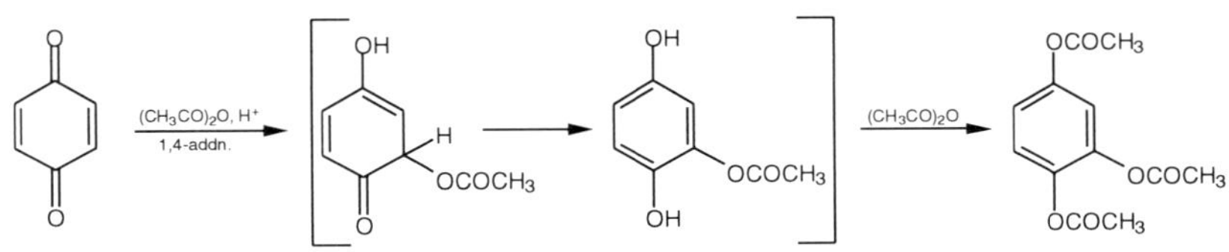

Using Thiele acetylation, 1,2,4-triacetoxybenzene (1.21) and 1,2,4-triacetoxynaphthalene (1.37) have been prepared.

Elb's Persulphate Reaction

A monohydric phenol can be oxidised to a dihydric phenol by alkaline potassium persulphate. Hydroxylation normally occurs in the p-position. However, if the para position is occupied, then hydroxylation occurs in ortho position.

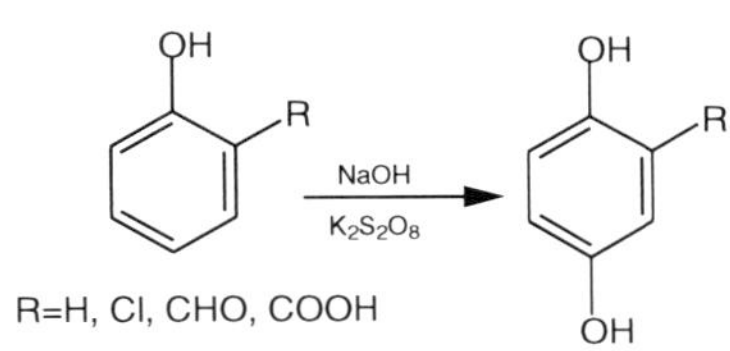

Mechanism

This is a general method of introducing a hydroxyl group in the p-position of the hydroxy compound. Using this method the preparation of following compounds is described. Partial demethylation of aryl methyl ethers.

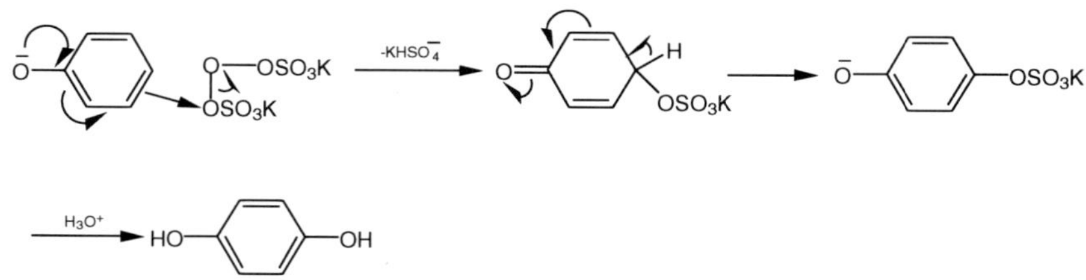

NaSCH$_2$Me in dimethylformamide is a powerful demethylating agent for the partial demethylation of aryl methyl ethers. The reagent NaSCH$_2$Me is prepared in situ by the action of sodium hydride on ethane thiol. Thus 1,2–dimethoxybenzene on stirring with ethane diol and sodium hydride in N,N-dimethylformamide gives o-methoxyphenol (1.1) in 97% yield.

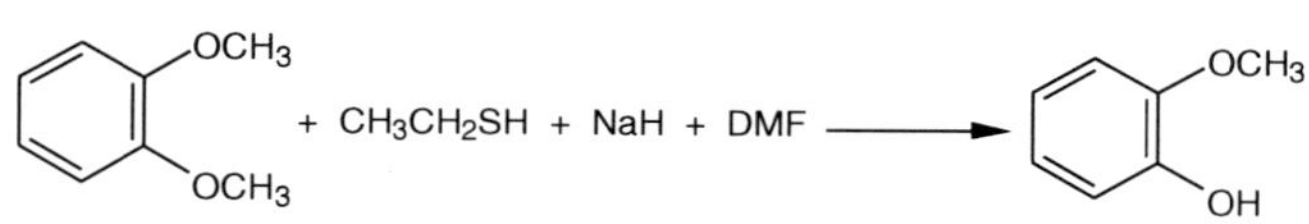

This method has been used for the preparation of 4-methoxy phenol (1.5), 2,3–dihydroxyphenol (1.7), 3,5-dimethoxyphenol (1.13) and 3-methoxy–5-methylphenol (1.15).

Methylation

Phenol can be conveniently methylated with sodium hydroxide and dimethyl sulphate to give partial or completely o-methylated products.

$$C_6H_5OH + NaOH + (CH_3)_2\,SO_4 \rightarrow C_6H_5OCH_3 + Na\,(CH_3)\,SO_4 + H_2O$$

The formation of partial or completely o-methylated product depends on the amount of dimethyl sulphate used. Using this method a number of o-methyl ethers, viz 1.2, 3, 4, 5, 6 and 7 have been prepared. This method is called the wet methylation.

Alternatively, methylation can be carried out under anhydrous conditions. In this method the phenolic compound is refluxed in dry acetone with dimethylsulphate in presence of anhydrous potassium carbonate. Using this method a number of o-methyl ethers , viz. 1.14, 15, 16 and 36 have been prepared.

Partial methylation of phenol can also be carried out by refluxing the phenolic compound in anhydrous methanol while dry hydrogen chloride gas is bubbled through the solution. Using this method phloroglucinol monomethyl ether (1.12) and phloroglucinol dimethyl ether (1.13) have been prepared.

1.1 o-METHOXYPHENOL (CATECHOL MONOMETHYL ETHER) (GUAIACOL)

Method (i)

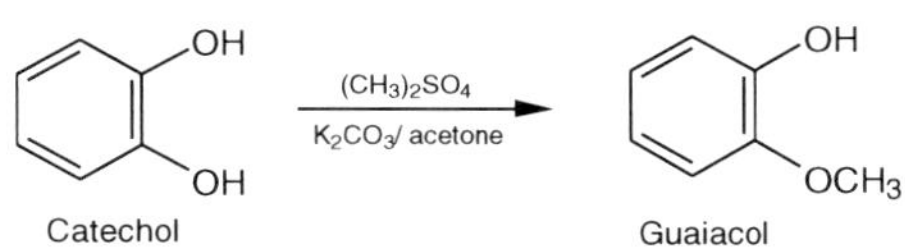

A mixture of catechol (11 g, 0.1 mole) and freshly ignited potassium carbonate (5 g) in dry acetone (100 ml) is refluxed under anhydrous conditions. Dimethyl sulphate (13.87 g, 10.4 ml, 0.11 mole) in dry acetone (20 ml) is added to the above refluxing solution in three instalments after every one hour The mixture is refluxed for additional 2 hours It is filtered and the inorganic salts washed with hot acetone (2 × 20 ml). The solvent is removed and the residual product is chromatographed over silica gel column (4 × 40 cm). Elution with petroleum ether gives the required compound as pale yellow oily product. Yield 8 g (64.5%). B.p. 202-205° (lit.[1] b.p. 53-55°/4 mm).

Method (ii)

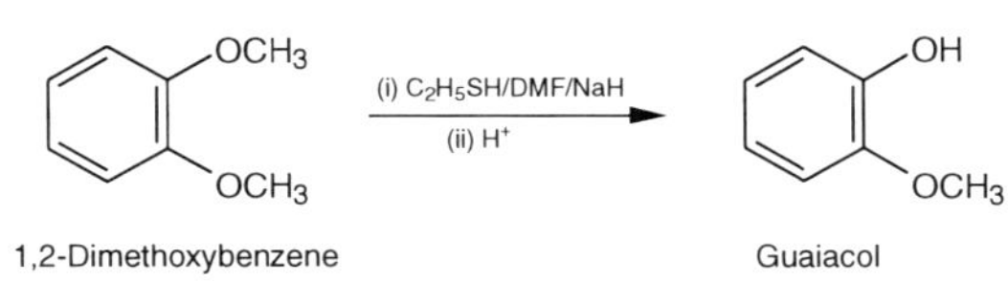

Ethanethiol (0.625 g, 0.56 ml, 0.01 mole) dissolved in N,N-dimethyl formamide (10 ml) is added to a suspension of sodium hydride (0.5 g of a 50% oil dispersion) in dry DMF (20 ml) under an atmosphere of nitrogen. The mixture is stirred for 5 min under anhydrous conditions. A solution of 1,2–dimethoxybenzene (veratrol) (1.2) (0.5 g, 0.46 ml, 0.0036 mole) in dry DMF (5 ml) is added and the mixture is refluxed for 1.5 hr. It is cooled and acidified with hydrochloric acid (10 %) and extracted with ether (2 × 50 ml). The ether extract is washed with water and extracted with sodium hydroxide solution (5%). The clear alkaline extract is acidified with hydrochloric acid (10%) and re-extracted with ether (2 × 50 ml). The ether extract is washed with water, dried (anhydrous sodium sulphate) and distilled. o-Methoxyphenol is obtained as yellow liquid. Yield 0.43 g (97%). b.p. 202-205° (lit[1] b.p. 53-55°/4 mm).

Note: The method described is a modification of lit.[2] procedure.

1.2 1,2–DIMETHOXYBENZENE (CATECHOL DIMETHYL ETHER) (VERATROLE)

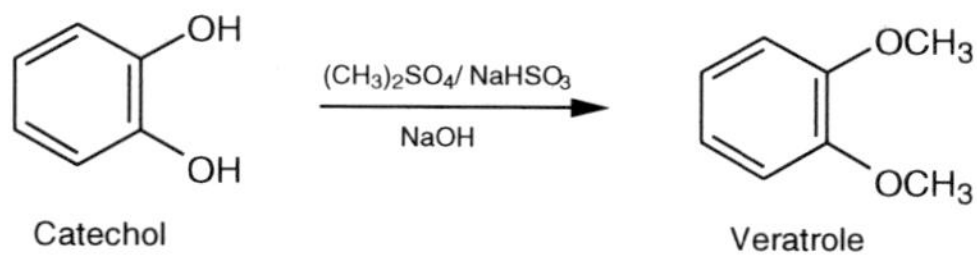

Aqueous sodium hydroxide (40%, 175 ml) is added dropwise during 3 hr at 18-22° to a well stirred mixture of catechol (50 g, 0.45 mole), sodium hydrosulphite (20 g, 0.115 mole), dimethyl sulphate (110 ml, 146.63 g, 1.16 mole) and alcohol (150 ml). Ice (1 kg) is added to the reaction mixture and left overnight at room temperature. It is extracted with ether (2 × 200 ml). The ether extract is washed with water, dried (anhydrous sodium sulphtate) and distilled. Veratrole is obtained as oil. Yield 52.75 g (84.1%). B.p. 207° (lit.[3] b.p. 207°).

1.3 3-METHOXYPHENOL (RESORCINOL MONOMETHYL ETHER)

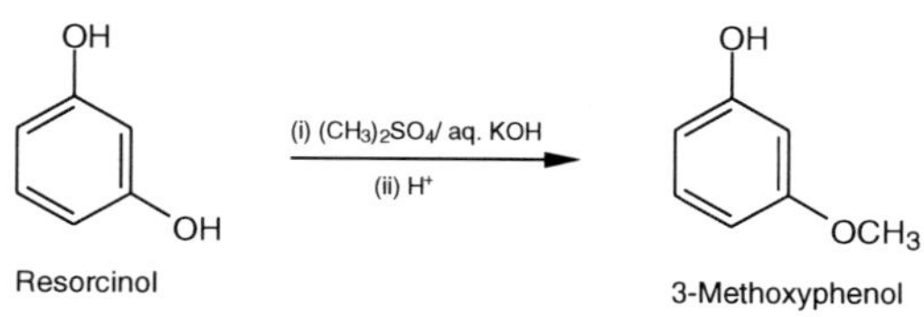

To a well stirred solution of resorcinol (150 g, 1.36 mole) and dimethyl sulphate (140 ml, 186.6 g, 1.48 mole) is added dropwise (separatory funnel) an aqueous solution of potassium hydroxide (40%, 300 ml) during 1 hr at room temperature. The mixture is stirred for 2 hr at room temperature and heated for 30 min in a boiling water bath. It is cooled, acidified with hydrochloric acid and extracted with benzene (3 × 100 ml). The benzene extract is washed with water (to remove unreacted resorcinol) and then extracted with sodium hydroxide solution (10%, 300 ml). The clear alkaline solution is acidified with hydrochloric acid and extracted with ether (2 × 100 ml). The ether solution is washed with a saturated solution of sodium chloride dried (anhydrous sodium sulphate) and ether distilled off. The oily product obtained is distilled and the fraction boiling at 242-243° is collected (Lit[4] b.p. 242-43°). Yield 60 g (35.5%). In case the product is still a mixture (as seen by TLC), it is purified by chromatography over silica gel column and elution with petroleum ether.

1.4 1,3-DIMETHOXYBENZENE (RESORCINOL DIMETHYL ETHER)

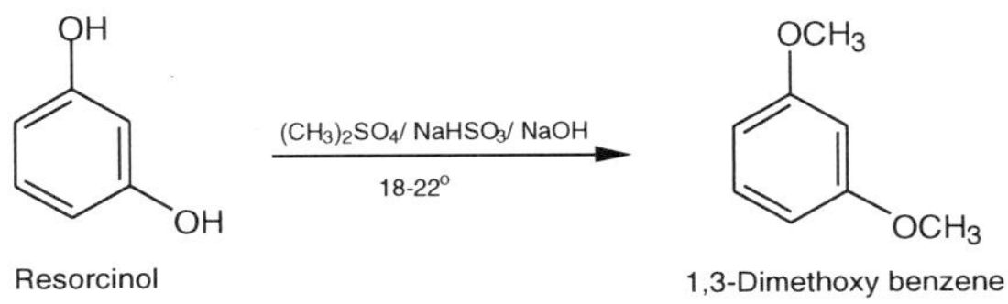

Aqueous sodium hydroxide (40%, 175 ml) is added dropwise during 3-4 hr to a well stirred mixture of resorcinol (50 g, 0.45 mole) sodium hydrosulphite (20 g, 0.115 mole), dimethyl sulphate (110 ml, 146.63 g, 1.16 mole) and alcohol (150 ml) kept at 18-22°. Ice (1 kg) is added to the reaction mixture and it is left overnight. It is extracted with ether (2 × 200 ml). The ether extract is washed with water, dried (anhydrous sodium sulphate) and distilled. The oily residue is distilled and the fraction distilling between 216-218° is collected. Yield 50 ml (52.75 g) (84.1%) (lit.[5] b.p. 216-18°).

1.5 4-METHOXYPHENOL (HYDROQUINONE MONOMETHYL ETHER)

Method (i)

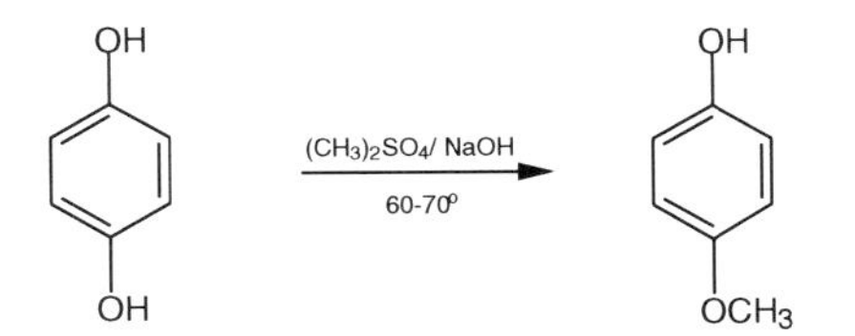

To a well stirred solution of hydroquinone (11 g, 0.1 mole) in aqueous sodium hydroxide (25%, 40 ml) is added dropwise during 1 hr, dimethyl sulphate (10.4 ml, 13.87 g, 0.11 mole) at 60-70°. The mixture is heated in boiling water bath for 30 min. It is cooled, extracted with ether (2 × 70 ml). The ether extract is washed with water, dried (sodium sulphate) and distilled. The oily residue is purified by column chromatography over silica gel (2 × 30 cm). The column is eluted with petroleum ether. 4-Methoxy phenol is obtained from the eluate by distillation. Yield 7.5 g (60.5%). It is crystallised from petroleum ether as colourless shining needles. M.p. 53° (lit.[1,6] m.p. 53°).

Method (ii)

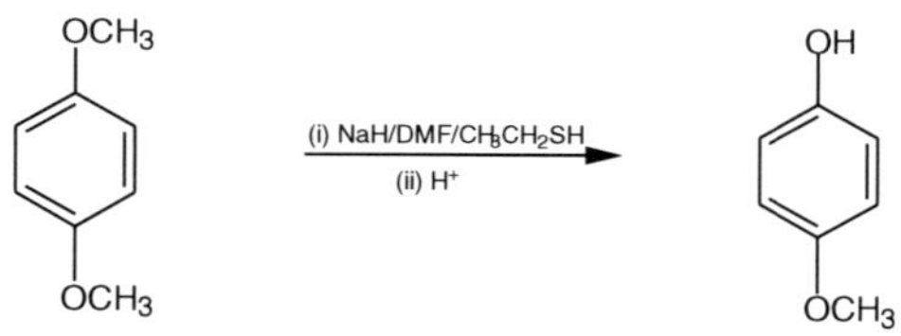

p-Dimethoxybenzene 4-Methoxyphenol

Ethanethiol (0.625 g, 0.56 ml, 0.01 mole) dissolved in N,N-dimethyl formamide (20 ml) is added to a suspension of sodium hydride (0.5 g of a 50% oil dispersion) in dry DMF (10 ml) under an atmosphere of nitrogen. The mixture is stirred for 5 min under anhydrous conditions. A solution of p-dimethoxybenzene (1.6) (0.5 g, 0.0036 mole) in dry DMF (10 ml) is added and the mixture refluxed for 3 hr. The reaction mixture is cooled and acidified with hydrochloric acid (10%). It is extracted with ether (3 × 50 ml). The ether layer is washed with water and extracted with sodium hydroxide solution (5%). The clear alkaline extract is acidified with hydrochloric acid. The mixture is extracted with ether. The ether layer is washed with water, dried (anhydrous sodium sulphate) and distilled. 4-Methoxy phenol is obtained as pale yellow oil. Yield 0.44 g (98%). It is crystallised from benzene -petroleum ether as needles. M.p. 54° (lit.[6] m.p. 53°).

Note: The method described is a modification of lit.[2] procedure.

1.6 1,4-DIMETHOXYBENZENE (HYDROQUINONE DIMETHYL ETHER)

An aqueous solution of sodium hydroxide (40%, 175 ml) is added dropwise during 3 hr at 18-22° to a well stirred mixture of hydroquinone (50 g, 0.45 mole), sodium hydrosulphite (2.0 g, 0.115 mole), dimethyl sulphate (110 ml, 146.63 g, 1.16 mole) and alcohol (150 ml). Ice (1 kg) is added and the

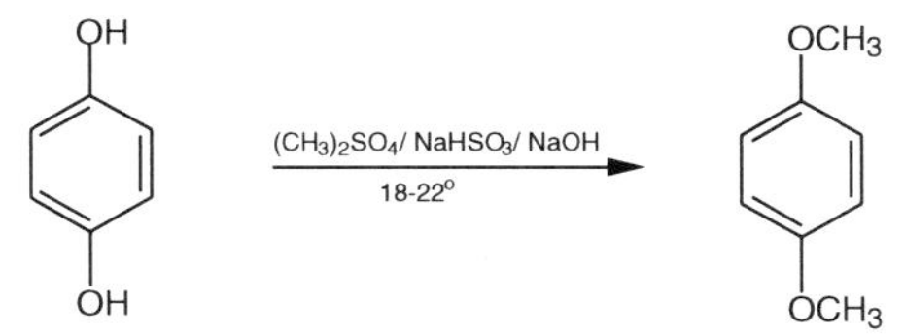

Hydroquinone 1,4-Dimethoxybenzene

reaction mixture left overnight at room temperature. It is extracted with ether (2 × 200 ml). The ether extract is washed with water, dried (anhydrous sodium sulphate) and distilled. 1,4–Dimethoxy benzene is obtained as an oil. Yield 52.75 g (84.1%). It solidifies on cooling and is crystallised from ether-petroleum ether as shining needles. M.p. 55-56° (lit.[7] m.p. 56°).

1.7 2,3 – DIHYDROXYANISOLE (PYROGALLOL MONOMETHYL ETHER)

Method (i)

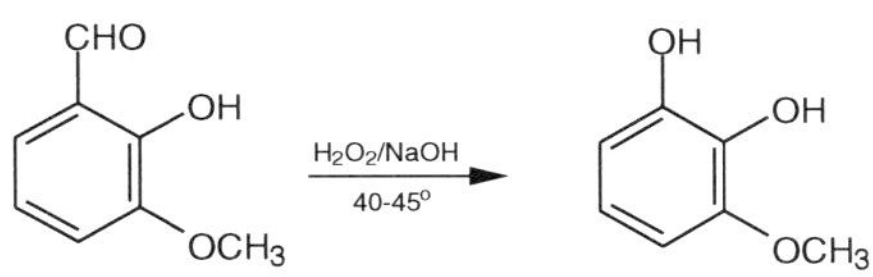

2-Hydroxy-3-Methoxy benzaldehyde 2,3-Dihydroxyanisole

To a solution of 2-hydroxy-3-methoxybenzaldehyde (o-vanillin) (20.3 g, 0.134 mole) in aqueous sodium hydroxide (2N, 66.6 ml, 0.133 mole) is added dropwise hydrogen peroxide (6%, 95 ml, 0.167 mole) during 1 hr at 40-45° (the solution is cooled if necessary). After the reaction is over, the reaction mixture is allowed to cool to room temperature. It is saturated with sodium chloride and then extracted with ether (3 × 70 ml). The ether solution is dried (sodium sulphate) and distilled. The residual product is distilled and the fraction boiling at 136-38°/22 mm is collected. Yield 11 g (58.8%) (lit.[8] b.p. 136-38°/22 mm).

Method (ii)

Ethanethiol (1g, 0.84ml, 0.015mole) dissolved in N,N- dimethyl-formamide (20 ml) is added to a suspension of sodium hydride (0.5 g of a 50% oil dispersion) in dry DMF (10 ml) under nitrogen atmosphere. The mixture is stirred for 5 min under anhydrous conditions. A solution of

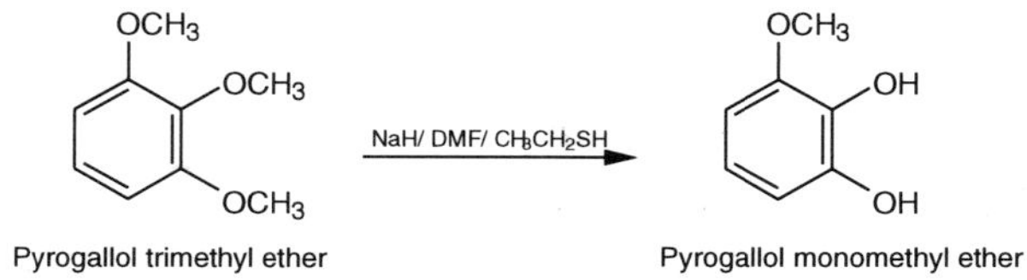

pyrogallol trimethyl ether (1.9) (0.5 g, 0.003 mole) in dry DMF (10 ml) is added and the mixture refluxed for 3 hr. The reaction mixture is cooled and acidified with hydrochloric acid (10%). It is extracted with sodium hydroxide solution (5%). The clear alkaline solution is acidfied with hydrochloric acid and extracted with ether. The ether extract is washed with water, dried (anhydrous sodium sulphate) and distilled. Pyrogallol monomethyl ether is obtained as an oil. Yield 0.4 g (96.8%). B.p. 136-38°/ 22 mm.

Note: The method described is a modification of lit.[2] procedure.

1.8 2,3- DIMETHOXYPHENOL (PYRAGALLOL-2,3–DIMETHYL ETHER)

Method (i)

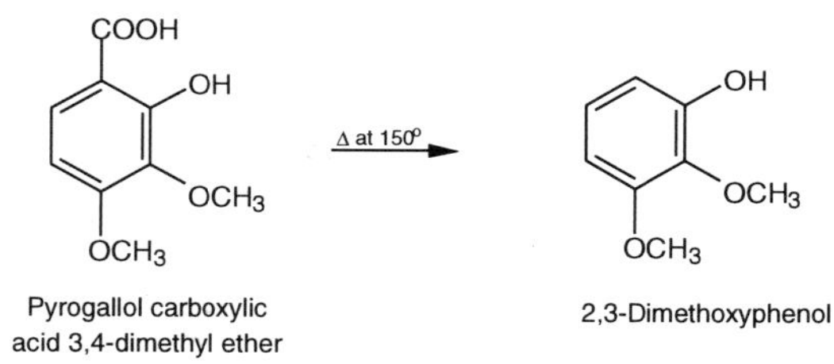

Pyrogallol carboxylic acid 3,4–dimethyl ether (3.25) (40 g, 0.2 mole) is heated in an oil bath at 150° under reduced pressure. The acid decarboxylates and pyrogallol dimethyl ether distils at 120-25°/17 mm. The product is redistilled at normal pressure and the fraction boiling at 232-234° is collected. Yield 15 g (48.2%) (lit.[9] b.p. 232-34°).

Method (ii)

A mixture of 2,3–dimethoxy benzaldehyde[10] (o-veratraldehyde) (1.66 g, 0.01 mole), hydrogen peroxide (30%, 2.6 ml, 0.023 mole), p-toluenesulphonic acid (6.88 g, 0.036 mole) and methanol (5 ml) is stirred at room temperature. When the reaction is complete (monitored by TLC), the reaction mixture is diluted with water (50 ml) and extracted with

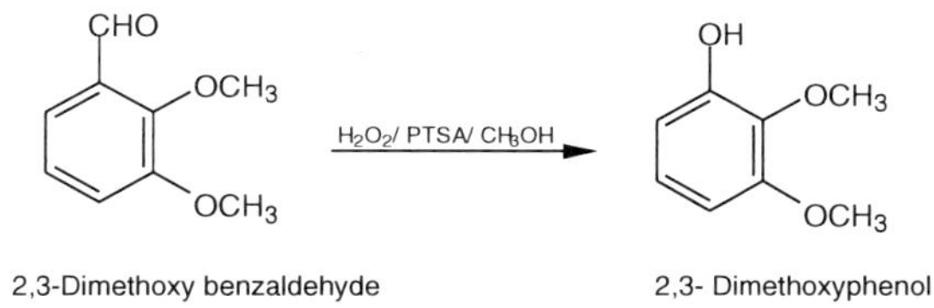

dichloromethane (5 × 50 ml). The extract is dried (anhydrous sodium sulphate) and distilled. The crude product is purified by column chromatography over silica gel (2 × 30 cm) and eluted with benzene. Pyrogallol 2,3-dimethyl ether is obtained as an oil. Yield 1.16 g (75.3%) b.p. 232-34°.

Note: The method described is a modification of lit.[11] procedure.

1.9 1,2,3 – TRIMETHOXYBENZENE (PYROGALLOL TRIMETHYL ETHER)

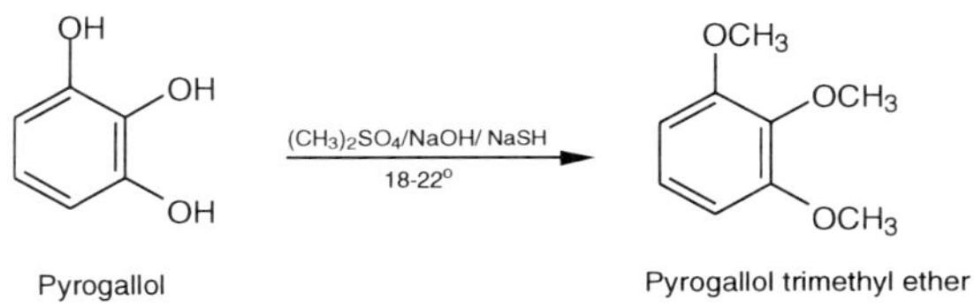

Aqueous sodium hydroxide (40%, 150 ml) is added dropwise during 3-4 hr to a stirred mixture of pyrogallol (42 g, 0.33 mole), sodium hydrosulphite (10 g, 0.057 mole), dimethyl sulphate (146 ml, 194.60 g, 1.54 mole) kept at 18-22°. Ice (1 kg) is then added and the reaction mixture left overnight. The separated solid is filtered and washed with water. It is crystallised from methanol as colourless prisms. Yield 37.5 g (66.9%). M.p. 46-47°. Some more product (4 g, 7.1 %) is obtained by extraction of the mother liquor with ether (lit.[12] m.p. 46-47°).

1.10 2,4–DIHYDROXYANISOLE (4- METHOXY RESORCINOL)

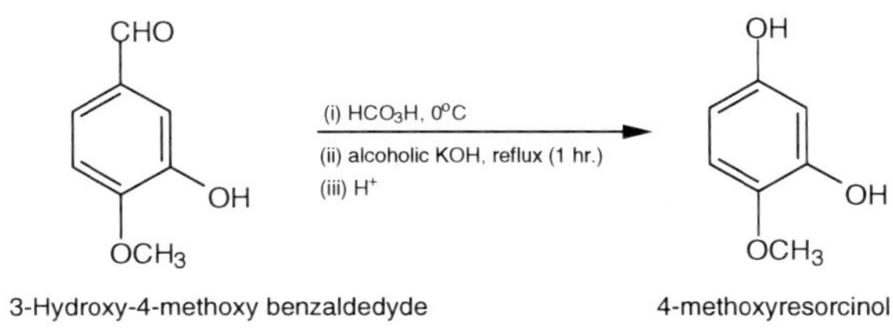

Isovanillin (3-hydroxy-4-methoxybenzaldehyde) (10 g, 0.066 mole) is dissolved in dry chloroform (100 ml) and the solution cooled to 0°. Performic acid (obtained as given in **1.36**) (50 ml) is added dropwise to the vigorously stirred solution. Addition of performic acid is stopped as soon as an aliquot showed the absence of isovanillin on TLC. The mixture is stirred for 3 hr and the solvent removed in vacuo. The residue consisting of the formate of isovanillin is treated with alcoholic potassium hydroxide solution (10%) till the solution is alkaline. Finally, the reaction mixture is refluxed for 1 hr with a solution of potassium hydroxide (10%, 100 ml) in aqueous ethanol (80%). The solvent is removed in vacuo and the remaining solution is acidified with dilute hydrochloric acid. It is extracted with ether (4 × 100 ml). The ether extract is dried (anhydrous sodium sulphate) and distilled. The required product is obtained as an oil. Yield 7 g (76%), b.p. 132-5°/4 mm. It solidifies on cooling, m.p. 66-7° (lit.[13] m.p. 66-7°).

1.11 3,4-DIMETHOXYPHENOL

Method (i)

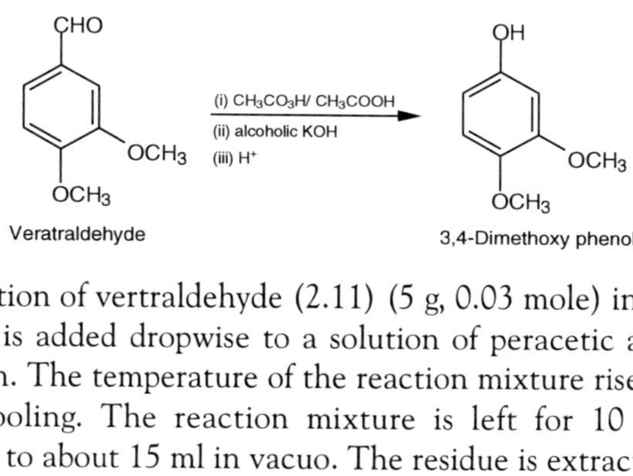

Veratraldehyde 3,4-Dimethoxy phenol

To a solution of vertraldehyde (2.11) (5 g, 0.03 mole) in glacial acetic acid (30 ml) is added dropwise to a solution of peracetic acid[15] (15 ml) during 30 min. The temperature of the reaction mixture rises, it is kept at 40-45° by cooling. The reaction mixture is left for 10 hr and then concentrated to about 15 ml in vacuo. The residue is extracted with ether (2 × 20 ml). The ether layer is distilled. The formate ester of 3,4–dimethoxyphenol thus obtained is hydrolysed by refluxing with potassium hydroxide (10 g) in aqueous alcohol (1:4, 100 ml) for 1 hr. The reaction mixture is concentrated in vacuo almost to dryness. It is dissolved in water (20 ml), the solution rendered acidic with dilute sulphuric acid and extracted with ether. The ether extract is dried (sodium sulphate) and distilled. The oily residue thus obtained is subjected to column chromatography over silica gel (2 × 30 cm). Elution with benzene gave 3,4

–dimethoxyphenol as yellow solid. Yield 3 g (65.2%). It is crystallised from benzene as yellow crystalline needles. M.p. 78-80° (lit.[14] m.p. 78-80°).

Method (ii)

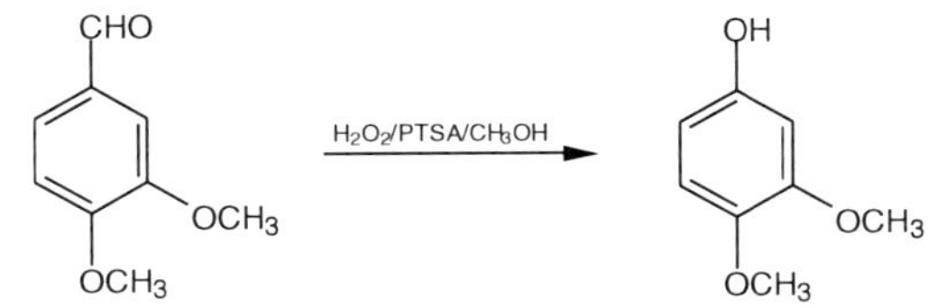

Veratraldehyde   3,4-Dimethoxyphenol

A mixture of veratraldehyde (2.11) (1.66 g, 0.01 mole), hydrogen peroxide (30%, 2.6 ml, 0.023 mole), p-toluenesulphonic acid (6.88 g, 0.036 mole) and methanol (5 ml) is stirred at room temperature. When the reaction is complete (as monitored by TLC) the reaction mixture is diluted with water (50 ml) and extracted with dichloromethane (5 × 50 ml). The extract is dried (anhydrous sodium sulphate) and purified by column chromatography over silica gel (2 × 30 cm). Elution with benzene gave 3,4-dimethoxyphenol as an yellow solid. Yield 1.16 g (75.3%). M.p. 78-80°.

1.12 1,3-DIHYDROXY-5-METHOXYBENZENE (PHLOROGLUCINOL MONOMETHYL ETHER)

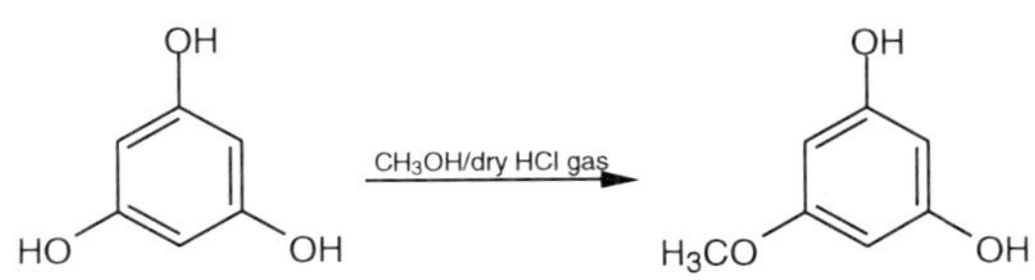

Phloroglucinol   1,3-Dihydroxy-5-methoxy benzene

Anhydrous phloroglucinol (25 g, 0.198 mole) is dissolved in absolute methyl alcohol (100 ml) and the solution cooled to 18°. Dry hydrogen chloride gas is passed through the solution for 25 min at such a rate that the solution does not get saturated (approx. 200 bubbles/min). The mixture is then refluxed for 15 min, cooled immediately to 18° and dry hydrogen chloride gas passed again for 25 min. The reaction mixture is allowed to stand for 12 hr and the solvent is removed by distillation in vacuo. Great care is taken to remove as much of the solvent as possible without applying any heat (preferably in rotatory evaporator). The residue is poured into water (500 ml) and extracted with ether (4 × 75 ml). The ether solution is dried (anhydrous sodium sulphate) and then distilled. The residual oily product is distilled in vacuo. A small quantity of o-dimethyl phloroglucinol (5 g, 16.4 %) distils over first at 185-190°/ 25 mm and then o-monomethyl phloroglucinol (15 g, 54.2%) distils over at 210° as a viscous oil. The viscous oil solidifies on standing for 2-3 days. M.p. 78°. The yields are not consistent and vary according to the extent of saturation (lit.[16,17] m.p. 78-81°).

1.13 3,5-DIMETHOXYPHENOL (PHLOROGLUCINOL DIMETHYL ETHER)

Method (i)

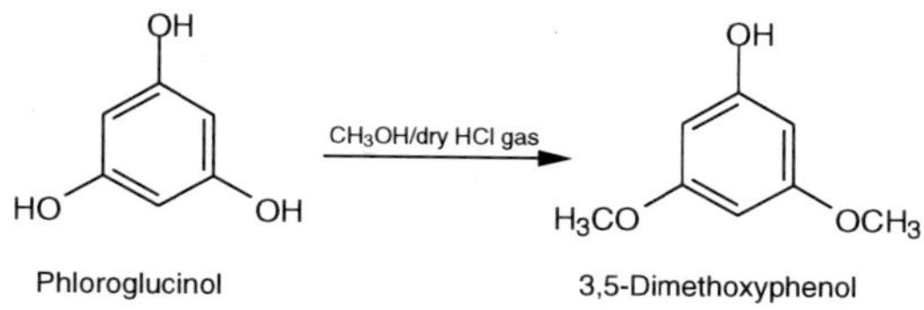

Phloroglucinol (20g, 0.159 mole) (dehydrated at 120° for 3 hr) is dissolved in acetone free absolute methanol (150 ml) and dry hydrogen chloride gas is passed into it at room temperature at such rate that the solution is saturated within 1 hr. The temperature of the solution rises to 65° towards the end of the reaction. The solution is refluxed for 1 hr on a water bath. It is cooled to room temperature and dry hydrogen chloride gas is passed again for 1 hr. The solution is left for 24 hr at room temperature and alcohol is removed by distillation. The residue is treated with water (350 ml). A heavy brown oil separates out. The reaction mixture is extracted with ether (2 × 100 ml). The ethereal solution is washed with water, dried (anhydrous sodium sulphate) and distilled. The oily product is distilled under reduced pressure. The dimethyl ether distils at 195-200°/30 mm as pale yellow viscous oil. Yield 10 g (40.9%) (Lit[18] b.p. 198-200°/30 mm).

Method (ii)

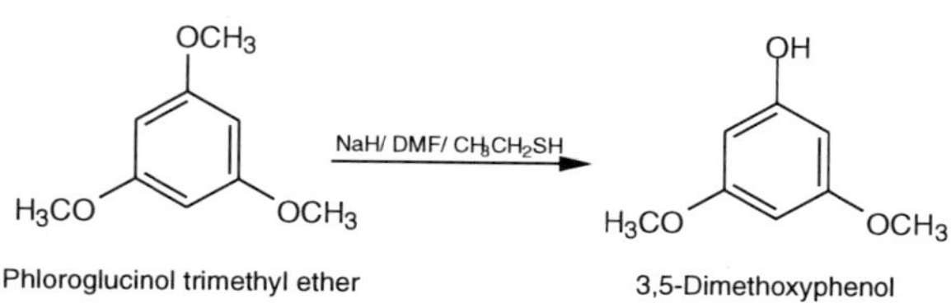

Ethanethiol (1.25g, 1.5ml, 0.02mole) dissolved in N,N- dimethylformamide (20 ml) is added to a suspension of sodium hydride (1 g of 50% oil dispersion) in dry DMF (10 ml) under an atmosphere of nitrogen. The mixture is stirred for 5 min under anhydrous conditions. A solution of pholroglucinol trimethyl ether (1.14) (1 g, 0.006 mole) in dry DMF (10 ml) is added and the mixture refluxed for 3 hr. The reaction mixture is cooled and acidified with hydrochloric acid (10%). It is extracted with ether (3 × 50 ml). The ether extract is washed with water and extracted with sodium hydroxide solution (5%). The clear alkaline extract is acidified with hydrochloric acid. The mixture is extracted with ether. The ether extract is washed with water, dried (anhydrous sodium sulphate) and distilled. 3,5–Dimethoxyphenol is obtained as an oil. Yield 0.9 g (98.1%). B.p. 198-199°/30 mm.

Note: The method described is a modification of lit[2] procedure.

1.14 1,3,5- TRIMETHOXYBENZENE (PHLOROGLUCINOL TRIMETHYL ETHER)

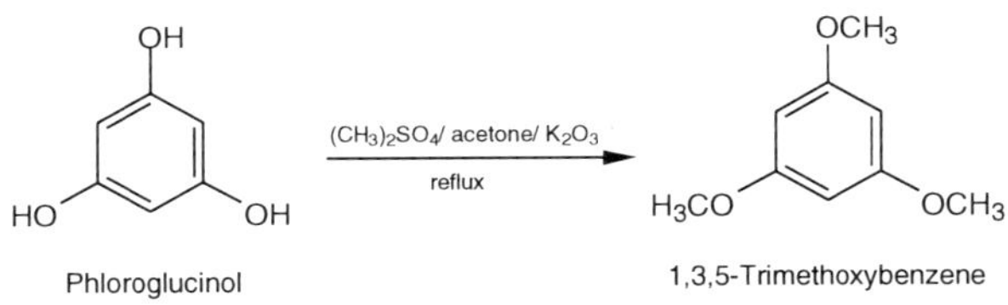

A solution of anhydrous phloroglucinol (6.3 g, 0.05 mole) in dry acetone (100 ml) is refluxed for 6 hr under anhydrous conditions with dimethyl sulphate (15. 6 ml, 20.8 g, 0.165 mole) in presence of ignited potassium carbonate (40 g). The mixture is filtered, and the inorganic salts washed with hot acetone (2 × 20 ml). The combined acetone solution is distilled and the residue macerated with crushed ice. It is extracted with ether, the ether extracts washed with sodium hydroxide solution (5%), water, dried (sodium sulphate) and distilled. Yield 6 g (71.4 %). **M.p** 52-3° (lit.[19] m.p. 52-53°).

1.15 3-METHOXY-5-METHYLPHENOL (ORCINOL MONOMETHYL ETHER)

Method (i)

A solution of anhydrous orcinol (12.4 g, 0.1 mole) in dry acetone (100 ml) is refluxed for 4 hr with dimethyl sulphate (10.63 ml, 14.2 g, 0.112 mole)

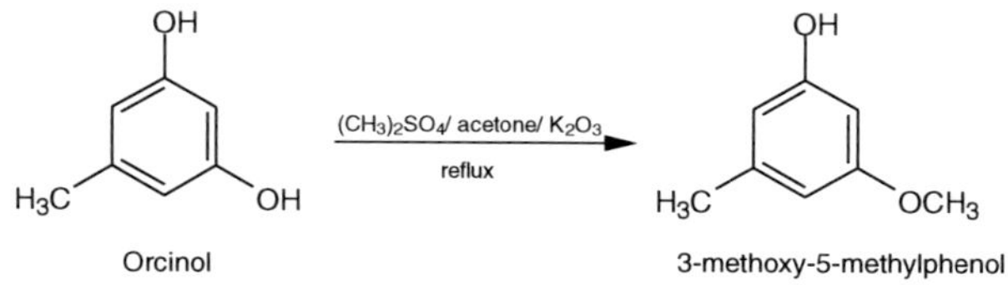

and ignited potassium carbonate (20 g) under anhydrous conditions. The mixture is filtered in hot and the inorganic salts washed with hot acetone (25 ml). The combined acetone solution is distilled. The oily product is purified by column chromatography over silica gel using chloroform as an eluant. Orcinol monomethyl ether is obtained as oil. Yield 4.2 g (30.4%). B.p. 89-90° (0.2 mm) m.p. 61-2° (lit.[20] b.p. 256°).

Method (ii)

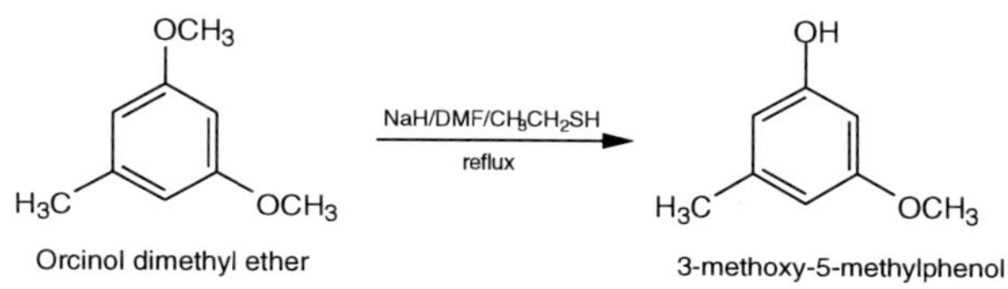

Dry N,N- dimethylformamide (125 ml) and sodium hydride (11 g, 0.46 mole) (60 % oil dispersion) is placed under nitrogen atmosphere in a R.B. flask fitted with a reflux condenser and calcium chloride drying tube. The flask is cooled in ice bath and a solution of ethanethiol (16.5 g, 19.7 ml, 0.266 mole) in N,N- dimethylformamide (75 ml) is added slowly through a dropping funnel during 20-25 min. The reaction mixture is kept stirring (magnetic stirrer). Stirring is continued for 15 min more and a solution of orcinol dimethyl ether (1.16) (18.25 ml) in dry N,N-dimethylformamide (50 ml) is added in one lot . The mixture is refluxed in nitrogen atmosphere for 3 hr. It is cooled and poured into cold water (500 ml) and extracted with petroleum ether (2 × 100 ml) (which is discarded). The aqueous layer is acidified with ice-cold hydrochloric acid (4 N, 165 ml) and extracted with ether (3 × 100 ml). The combined ether extract is washed with saturated aqueous sodium chloride (50 ml), dried (magnesium sulphtate) and distilled. Orcinol monomethyl ether is obtained as an oil. Yield 28 g (81%). B.p. 89-90° (0.2 mm). The distilled product solidifies after 6 hr. It is crystallised from petroleum ether as off-white prisms. M.p. 61-62°.

Note: The method described is a modification of lit.[2] procedure.

1.16 3,5–DIMETHOXYTOLUENE (ORCINOL DIMETHYL ETHER)

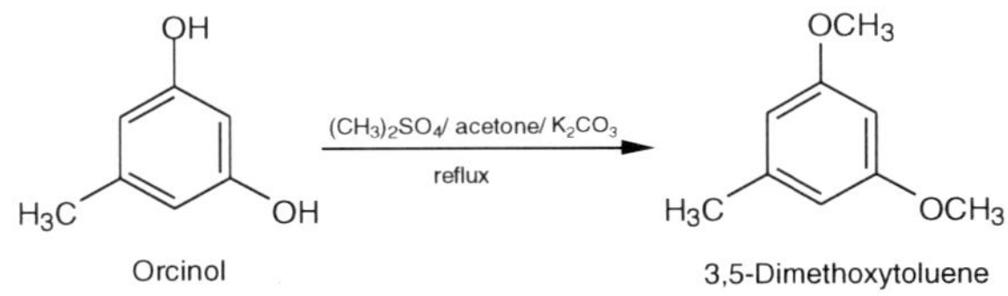

A solution of dry orcinol (14.2 g, 0.115 mole) in dry anhydrous acetone (150 ml) containing anhydrous potassium carbonate (45 g) is refluxed with dimethyl sulphate (23.6 ml, 31.46 g, 0.25 mole) for 4 hr. The dimethyl sulphate is added in small lots of 5 ml. After refluxing, the solution is filtered and the inorganic residue washed with hot acetone (2 × 25 ml). The combined acetone solution is distilled and the residue macerated with crushed ice. It is extracted with ether, the ether extract dried (anhydrous sodium sulphate) and distilled. Yield 14.2 g (81.6%). B.p. 102° (8 mm) (lit.[21], b.p. 102°/8 mm).

1.17 2,6-DIHYDROXYTOLUENE (2- METHYL RESORCINOL)

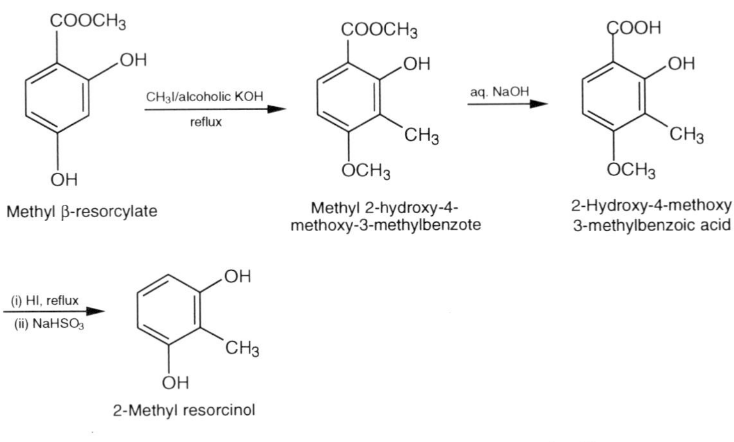

Step (i) Methyl 2-hydroxy-4-methoxy-3-methylbenzoate

To a cooled solution (0°) of methyl β-resorcylate (β-resorcylic acid methyl ester) (3.7) (10 g, 0.06 mole) in methanolic potassium hydroxide (10 g in 100 ml methanol) is added to cold methyl iodide (20 ml, 45.6 g, 0.32 mole) in one lot. The reaction mixture is shaken for 5-10 min and kept for 24 hr at room temperature. It is refluxed for 4 hr. Methanol and excess methyl iodide are removed in vacuo and the residue treated with crushed ice. The solid product is filtered and washed with water. It is crystallised from dilute methanol as colourless prismatic needles. Yield 2 g (17.2%). M.p. 77° (lit.[22] m.p. 77°).

Step (ii) 2- Hydroxy-4- methoxy-3-methylbenzoic acid

The above methyl ester (10 g, 0.05 mole) is refluxed with sodium hydroxide solution (10%, 100 ml) for 15 min. The clear alkaline solution on acidification gave the required product. It is crystallised from methanol as white crystalline needles. Yield 8 g (86.2%). M.p. 213° (lit.[22] m.p. 213°).

Step (iii) 2- Methylresorcinol

The above 2-hydroxy-4-methoxy-3-methylbenzoic acid (4 g, 0.022 mole) is refluxed for 5 hr with hydriodic acid (80 ml, d = 1.87). The mixture is diluted with water (100 ml) and filtered. The clear solution is treated with sodium bisulphite till it becomes colourless. It is saturated with sodium chloride and extracted with ether (4 × 75 ml). The ether extract is dried (magnesium sulphate) and distilled. The residual product is crystallised from benzene–petroleum ether as colourless elongated prisms. Yield 2 g (74%). M.p. 118-19° (lit.[22] m.p. 119-121°).

1.18 2,4-DIHYDROXYTOLUENE (4-METHYL RESORCINOL)

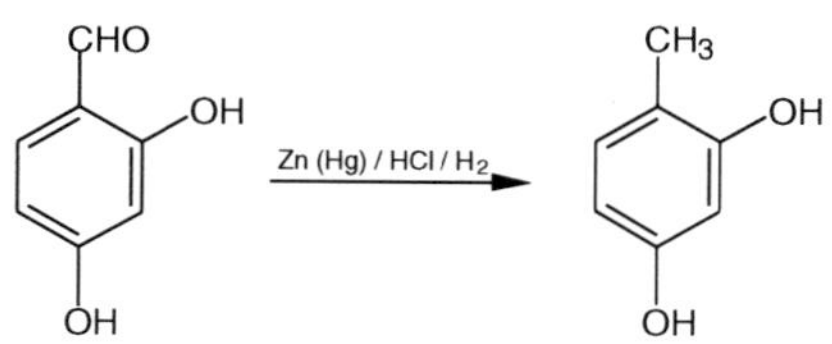

β-Resorcylaldehyde 4-Methylresorcinol

A mixture of granulated zinc (50 g, 0.76 mole), mercuric chloride (3.8 g, 0.014 mole), concentrated hydrochloric acid (2.5 ml) and water (65 ml) is shaken vigorously for 5 min. The aqueous solution is decanted and the amalgamated zinc is covered with water (20 ml) and concentrated hydrochloric acid (40 ml). β- resorcylaldehyde (7.5) (20 g, 0.145 mole) is then added immediately and a current of hydrogen is passed through the reaction mixture with occasional shaking for 30 min. The reaction mixture is diluted with water (25 ml) and extracted with ethyl acetate (3 × 100 ml).

The combined ethyl acetate extract is washed with water, dried (anhydrous potassium sulphate) and distilled. The dark red product thus obtained is chromatographed over silica gel column (4 × 40 cm). Elution with chloroform-methanol (95:5) gave the required 2,4–dihydroxytoluene. Yield 6.5 g (36.2%). M.p. 104-105° (lit.[23] m.p. 104-105°).

1.19 2,4-DIHYDROXY-1-ETHYLBENZENE

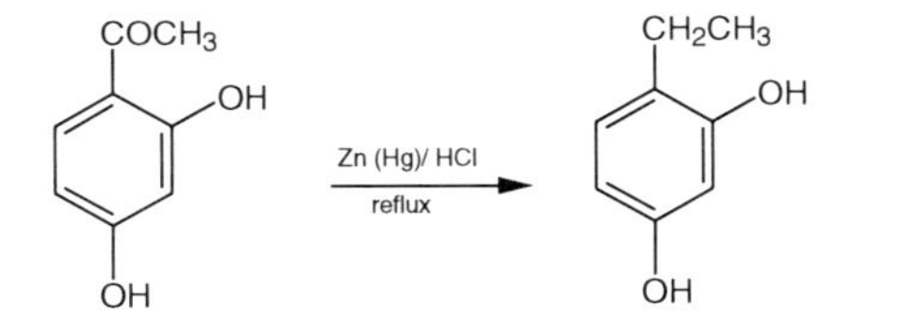

β-Resacetophenone 2,4-Dihydroxy-1-ethylbenzene

2,4-Dihydroxyacetophenone (β-resacetophenone) (4.A.6) (10 g, 0.066 mole) is added to zinc amalgam [obtained by shaking vigorously for 5 min a mixture of granulated zinc (50 g, 0.76 mole), mercuric chloride (3.8 g, 0.014 mole), concentrated hydrochloric acid (2.5 ml) and water (65 ml); the aqueous solution is decanted and the formed amalgamated zinc is covered with water and concentrated hydrochloric acid (100 ml)]. The mixture is refluxed for 3-4 hr, concentrated hydrochloric acid (5 ml) is added every hour. After the completion of the reaction (as indicated by TLC and negative ferric reaction), the supernatant liquid is decanted from the unreacted amalgam. The solution is saturated with sodium chloride and extracted with ether (4 × 50 ml). The ether extract is washed with water, dried (anhydrous sodium sulphate) and distilled. The residual product is crystallised from benzene as colourless crystals. Yield 7.2 g (79.3%). M.p. 97-8° (lit.[24] m.p. 97-8°).

1.20 P-BENZOQUINONE

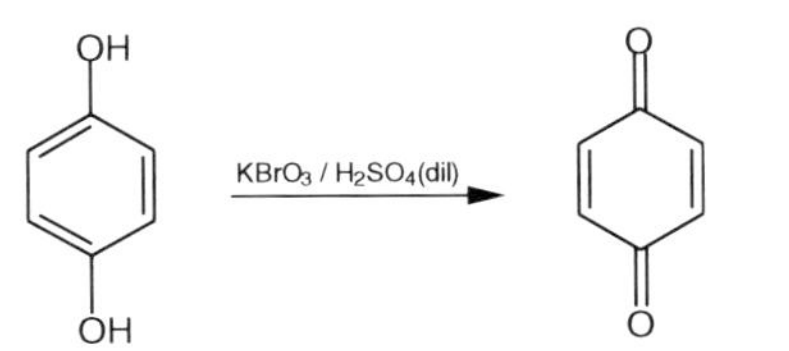

Hydroquinone p-Benzoquinone

A mixture of hydroquinone (10 g, 0.09 mole), potassium bromate (5.6 g, 0.033 mole), sulphuric acid (1N, 5 ml) and water (100 ml) is slowly heated to 60° in a water bath. After a few minutes, the temperature is raised to 80° and maintained for 10 min. The reaction mixture is cooled in ice-bath. The separated quinone is filtered, washed with cold water and dried in vacuum desiccator over fused calcium chloride. Yield 7 g (71.4%). M.p. 115°. It can be crystallised from petroleum ether (lit.[25] m.p. 117°).

1.21 1,2,4-TRIACETOXYBENZENE (HYDROXYHYDROQUINONE TRIACETATE)

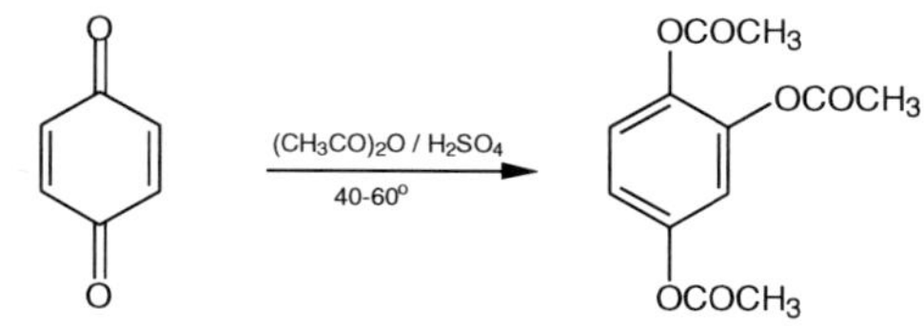

p-Benzoquinone   1,2,4-triacetoxybenzene

Concentrated sulphuric acid (12 ml) is added to acetic anhydride (180 g, 166 ml, 1.76 mole) in a beaker after which p-benzoquinone (1.20) (60 g, 0.55 mole) is added in small portions with constant mechanical stirring; the temperature of the reaction mixture is maintained between 40-60° throughout the addition. The reaction mixture is cooled and poured over crushed ice (500 g). The separated triacetate is filtered and washed with water. It is crystallised from dilute alcohol as rectangular prisms. Yield 120 g (85.7%) m.p. 96-97° (lit.[26,27] m.p. 96-97°).

1.22 1,2,4-TRIMETHOXYBENZENE

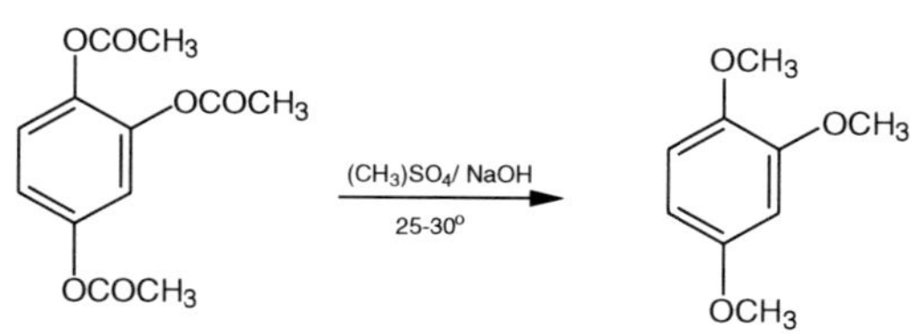

1,2,4-Triacetoxybenzene   1,2,4-Trimethoxybenzene

To a cooled solution of 1,2,4-triacetoxybenzene (1.21) (12.5 g, 0.05 mole) in methanol (50 ml) and dimethyl sulphate (45 ml, 60 g, 0.475 mole) is added dropwise during 2 hr, a solution of sodium hydroxide (37.5 g in 37.5 ml water) with efficient stirring. The temperature of the reaction mixture is kept between 25–30° throughout this addition. The reaction mixture is stirred for 30 min more and left overnight. It is extracted with ether (3 × 70 ml). The ether extract is washed with water, dried (sodium sulphate) and distilled. On distillation the required 1,2,4-trimethoxybenzene is obtained as colourless oil. Yield 6 g (72.3%). B.p. 247° (lit.[27] b.p. 247°).

1.23 1,2,4-TRIHYDROXYBENZENE (HYDROXYQUINOL)

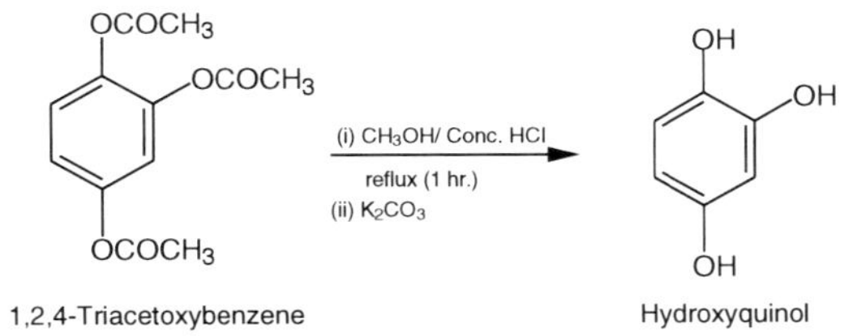

A mixture of 1,2,4-triacetoxybenzene (1.22) (50 g, 0.198 mole), methyl alcohol (100 ml) and concentrated hydrochloric acid (15 ml) is refluxed for 1 hr. The alcohol is distilled below 30° under reduced pressure until the dark green coloured liquid acquired a brown colour. Ether (150 ml) is added to the solution followed by addition of potassium carbonate to neutralise the acid. The ether is distilled off and the separated product is filtered. It is chromatographed over silica gel column. Elution with benzene-ethyl acetate (3:1) gives the trihydroxy benzene. It is crystallised from ether as small grey coloured crystals. Yield 20 g (80%). M.p. 140-41° (lit.[28] m.p. 140-41°).

1.24 2-METHOXY-1,4-DIHYDROXYBENZENE (o-METHOXYHYDROQUINONE)

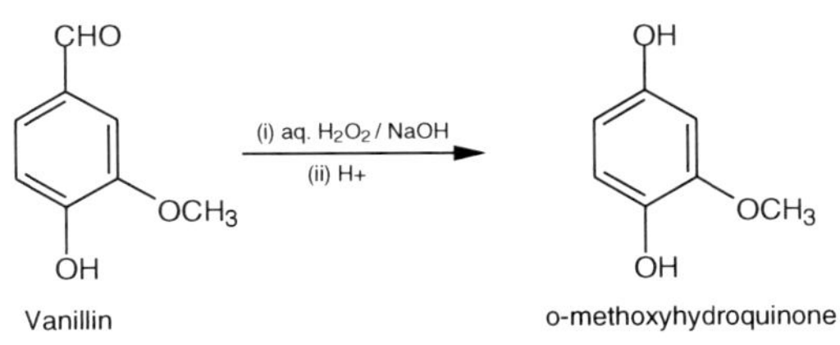

A solution of 3-methoxy-4-hydroxybenzaldehyde (vanillin) (1 g, 0.0065 mole) in aqueous sodium hydroxide (1N, 10 ml) is treated with aqueous hydrogen peroxide (6%, 10 ml, excess) at 0° in an atmosphere of nitrogen. The initial precipitate that forms gradually goes into solution and after 1 hr the reaction is complete. The reaction mixture is acidified with hydrochloric acid and extracted with chloroform (3 × 30 ml). The extract is dried (anhydrous sodium sulphate) and distilled. It is crystallised from benzene—petroleum ether as colourless needles. Yield 0.65 g (70.7%). M.p. 82° (lit.[29] m.p. 82° (dec.)).

1.25 2-METHOXY-p-BENZOQUINONE

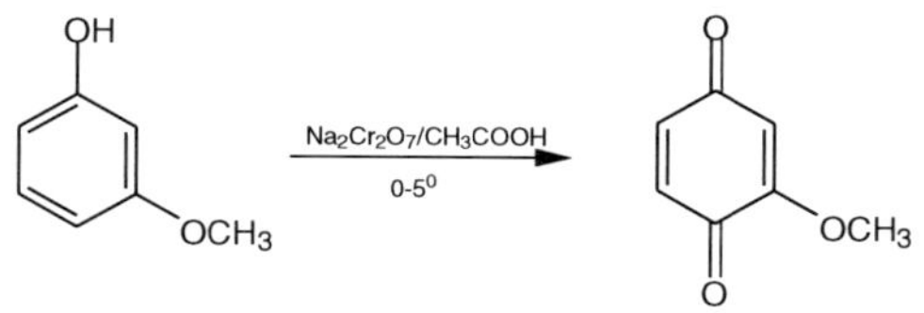

Resorcinol monomethyl ether   2-Methoxy-p-benzoquinone

Resorcinol monomethyl ether (1.3) (2 g, 0.016 mole) in acetic acid (20 ml) is treated with a solution of sodium dichromate in acetic acid (10%, 20 ml) during 4 hr; the solution is stirred at 0-5° throughout the addition. The mixture is stirred for 4 hr more and the temperature is allowed to rise to room temperature. The reaction mixture is diluted with water (100 ml) and extracted with ethyl acetate (4 × 50 ml). The extract is washed with water, dried (anhydrous sodium sulphate) and distilled. The quinone thus obtained is crystallised from hot water as orange yellow crystals. Yield 0.6 g (27%). m.p. 142-43° (lit.[30] m.p. 145°).

1.26 2,3-DIMETHOXY-p-BENZOQUINONE

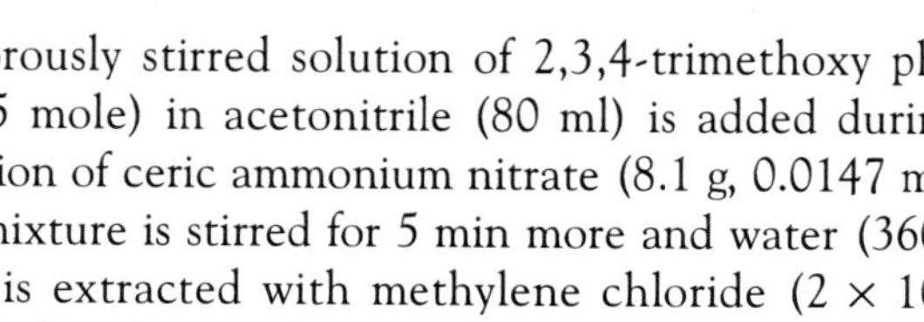

2,3,4-Trimethoxy phenol   2,3-Dimethoxy-p-benzoquinone

To a vigorously stirred solution of 2,3,4-trimethoxy phenol (1.40) (1.2 g, 0.0065 mole) in acetonitrile (80 ml) is added during 5 min an aqueous solution of ceric ammonium nitrate (8.1 g, 0.0147 mole in 80 ml water). The mixture is stirred for 5 min more and water (360 ml) added. The mixture is extracted with methylene chloride (2 × 100 ml). The extract is washed with water (2 × 50 ml), sodium bicarbonate solution (5%, 2 × 50 ml), dried (anhydrous sodium sulphate) and concentrated. The separated 2,3–dimethoxybenzoquinone is filtered. Yield 0.7 g (64.2%). M.p. 68° (lit.[31] m.p. 66-7°)

1.27 2,5-DIMETHOXY-1,4-BENZOQUINONE

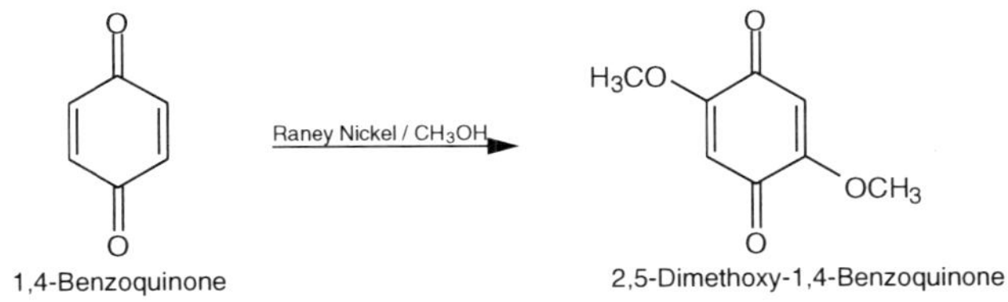

1,4- Benzoquinone (1.20) (1.08 g, 0.01 mole) is dissolved in dry methanol (100 ml) in a R.B.flask. To this solution small amount of Raney Nickel (6.35) (0.5 g) is added and the mixture stirred at room temperature. The progress of the reaction is monitored by TLC. After the completion of the reaction, it is filtered and the solvent is removed in vacuo. The residual product is subjected to column chromatography over silica gel (2 × 30 cm). Elution with petroleum ether-benzene (95:5) gives the required product. Yield 1 g (59.5%). M.p. 250° (lit.[32] m.p. 250°).

1.28 3,4,5-TRIHYDROXYTOLUENE (5- METHYL PYROGALLOL)

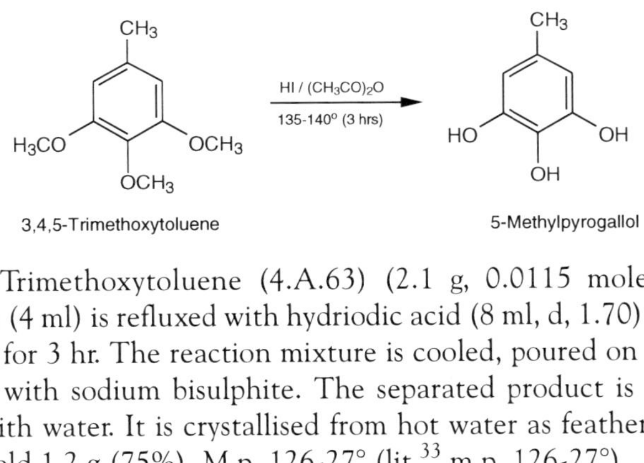

3,4,5-Trimethoxytoluene (4.A.63) (2.1 g, 0.0115 mole) in acetic anhydride (4 ml) is refluxed with hydriodic acid (8 ml, d, 1.70) at 135-140° (oil bath) for 3 hr. The reaction mixture is cooled, poured on to ice water saturated with sodium bisulphite. The separated product is filtered and washed with water. It is crystallised from hot water as feathery colourless blades. Yield 1.2 g (75%). M.p. 126-27° (lit.[33] m.p. 126-27°).

1.29 2-BENZYL-5-METHOXY-3-METHYLPHENOL[34]

2-Hydroxy-4-methoxy-6-methylbenzophenone (4.E.116) (0.1 g, 0.0004 mole) in methanol (25 ml) is refluxed for 20 min with amalgamated zinc [prepared from zinc dust (0.1 g), mercuric chloride (0.01 g) and hydrochloric acid (0.01 ml)] and hydrochloric acid (1 ml). The solution is

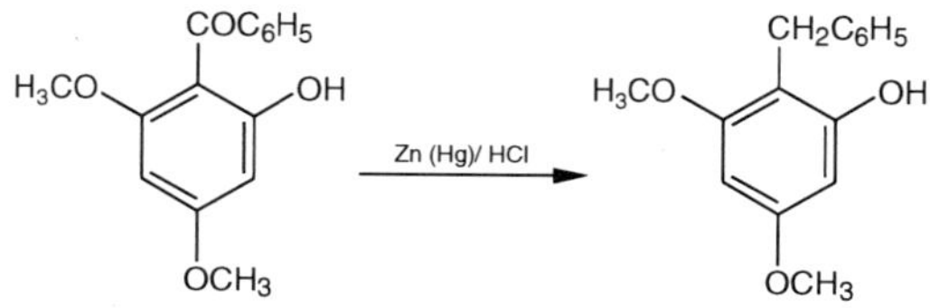

2-Hydroxy-4,6-dimethoxybenzophenone 2-Benzyl- 3,5-dimethoxyphenol

filtered, methanol removed in vacuo and water (10 ml) added. It is extracted with ether (2 × 25 ml), ether extract dried (anhydrous sodium sulphate) and distilled. TLC examination showed it to be a mixture of two products. These are separated by preparative TLC (benzene-acetone, 19:1) over silica gel plates. The upper band is eluted with ether to give the required 2-benzyl -5-methoxy-3-methyl phenol as colourless oil. Yield 0.05 g (53.2%). The lower band on elution with ether gave orcinol monomethyl ether (0.03 g) as an oil.

1.30 2-BENZYL-3,5-DIMETHOXYPHENOL[34]

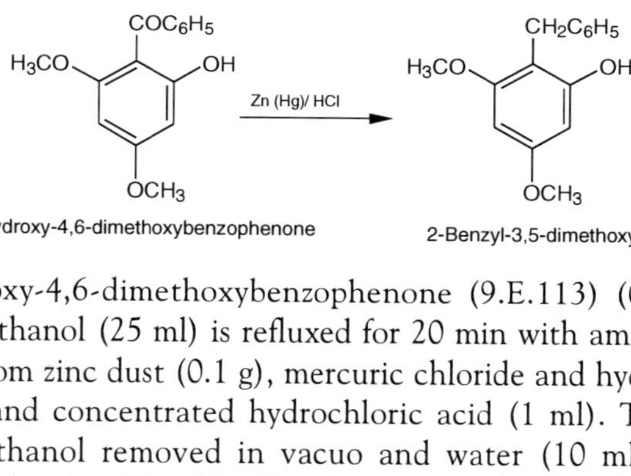

2-Hydroxy-4,6-dimethoxybenzophenone 2-Benzyl-3,5-dimethoxyphenol

2-Hydroxy-4,6-dimethoxybenzophenone (9.E.113) (0.1 g, 0.00038 mole) in methanol (25 ml) is refluxed for 20 min with amalgamated zinc [prepared from zinc dust (0.1 g), mercuric chloride and hydrochloric acid (0.01 ml)] and concentrated hydrochloric acid (1 ml). The solution is filtered, methanol removed in vacuo and water (10 ml) added. It is extracted with ether (2 × 25 ml), ether extract dried (anhydrous sodium sulphate) and distilled. TLC examination showed it to be a mixture of two products. These are separated by preparative TLC on silica gel plates using benzene-acetone (47:3). The upper band on elution with ether gave the required 2-benzyl-3,5-dimethoxyphenol as colourless oil. Yield 0.05 g (52.6%). The lower band is eluted with ether to give phloroglucionol dimethyl ether (0.04 g) as an oil.

1.31 2,4,6-TRIHYDROXYTOLUENE (C-METHYL PHLOROGLUCINOL)

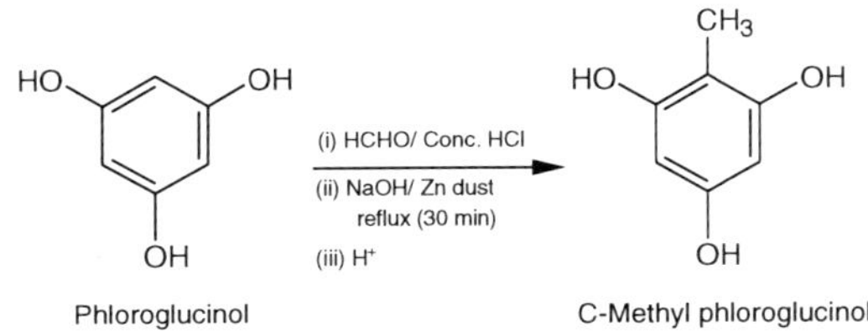

A solution of phloroglucinol (43.2 g, 0.34 mole) in water (2 litre) is added with stirring to a mixture of formalin (30 ml, 40%, 0.4 mole) and concentrated hydrochloric acid (100 ml, sp. gr. 1.12). The reaction mixture is stirred vigorously for 5-10 min and then allowed to stand undisturbed for 2 hr. The hexahydroxy diphenyl methane which separates as a white flocculent mass is filtered and pressed as dry as possible. The crude product is dissolved in sodium hydroxide solution (250 ml, 15%) and to the reddish solution so obtained is added zinc dust (100 g). The mixture is refluxed for 30 min and filtered; the residual zinc dust is washed with hot water (2 × 75 ml). The combined filtrate (total volume approx. 600 ml) is acidified to congo red with hydrochloric acid and extracted with ether (5 × 200 ml). The ether extract is distilled and the crude product thus obtained is found to be a mixture of C-methyl phloroglucinol and phloroglucinol. The crude product is dissolved in water (50 ml), treated with decolourising carbon, filtered and cooled in a refrigerator (24 hr). The separated phloroglucinol is filtered and the filtrate extracted with ether (3 × 75 ml). The ether extract is dried (anhydrous sodium sulphate) and distilled. The product obtained is dissolved in ethyl acetate (50 ml) and impurities precipitated by adding xylene. The clear solution on cooling gave C-methylphloroglucinol. Yield 10 g (20.8%). m.p. 213-215° (lit.[35] m.p. 213-15°)

1.32 2,6-DIMETHOXY-4-HYDROXYTOLUENE

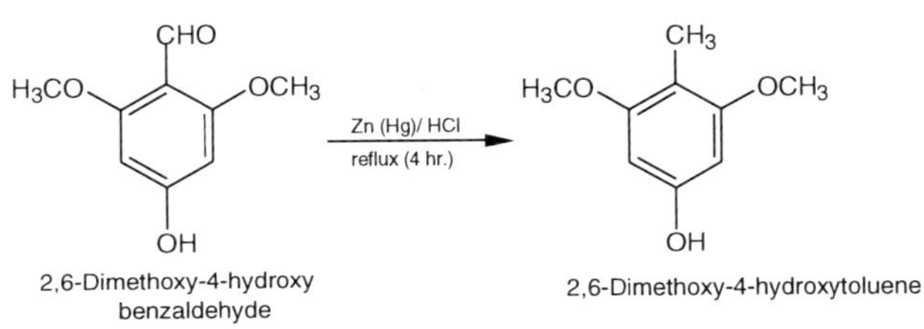

A solution of 2,6-dimethoxy-4-hydroxybenzaldehyde (2.15) (9 g, 0.0494 mole) in glacial acetic acid (80 ml) is added to zinc amalgam (prepared from 40 g zinc dust). Hydrochloric acid (4N, 100 ml) is added and the reaction mixture refluxed for 4 hr. It is filtered and the zinc dust washed with acetic acid (2 × 20 ml). The combined filtrate is concentrated in vacuo. The product obtained is crystallised from acetic acid as colourless needles. Yield 6 g (72.3%). M.p. 149-50° (lit.[36] m.p. 149-50°).

1.33 2,4,6-TRIMETHOXYTOLUENE(C-METHYL PHLOROGLUCINOL TRIMETHYL ETHER)

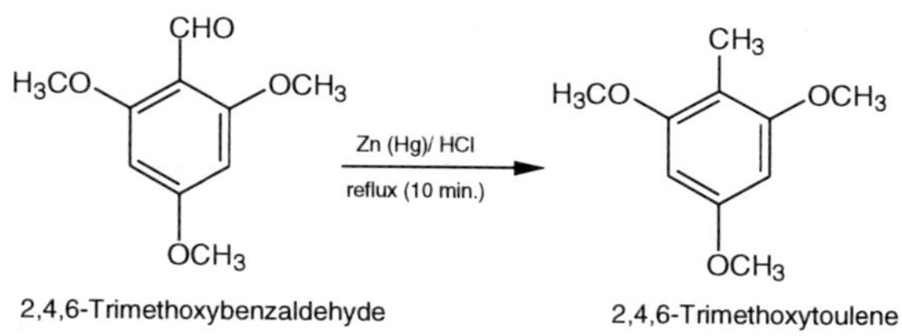

A solution of 2,4,6-trimethoxybenzaldehyde (2.18) (4.5 g,0.023 mole) in ethyl acetate (50 ml) is added to zinc amalgam (prepared from 20 g zinc dust, mercuric chloride and hydrochloric acid). Hydrochloric acid (30 ml) is added and the reaction mixture refluxed for 10 min. It is filtered in hot and the residual zinc washed with hot ethyl acetate (2 × 15 ml). The organic layer is separated and the aqueous solution extracted with ethyl acetate (2 × 20 ml). The combined ethyl acetate solution is washed with water, dried (anhydrous sodium sulphate) and distilled. The residual product is purified by column chromatography over silica gel column (2 × 30 cm). Elution with petroleum ether gave the required product as oil. Yield 3 g (71.9%). B.p. 140-42°/18 mm (lit.[37] b.p. 140-42°/18 mm).

1.34 1,3-DIHYDROXY-2,5-DIMETHOXYBENZENE

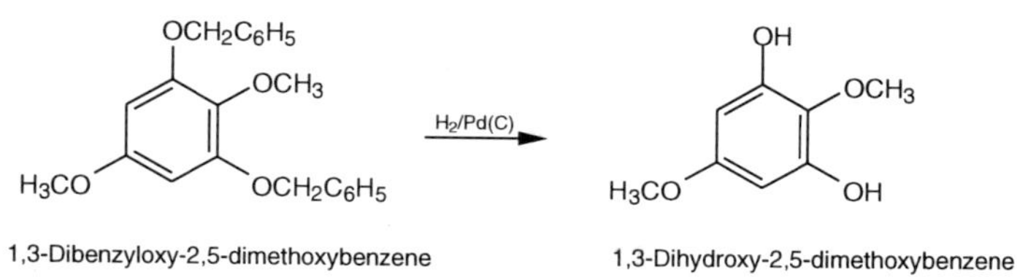

Palladium charcoal (10%, 1 g) in ethyl acetate (30 ml) is saturated with hydrogen gas with stirring. A solution of 1,3- dibenzyloxy-2,5-dimethoxybenzene (4.B.80) (1 g, 0.0028 mole) is then added and hydrogenation continued till absorption of hydrogen ceases. The reaction mixture is distilled to give the required product. Yield 0.3 g (61.8%). It crystallised from methanol as colourless needles. M.p. 62-63° (lit.[38] m.p. 61-2°).

1.35 1,2-DIHYDROXY-3,5-DIMETHOXYBENZENE

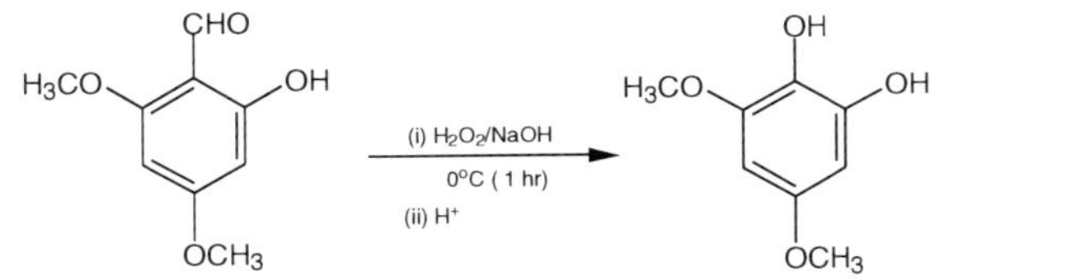

2-Hydroxy-4,6-dimethoxybenzaldehyde 1,2-Dihydroxy-3,5-dimethoxybenzene

A solution of 2-hydroxy-4,6-dimethoxybenzaldehyde (2.16) (0.9 g, 0.005 mole) in aqueous sodium hydroxide (N, 5 ml) is treated dropwise with hydrogen peroxide (6%, 5 ml, excess) at 0° in an atmosphere of nitrogen. The initial precipitate formed gradually dissolves. After 1 hr the reaction is complete. The reaction mixture is acidified with hydrochloric acid and extracted with chloroform (3 × 50 ml). The chloroform extract is washed with water, dried (anhydrous sodium sulphate) and distilled to give the required product as an oil (TLC showed it to be a single product). Yield 0.5 g (60%). It is characterized as its diacetate, which is prepared by acetic anhydride–pyridine (6 hr room temperature). The diacetate crystallised from ether-petroleum ether as colourless crystals, m.p. 101-102° [lit.[39] m.p. 101-102°].

1.36 1,2,3,4-TETRAMETHOXYBENZENE

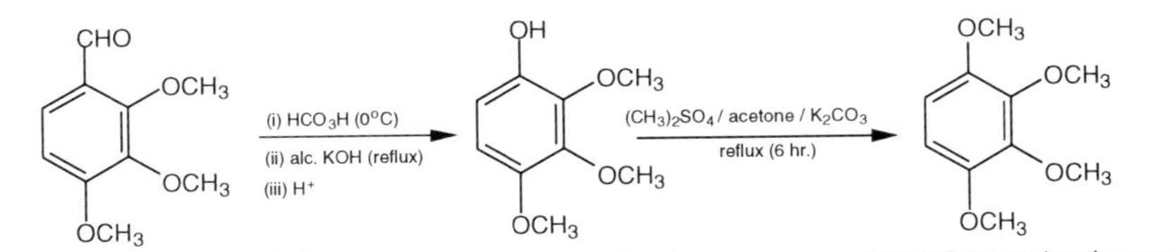

2,3,4-Trimethoxybenzaldehyde 2,3,4-Trimethoxyphenol 1,2,3,4-Tetra-methoxybenzene

Step (i) Performic Acid

Hydrogen peroxide (15 ml, 30%, 0.13 mole) is added dropwise to formic acid (75 ml, 98-100%, 1.98 mole) and the mixture is warmed to 60° for 2 min. The reaction mixture is then left at room temperature and stored in a refrigerator.

Step (ii) 2,3,4-Trimethoxyphenol

To a well stirred solution of 2,3,4-trimethoxybenzaldehyde (2.21) (10 g, 0.05 mole) in dry chloroform (100 ml) kept at 0° is added dropwise a solution of performic acid (50 ml, 0.8 mole). Addition of performic acid is stopped as soon as the absence of aldehyde is shown on TLC. The mixture is continuously stirred for 3 hr. It is treated with sodium bisulphite (2.5 g). Perfromic acid is removed by distillation in vacuo. The residual solution is rendered alkaline by adding alcoholic potassium hydroxide solution (10%). The mixture is refluxed for 1 hr with potassium hydroxide solution (10%, 100 ml). Alcohol is removed by distillation and the residue is acidified with hydrochloric acid. It is extracted with ether (3 × 50 ml). The ether extract is dried (anhydrous sodium sulphate) and distilled to give the required compound as an oil. Yield 6.1 g (65%). It is used as such for the next step.

Step (iii) 1,2,3,4-Tetramethoxybenzene

The above 2,3,4-trimethoxy phenol (5 g, 0.027 mole) in dry acetone (100 ml) is refluxed for 6 hr under anhydrous conditions with dimethyl sulphate (4 ml, 5.33 g, 0.042 mole) and ignited potassium carbonate (15 g). The mixture is filtered and the inorganic salts washed with hot acetone (2 × 25 ml). The combined acetone solution is distilled and the residue macerated with ice. The separated product is filtered and washed with water. It is crystallised from methanol as colourless needles. Yield 4.2 g (78%). M.p. 89-90° (lit.[40] m.p. 89°).

1.37 1,2,4-TRIACETOXYNAPHTHALENE

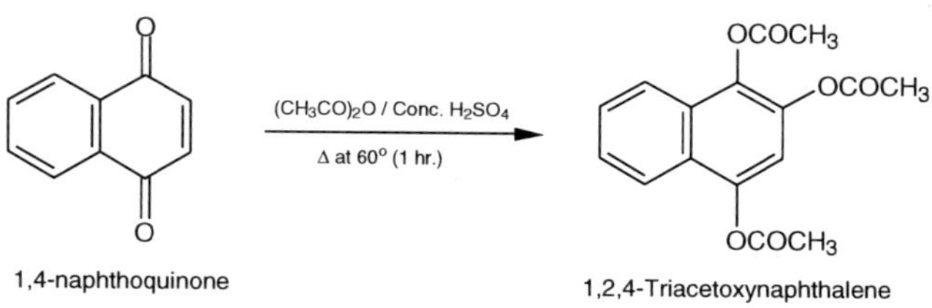

Acetic anhydride (4 ml) containing small amount (2 drops) of concentrated sulphuric acid is added to 1,4-naphthoquonone (6.E.48) (0.3 g, 0.0019 mole) (naphthoquinone is purified by crystallisation from alcohol). The mixture is heated on a water bath at 60° for 1 hr. It is poured over crushed ice. The separated product is filtered, washed with water and crystallised from alcohol as colourless needles. Yield 0.25 g (44%). M.p. 134-35° (lit.[41] m.p. 134-35°).

1.38 3,4-METHYLENEDIOXYPHENOL

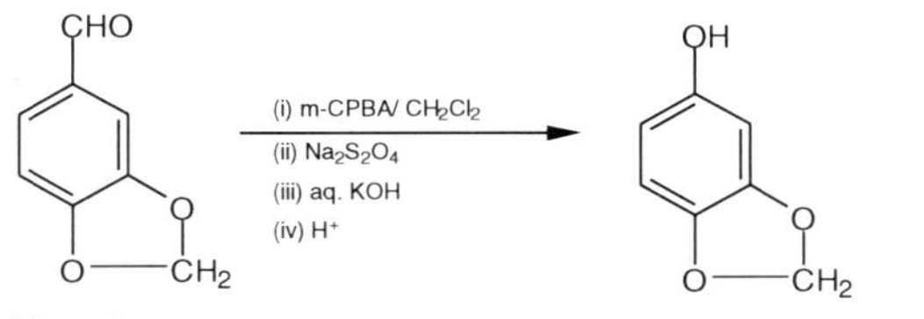

To a solution of pipronal (3,4-methylenedioxy benzaldehyde) (1 g, 0.0066 mole) in methylene chloride (30 ml) is added m-chloroperbenzoic acid (6.A.14) (2.4 g, 0.0139 mole) and the mixture is stirred for 2.5 hr. An aqueous solution of sodium thiosulphate (10%, 15 ml) is added and the mixture stirred for 45 min more. The reaction mixture is transferred to a separating funnel and shaken vigorously for 15 min with another portion of sodium thiosulphate (10%, 15 ml). The organic phase is separated and the aqueous solution extracted with methylene chloride (2 × 20 ml). The combined methylene chloride extract is washed successively with aqueous sodium thiosulphate (10%, 2 × 50 ml), brine solution (10%, 2 × 50 ml), dried (anhydrous sodium sulphate) and concentrated to give the intermediate formate ester.

A solution of the above formate ester in a mixture of ethanol and tetrahydrofuran (1:1, 25 ml) is cooled to 10° and treated with an aqueous solution of potassium hydroxide (10%, 8 ml) added slowly. The reaction mixture is kept for 30 min at 10°. Ice-cold hydrochloric acid (5%, 15 ml) is added in three instalments during 5 min with shaking. The reaction mixture changes its colour and acquires pH 1.0. It is extracted with methylene chloride (3 × 25 ml). The extract is washed with water (2 × 25 ml), brine (2 × 25 ml), dried (anhydrous sodium sulphate) and concentrated in vacuo. The residual brownish liquid on cooling gives the required compound. Yield 0.8 g (86.9%). It is crystallised from petroleum ether. M.p. 65-66° (lit.[42] m.p. 65°).

1.39 6-ALLYLGUAIACOL (EUGENOL)

Step(i) Guaiacol allyl ether

Guaiacol (1.1) (7.5 g, 0.06 mole) and freshly ignited potassium carbonate (20 g) are taken in acetone (100 ml) and refluxed under anhydrous conditions. A solution of allyl bromide (8 g, 5.7 ml, 0.066 mole) in dry

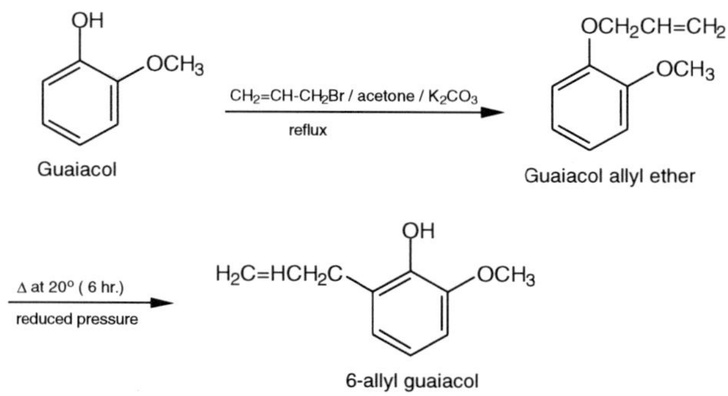

acetone(15 ml) is added in three instalments at an interval of 1 hr. The mixture is refluxed for 2 hr more. The acetone is filtered and the inorganic residue washed with hot acetone (2 × 20 ml). The combined acetone solution is distilled. The residual product is purified by passing over a small column of silica gel (2 × 20 cm) and elution with petroleum ether. Guaiacol allyl ether is obtained as an oily liquid. Yield 5.5 g (55.5%). B.p. 110°/12 mm.

Step (ii) 6-Allylguaiacol (Eugenol)

Guaiacol allyl ether (5 g, 0.03 mole) is heated for 6 hr at 200° (oil bath) under reduced pressure. A dark viscous liquid is obtained. It is cooled and purified by column chromatography over silica gel (2 × 40 cm). Elution with petroleum ether gives the required 6-allyl-2-methoxyphenol (eugenol). Yield 2.3 g (46%). B.p. 250-52° (lit.[43] B.p. 250-52°).

1.40 METHOXYPHENOLS: GENERAL METHOD OF PREPARATION

General Procedure[11]

A mixture of the appropriate benzaldehyde (0.001 mole), hydrogen peroxide (30%, 0.0025 mole), p-toluenesulphonic acid (0.0004 mole) and methanol (5 ml) is stirred at room temperature. After completion of the reaction (monitored by TLC) the reaction mixture is diluted with water (50 ml) and extracted with dichloromethane (5 × 50 ml). The extract is dried over anhydrous sodium sulphate and distilled. The crude material is purified by column chromatography over silica gel and elution with benzene.

Starting aldehyde	*Methoxyphenyl obtained*	*Yield (%)*
2,4,6-Trimethoxybenzaldehyde (2.18)	2,4,6-Trimethoxyphenol[a]	92
2,3,4-Trimethoxybenzaldehyde (2.21)	2,3,4-Trimethoxyphenol[b]	92
3,4,5- Trimethoxybenzaldehyde (2.26)	3,4,5-Trimethoxyphenol[c] (Antiarol), m.p. 148°	65
2,4,5-Trimethoxybenzaldehyde (2.25)	2,4,5-Trimethoxyphenol[d]	80
2,4-Dimethoxybenzaldehyde (2.7)	2,4-Dimethoxyphenol[e]	70
2,6-Dimethoxybenzaldehyde (2.10)	2,6-Dimethoxyphenol[f] M.p. 55-6°; B.p. 262-7°	75

a. Obtained as oil, on complete methylation it gives 1,2,3,5-tetramethoxybenzene, m.p. 47°, B.p. 271° (Dictionary of organic compounds by Chapman & Hall, 5th Edition, B-00218).

b. Obtained as oil. On complete methylation it gives 1,2,3,4 -tetramethoxybenzene, m.p. 89° (Dictionary of organic compounds by Chapman & Hall, 5th Edition, B-00217).

c. Dictionary of organic compounds by Chapman & Hall, 5th Edition, B-00218.

d. On complete methylation it gives 1,2,4,5-tetramethoxybenzene, m.p. 103° (Dictionary of organic compounds by Chapman & Hall, 5th Edition, B-00219).

e. On complete methylation it gives 1,2,4-trimethoxybenzene. b.p. 247° (Dictionary of organic compounds by Chapman & Hall, 5th Edition, B-00230).

f. Dictionary of organic compounds by Chapman & Hall, 5th Edition, Supplement –2, B-20017.

References

1. M. William, J. Org Chem. 1962 , 27, 627.
2. G.I. Feutrill and R.N. Mirrington, Aust. J. Chem., 1972, 25, 1727.
3. Dictionary of organic compounds by Chapman & Hall, 5th Edition, B- 00180.
4. Dey , Rao and Seshadri, J. Indian Chem. Soc., 1935, 12, 140.
5. Dictionary of organic compounds by Chapman & Hall, 5th Edition, B-00181.
6. Dictionary of organic compounds by Chapman & Hall, 5th Edition, B-00182.
7. E. Konz & R. Pistorius, Synthesis, 1979, 603.
8. Baker and Savage, J. Chem Soc., 1938, 1607.
9. J. Herzig and J. Pollak, Ber., 1903, 662.
10. 2,3-Dimethoxybenzaldehyde (o-verataldehyde), m.p. 54° is obtained by the wet methylation of o-vanillin as in the case of o-methoxybenzaldehyde (II.1).

11. R.N. Baruah, Indian J. Chem., 1994, 33B,, 1103.
12. W. Baker, J. Chem. Soc., 1941, 665.
13. Dictionary of organic compounds by Chapman & Hall, 5th Edition, B-00230.
14. R.I. Meltzer and J. Doczi, J. Am. Chem Soc., 1950, 72, 4986.
15. Peracetic acid solution is prepared in situ by adding 1 part of hydrogen peroxide (30%) to glacial acetic (3 parts) in presence of catalytic amount of sulphuric acid.
16. H. Weidel and J. Pollak, Monatsch, 1900, 21, 22; H. Herzig and F. Wenzcel, Monatsch, 1906, 27, 785.
17. Dictionary of organic compounds by Chapman & Hall, 5th Edition B-20019.
18. Pratt and Robinson, J. Chem. Soc., 1924, 125, 193.
19. K. Aghormurthy, N. Narascimhachari and T.R. Seshadri, Proc. Indian Acad. Sci., 1951, 33A, 257.
20. F. Henrich and G. Nachtigall, Ber., 1903, 36, 889.
21. Dictionary of organic compounds by Chapman & Hall, 5th Edition, M-00947.
22. S. Rangaswami and T.R. Seshadri, Proc. Indian Acad. Sci., 1938, 8A, 214; E.T. Jones and A. Robertson, J. Chem. Soc., 1932, 1690.
23. I. Meirovus, V. Tilika, J. Linaberg and O. Neilands, Chem. Abstr., 1970, 72, 110943.
24. T.B. Johnson and F. W. Lane, J. Am. Chem. Soc., 1921, 43, 348.
25. R. Nietzki, Ber., 1877, 10, 1934; 1886, 19, 1468; Dictionary of organic compounds by Chapman & Hall, 5th Edition, B-00515.
26. Thiele, Ber., 1898, 31, 1247.
27. T.R. Govindachari, K. Nagarajan and P. C. Parthasarthy, J. Chem. Soc., 1957, 550.
28. Dictionary of organic compounds by Chapman & Hall, 5th Edition, B-00230.
29. A Chatterjee, D. Ganguly and R. Sen, Tetrahedron, 1976, 32, 2407.
30. D.S. Deorha, V.K. Mahesh and S.P. Sareen, J. Ind. Chem. Soc., 1965, 42, 315.
31. C. Bogentoft, A.V. Klandy and K. Folkers, J. Med. Chem; 1972, 15, 1135.
32. E. Knovenagel and C. Buckel, Ber., 1901, 34, 3993; Dictionary of organic compounds by Chapman & Hall, 5th Edition, D-04301.
33. S. Goodwin and B. Witkop, J. Amer . Chem. Soc., 1957, 79, 179.
34. V.K. Ahluwalia, Devendra Kumar and M.C. Gupta, Indian. J. Chem., 1978, 16B, 574.

35. A . McGookin, A. Robertson and T.H. Simpson, J. Chem. Soc., 1951, 2021; A. Robertson and W.B. Whalley, J. Chem . Soc., 1951, 3355.
36. A.C. Jain, Pyare Lal & T.R. Seshadri, Indian J. Chem., 1968, 6, 486.
37. Dictionary of organic compounds by Chapman & Hall, 5th Edition, M-00973.
38. W. Baker, R. Nodzu and R. Robinson, J. Chem. Soc., 1929, 74.
39. S. K. Mukerjee, T. Saroja and T.R. Seshadri, Tetrahedron, 1968, 24, 6527.
40. Einhorn, Cobliner and Pfeifler, Ber., 1904, 37, 119.
41. J. Thiele, Justus Liebiggs Ann. Chem., 1911, 311, 345.
42. Y. Takata and T. Matsuda, J. Pharm. Soc. Jpn, 1954, 74, 693.
43. E. D. Laskina et al., Zh. Org. Khim., 1972, 8, 611; Dictionary of organic compounds by Chapman & Hall, 5th Edition, M-00716.

2

Aldehydes

Aldehydes (aromatic) are prepared by the following methods.

GATTERMANN FORMYLATION

This reaction is also known as Gatterman Aldehyde synthesis. It consists in the treatment of aromatic substance (phenols and phenolic ether) with hydrogen cyanide in the presence of lewis acid catalysts (zinc chloride, aluminium chloride). The reaction is believed to proceed as follows.

$$HCN + HCl \xrightarrow{AlCl_3} HN{=}CHCl$$

$$C_6H_6 + HN{=}CHCl \xrightarrow{AlCl_3} HCl + C_6H_5CH{=}\overset{+}{N}H$$

$$C_6H_5CH{=}\overset{+}{N}H \xrightarrow{H_2O} C_6H_5CHO + NH_3$$

In this reaction hydrogen cyanide is generated in situ by passing dry hydrogen chloride gas into a mixture of the organic compound, zinc cyanide and or aluminium chloride in anhydrous ether.

Following aldehydes have been prepared by this method 2.2, 5, 14, 17, 19, 20, 23, 27, 28, 30, 35.

VILSMEIER-HAAK REACTION

This method is used for the preparation of aromatic aldehydes. It consists in heating the substrate (phenolic ethers) with dimethyl formamide in the presence of phosphorous oxychloride. The formed complex ArCH $=$ $N^+(CH_3)_2Cl^-$ on heating with aqueous sodium acetate give the aldehyde in good yield.

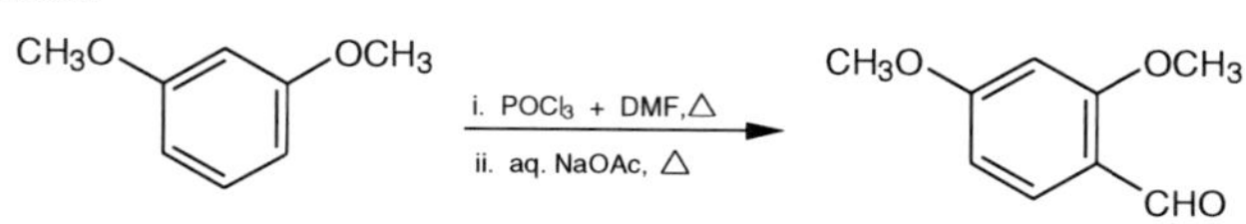

Mechanism

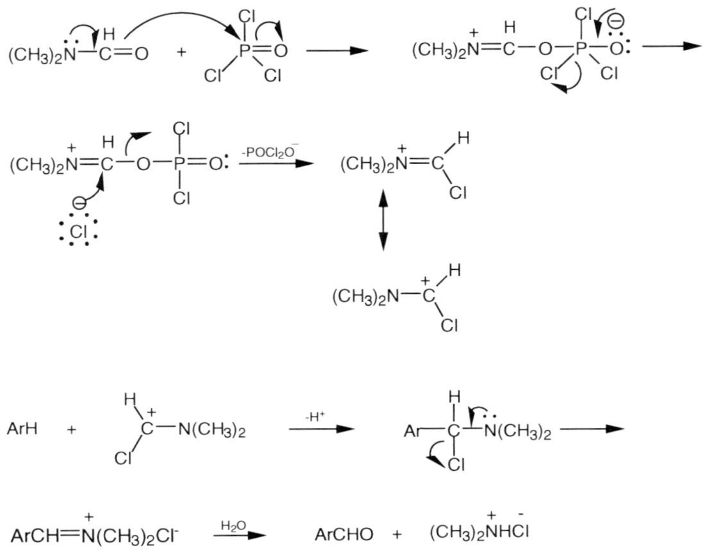

Following aldehydes have been prepared by this method 2.7, 15, 16, 18, 21, 25, 31.

3. Partial methyl ethers of some of the aromatic aldehydes are obtained by partial methylation of hydroxyaldehydes or by Elbs persulphate oxidation followed by methylation (for mechanisms of these reactions see Chapter 1 page 3 and 4).

2.1 o-METHOXYBENZALDEHYDE

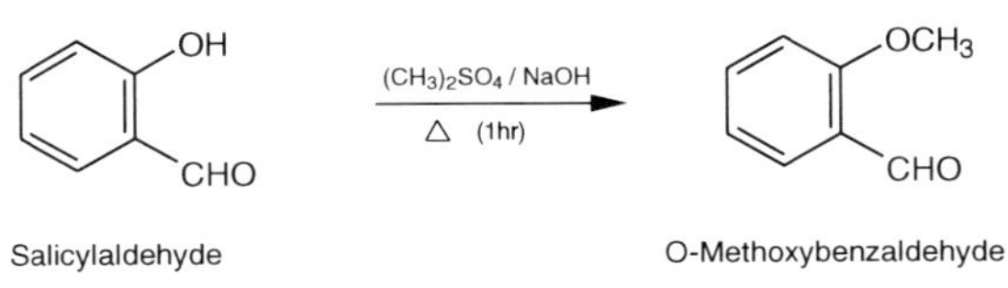

A mixture of salicylaldehyde (25g, 21.8 ml, 0.205 mole) aqueous sodium hydroxide (100 ml, 2N) and dimethyl sulphate (34.66 g, 26 ml, 0.275 mole) is heated on a water bath for 1 hr with occasional shaking. The mixture is cooled and extracted with ether. The ether solution is washed with sodium hydroxide solution (5%, 3 × 25 ml), dried (anhydrous calcium chloride). The ether is removed by distillation. The residual product is distilled and o-methoxybenzaldehyde is collected, at 241-242.° yield 13.5 g (48.4%) (lit.[1] b.p. 242-45°).

2.2 p-METHOXYBENZALDEHYDE (ANISALDEHYDE)

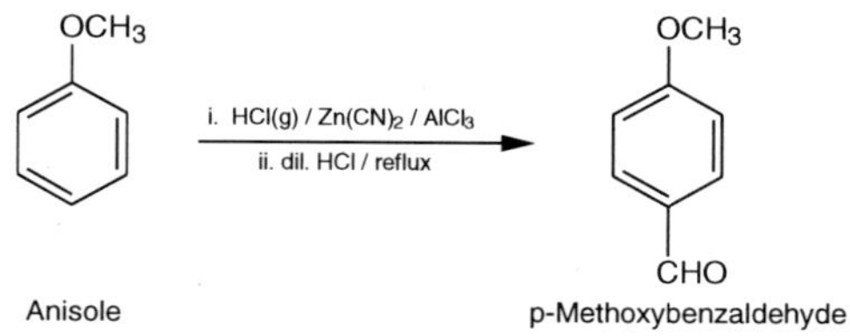

Dry hydrogen chloride gas is passed for 1 hr through a cold (5°) and stirred suspension of anisole (30 ml, 0.28 mole), anhydrous benzene (70 ml) and powdered dry zinc cyanide (52 g, 0.44 mole). Anhydrous aluminium chloride (45 g) is added slowly to the reaction mixture (the supply of hydrogen chloride is disconnected during the addition). A slow stream of hydrogen chloride gas is again passed through the stirred mixture for 3-4 hr at 40-45°. The cooled reaction mixture is poured into excess dilute hydrochloric acid. A solid imine hydrochloride is separated. The whole mixture is refluxed for 30 min to decompose the imine hydrochloride. The organic layer is separated, dried (magnesium sulphate) and is distilled to give p-methoxy benzaldehyde; it is purified by distillation, b.p. 246-248°. Yield 35 g (92%) (lit.[2] b.p. 241-250°).

2.3 m-HYDROXYBENZALDEHYDE

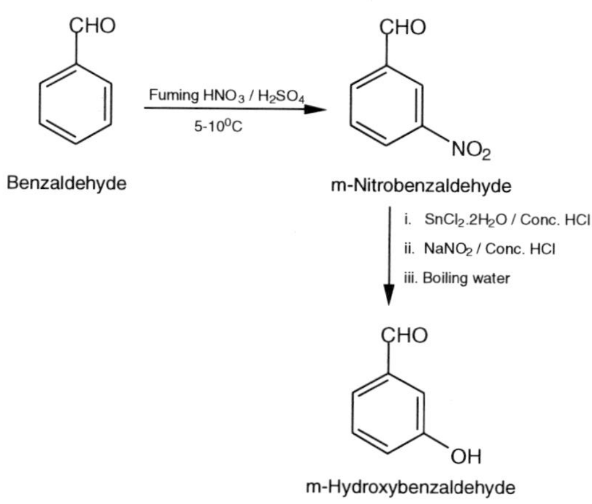

Step (i) m-Nitrobenzaldehyde

Fuming nitric acid (125 g, 83.7 ml, d, 1.49-1.50) is added fairly rapidly to stirred concentrated sulphuric acid (650 ml, d, 1.84); the mixture is cooled by ice-bath so that the temperature of the mixed acid remains $< 10°$. The mixture is stirred at 5-10° and benzaldehyde (101.5 g, 1 mole) is added during 2-3 hr. After the addition is complete the reaction mixture is allowed to stand overnight at room temperature. The mixture is poured carefully into ice (1.5 kg) with stirring. The separated product is filtered, washed with cold water and pressed as dry as possible. A small amount of oily liquid consisting of the ortho- and meta-isomers passes through the filter and is discarded. Further removal of water is accomplished by dissolving in warm benzene (75 ml) in a separatory funnel. The clear benzene solution is concentrated on a steam bath. The residual m-nitrobenzaldehyde is sufficiently pure. Yield 135.2 g (82.5 %).

Pure m-nitrobenzaldehyde is obtained by diluting the above benzene solution with additional benzene (total volume approx. 200 ml) and washing with aqueous sodium bicarbonate solution until the washings are alkaline. Incomplete washing of the product results in explosion during distillation; all remaining acid must be removed before distillation is attempted. The benzene solution is dried (sodium sulphate) and distilled. The residual oily product is distilled at 119-123°/4 mm. Yield 123 g (75%) (lit.[3] b.p. 119-23°/4 mm).

Step (ii) m-Hydroxybenzaldehyde

m-Nitrobenzaldehyde (50 g, 0.33 mole) is added in one lot to a cooled and stirred mixture (5°, ice-salt bath) of powdered stannous chloride dihydrate (250 g, 1 mole) and concentrated hydrochloric acid (300 ml). After the addition, the mixture is removed from the cooling mixture of ice-salt and stirring continued. The temperature rises slowly to 25-30° in about 5 minutes; it rises further to about 100°. At this stage the reaction mixture is again cooled (ice-salt). The stirring is continued. The formed m-aminobenzaldehyde dissolves and a red solution is obtained. The solution is cooled with slow stirring in ice-salt mixture for 2-2.5 hr. The orange red paste of stannic chloride and m-aminobenzaldehyde is filtered preferably using a sintered glass funnel. It is used as such for the next step.

To a cooled and stirred suspension of the above product in concentrated hydrochloric acid (75 ml) (ice-salt bath) is added a solution of sodium nitrite (23 g in 75 ml water) through a separating funnel, with its stem below the surface of the suspension; the temperature of the mixture is kept at 4-5°. After the addition of sodium nitrite solution is complete (1 hr), stirring is continued for 1 hr more. During this time the stannic chloride of the diazonium salt crystallises out. It is filtered using a sintered glass funnel.

The above product is added in small portions during 40 min to boiling water (900 ml). The reaction is treated with decolourising charcoal (4 g), boiled for 3-4 min, filtered and cooled (10-12 hr in ice box). The separated m-hydroxybenzaldehyde is filtered. Yield 48 g (59%). M.p. 99-101°. It is finally crystallised from benzene using decolourising charcoal as light tan crystals. M.p. 101-102°. (lit.[4] m.p. 101-102°).

2.4 m-METHOXYBENZALDEHYDE

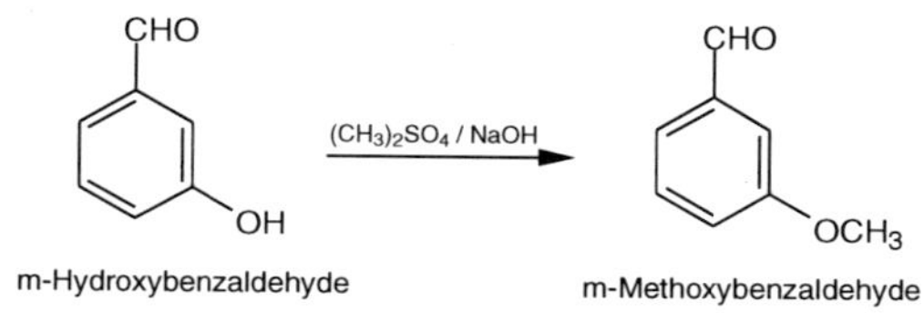

m-Hydroxybenzaldehyde (2.3) (60.4 g, 0.4 mole) is dissolved in sodium hydroxide solution (2N, 225 ml). To the dark coloured stirred solution is added dropwise dimethyl sulphate (63 g, 47.5 ml, 0.5 mole) and the temperature of the reaction mixture is maintained at 40-45°. After the addition is over, the mixture is stirred for 10 min more. Sodium hydroxide solution (2N, 137 ml) is added in one lot and dimethyl sulphate (31.5 g, 23.8 ml) is added as before, this time the temperature of the reaction mixture is allowed to rise to 50°. The mixture is stirred for 30 min at 50°, cooled and extracted with ether (2 × 150 ml). The ether solution is dried over anhydrous sodium sulphate (6 hr), filtered and distilled. m-Methoxybenzaldehyde is obtained as pale yellow liquid. Yield 86 g (63.0%). B.p. 88-90°/3 mm. (lit.[5] b.p. 88-9°/3 mm).

2.5 2,4-DIHYDROYXBENZALDEHYDE (β-RESORCYLIC ALDEHYDE)

An ice cold solution (0-10°) of dry resorcinol (10 g, 0.09 mole) and zinc cyanide (40 g, 0.34 mole) in dry ether (175 ml) is saturated with dry hydrogen chloride gas for 4-5 hr under anhydrous conditions. The separated imine hydrochloride is allowed to stand for 3 hr. Ether is

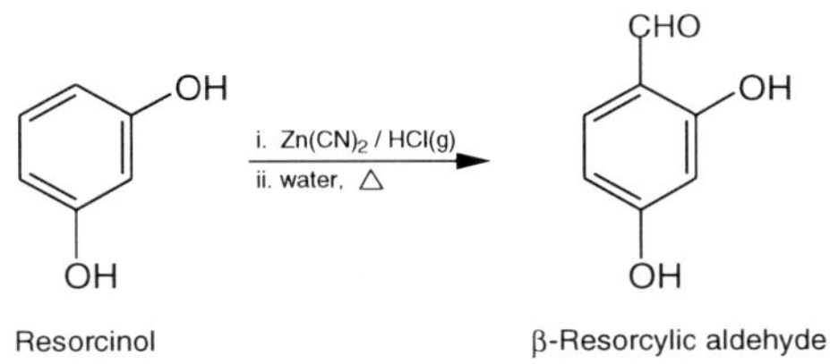

decanted and the solid product washed with dry ether (2 × 50 ml) and is hydrolysed by heating with water (70 ml) on a water bath at 60-75° for 15 min. and the solution cooled. The separated product is filtered. It is crystallised from water as colourless needles. Yield 9 g (71.8%). M.p. 136-7° (lit.[6] m.p. 136-7°).

2.6 2-HYDROXY-4-METHOXYBENZALDEHYDE

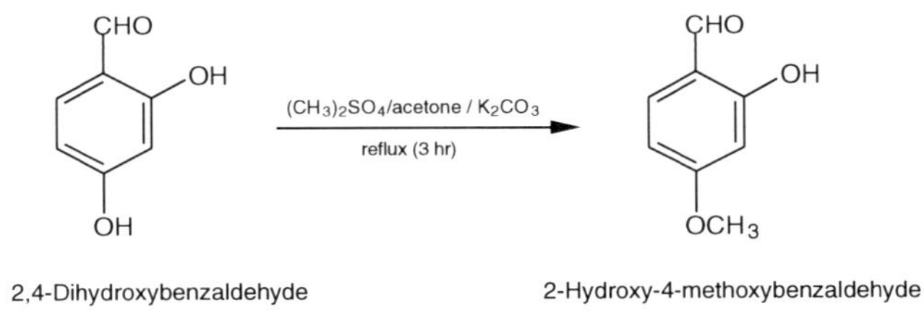

2,4–Dihydroxybenzaldehyde (2.5) (10 g, 0.072 mole) is dissolved in dry acetone (150 ml) and refluxed for 3 hr (water bath) with dimethyl sulphate (12.61 g, 9.5 ml, 0.1 mole) and ignited potassium carbonate (25 g) under anhydrous conditions. The potassium salts are filtered and the residue washed with hot acetone (2 × 25 ml). The combined acetone is distilled. 2-Hydroxy-4-methoxybenzaldehyde is obtained as the steam volatile product. It is crystallised from alcohol as colourless needles. Yield 2.5 g (22.7%). M.p. 41° (lit.[7] m.p. 40-2°).

2.7 2,4-DIMETHOXYBENZALDEHYDE

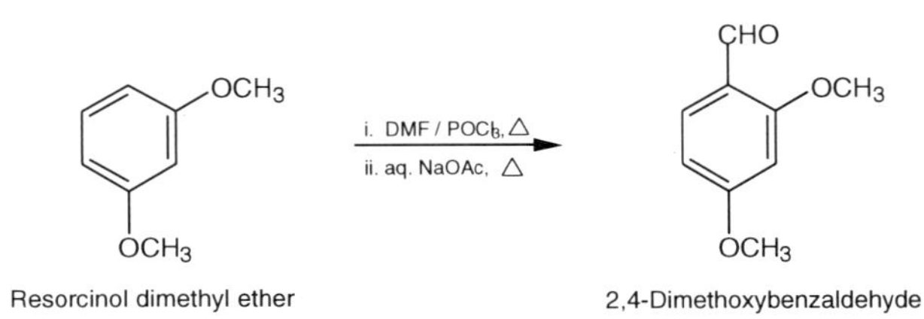

Resorcinol dimethyl ether (2.4) (14 g, 0.10 mole) is added under anhydrous conditions to N,N-dimethylformamide (10 ml, 9.44 g, 0.13 mole) and phosphorus oxychloride (10 ml, 16.45 g, 0.11 mole) is added in small amounts while cooling the reaction mixture in ice-bath. The resulting reddish-orange viscous liquid is heated at 100° (water-bath) for 6 hr under anhydrous conditions. Excess phosphorus oxychloride is removed under reduced pressure and mixture is cooled and saturated solution of sodium acetate (30 g) added. The mixture is refluxed for 30 min. It is poured into ice-water. The separated product is filtered. It is crystallised from dilute methanol as colourless plates. Yield 10 g (59.4%). M.p. 68-70° (lit.[8] m.p. 69-70°).

2.8 2,5 DIHYDROXYBENZALDEHYDE (GENTISIC ALDEHYDE)

Method (i)

Aqueous potassium persulphate (150 g, 0.55 mole in 1500 ml water) and sodium hydroxide solution (40%, 200 ml) are simultaneously added to a solution of 2-hydroxybenzaldehyde (61 g, 53.2 ml, 0.5 mole) and sodium hydroxide (25 g) in water (400 ml) with vigorous stirring at 30-35° during 90 min at such a rate that the mixture always remains alkaline. After the addition, the mixture is stirred for 1 hr more and kept at room temperature for 36 hr. The solution is cooled and acidified to congo red with hydrochloric acid. The unchanged 2-hydroxybenzaldehyde (35 g) is extracted with ether (2 × 100 ml). The aqueous solution is strongly acidified with concentrated hydrochloric acid (350 ml) and heated slowly to 70° during 1 hr. The dark–brown amorphous precipitate is filtered. 2,5–Dihydroxybenzaldehyde is extracted from the filtrate with ether (3 × 100 ml). The extract is dried (anhydrous sodium sulphate) and distilled. The residue is purified by column chromatography on silica gel and eluting with benzene. Yield 13 g (18.8%). m.p. 97-98° (lit.[9] m.p. 98-99°).

Method (ii)

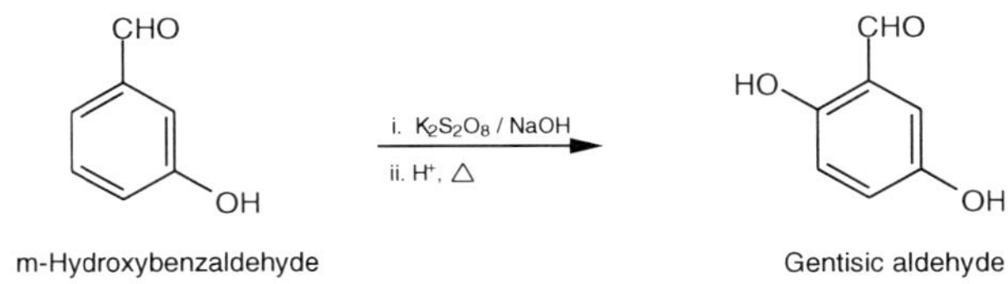

Aqueous potassium persulphate (150 g, 0.55 mole in 1500 ml water) and sodium hydroxide (40%, 200 ml) are added simultaneously to a solution of 3-hydroxybenzaldehyde (2.3) (61 g, 0.5 mole) in sodium hydroxide (25 g, 400 ml water) with vigorous stirring at 30-35° during 90 min at such a rate that the mixture is always alkaline. The reaction mixture is stirred for 1 hr more and kept at room temperature for 36 hr. The aqueous solution is acidified to congo red with hydrochloric acid and the unchanged m-hydroxybenzaldehyde is extractecd with ether (2 × 100 ml). The aqueous solution is acidified slowly with concentrated hydrochloric acid (350 ml) and heated slowly to 70° for 1 hr. The dark brown amorphous precipitate is filtered off and the filtrate is extracted with ether (3 × 100 ml). The ether solution is dried (anhydrous sodium sulphate) and distilled. The residual product is purified by column chromatography on silica gel and eluting with benzene. Yield 15 g (21.7%). M.p. 97-98°.

2.9 2,5-DIMETHOXYBENZALDEHYDE

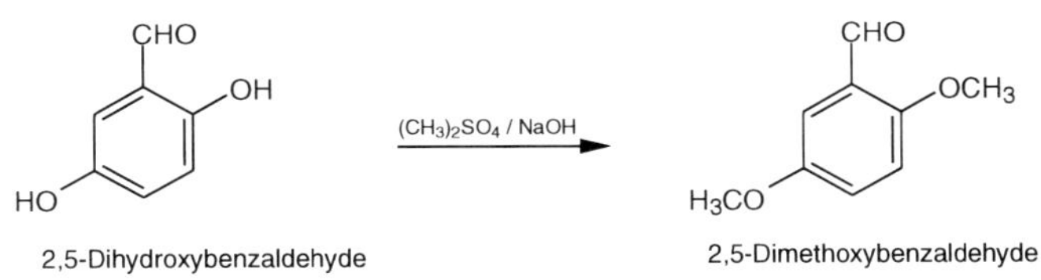

To a stirred solution of 2,5-dihydroxybenzaldehyde (2.8) (10 g, 0.072 mole) in aqueous sodium hydroxide (10%, 200 ml) is added slowly dimethyl sulphate (20 g, 15 ml, 0.158 mole). The reaction mixture is stirred for 2 hr more. It is extracted with ethyl acetate (2 × 100 ml), extract is washed with water, dried (anhydrous sodium sulphate) and solvent removed under reduced pressure. The residue is crystallised from alcohol. Yield 6.8 g (56.6%). M.p. 50° (lit.[10] m.p. 53°).

2.10 2,6 DIHYDROXYBENZALDEHYDE (γ-RESORCYLIC ALDEHYDE)

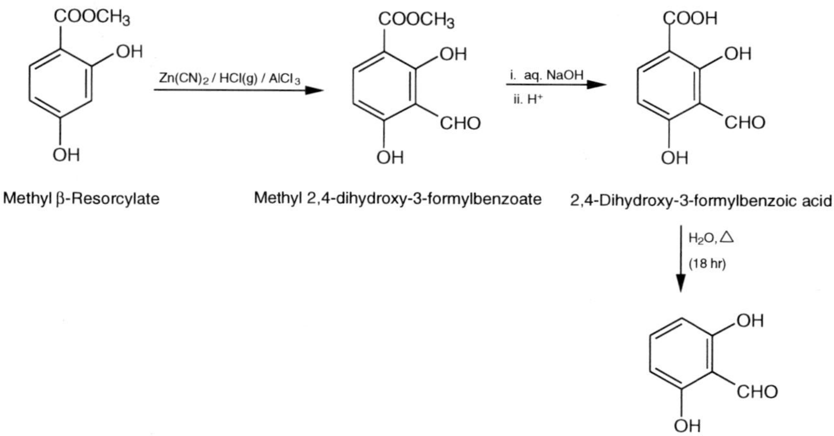

Methyl β-Resorcylate — Methyl 2,4-dihydroxy-3-formylbenzoate — 2,4-Dihydroxy-3-formylbenzoic acid — 2,6-Dihydroxybenzaldehyde

Step (i) Methyl 2,4-dihydroxy-3-formylbenzoate

To a cooled solution of anhydrous aluminium chloride (8 g, 0.06 mole) in dry ether (100 ml) is added methyl β-resorcylate (3.7) (5 g, 0.03 mole) and anhydrous zinc cyanide (7 g, 0.06 mole). The mixture is cooled to 0° (ice–salt mixture) and dry hydrogen chloride gas is passed through the stirred suspension for 5 hr under anhydrous conditions. The ether is decanted from the separated oily product and washed with more ether (2 × 50 ml). The remaining oily product is treated with water (70 ml) (caution) and the mixture is heated for 30 min at 100° (water bath) with occasional shaking. The separated product is filtered while hot. It is crystallised from alcohol as colourless needles. Yield 3 g (51.5%). M.p. 138-140° (lit.[11] m.p. 138-140°). The aqueous filtrate on cooling gave unchanged methyl β-resorcylate (0.5 g).

Step (ii) 2,4-Dihydroxy-3-formyl Benzoic Acid

The above ester (5 g, 0.025 mole) is dissolved in aqueous sodium hydroxide (N, 84 ml) and left at room temperature for 45 hr. The brownish solution is acidified with hydrochloric acid. It is extracted with ether and the ether distilled. The oily product is titurated with benzene (50 ml) to remove the unchanged ester. It is crystallised from the aqueous methanol as colourless needles. Yield 3.5 g (75.4%). M.p. 193-94° (lit.[11] m.p. 193-94°).

Step (iii) 2,6- Dihydroxybenzaldehyde

2,4-Dihydroxy-3-formyl benzoic acid (2 g, 0.011 mole) is heated with water (30 ml) at 100° (water bath) for 18 hr in a sealed tube. The mixture is cooled and extracted once with ether. The remaining reddish oily residue is separated. It is crystallised from boiling water as colourless needles. Yield 0.3 g (19.8%). M.p. 155-56° (lit.[11] m.p. 155-56°).

2.11 3,4-DIMETHOXYBENZALDEHYDE (VERATRALDEHYDE)

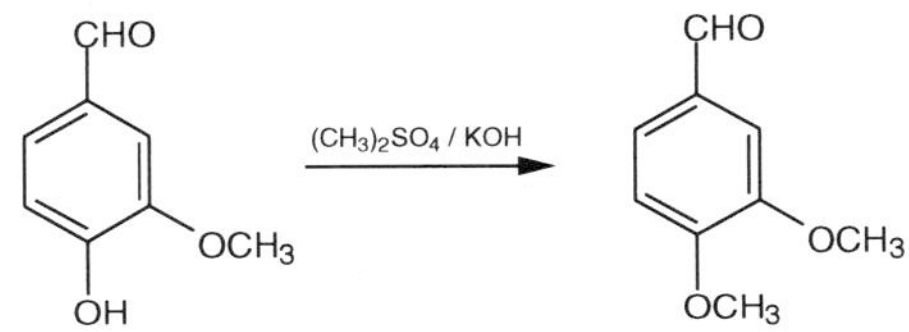

Vanillin 3,4-Dimethoxybenzaldehyde

Vanilin (76 g, 0.5 mole) is melted by warming on a steam bath in a three necked flask (500 ml capacity) fitted with a mechanical stirrer and two separatory funnels. A solution of potassium hydroxide (46 g, 0.75 mole) in water (75 ml) is added dropwise through a separatory funnel. Simultaneously dimethyl sulphate (80 g, 60 ml, 0.063 mole) is added dropwise through the second separatory funnel to the vigorously stirred mixture. The external heating is stopped after a few minutes and the mixture continues to reflux by the heat of the reaction. The addition of both the reagents is completed in 20-25 min. It is ascertained that the reaction mixture remains alkaline throughout the reaction. The reaction mixture is transferred to a large beaker and allowed to cool over night. The hard crystalline mass of veratraldehyde is removed, ground in a mortar with ice- cold water (200 ml) and filtered. It is crystallised from petroleum ether as colourless needles. Yield 79 g (95%). M.p. 42-43° (lit.[12] m.p. 43°).

2.12 3,4 DIHYDROXYBENZALDEHYDE (PROTOCATECHUALDEHYDE)

Phosphorous pentachloride (113 g, 0.54 moles) is added in small lots during 30 min under anhydrous conditions to piperonal (27 g, 0.18 mole) in 1 L R.B. flask. The reaction is vigorous in the beginning, the flask is cooled (ice) during the initial stage. The resulting mixture is gently heated over a small flame for about 45 min to expel hydrogen chloride. The

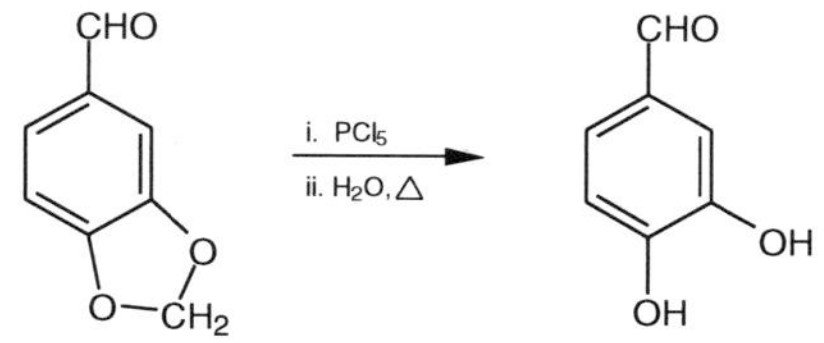

Piperonal Protocatechualdehyde

mixture is heated for 30 min on a steam bath under reduced pressure to remove the volatile material. It is poured into water (1000 ml) with caution. An oily product is formed, which becomes solid after 1-2 hr. The mixture is left overnight and heated gently for 3 hr. It is charcoalised and the clear solution concentrated in vacuo to about 150 ml. The solution is allowed to stand overnight at 0°. The separated protocatechualdehyde is filtered and washed with water. It is crystallised from water, yield 15 g (60-61%), m.p. 153-54° (decomp.) (lit.[13] m.p. 153-54°).

2.13 4,5-DIHYDROXY-3-METHOXYBENZALDEHYDE

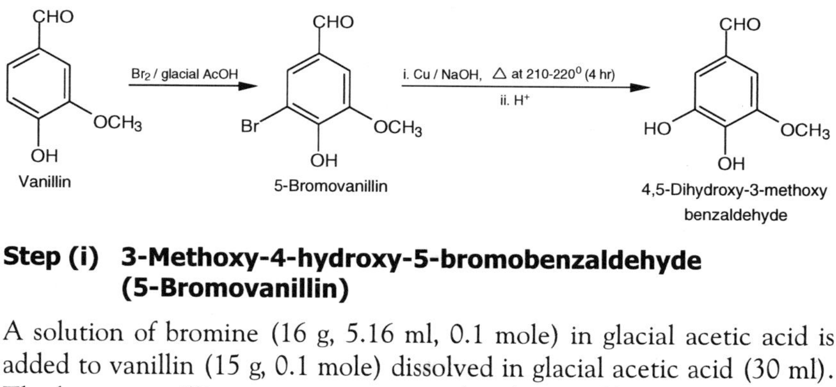

Vanillin 5-Bromovanillin 4,5-Dihydroxy-3-methoxy benzaldehyde

Step (i) 3-Methoxy-4-hydroxy-5-bromobenzaldehyde (5-Bromovanillin)

A solution of bromine (16 g, 5.16 ml, 0.1 mole) in glacial acetic acid is added to vanillin (15 g, 0.1 mole) dissolved in glacial acetic acid (30 ml). The bromo vanillin separates out immediately. It is filtered, washed with water and crystallised from alcohol as cubic crystals. Yield 16 g (70.2%). M.p. 163-64° (lit.[14] m.p. 164°).

Step (ii) 4,5 Dihydroxy-3-methoxybenzaldehyde

A mixture of 5- bromovanillin (9.2 g, 0.04 mole), copper bronze (3.6 g) and aqueous sodium hydroxide (8%, 140 ml) is heated for 4 hr in a sealed tube at 210-220° (oil bath). The mixture is cooled, filtered and acidified with dilute sulphuric acid. The clear filtrate is concentrated in vacuo and extracted with ether (5 × 50 ml). The ether solution is distilled and the crude product is purified by column chromatography over silica gel column (2 × 20 cm) and elution with benzene. The benzene elute is distilled. The separated product is crystallised from ethyl acetate-petroleum ether as colourless needles. Yield 3. 5 g (52.3%). M.p. 131-32° (lit.[15] m.p. 132-34°).

2.14 2,4,6-TRIHYDROXYBENZALDEHYDE (PHLOROGLUCINALDEHYDE)

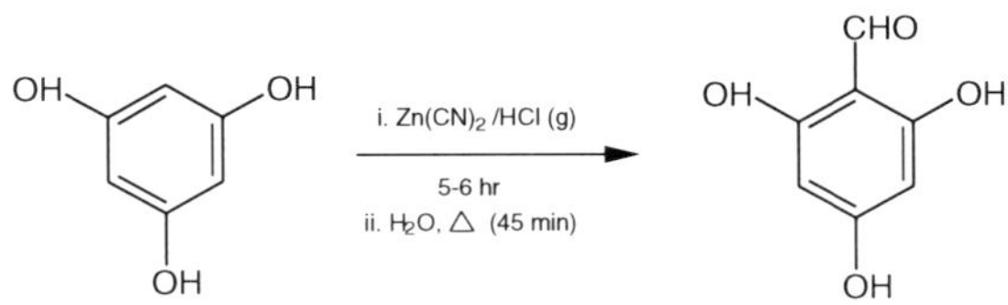

Phloroglucinol   2,4,6-Trihydroxybenzaldehyde

Into a cooled solution (0-5°) of anhydrous phloroglucinol (10 g, 0.08 mole) in dry ether (150 ml) containing dry zinc cyanide (29 g, 0.25 mole), a constant stream of dry hydrogen chloride gas is passed for 5-6 hr (under anhydrous conditions) until the oil formed becomes solid. The reaction mixture is allowed to stand overnight at room temperature. The ether is decanted and the solid washed with anhydrous ether (2 × 50 ml). The remaining imine salt is dissolved in water (200 ml) and heated on a water bath for 45 min and the solution cooled. The solid product is filtered and dissolved in ether. The clear ether solution is distilled. The product obtained is crystallised from water containing a trace of sulphur dioxide as cream coloured long needles. Yield 8 g (65.5%) it has no definite m.p.

Note: The procedure described is a modification of literature procedure[16].

2.15 4-HYDROXY-2,6-DIMETHOXYBENZALDEHYDE AND

2.16 2-HYDROXY-4,6-DIMETHOXYBENZALDEHYDE

Phosphorous oxychloride (6.5 ml, 10.7 g, 0.07 mole) is added to a mixture of di-o-methlphloroglucinol (1.13) (10 g, 0.065 mole) and dimethyl formamide (6.5 ml, 6.15 g, 0.084 mole), cooled to 0° during 30 min. The mixture is left for 2hr at room temperature with occasional shaking. The reaction mixture is treated with water (65 ml) and extracted with ether. The aqueous portion is saturated with sodium acetate and left overnight. The separated orange colour solid is filtered and stirred with ethyl acetate (60 ml) and filtered (mother liquor M). The insoluble solid product is

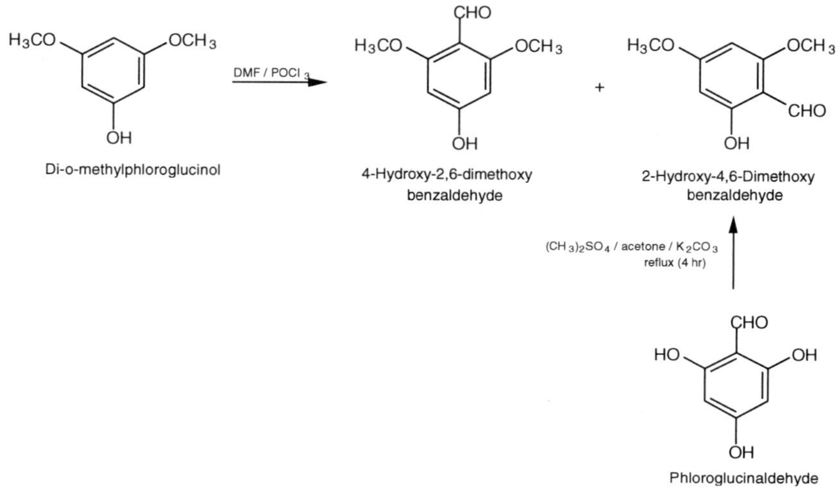

crystallised from methanol as colourless needles of 4-hydroxy-2,6-dimethoxybenzaldehyde. Yield 4 g (38%). M.p. 223-24° (lit.[17] m.p. 223-24°).

The ethyl acetate extract (M) is dried (sodium sulphate). It is filtered and treated with petroleum ether to remove the impurities. The clear solution is evaporated. The residual product is crystallised from benzene-petroleum ether as colourless needles of 2-hydroxy-4,6-dimethoxybenzaldehyde. Yield 4.1 g (34.7%). M.p. 69-70° (lit.[17] m.p. 69-70°).

2-Hydroxy-4,6-dimethoxybenzaldehyde (2.16) is also obtained by refluxing phloroglucinaldehyde (5 g, 0.03 mole) in dry acetone (100 ml) with ignited potassium carbonate (20 g) and dimethyl sulphate (10 g, 7.5 ml, 0.08 mole) for 4 hr. The mixture is worked up in the usual way. The product obtained is crystallised from methanol as colourless needles. Yield 4 g (67.8%). M.p. 69-70°.

2.17 2,4-DIHYDROXY-6-METHOXYBENZALDEHYDE

A mixture of phloroglucinol monomethyl ether (1.12) (10g, 0.07 mole), anhydrous zinc cyanide (10 g, 0.085 mole) and dry ether (100 ml) is saturated with dry hydrogen chloride gas for 4 hr at 0° (under anhydrous conditions). The reaction mixture is left overnight at room temperature. The ether layer is decanted and the remaining semi-solid viscous mass washed with dry ether (2 × 50 ml). Cold water (100 ml) is added to the oily

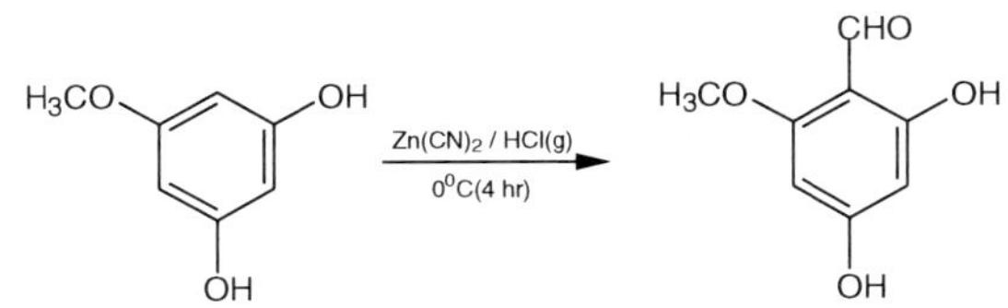

Phloroglucinol monomethyl ether — 2,4-Dihydroxy-6-Methoxy benzaldehyde

residue and the solution is heated in a water bath at 60° for 2 hr. The solution is cooled and the separated product filtered. Yield 10 g (83.3%). M.p. 200-2° (lit.[18] m.p. 200-2°).

2.18 2,4,6-TRIMETHOXYBENZALDEHYDE

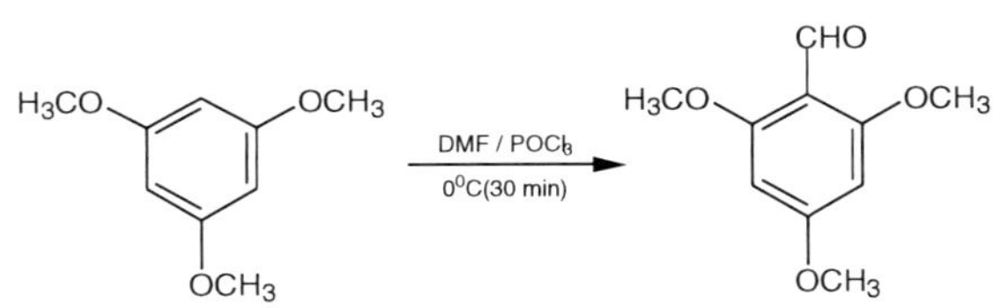

O-Trimethylphloroglucinol — 2,4,6-Trimethoxybenzaldehyde

Phosphorous oxychloride (13 ml, 21.4 g, 0.14 mole) is added to a cooled mixture (0°) of O-trimethylphloroglucinol (1.14) (20 g, 0.12 mole) and dimethyl formamide (12.4 ml, 11.7 g, 0.16 mole) during 30 min under anhydrous conditions. The mixture is allowed to stand for 1 hr at room temperature. The resulting reddish orange viscous mass is treated with water (70 ml) and extracted with ether (30 ml). The aqueous portion is saturated with sodium acetate and left overnight and the separated product filtered. It is crystallised from ethyl acetate- petroleum ether as colourless plates. Yield 10 g (42.9%). M.p. 119-20° (lit.[19] m.p. 118°).

2.19 2,4,6-TRIMETHYLBENZALDEHYDE (MESITALDEHYDE)

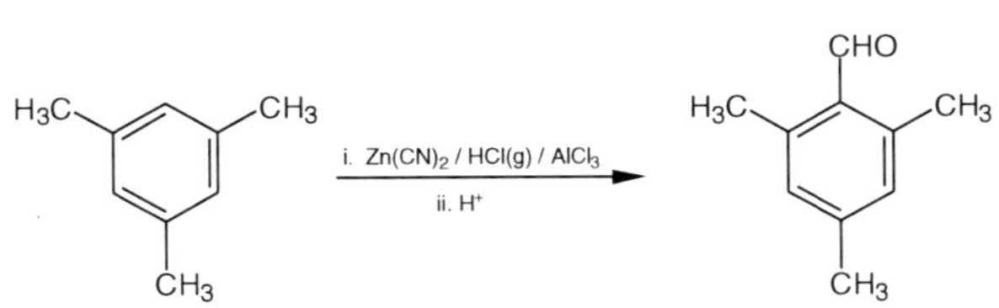

Mesitylene — Mesitaldehyde

A current of dry hydrogen chloride gas is passed through a stirred mixture of mesitylene (20.4 g, 23.6 ml, 0.17 mole), zinc cyanide (29.4 g, 0.25 mole) and carbon tetrachloride (60 ml) under anhydrous conditions contained in a three necked R.B. flask (1 litre capacity), fitted with a stirrer, a reflux condenser, an inlet tube and a thermometer. The passage of hydrogen chloride gas is continued for 2-3 hr till the zinc cyanide is decomposed (all the operations must be done in an efficient fume cupboard). The flask is now immersed in ice-bath, the inlet tube for the supply of hydrogen chloride removed and finely powdered aluminium chloride (47.8 g, 0.36 mole) is added to the mixture with vigorous stirring. The reaction is exothermic; the mixture attains a temperature of 70° at the end of an hour. It is stirred for 2 hr more at 68-70°. The cooled mixture is poured cautiously with stirring into crushed ice (about 400 g) to which has been added concentrated hydrochloric acid (20 ml). It is allowed to stand overnight and refluxed for 2.5 hr. The total mixture is steam- distilled. The first 120-130 ml of the distillate is kept separately for recovery of the solvent; the subsequent portion is collected as long as oily drops are observed (3.5- 4 hr). The steam distillate is extracted with benzene (2 × 150 ml). The solvent is removed on steam bath and the residual oily product distilled. The fraction boiling at 118-121°/16 mm is collected (lit.[20] b.p. 118-21°/16 mm). Yield 20.4 g (81%).

2.20 2,3,4-TRIHYDROXYBENZALDEHYDE (PYROGALLOL ALDEHYDE)

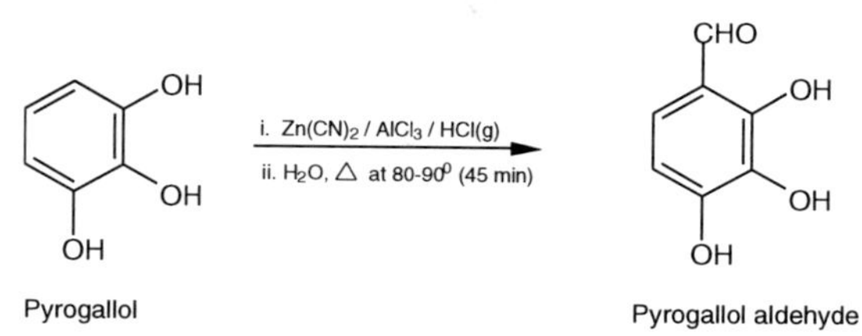

Pyrogallol (20 g, 0.16 mole) is dissolved in dry ether (200 ml) zinc cyanide (40 g, 0.34 mole) and powdered anhydrous aluminium chloride (5 g, 0.037 mole) is added at 0°. Dry hydrogen chloride gas is passed through the mixture under anhydrous conditions (0°). The separated oily aldimine hydrochloride solidifies after leaving the reaction mixture in ice-chest. The reaction mixture is allowed to stand for 2 days at room temperature. The clear ether solution is decanted and the aldimine hydrochloride washed with fresh dry ether (2 × 50 ml). It is hydrolysed by heating with water (100 ml) at 80-90° in a water bath for 45 min. The dark solution is cooled and extracted with ether. The ether extract is dried (anhydrous magnesium sulphate) and distilled to give a crystalline solid. Yield 7.2 g (29.5%). M.p. 161° (lit.[21] m.p. 161°).

2.21 2,3,4-TRIMETHOXYBENZALDEHYDE

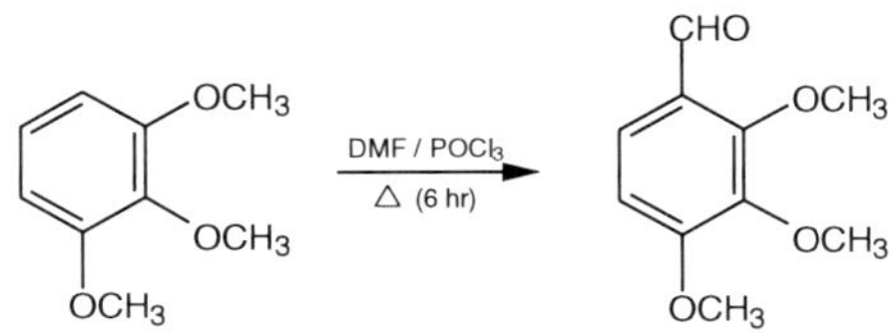

Pyrogallol trimethyl ether 2,3,4-Trimethoxybenzaldehyde

A solution of pyrogallol trimethyl ether (1.9) (15 g, 0.09 mole) in dimethyl formamide (10.5 g, 11.12 ml, 0.14 mole) and phosphorous oxychloride (15 ml, 24.7 g, 0.16 mole) is heated in a boiling water bath for 6 hr under anhydrous conditions. The mixture is treated with a saturated solution of sodium acetate (50 ml) and heating continued for 30 min more in boiling water bath. The reaction mixture is cooled, diluted with water (50 ml) and extracted with ether. The ether extract is washed with water, dried (magnesium sulphate) and distilled. The residual oily product is crystallised from petroleum as tiny needles. Yield 12 g (68.5%). M.p. 37° (lit.[22] m.p. 37°).

2.22 2-HYDROXY-3,4-DIMETHOXYBENZALDEHYDE

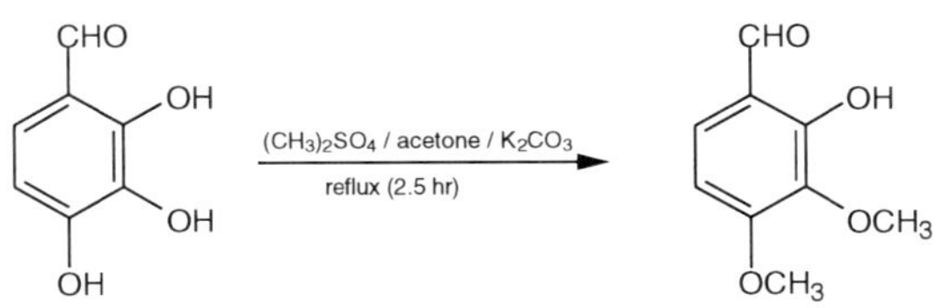

Pyrogallol aldehyde 2-Hydroxy-3,4-dimethoxybenzaldehyde

A solution of pyrogallol aldehyde (2.20) (1g, 0.0065 mole) in dry acetone (50 ml) is refluxed for 2.5 hr in a water bath with dimethyl sulphate (1.3 ml, 1.76 g, 0.014 mole) in the presence of anhydrous potassium carbonate (2 g). The inorganic salts are filtered and the residue washed with hot acetone (20 ml). The combined acetone solution is distilled and the residual product is crystallised from petroleum ether as colourless prisms. Yield 0.74 g (62.7%). M.p. 73-74° (lit.[23] m.p. 74°).

2.23 2,4,5-TRIHYDROXYBENZALDEHYDE

Method (i)

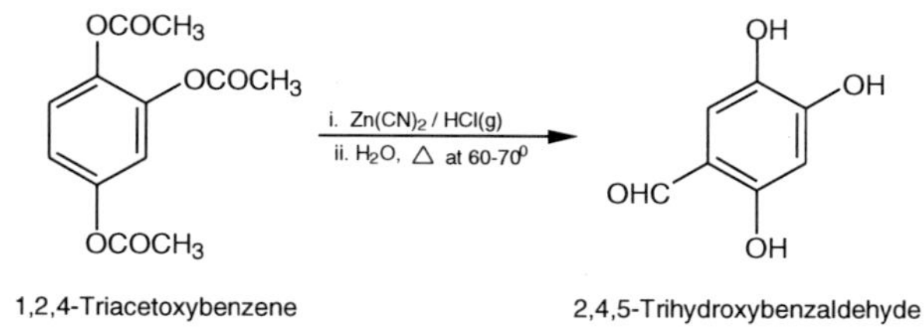

A current of dry hydrogen chloride gas is passed through a cooled solution (0°) of 1,2,4-triacetoxybenzene (1.21) (10 g, 0.04 mole) in dry ether (100 ml) containing zinc cyanide (14 g, 0.12 mole) under anhydrous conditions. After 5 hr, the current of hydrogen chloride gas is stopped and the reaction mixture left for 2-3 hr at 0°. The clear ether solution is decanted and the remaining oily product washed with ether (2 × 25 ml). It is dissolved in water (40 ml) and the solution heated at 60-70° for 15 min. The cooled solution is extracted with ethyl acetate (5 × 25 ml). The combined extract is dried (sodium sulphate) and distilled. The residual product is crystallised from water containing a little dilute hydrochloric acid as pink plates. Yield 4.0 g (65.6%). M.p. 221-22° (lit.[24] m.p. 223°).

Method (ii)

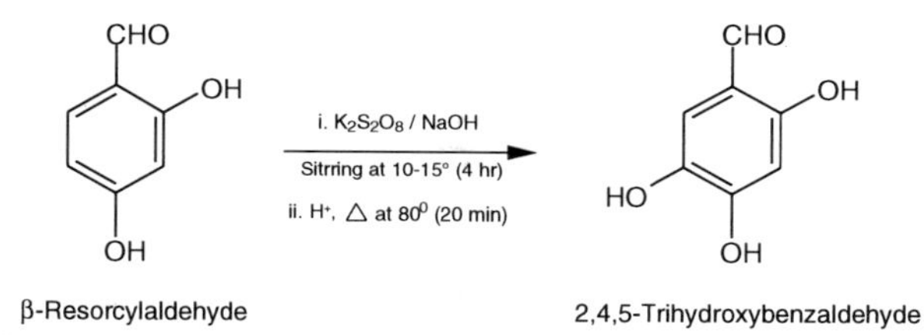

A solution of potassium persulphate (27.2 g, 0.1 mole in 500 ml water) is added dropwise to a cooled solution (5-10°) of β-resorcylaldehyde (2.5) (13.8 g, 0.1 mole) in aqueous sodium hydroxide (10%, 200 ml). The mixture is stirred for 4 hr at 10-15° and allowed to stand for 24 hr at room temperature. The mixture is cooled and acidified to congo red with dilute hydrochloric acid. The unreacted aldehyde is removed by ether extraction

(3 × 100 ml) and the remaining solution is heated for 20 min at 80° with concentrated hydrochloric acid (175 ml) and sodium sulphate (10 g). The cooled solution is extracted with ether (5 × 100 ml). The extract is dried (sodium sulphate) and distilled. The 2,4,5-trihydroxybenzaldehyde is crystallised from water as light pink plates. Yield 1.2 g (7.8%). M.p. 220°.

2.24 2-HYDROXY-4,5-DIMETHOXYBENZALDEHYDE

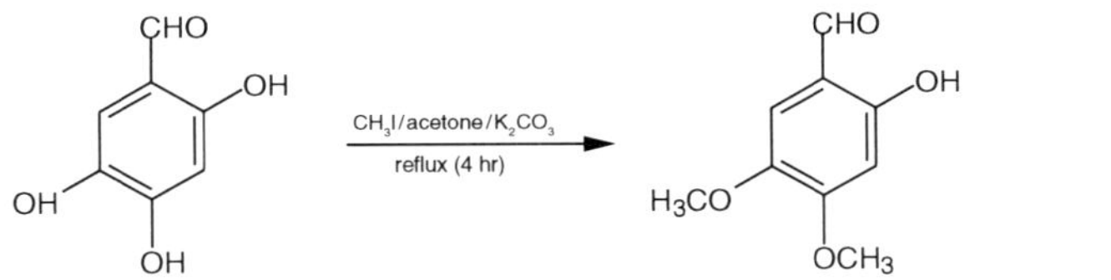

2,4,5-Trihydroxybenzaldehyde   2-hydroxy-4,5-dimethoxybenzaldehyde

A solution of 2,4,5-trihydroxybenzaldehyde (2.23) (1g, 0.0065 mole) in dry acetone (50 ml) containing ignited potassium carbonate (3 g) and methyl iodide (1.5 ml, 3.42 g, 0.024 mole) is refluxed for 4 hr till the starting compound disappeared (TLC on silica gel plates; benzene-ethyl acetate, 47:3). The acetone is filtered, inorganic residue washed with hot acetone (20 ml). The combined acetone solution is distilled. The residual product is crystallised from alcohol as cream coloured needles. Yield 0.6 g (50.8%). M.p. 104-5° (lit.[25] m.p. 105°).

2.25 2,4,5-TRIMETHOXYBENZALDEHYDE (ASARLALDEHYDE)

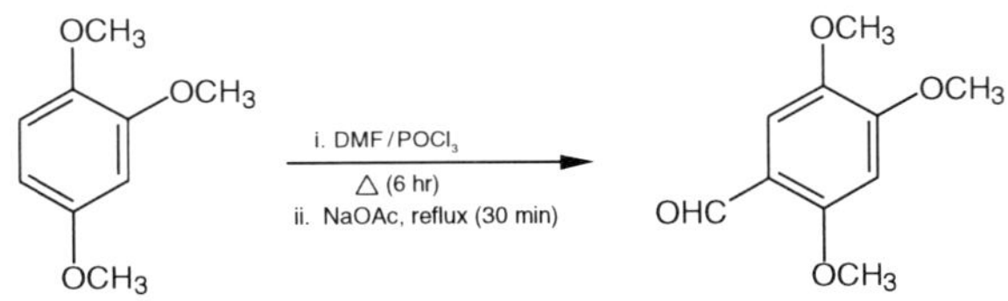

1,2,4-Trimethoxybenzene   2,4,5-Trimethoxybenzaldehyde

A mixture of 1,2,4-trimethoxybenzene (1.22) (15 g, 0.09 mole) dimethylformamide (10.5 g, 11.12 ml, 0.14 mole) and freshly distilled phosphorous oxychloride (15 ml, 24.7 g, 0.165 mole) is heated on a boiling water bath for 6 hr under anhydrous conditions. A saturated solution of sodium acetate (50 ml) is then added and the mixture refluxed for 30 min. The solution is cooled and the separated product is filtered and washed with cold water. It is crystallised from boiling water as colourless slender needles. Yield 10.5 g (60%). M.p. 113-114° (lit.[26,27] m.p. 115°).

2.26 3,4,5-TRIMETHOXYBENZALDEHYDE (TRI-O-METHYLGALLALDEHYDE)

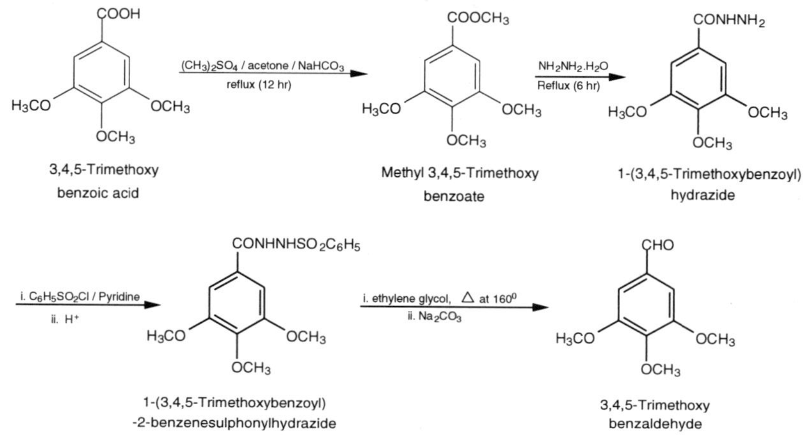

Step (i) Methyl 3,4,5-trimethoxybenzoate (See 3.28)

Step (ii) 1-(3,4,5-Trimethoxybenzoyl) hydrazide

A mixture of methyl 3,4,5-trimethoxybenzoate (44 g, 0.19 mole) and hydrazine hydrate (14 ml, excess) is refluxed for 6 hr. Excess hydrazine is removed under reduced pressure. The residue is treated with water and the separated product is filtered and washed with water. It is crystallised from alcohol as colourless shining needles. Yield 42 g (95.5%). M.p. 159-60° (lit.[20] m.p. 159-60°).

Step (iii) 1-(3,4,5-Trimethoxybenzoyl)-2-benzene sulfonylhydrazide

The above hydrazide (41 g, 0.18 mole) is dissolved in pyridine (50 ml) and treated with benzene sulphonyl chloride (34.5 g, 0.18 mole) with stirring. The mixture is stirred for 2 hr. It is poured over a mixture of ice and hydrochloric acid (the overall mixture should be acidic). The separated product is filtered and washed with water. It is crystallised from alcohol as creamish coloured plates. Yield 56 g (81.2%). M.p. 243-44°. (lit.[28] m.p. 243-44°).

Step (iv) 3,4,5-Trimethoxybenzaldehyde

The above benzene sulphonyl hydrazide (55 g, 0.14 mole) in ethylene glycol (185 ml) is heated in an oil-bath at 160°. To this hot solution, sodium carbonate (33 g) added in lots. A heavy frothing is produced. The mixture is cooled and dried (sodium sulphate) and distilled. The product is crystallised from methanol as colourless needles. Yield 13. 5 g (47.6%). M.p. 77° (lit.[28] m.p. 77°).

Note: 3,4,5-trimethoxybenzoyl chloride is also obtained by Rosenmund reduction of 3,4,5-trimethoxybenzoyl chloride with hydrogen (Pd-C). This procedure needs very elaborate arrangements (Org. Syn. Coll. Vol. VI, p. 1007).

2.27 2,4-DIHYDROXY-3,6-DIMETHOXYBENZALDEHYDE

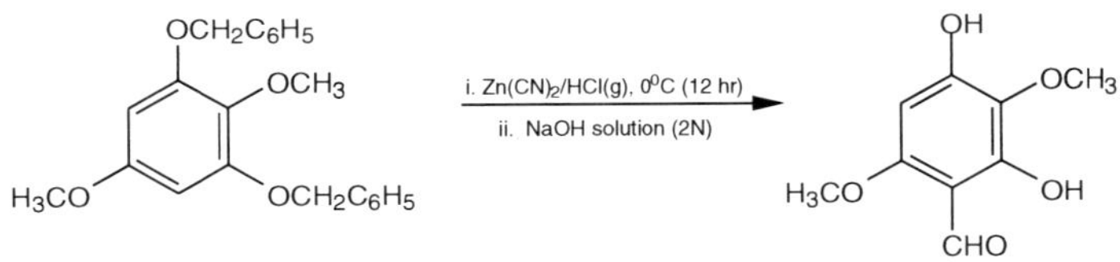

1,3-Dibenzyloxy-2,5-dimethoxybenzene

2,4-Dihydroxy-3,6-dimethoxy benzaldehyde

A mixture of 1,3-dibenzyloxy-2,5-dimethoxybenzene (4.B.80) (20g, 0.06 mole), zinc cyanide (35 g, 0.3 mole) and dry ether (250 ml) is saturated with dry hydrogen chloride gas at 0° (ice-salt mixture) for 12 hr under anhydrous conditions. The mixture is left for 48 hr at room temperature. The clear ether solution is decanted from the residual oily product and the residue is washed with more dry ether (2 × 50 ml). The remaining oily product is dissolved in water (25 ml) and the solution treated with sodium hydroxide solution (2N) until just acidic to congo red and the solution is heated for 1 hr (boiling water bath). The solution is cooled and the separated product filtered. It is crystallised twice from methanol as golden yellow needles. Yield 9 g (79.6%) m.p. 198° (lit.[29] m.p. 198°).

Note: Sometimes, a partially debenzylated compound (m.p. 154°) is obtained especially if the reaction mixture is kept for shorter duration. This is sparingly soluble in alcohol and can be separated from the required product. However, the partially debenzylated product could be smoothly debenzylated in glacial acetic acid in presence of Pd-C as catalyst in presence of hydrogen gas to yield the same required aldehyde.

2.28 2,4-DIHYDROXY-6-METHYLBENZALDEHYDE (β-ORCINALDEHYDE)

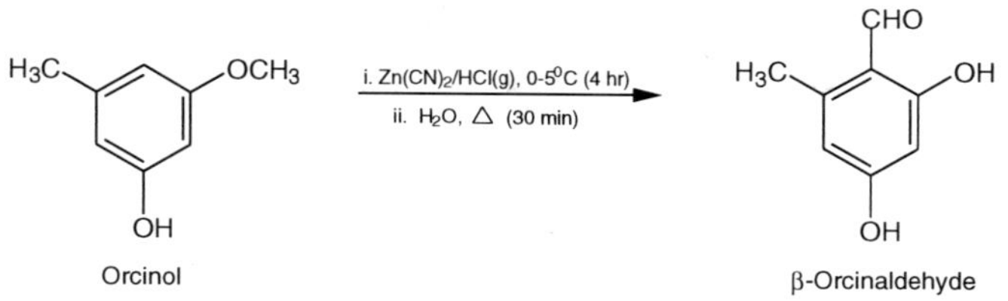

A solution of orcinol (6 g, 0.048 mole) in dry ether (100 ml) containing zinc cyanide (15 g, 0.13 mole) is saturated with dry hydrogen chloride gas at 0-5° for 4 hr under anhydrous conditions; the mixture is stirred gently during saturation. The oily aldimine hydrochloride is obtained first as a thick oil; it slowly solidifies after about 1 hr. The mixture is left in an ice chest for 2 days. The ether solution is decanted and the solid product washed with dry ether (2 × 30 ml). It is dissolved in water (100 ml) and heated on a water bath for 30 min. The resulting solution is cooled and the separated product is filtered and washed with water. It is crystallised from alcohol as colourless needles. Yield 3.5 g (47.6%). M.p. 182-83° (lit.[24,30] m.p. 180°).

2.29 2-HYDROXY-4-METHOXY-6-METHYLBENZALDEHYDE (EVERNINALDEHYDE)

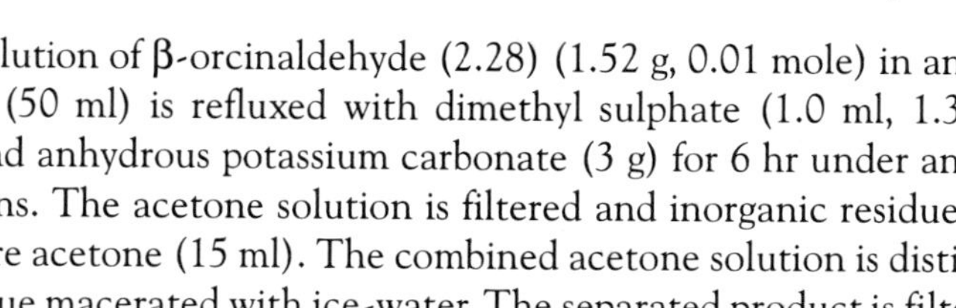

A solution of β-orcinaldehyde (2.28) (1.52 g, 0.01 mole) in anhydrous acetone (50 ml) is refluxed with dimethyl sulphate (1.0 ml, 1.3 g, 0.11 mole) and anhydrous potassium carbonate (3 g) for 6 hr under anhydrous conditions. The acetone solution is filtered and inorganic residue washed with more acetone (15 ml). The combined acetone solution is distilled and the residue macerated with ice-water. The separated product is filtered and washed with water. It is crystallised from methanol as colourless needles. Yield 1.4 g (84.3%). M.p. 63-64° (lit.[31] m.p. 65°).

2.30 2,4-DIHYDROXY-5-METHYLBENZALDEHYDE

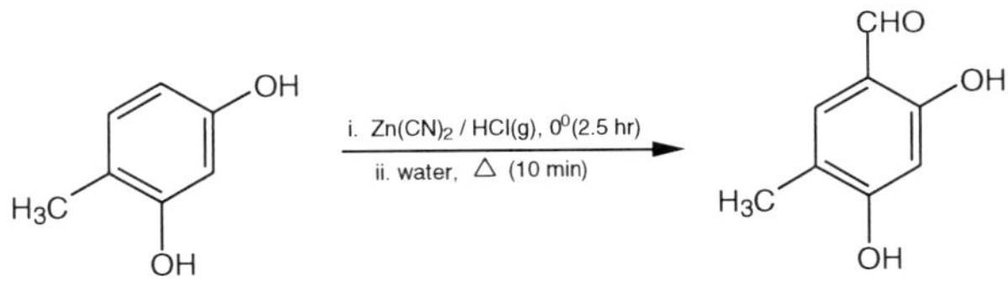

4-Methyl resorcinol 2,4-Dihydroxy-5-methylbenzaldehyde

A rapid stream of dry hydrogen chloride gas is passed through a solution of 4-methyl resorcinol (1.18) (6 g, 0.048 mole) in dry ether (75 ml) containing dry zinc cyanide (12 g, 0.10 mole) at 0° for 2.5 hr. The mixture is allowed to stand overnight. The clear ether solution is decanted and the oily residue washed with dry ether (25 ml). The remaining oily product is dissolved in water (50 ml) and heated on a boiling water bath for 10 min. The solution is cooled and the separated product is filtered. It is crystallised from chloroform as colourless shining needles. Yield 4.5 g (61.2%). M.p. 144-45° (lit.[32] m.p. 146-47°).

2.31 2,4-DIHYDROXY-3-METHYLBENZALDEHYDE

Method (i)

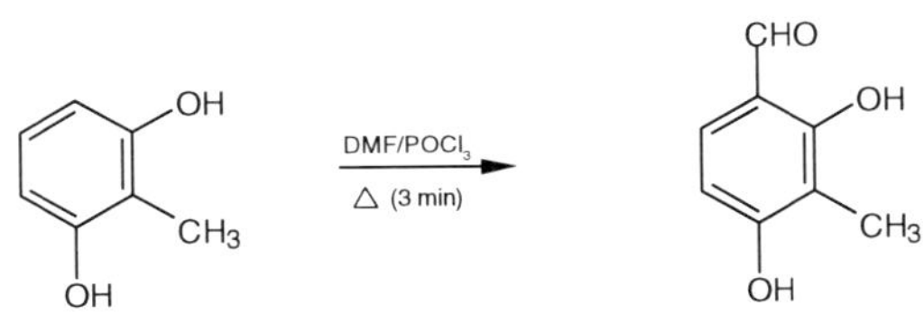

2-Methyl resorcinol 2,4-Dihydroxy-3-methylbenzaldehyde

To a stirred solution of 2-methyl resorcinol (14 g, 0.11 mole) in dimethyl formamide (10 ml, 9.44 g, 0.13 mole) is added phosphorus oxychloride (10 ml, 16.45 g, 0.11 mole) in small portions. The mixture is heated on a water bath for 3 hr under anhydrous conditions. It is cooled, water (50 ml) added and the mixture is extracted with ether. The ether extract is washed with water, dried (magnesium sulphate) and distilled. The residual product is crystallised from methanol as colourless needles. Yield 10.5 g (61.2%). M.p. 145-46° (lit.[33,34] m.p. 145-46°).

Method (ii)

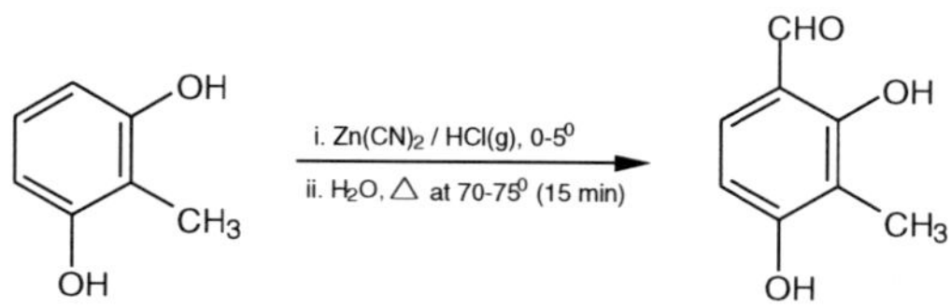

2-Methyl resorcinol 2,4-Dihydroxy-3-methylbenzaldehyde

A rapid stream of dry hydrogen chloride gas is passed for 5-6 hr through a cooled solution (0-5°) of 2-methyl resorcinol (1.17) (5 g, 0.04 mole) in dry ether (100 ml) containing anhydrous zinc cyanide (7.5 g, 0.064 mole) until the imine hydrochloride formed as a thick oil solidifies. The mixture is left for 24 hr in an ice chest. The ether is decanted and the residual product washed with dry ether (2 × 20 ml). It is dissolved in water (70 ml) and heated on a water-bath (70-75°) for 15 min. The solution is cooled and separated product filtered. It is crystallised from methanol as colourless needles. Yield 3.8 g (61.99%). M.p. 146-47°.

2.32 2-HYDROXY-4-METHOXY-3-METHYLBENZALDEHYDE

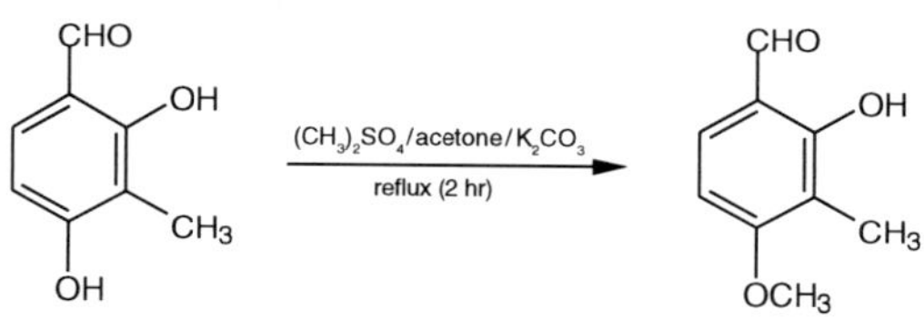

2,4-Dihydroxy-3-methyl benzaldehyde 2-Hydroxy-4-methoxy-3-methyl benzaldehyde

A solution of 2,4-dihydroxy-3-methyl benzaldehyde (2.31) (1 g, 0.0065 mole) in dry acetone (50 ml) is refluxed for 2 hr with dimethyl sulphate (0.66 ml, 0.88 g, 0.007 mole) in presence of anhydrous potassium carbonate (2 g). The mixture is filtered and the inorganic salts washed with hot acetone (20 ml). The combined acetone solution is distilled and the residue treated with ice-water. The separated product is filtered. It is crystallised from benzene-petroleum ether as colourless needles. Yield 0.84 g (77.1%). M.p. 66-67° (lit.[35] m.p. 66-67°).

2.33 2,5-DIHYDROXY-3-METHOXYBENZALDEHYDE

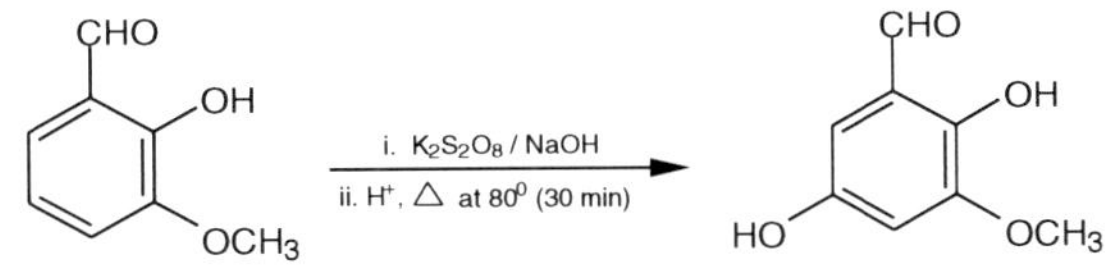

o-vanillin | 2,5-Dihydroxy-3-methoxy-benzaldehyde

A solution of potassium persulphate (36 g, 0.133 mole) in water (720 ml) is added dropwise during 3-4 hr at 10-15° to a stirred solution of o-vanillin (20 g, 0.13 mole) in water (160 ml) containing sodium hydroxide (32 g). The mixture is stirred for 1 hr more and allowed to stand for 24 hr at room temperature. The reaction mixture is cooled and acidified to congo red with hydrochloric acid. The separated unchanged o-vanillin is filtered and the solution extracted with ether (2 × 100 ml) to remove the starting material completely. The solution is heated at 80° for 30 min with concentrated hydrochloric acid (250 ml) and sodium sulphite (10 g). The solution is cooled and extracted with ether (5 × 100 ml). The ether extracted is dried (sodium sulphate) and distilled. The crude aldehyde thus obtained is crystallized from benzene as pale yellow needles. Yield 7 g (31.6%). M.p. 144° (lit.[36] m.p. 144°).

2.34 2,4,5-TRIHYDROXY-6-METHYLBENZALDEHYDE

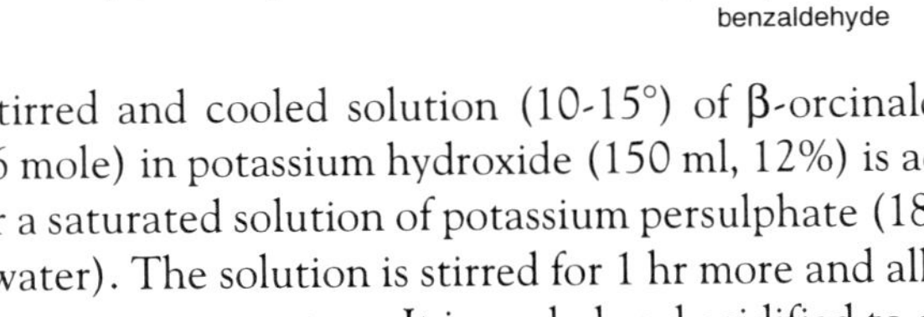

β-Orcinaldehyde | 2,4,5-Trihydroxy-6-methyl benzaldehyde

To a stirred and cooled solution (10-15°) of β-orcinaldehyde (2.28) (10g, 0.066 mole) in potassium hydroxide (150 ml, 12%) is added dropwise during 4 hr a saturated solution of potassium persulphate (18 g, 0.067 mole in 300 ml water). The solution is stirred for 1 hr more and allowed to stand overnight at room temperature. It is cooled and acidified to congo red with hydrochloric acid. The solution is extracted with ether (3 × 100 ml) to remove the unreacted aldehyde. The aqueous solution is refluxed under a layer of benzene (100 ml) for 3 hr with concentrated hydrochloric acid (100 ml) and sodium bisulphite (4 g). The solution is cooled and the benzene layer separated. It is dried (anhydrous sodium sulphate) and distilled. The residue is crystallised from water containing a little sulphur dioxide as brown prisms. Yield 0.4 g (3.61 %). M.p. 208°. (lit.[37] m.p. 208).

2.35 2,5-DIHYDROXY-4,6-DIMETHOXYBENZALDEHYDE

Method (i)

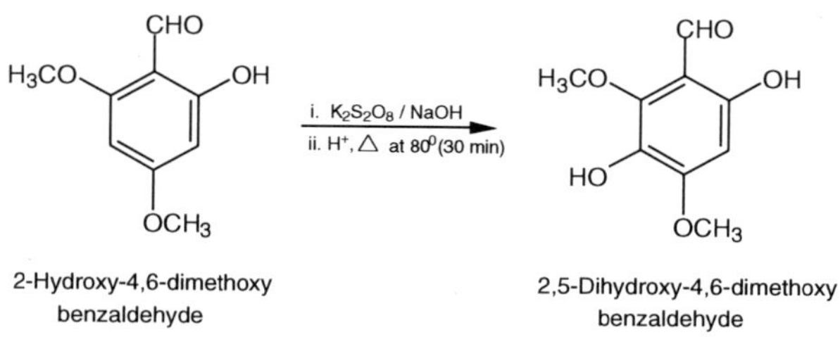

A solution of 2-hydroxy-4,6-dimethoxybenzaldehyde (2.16) (10 g, 0.055 mole) in sodium hydroxide solution (10.1 g, 0.25 mole in 110 ml water) is treated dropwise during 3 hr at 10-15° with a solution of potassium persulphate (16.5 g, 0.06 mole in 300 ml water) with continuous stirring. The solution is allowed to stand overnight, cooled and acidified to congo red with hydrochloric acid. It is extracted with ether (2 × 100 ml) to remove the unchanged aldehyde. The remaining solution is heated for 30 min at 80° with concentrated hydrochloric acid (100 ml) and sodium sulphite (10g). The mixture is cooled and extracted with ether (5 × 100 ml). The ether extract is dried (sodium sulphate) and distilled. The residual product is crystallised from hot water containing sulphur dioxide. Yield 0.9 g (8.27%). M.p. 135-36° (lit.[38,39] m.p. 135-36°).

Method (ii)

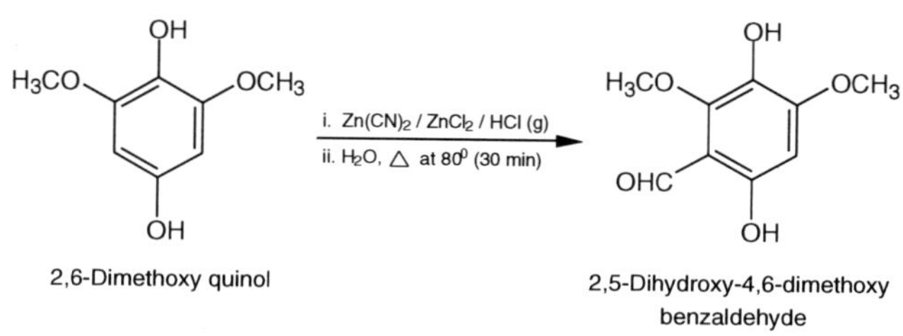

2,6-Dimethoxy quinol (4.A.54) (5 g, 0.029 mole) is dissolved in dry ether (250 ml) and dry zinc cyanide (10 g, 0.085 mole) and fused zinc chloride (1 g, 0.007 mole) are added. The mixture is cooled (0°, ice-salt mixture) and saturated with dry hydrogen chloride gas for 4-5 hr under anhydrous conditions. The mixture is allowed to stand overnight at room temperature and the clear ether solution decanted off. The remaining aldimine hydrochloride is washed with dry ether (2 × 50 ml). It is hydrolysed by heating for 30 min at 80° with water (50 ml). The solution is cooled and the separated product filtered. It is crystallised from water containing sulphur dioxide. Yield 1.5 g (25.7%). M.p. 135-36°.

2.36 2,5-DIHYDROXY-4-METHOXY-6-METHYLBENZALDEHYDE

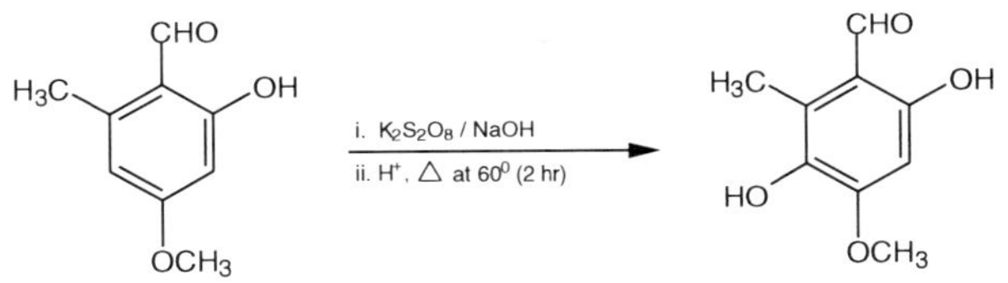

2-Hydroxy-4-methoxy-6-methylbenzaldehyde

2,5-Dihydroxy-4-methoxy-6-methylbenzaldehyde

2-Hydroxy-4-methoxy-6-methylbenzaldehyde (2.29) (5.7 g, 0.34 mole) is dissolved in pyridine (10 ml) and aqueous sodium hydroxide solution (7 g, 0.175 mole in 70 ml water). An aqueous solution of potassium persulphate (10 g, 0.37 mole in 250 ml water) is added dropwise during 4 hr at 10-15° with stirring. The solution is left overnight at room temperature. It is cooled and acidified to congo red with hydrochloric acid. The solution is extracted with ether (3 × 50 ml) to remove the unreacted aldehyde. The clear solution is heated at 60° for 2 hr with concentrated hydrochloric acid (40 ml) and sodium sulphite (2 g). The solution is cooled and the separated product filtered. It is crystallised from alcohol as golden yellow prisms. Yield 1.5 g (24.1%). M.p. 162-63° (lit.[40] m.p. 162-63°).

2.37 2-METHOXY-4,5-METHYLENEDIOXYBENZALDEHYDE

Step (i) 6-Nitropiperonal

Piperonal (50 g, 0.33 mole) is added in small lots to well stirred concentrated nitric acid (250 ml, sp. 1.38) at 40°. After the addition is over, the mixture is stirred for 1 hr at 40°. It is poured over crushed ice. The

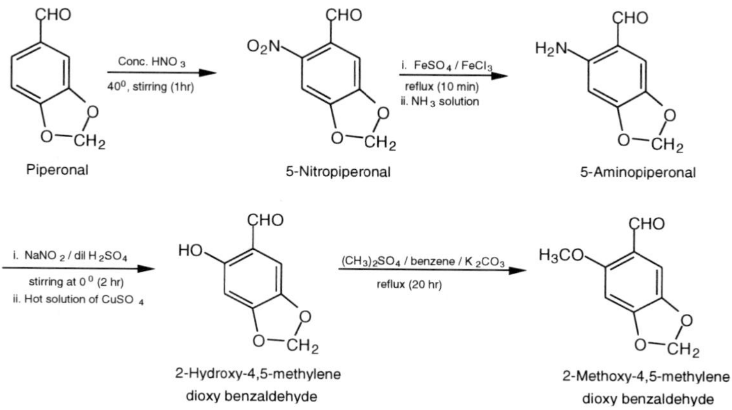

separated product is filtered, washed with cold water and then with sodium bicarbonate solution (to remove excess of nitric acid) and finally with cold water. It is crystallised from ethanol as pale yellow needles. Yield 30 g (46.2%). M.p. 97-98° (lit.[41] m.p. 97-98°).

Step (ii) 6-Aminopiperonal

To a boiling solution of ferrrous sulphate (100 g in 500 ml water) containing little ferric chloride (1.5 g) is added with stirring during 10 min, a solution of 6-nitropiperonal (10 g, 0.05 mole) in alcohol (500 ml, 50%). After the addition the reaction mixture is refluxed for 10 min and then a strong solution of ammonia (150 ml of 50% liquor ammonia) is added in lots of 10 ml with vigorous stirring. The solution is filtered and the residual product washed with boiling water (250 ml). The combined filtrate and washings on cooling gave a solid product. It is filtered and crystallised from hot water as bright yellow needles. Yield 6.3 g (74.5%). M.p. 107-108° (lit.[42] m.p. 107-108°).

Step (iii) 2-Hydroxy-4,5-methylenedioxybenzaldehyde

6-Aminopiperonal (10 g, 0.06 mole) is suspended in water (300 ml), cooled to 0° and dilute sulphuric acid (10 ml concentrated sulphuric acid in 60 ml water) is added dropwise with shaking. The mixture is stirred and cooled (0°). It is treated with a solution of sodium nitrite (5 g, 0.72 mole in 50 ml water) added dropwise during 30 min. The mixture is stirred for 2 hr at 0° and the reaction mixture is left overnight at room temperature. It is diluted with an equal volume of water and then added slowly to a hot solution of copper sulphate (150 g in 150 ml water). The solution is steam distilled (10 hr). The collected distillate (approx. 4.5 litres) is extracted with ether (4 × 250 ml). The ether extract is concentrated and extracted with sodium hydroxide solution (10%, 3 × 100 ml). The combined clear alkaline extract is acidified and the separated product filtered. It is crystallised from dilute alcohol as colourless needles. Yield 2.5 g (24.9%). M.p. 125-26° (lit.[42] m.p. 125-26°).

Step (iv) 2-Methoxy-4,5-methylenedioxybenzaldehyde

A mixture of the above 2-hydroxy-4,5-methylenedioxybenzaldehyde (6 g, 0.036 mole) in dry benzene (100 ml) containing anhydrous potassium carbonate (20 g) and dimethyl sulphate (9 ml, 12 g, 0.095 mole) is refluxed for 20 hr. The mixture is filtered and the inorganic salts washed with hot benzene (2 × 25 ml). The combined benzene solution is concentrated in vacuo and the residue treated with cold water. The separated product is filtered. It is crystallised from hot water as colourless needles. Yield 4.5 g (69.2%). M.p. 112° (lit.[42] m.p. 112°).

2.38 2,3,4,5- TETRAMETHOXYBENZALDEHYDE[43]

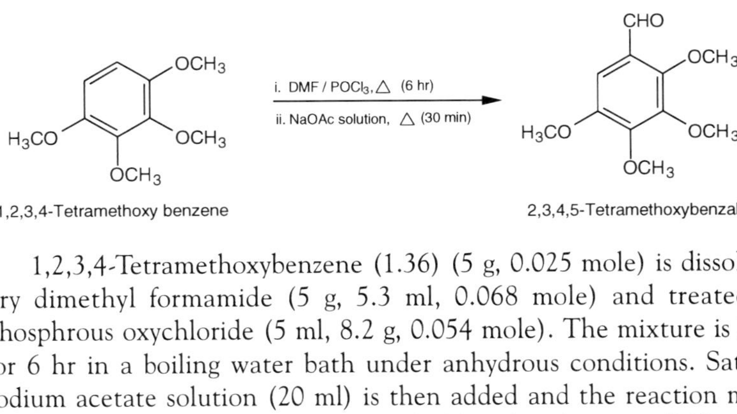

1,2,3,4-Tetramethoxy benzene   2,3,4,5-Tetramethoxybenzaldehyde

1,2,3,4-Tetramethoxybenzene (1.36) (5 g, 0.025 mole) is dissolved in dry dimethyl formamide (5 g, 5.3 ml, 0.068 mole) and treated with phosphrous oxychloride (5 ml, 8.2 g, 0.054 mole). The mixture is heated for 6 hr in a boiling water bath under anhydrous conditions. Saturated sodium acetate solution (20 ml) is then added and the reaction mixture heated for 30 min more. It is cooled, diluted with water (100 ml) and extracted with ether (3 × 100 ml). The combined ether extract is washed with water, dried (magnesium sulphate) and distilled. 2,3,4,5-Tetramethoxybenzene is obtained as an oil. Yield 4g (70.2%).

2.39 2-HYDROXY-3,4,5,6-TETRAMETHOXYBENZALDEHYDE

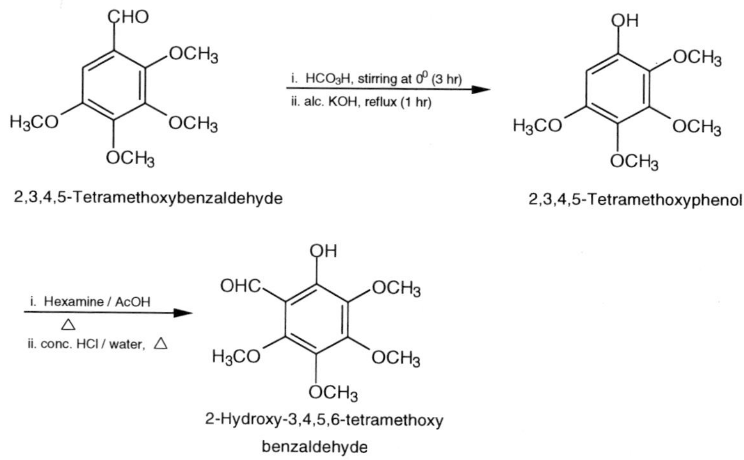

Step (i) 2,3,4,5-Tetramethoxyphenol[43]

2,3,4,5-Tetramethoxybenzaldehyde (2.38) (5 g, 0.022 mole) is dissolved in dry chloroform (50 ml) and treated dropwise with performic acid (1.36) (25 ml) with constant stirring; the temperature of the reaction mixture is maintained at 0°. The reaction mixture is stirred for 3 hr and treated with sodium bisulphite (1 g). Performic acid is removed in vacuo and the residue rendered alkaline with alcoholic potassium hydroxide (10%). The solution is refluxed for 1 hr with potassium hydroxide solution (50 ml, 10%). The alcohol is removed in vacuo and the residue acidified with hydrochloric acid. The reaction mixture is extracted with ether, ether extract dried (sodium sulphate). The ether is distilled and 2,3,4,5-tetramethoxyphenol is obtained as an oil. Yield 2 g (42.3%).

The above phenol (1 g, 0.0047 mole) is refluxed for 4 hr in dry acetone (50 ml) dimethyl sulphate (1 ml, 1.3 g, 0.01 mole) in the presence of anhydrous potassium carbonate (5 g). The reaction mixture is worked up in the usual way. 1,2,3,4,5- Pentamethoxybenzene is crystallised from petroleum ether as colourless needles. Yield 1 g (94.3%). M.p. 58-59° (lit.[44], m.p. 58-59°).

Step (ii) 2-Hydroxy-3,4,5,6- tetramethoxybenzaldehyde[43]

A mixture of the above 2,3,4,5- tetramethoxyphenol (2 g, 0.009 mole), glacial acetic acid (20 ml) and hexamine (6 g) is heated in a boiling water bath under anhydrous conditions. Concentrated hydrochloric acid (10 ml) and water (20 ml) is then added to the reaction mixture and heating continued for 30 min more. The reaction mixture is cooled, diluted with water and extracted with ether. The ether solution is washed with water, dried (sodium sulphate) and distilled. 2-Hydroxy-3,4,5,6-tetramethoxybenzaldehyde is obtained as an oil. Yield 0. 5 g (22.4%) (lit.[44] oil).

2.40 2-METHOXY-3-NITROBENZALDEHYDE

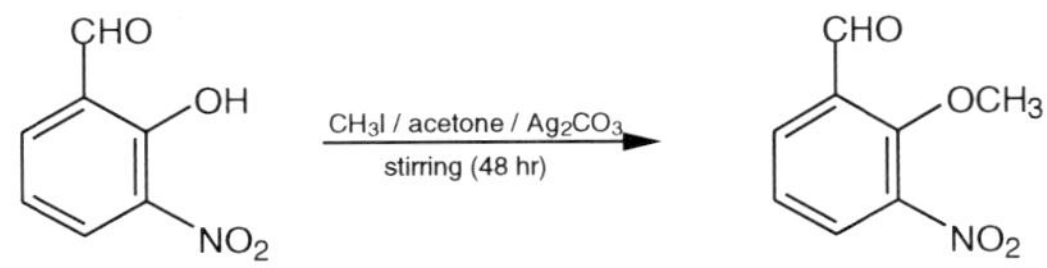

3-Nitrosalicylaldehyde 2-Methoxy-3-nitrobenzaldehyde

To a solution of 3-nitrosalicylaldehyde (2.42) (2.2 g, 0.013 mole) in pure dry acetone (70 ml) is added air-dried silver carbonate (5.4 g) and methyl iodide (2.5 ml, 5.7 g, 0.04 mole). The mixture is stirred for 48 hr, allowed to stand for 6 days at room temperature and filtered. The filtrate is concentrated in vacuo. The solid product is taken up in ether, the ethereal solution washed with sodium hydroxide solution (5%), water and then distilled. The residual product is crystallised from alcohol as almost colourless needles with an yellow tinge. Yield 2 g (84%). M.p. 97-99° (lit.[45] m.p. 102°).

2.41 2-METHOXY-5-NITROBENZALDEHYDE

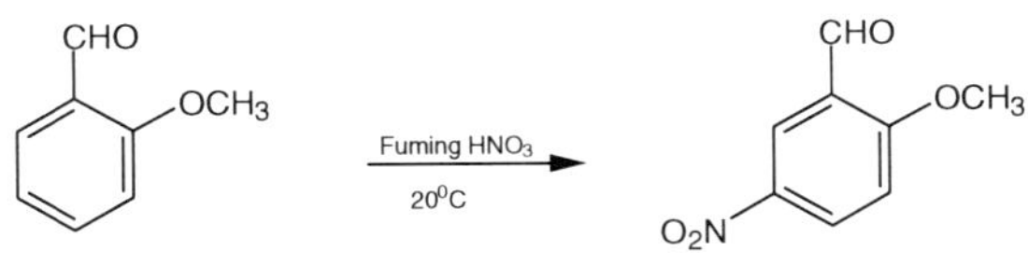

o-Methoxybenzaldehyde 2-Methoxy-5-nitrobenzaldehyde

o-Methoxybenzaldehyde (2.1) (5 ml, 5.6 g, 0.04 mole) is slowly added to cold fuming nitric acid (25 ml, d, 1.5) at such a rate that temperature did not rise above 20° during the addition. The deep yellow solution is allowed to stand at 20° for 15-20 min and poured into ice-water (200 ml). The separated product is filtered and washed with cold water. It is crystallised from large volume of water as nearly colourless slender long needles. Yield 4 g (53.7%). M.p. 86-88° (lit.[46] m.p. 89-90°).

2.42 3,5-DINITROSALICYLALDEHYDE

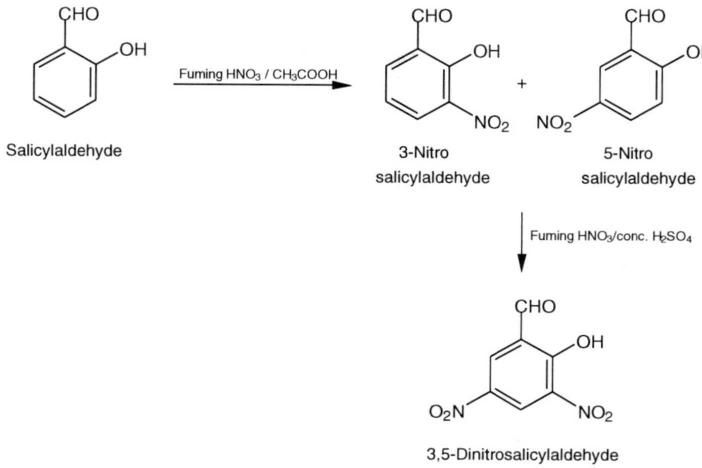

Step (i) 3- and 5-Nitrosalicylaldehydes

Fuming nitric acid (20 ml, d, 1.2) is added slowly to a stirred and cooled (25°) mixture of glacial acetic acid (100 ml) and salicylaldehyde (25 g, 21.8 ml, 0.205 mole). After the addition of first 5 ml of nitric acid, the mixture is kept at 15° till the addition is over. The temperature is allowed to rise during 40-45 min to 45°. The reaction mixture is poured into ice. After 4-5 hr at room temperature, the product is filtered and washed with water.

The above mixture of 3- and 5-nitrosalicylaldehyde (20 g) is dissoved in aqueous sodium hydroxide solution (5 g in 150 ml water) by warming, solution filtered and the clear solution allowed to stand overnight. The 5- nitrosalicylaldehyde separates out as sodium salt. It is filtered, suspended in water and acidified with hydrochloric acid. The separated 5-nitrosalicylaldehyde is filtered and washed with water. It is crystallised from alcohol as tan coloured needles. Yield 7 g (20.4%). M.p. 126° (lit [47,48] m.p. 126°). The red coloured filtrate (after separation of sodium salt of 5-nitrosalicyldehyde) is acidified. The separated 3-nitrosalicylaldehyde is filtered. It is purified by dissolving in slight excess of alcohol and slowly precipitated by addition of water. It is finally crystallised from acetic acid as small yellow prisms. Yield 3. 5 g (10.2%). M.p. 109-110° (lit.[49] m.p. 109-110°).

Step (ii) 3,5-Dinitrosalicylaldehyde

The mixture of 3- and 5- nitrosalicylaldehyes obtained above (3.3 g, 0.02 mole) is cautiously added to an ice-cold mixture of concentrated sulphuric acid (4 ml) and fuming nitric acid (2 ml, d, 1.5). The mixture is allowed to stand for 30 min with occasional shaking. It is poured into cold water and the separated yellow product filtered and washed with water. It is crystallised twice from benzene as lemon yellow prisms. Yield 1.6 g (38.2%). M.p. 60-61° (lit.[50] m.p. 62-63°).

2.43 2,4-DIHYDROXY-5-NITROBENZALDEHYDE (5-NITRORESORCYLIC ALDEHYDE)

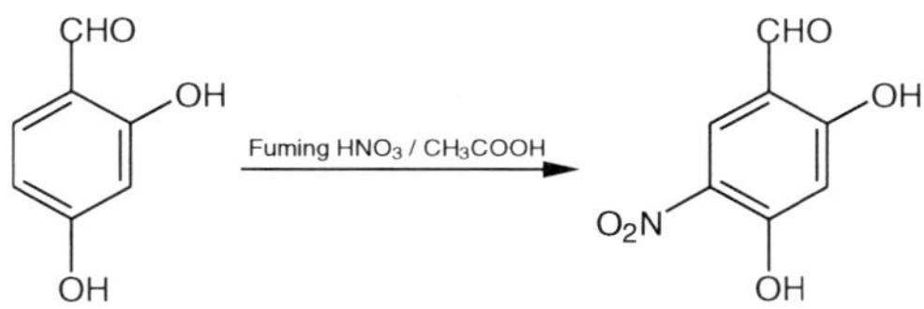

β-resorcyclic aldehyde 2,4-Dihydroxy-5-Nitrobenzaldehyde

A cold solution of β-resorcylic aldehyde (2.5) (5 g, 0.036 mole) in glacial acetic acid (10 ml) is added to cold solution of fuming nitric acid (2.1 ml, d, 1.5) in glacial acetic acid (5 ml). As soon as the reaction subsided, the reaction mixture is heated for 10-15 min in a boiling water bath and then poured into water (30 ml). The separated product is filtered and washed with water. It is crystallised from benzene as brownish yellow prisms. Yield 3 g (45.2%). M.p. 148-49° (lit.[51] m.p. 148-49°).

2.44 2-HYDROXY-3-METHOXY-5-NITROBENZALDEHYDE (5-NITRO-o-VANILLIN)

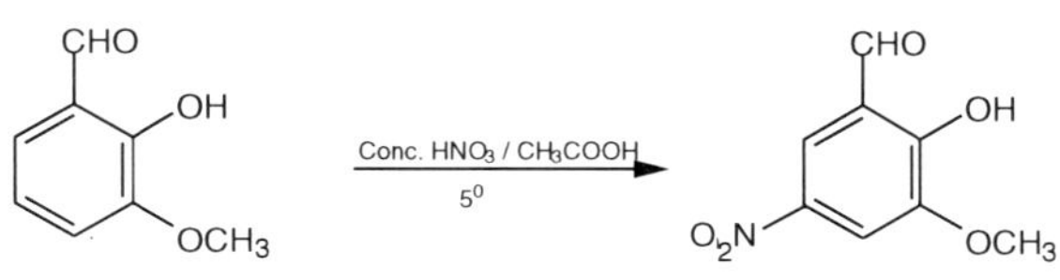

o-vanillin 2-Hydroxy-3-methoxy-5-Nitrobenzaldehyde

Concentrated nitric acid (1.5 ml, d, 1.5) is added dropwise with stirring to a cooled (5°) solution of o-vanillin (3 g, 0.002 mole) in glacial acetic acid (1.2 ml). The reaction mixture is allowed to stand for 30 min and then poured into water. The separated product is filtered and washed with water. It is crystallised from acetic acid as yellow needles. Yield 3 g (77.3%). M.p. 140-142° (lit.[52] m.p. 142°).

2.45 2-HYDROXY-3-METHOXY-4-NITROBENZALDEHYDE (4-NITRO-o-VANILLIN)

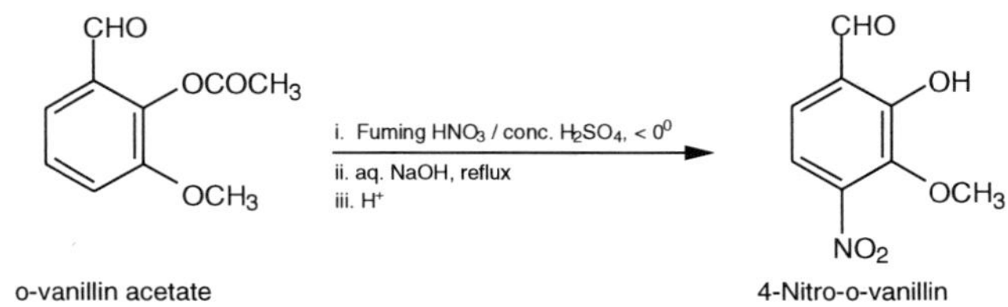

To a previously cooled mixture (–6°) of fuming nitric acid (10 ml, d, 1.52) and concentrated sulphuric acid (2 ml), o-Vanillin acetate (3 g, 0.015 mole) is added in small lots with brisk stirring. During the reaction and subsequent 15 min, the reaction mixture is kept stirred (<0°). It is poured into crushed ice and the separated viscous oil taken up in benzene. The benzene extract is washed with ice-cold water (3 × 100 ml), cold aqueous sodium bicarbonate (5 g) and again with water. The deep red bicarbonate solution on acidification gave the 5-nitro isomer (0.2 g), m.p. 140-42° (after crystallisation from acetic acid). The remaining benzene solution is distilled and the residual oily product is refluxed with aqueous sodium hydroxide (20 ml, 2%) for 10-15 min. The clear solution is cooled and acidified. The separated product is filtered and washed with water. It is crystallised from acetic acid as yellow needles of 4-nitro-o-vanillin. Yield 0.8 g (26.3%). M.p. 92-93°. (lit.[52] m.p. 92-93°).

2.46 3,4-DIMETHOXY-6-NITROBENZALDEHYDE (6-NITROVERATRALDEHYDE)

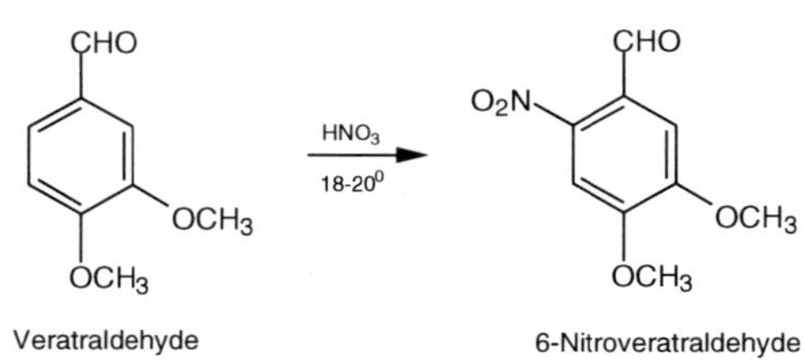

Powdered veratraldehyde (2.11) (7 g, 0.042 mole) is added in very small quantities to a vigorously stirred nitric acid (35 ml, sp. gr. 1.4). The temperature of the reaction mixture is maintained between 18-20° and the addition of veratraldehyde should take about 45 min. The mixture is stirred for 10 min more and then it is poured into cold water (400 ml) with stirring (the mixture be now onwards protected from light). The separated product is filtered and stirred again in cold water (10 ml) and filtered. It is dried as much as possible (50°, 8 hr). It is crystallised from alcohol. Yield 6.5 g (73%). M.p. 132-133° (lit.[53] m.p. 132-133°).

References

1. W.P. Dickison and P.G. Marshall, J. Chem. Soc., 1929, 1496.
2. L.E. Hinkel, E.A. Ayling and W.H. Morgan, J. Chem. Soc., 1932, 2793.
3. Baker and Moffitt, J. Chem. Soc., 1931, 314.
4. Tiemann and Ludwig, Ber., 1882, 15, 2045.
5. Spatt, Monatsch, 1913, 34, 1998; Tieman and Ludwig, Ber., 1882, 15, 2043.
6. L. Gatterman and M. Kobner, Ber., 1899, 32, 280.
7. Dictionary of organic compounds by Chapman & Hall, 5th Edition, D-50424.
8. R. Adams and E. Montgomery, J. Am. Chem. Soc., 1924, 46, 1518; A. C. Jain, P.D. Sarpal and T.R. Seshadri, Indian J. Chem., 1966, 4, 226.
9. G. Casiraghi, G. Casnati and M. Cornia, J. Chem. Soc., P-1, 1978, 318.
10. Dictionary of organic compounds by Chapman & Hall, 5th Edition, D-04213.
11. R.C. Shah and M.C. Laiwalla , J. Chem. Soc., 1938, 1828.
12. Barger and Silberschmidt, J. Chem. Soc., 1928, 133, 2924.
13. Fittig and Remsen, Ann., 1871, 159, 144; Hoering and Baum, Ber., 1908, 41, 1914.
14. Dictionary of organic compounds by Chapman & Hall, 5th Edition, B-02262.
15. W. Bradley, R. Robinson and G. Schwarzenbach, J. Chem. Soc., 1930, 16, 793.
16. T. Malkin and M. Nierenstein, J. Am. Chem. Soc., 1931, 53, 241.
17. A. C. Jain, Pyrare Lal and T.R. Seshadri, Indian J. Chem., 1968, 6, 486.
18. Dictionary of organic compounds by Chapman & Hall, 5th Edition, T-03268.
19. V.K. Ahluwalia et al, unpublished results.

20. Fuson, Horning, Ward, Rowland and Marsh, J. Am. Chem. Soc., 1942, 64, 31.
21. L. Gattermann, Ber., 1898, 31, 1768.
22. Fr. Schaaf and A. Labouchere, Helv. Chim. Acta, 1924, 7, 357.
23. Dictionary of organic compounds by Chapman & Hal, 5th Edition, T-03264.
24. L. Gattermann and M. Kobner, Ber., 1899, 32, 278.
25. Dictionary of organic compounds by Chapmann & Hall, 5th Edition, T-03267.
26. T.R. Govindachari, K. Nagarajan and P.C. Parthsarathy, J. Chem. Soc., 1957, 550.
27. A.J. Birch, A.H. Jackson, P.V.R. Shannon and G.W. Stewart, J. Chem. Soc. Perkin, 1,1975, 2492.
28. A. Dornow and G. Petsch, Arch. Pharm., 1951, 284, 160 (Chem. Abstr., 1952, 46, 10126b).
29. R.A. Baxter, G.H. Ramage and J.A. Timson, J. Chem. Soc., 1949, 530.
30. Dictionary of organic compounds by Chapman and Hall, 5th edition, D-04652.
31. Robertson and Stephenon, J. Chem. Soc., 1932, 1392.
32. R.T. Foster, A. Roberston and A. Bushra, J. Chem. Soc., 1948, 2254.
33. V.D. Mehta, Ph.D. Thesis, Delhi University, 1987, p 183.
34. A. Gatterman, Ann. 1907, 357, 340.
35. K. Kiyoshi, N. Jun, K. Kiyonosin, I. Kiyoshi, S. Yugi, I. Katsuro, Euo. Pat. Appl., 11 839 (June 1980); Chem. Abstr., 1981, 91, 473354.
36. W. Baker, N.C. Brown and J.A. Scott., J. Chem. Soc., 1939, 1922.
37. L. Ponniah, Ph.D. Thesis, Delhi University, 1953, p 54.
38. L. Ponniah, Ph.D. Thesis, Delhi University, 1953, p 48.
39. Chapman, Perkin and Robinson, J. Chem. Soc., 1927, 3208.
40. K. Aghoramurthy and T.R. Seshadri, Proc. Indian Acad. Sci., 1952, 35A, 334.
41. Ekeley and Klemms, J. Am. Chem. Soc., 1928, 50, 2711.
42. Campbell et al., J. Org. Chem., 1951, 16, 1736.
43. V.K. Ahluwalia and Sunita, Indian J. Chem., 1977, 15B, 936.
44. Dictionary of organic compounds by Chapman & Hall, 5th Edition, B-00200.
45. W.V. Miller and F. Kinkelin, Ber., 1889, 22, 1706.
46. Dictionary of organic compounds by Chapman & Hall, 5th Edition, H-02770.

47. C.C. Hach, L.M. Liggett and Diehl, Iowa State, Coll. J. Sci., 1947, 21, 311 (Chem. Abstr., 1948, 42, 1240).
48. D.A. Denton and H. Suschitzky, J. Chem. Soc., 1963, 4741.
49. W.V. Miller, Ber., 1887, 20, 1927.
50. A.B.E. Lovett and E. Hoberts, J. Chem. Soc., 1928, 1975.
51. Dictionary of organic compounds by Chapman & Hall, 5th Edition, D-04823: Gattermann, Annalen, 1907, 357, 313.
52. B.B. Dey and V.A. Kutti, Proc. Nat. Inst. Sci. India, 1940, 6, 641.
53. Salway, J. Chem. Soc., 1909, 95, 1163.

3

Carboxylic Acids

Carboxylic acids (aromatic) are prepared by the following methods.

KOLBE-SCHMIDT REACTION

This reaction is useful for the introduction of a carboxyl group in aromatic compounds. The carboxyl group is introduced in ortho position to the phenolic group in the aromatic compound. The reaction consists in heating the organic compound with aqueous potassium bicarbonate while carbon dioxide gas is bubbled into it.

Mechanism

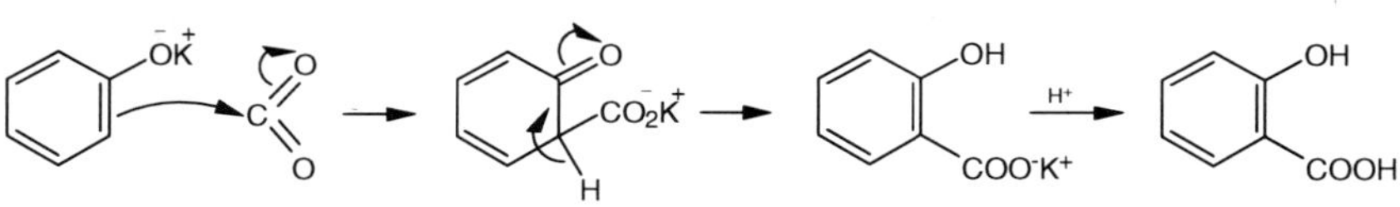

The reaction involves the high reactivity of phenoxide ion towards electrophilic substitution. In this case carbon dioxide, a relatively weak electrophile is the reactive species. The unstable intermediate undergoes proton shift to give the carboxylic acid.

This reaction is particularly facile with di- and trihydric phenols.

This method is used to synthesise o-hydroxy carboxylic acid, viz. 3.6, 23, 33, 38, 39.

FROM ALDEHYDES

Aromatic aldehydes can be oxidised to carboxylic acids either by potassium permanganate or by fusion with alkali.

a. Potassium permaganate oxidises aldehydes to carboxylic acid in alkaline or acidic medium. In alkaline medium, the oxidation takes place as given below:

$$3\ ArCHO + 2MnO_4^- + {}^-OH \rightarrow 3\ ArCOO^- + 2MnO_2 + 2H_2O$$

However the mechanism for oxidation in acid solution is as given below.

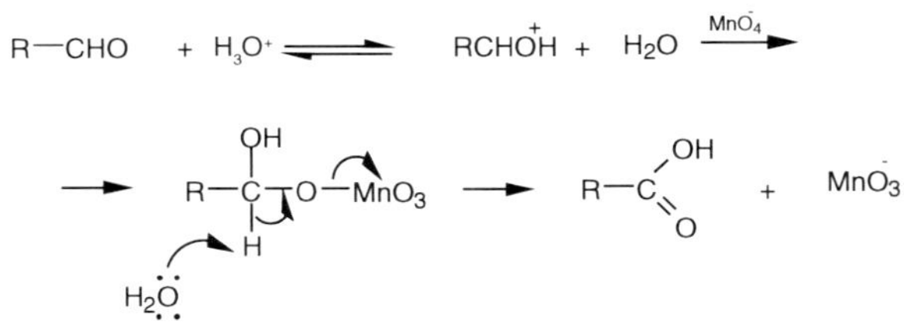

$$2MnO_3^- + H_2O \longrightarrow MnO_2 + MnO_4^- + 2OH^-$$

Potassium permaganate oxidation has been used to synthesise a number of carboxylic acids, viz. 3.17, 19, 20, 36.

Potassium permaganate has also been used to prepare keto acids by the oxidation of acetophenols. Thus, p-methoxyacetophenone on oxidation with potassium permanganate gives p-methoxyphenylglyoxalic acid (3.43).

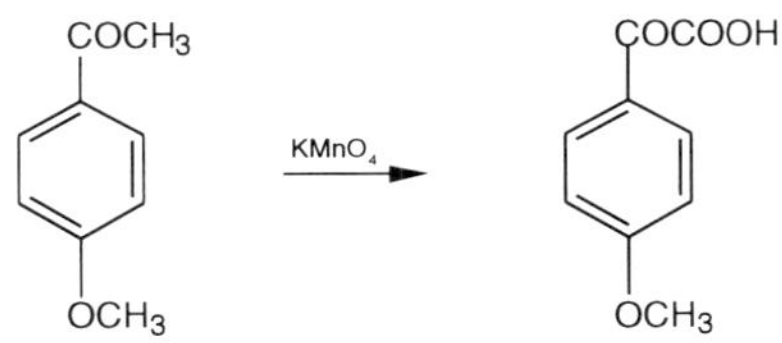

b. Fusion with Alkali

Aromatic aldehydes on fusion with alkali (viz. sodium hydroxide) give the corresponding carboxylic acid. Thus

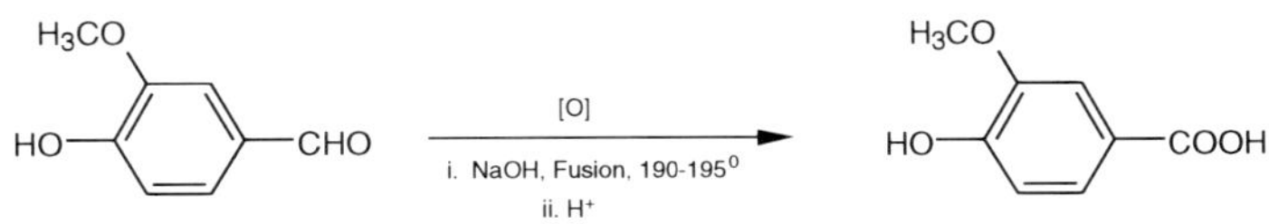

Vanillin — Vanillic acid

if a mixture of sodium hydroxide and potassium hydroxide is used, the oxidation is accompanied by demethylation of the methoxyl group to give the corresponding carboxylic acid.

This method is used to synthesize carboxylic acids, viz. 3.3, 13 and 14.

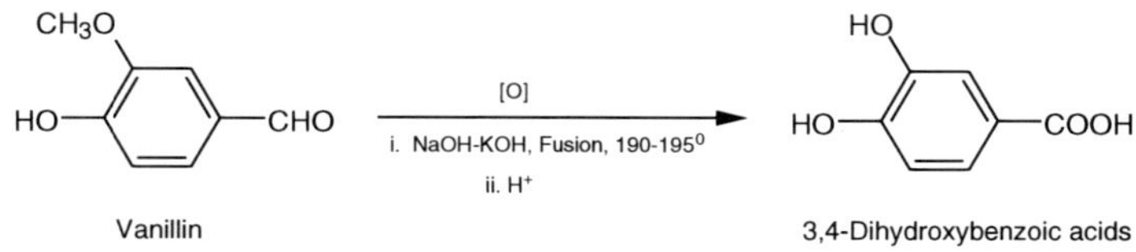

FROM KETONES

Ketones can be oxidised into carboxylic acid by oxidation with iodine-pyridine followed by hydrolysis of the formed complex. It is believed that iodine and pyridine react with ketones to give the pyridinium iodide in 95% yield. The salts are cleaved by heating with alkali to give carboxylic acids.

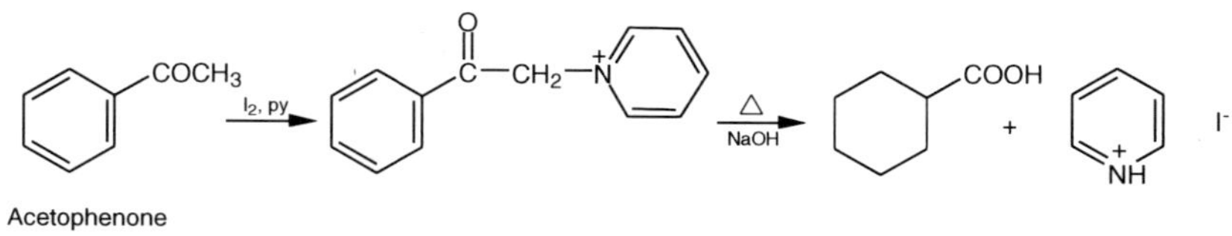

This reaction has been used for the preparation of 3.12, 34 and 42.

WILLGERDOT REACTION

This reaction is useful for the oxidation of aromatic ketones to the corresponding phenyl acetic acids. It consists in heating the aromatic ketones with sulphur and morpholine. The formed thioamide hydrolysed with alkali to give the corresponding phenyl acetic acid.

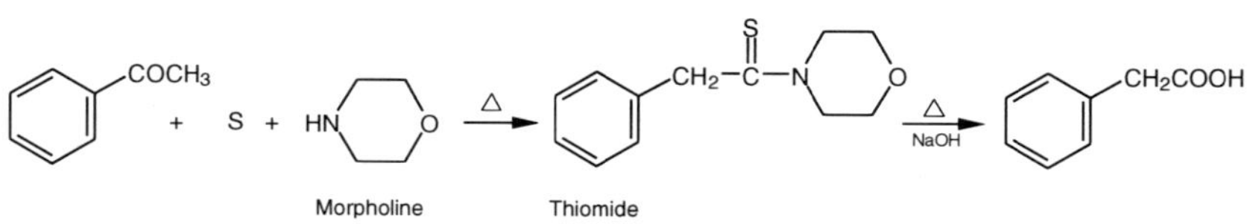

Using this method p-methoxyphenyl acetic acid (3.46) has been prepared.

KNOEVENAGEL CONDENSATION

In this method an aldehyde is condensed with malonic acid in the presence of a base (like piperidine) followed by heating with hydrochloric acid to give the corresponding cinnamic acid.

$$C_6H_5CHO + CH_2(COOH)_2 \xrightarrow[\text{ii. HCl, }\Delta]{\text{i. Piperidine, }\Delta} C_6H_5CH{=}CHCOOH$$

Benzaldehyde   Malonic acid   Cinnamic acid

Mechanism

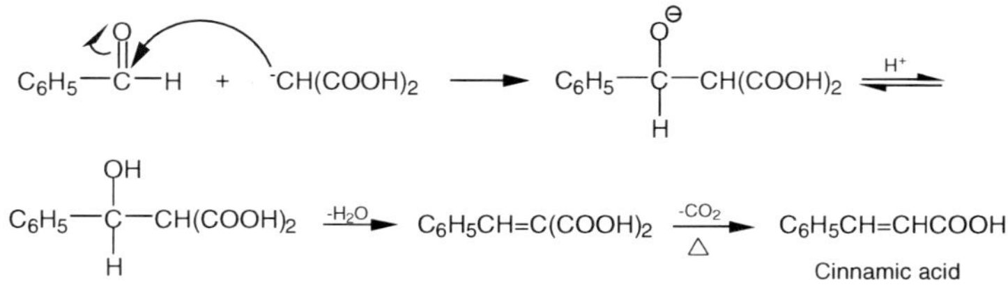

Knoevenagel condensation is a general method for the conversion of aldehyde into α, β-unsaturated carboxylic acid. Using this method p-hydroxycinnamic acid (3.52) has been prepared.

PERKIN REACTION

Like Knoevenagel reaction, the Perkin reaction is also used for the conversion of aromatic aldehydes into the corresponding α, β-unsaturated acids. It consists in heating the aromatic aldehyde with aliphatic acid anhydride in presence of sodium or potassium salt of the corresponding anhydride.

$$C_6H_5CHO + (CH_3CO)_2O + CH_3COO^-Na^+ \xrightarrow[140\text{-}150^0C]{\Delta} C_6H_5CH{=}CHCOOH$$

Mechanism

$$CH_3COONa \rightleftharpoons CH_3COO^- + Na^+$$

$$CH_3COO^- + H\text{-}CH_2\text{-}CO\text{-}O\text{-}CO\text{-}CH_3 \rightleftharpoons CH_3COOH + {}^-CH_2\text{-}CO\text{-}O\text{-}COCH_3$$

or

$(CH_3CO)_2O$

acetic anhydride

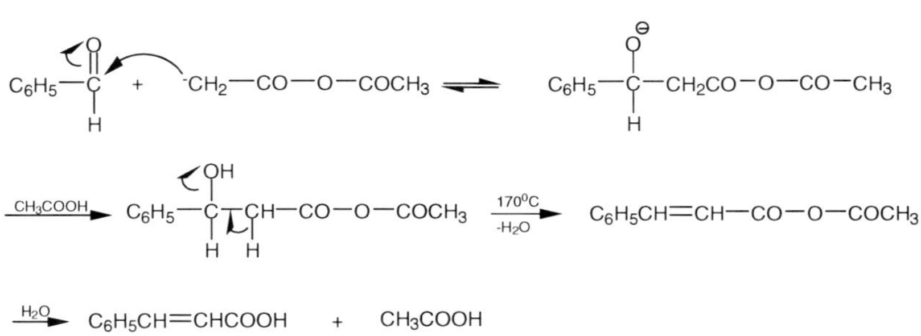

Perkin reaction is a general method for the synthesis of α, β-unsaturated acid (3.49).

SAPONIFICATION

If esters are conveniently available, these can be hydrolysed (saponification) with alkali to give the corresponding acids.

$$RCOOC_2H_5 \xrightarrow{HO^-} RCOOH + C_2H_5OH$$

Mechanism

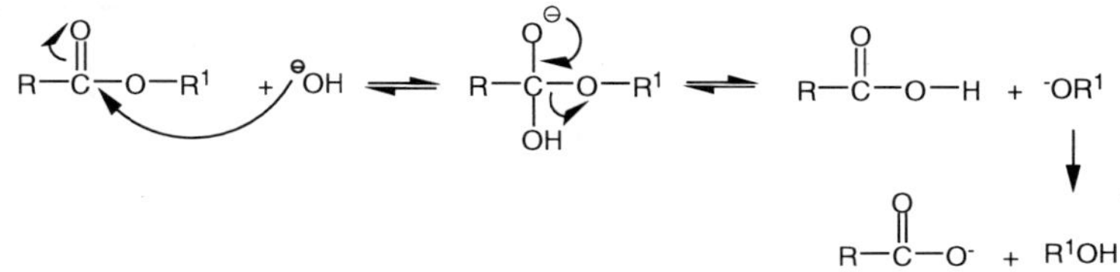

The special use of saponification is the preparation of o-methyl ethers of carboxylic acid. Thus, salicylic acid on methylation gives methyl salicylate, which on saponification gives o-methoxybenzoic acid (3.1).

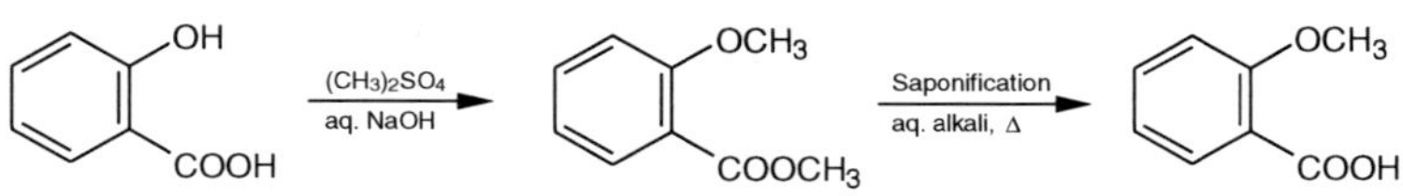

Salicylic acid   Methyl salicylate   o-Methoxybenzoic acid

3.1 o-METHOXYBENZOIC ACID

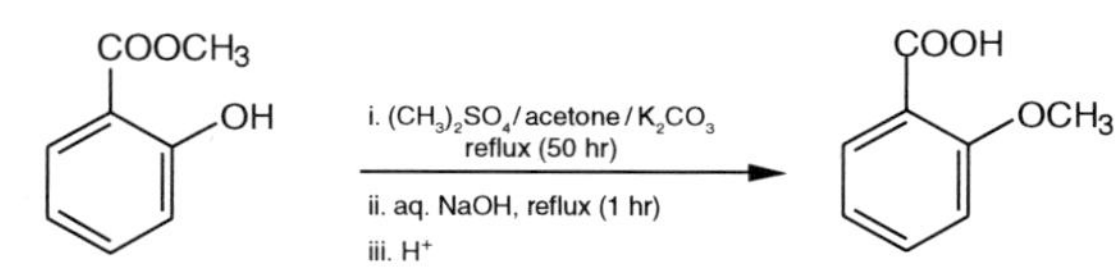

Methyl salicylate   o-Methoxybenzoic acid

Methyl salicylate (15 g, 0.99 mole) is dissolved in dry acetone (50 ml) and dimethyl sulphate (10.8 ml, 0.11 mole) and ignited potassium carbonate (25 g) added. The mixture is refluxed for 50 hr on a water bath under anhydrous conditions. The acetone is filtered and the potassium salts washed with hot acetone (2 × 50 ml). The combined acetone solution is distilled and the oily residue is hydrolysed by refluxing for 1 hr with aqueous sodium hydroxide (10%, 150 ml). The clear, cooled alkaline solution is acidified with hydrochloric acid. The separated product is filtered, washed with water and crystallised from dilute alcohol as colourless plates. Yield 12 g (80%). M.p. 101-102° (lit.[1], m.p. 101.5°).

Note: Methyl salicylate is prepared by refluxing salicylic acid (28 g, 0.2 mole), dry methanol (80 ml) and concentrated sulphuric acid (8 ml) for 5 hr (water bath). The alcohol is removed in vacuo and water (200 ml) added. It is extracted with ether (2 × 75 ml), ether extract washed with aqueous sodium bicarbonate (5%), water and dried (sodium sulphate). The ether is removed by evaporation and methyl salicylate is collected by distillation. B.p. 221-24°C, yield 25g (81%).

3.2 p-HYDROXYBENZOIC ACID

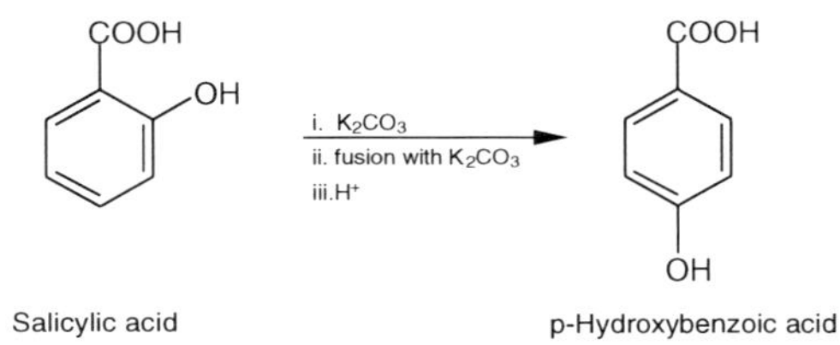

A suspension of salicylic acid (50 g, 0.362 mole) and water (80 ml) is treated with potassium carbonate (30 g, 0.215 mole). The solution is evaporated on a steam bath in a porcelain dish until a pasty residue is obtained. It is dried for 2 hr at 100-110° (air-oven). The solid product is powdered, dried for 2 hr more at 100-110°and again powdered. The dried mixture of potassium salicylate and potassium carbonate is heated in an oil bath in a R.B. flask. The oil bath is heated to 250° and maintained at this temperature for 2 hr. The mixture is stirred occasionally during this time. The hot reaction mixture is poured into hot water (500 ml) and the alkaline solution acidified (concentrated hydrochloric acid, about 40 ml). The solution is heated to boiling, treated with charcoal and filtered. The solution is cooled and the separated p-hydroxybenzoic acid filtered. More acid can be obtained by concentrating the filtrate to about 125 ml and cooling. Total yield is 20-25 g (70-80%). It is crystallised from water (can be charcoalised again if necessary), m.p. 211-12° (lit.[2], m.p. 211-12°).

3.3 2-HYDROXY-3-METHOXYBENZOIC ACID (o-VANILLIC ACID)

Potassium hydroxide (100 g) is moistened with water (3 ml) in a nickel dish and heated. To the molten mass, is added o-vanillin (10 g, 0.065 mole)

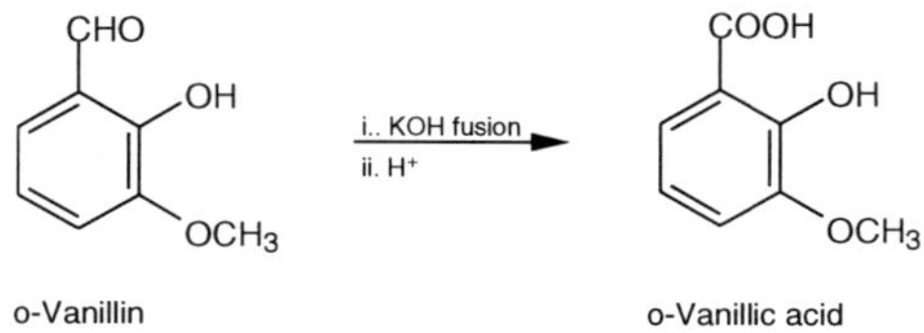

gradually with constant stirring. After the addition is over, the mass is cooled and dissolved in water (100 ml). The clear alkaline solution is acidified with hydrocholric acid. The o-vanillic acid first separates as an oil but it rapidly solidifies. It is filtered and washed with water. It is crystallised from benzene as colourless needles. Yield 9.5 g (85.9 %). M.P. 152° (lit.[3] m.p. 152°). A further quantity (3-4 g) is obtained by extracting the aqueous filtrate with ether.

3.4 2,3-DIHYDROXYBENZOIC ACID

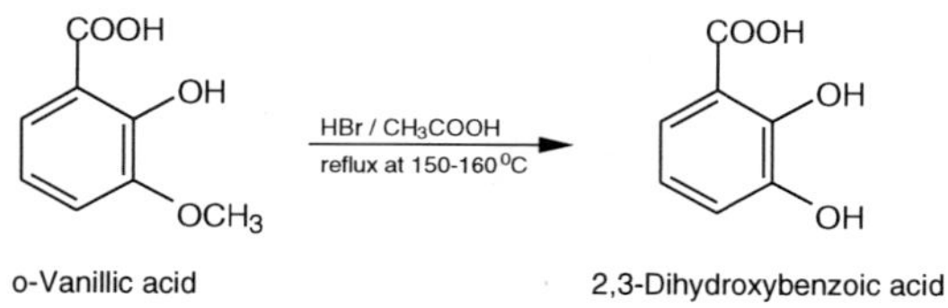

o-Vanillic acid (3.3) (20 g, 0.12 mole) is dissolved in glacial acetic acid (100 ml) and hydrobromic acid (100 ml, 47%) added. The mixture is refluxed at 150-160° (oil bath) for 3 hr and cooled. It is poured into cold water (50 ml). It is filtered and the clear filtrate allowed to stand in a refrigerator for 24 hr. The separated 2,3-dihydroxybenzoic acid is filtered. Yield 10.8 g (59.0%). M.p. 204-206°C (lit.[4], m.p. 204-206°C).

3.5 2,3-DIMETHOXYBENZOIC ACID

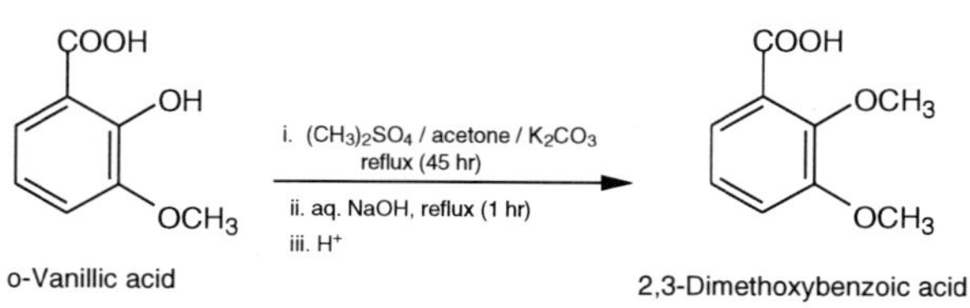

A solution of 2-hydroxy-3-methoxybenzoic acid (3.3) (20 g, 0.12 mole) in dry acetone (200 ml) is refluxed with excess dimethyl sulphate (45.3 ml, 0.48 mole) and ignited potassium carbonate (90 g) for 45 hr. The acetone is filtered and the inorganic salts washed with hot acetone (2 × 50 ml). The combined acetone solution is distilled and the oily product is refluxed for one hr with aqueous sodium hydroxide (5%, 200 ml). The clear alkaline solution is cooled and acidified with hydrochloric acid. The separated product is filtered and washed with water. It is crystallised from alcohol as colourless small prisms. Yield 20 g (92.3%). M.p. 124-25° (lit., m.p. 120-22°).

3.6 2,4-DIHYDROXYBENZOIC ACID (β-RESORCYCLIC ACID)

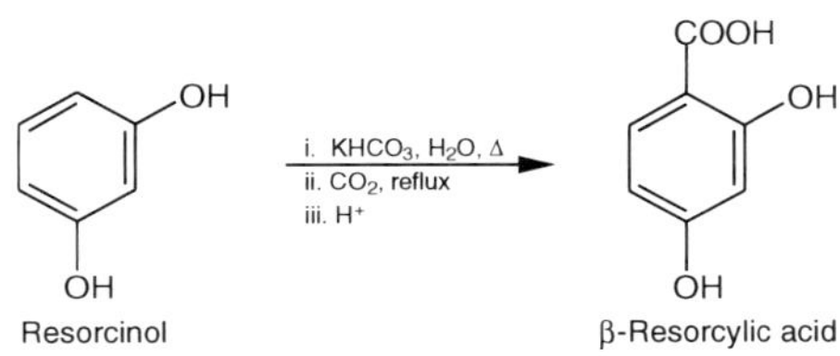

A mixture of resorcinol (50 g, 0.45 mole), potassium bicarbonate (250 g) and water (500 ml) is heated on a steam bath for 4 hr in a two litre three necked R.B. flask fitted with a reflux condenser. The solution is then vigorously refluxed over a flame for 30 min, a rapid stream of carbon dioxide gas is passed through the solution during refluxing. While still hot, the solution is acidified by adding concentrated hydrochloric acid (approx. 225 ml) from a separatory funnel with a tube delivering the acid to the bottom of the flask; this process prevents the formation of a layer of acid over the neutralised solution. The reaction mixture is allowed to cool to room temperature and then chilled in an ice-bath. The separated product is filtered and crystallised from water (charcoal) to give 2,4-dihydroxy benzoic acid as colourless product. Yield 55 g (78.6%). M.p. 217-218° (lit.[6] m.p. 217-18°). Some more of the acid (9 g) is obtained by ether extraction of the combined filtrate with ether. The acid is isolated from the ether extract by extracting with sodium bicarbonate solution and acidifying the clear bicarbonate extract.

3.7 METHYL 2,4-DIHYDROXYBENZOATE

2,4-Dihydroxybenzoic acid (3.6) (18 g, 0.12 mole) in dry acetone (250 ml) is refluxed for 4 hr with dimethyl sulphate (12.3 ml, 0.13 mole) in the presence of anhydrous sodium bicarbonate (25 g). The acetone is filtered and inorganic salts washed with hot acetone (2 × 50 ml). The combined

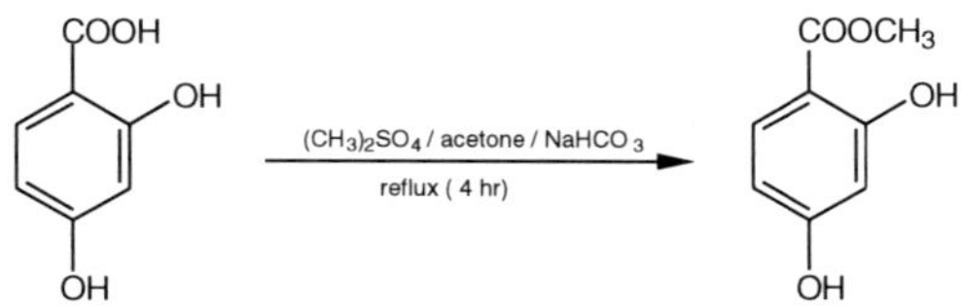

β-Resorcylic acid Methyl-2,4-Dihydroxybenzoate

acetone solution is distilled and the oily residue treated with ice-water. The solid product is filtered and washed with water. It is crystallised from dilute alcohol as colourless needles. Yield 16 g (81.6%). M.p. 119° (anhydrous sample) (lit.[7] m.p. 118-19°).

Note: The acid separated after acidification melts at 78-79° due to water of crystallization. It loses water of crystallization on drying in vacuum desiccator and the m.p. rises to 119°.

3.8 2-HYDROXY-4-METHOXYBENZOIC ACID

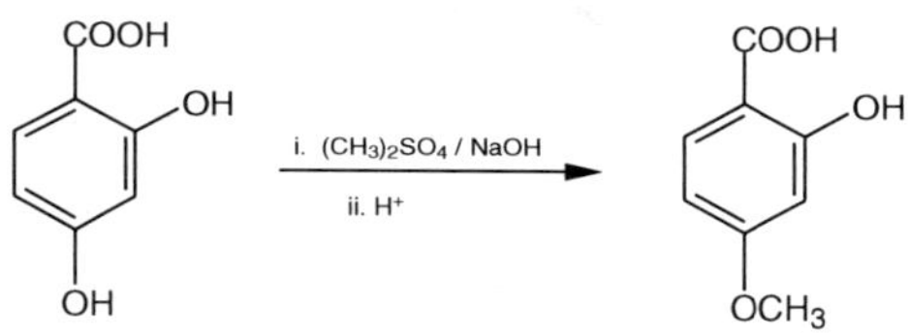

β-Resorcylic acid 2-Hydroxy-4-methoxybenzoic acid

2,4-Dihydroxybenzoic acid (3.6) (24 g, 0.16 mole) is dissolved in sodium hydroxide solution (80 ml, 20%) in a conical flask. The solution is stirred mechanically and dimethyl sulphate (50 ml, 0.53 mole) added dropwise while cooling the solution during 1-1.5 hr. It is allowed to stand overnight at room temperature and then heated on a water bath for 30 min. The solution is cooled and acidified with hydrocholoric acid. The separated product is filtered and washed with water. It is crystallised from dilute alcohol as colourless needles. Yield 20 g (76.4%). M.p. 160-61° (lit.[8] m.p. 160-61°).

3.9 2,5-DIHYDROXYBENZOIC ACID (GENTISIC ACID)

To a solution of salicylic acid (2 g, 0.14 mole) in aqueous sodium hydroxide (200 ml, 10%) cooled to 15-20° (ice-bath) is added dropwise during 4-5 hr, an aqueous solution of potassium persulphate (38 g, 0.14 mole in 400 ml water) with constant stirring. The reaction mixture is kept for 72 hr at room temperature. It is cooled and acidified to congo red with dilute

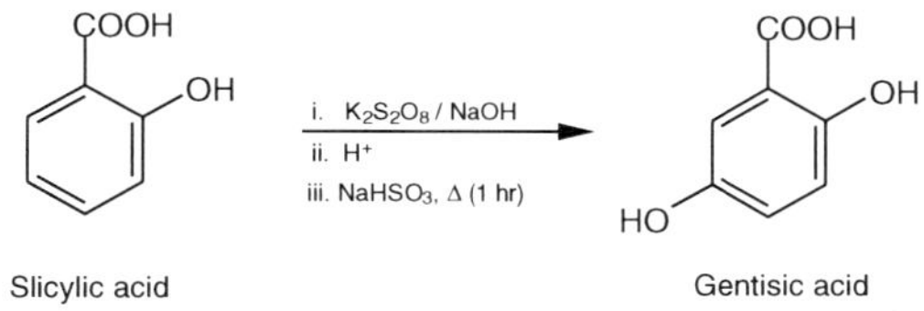

hydrochloric acid. The solution is extracted with ether (3 × 100 ml) to remove the unreacted salicylic acid. The remaining solution is heated on a boiling water bath for 1 hr with sodium bisulphite (15 g) and concentrated hydrocholoric acid (100 ml). It is cooled and repeatedly extracted with ethyl acetate (6 × 75 ml). The ethyl acetate extract is dried (anhydrous sodium sulphate) and distilled. The required 2,5-dihydroxybenzoic acid is crystallised from ethyl acetate as colourless needles. Yield 6 g (26.9%). M.p. 204-205° (lit.[9] m.p. 197-98°).

3.10 2,5-DIMETHOXYBENZOIC ACID

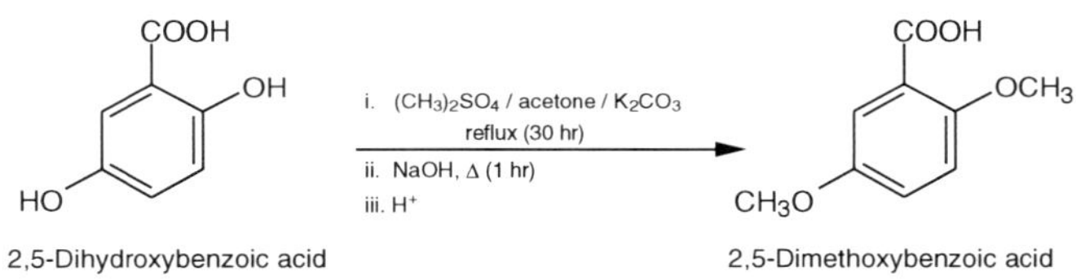

A solution of 2,5-dihydroxybenzoic acid (3.9) (6 g, 0.039 mole) in dry acetone (200 ml) containing dimethyl sulphate (12 ml, 0.13 mole) and anhydrous potassium carbonate (18 g) is refluxed on a water bath for 30 hr under anhydrous conditions. The acetone solution is filtered and the inorganic residue washed with hot acetone (2 × 50 ml). The acetone is distilled and the resulting methyl 2,5-dimethoxybenzoate is hydrolysed directly by heating for 1 hr in a boiling water bath with sodium hydroxide solution (40 ml, 10%). The resulting clear alkaline solution is acidified with hydrocholoric acid. The separated product is filtered and washed with water. It is crystallised from alcohol as colourless needles. Yield 4 g (56.4%). M.p. 76° (lit.[10] m.p. 76°).

3.11 2-HYDROXY-5-METHOXYBENZOIC ACID

2,5-Dihydroxybenzoic acid (3.9) (17.5 g, 0.11 mole) in dry acetone (350 ml) is refluxed with dimethyl sulphate (24 ml, 0.25 mole) in presence of anhydrous potassium carbonate (70 g) for 6 hr (water-bath). The hot mixture is filtered and the inorganic salts washed with hot acetone

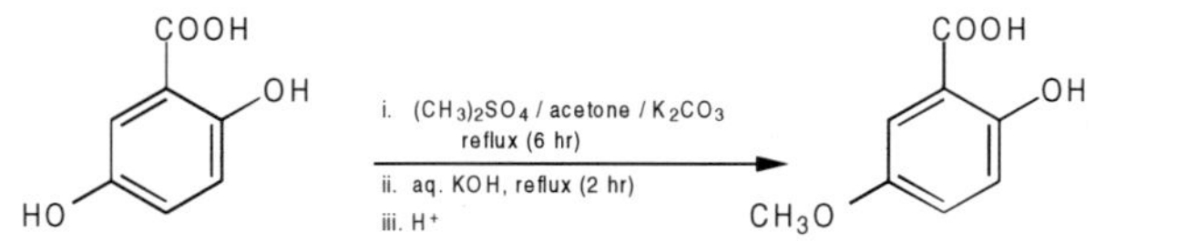

2,5-Dihydroxybenzoic acid → 2-Hydroxy-5-methoxybenzoic acid

(2 × 50 ml). The combined acetone solution is distilled and the resulting methyl 2-hydroxy-5-methoxybenzoate is hydrolysed by refluxing with aqueous potassium hydroxide (200 ml, 10%) for 2 hr. The clear and cooled alkaline solution is solidified and cooled. The separated 2-hydroxy-5-methoxybenzoic acid is obtained as colourless prisms. Yield 12 g (62.8%). M.p. 145-46° (lit.[11] m.p. 145-46°).

3.12 2-HYDROXY-6-METHOXYBENZOIC ACID

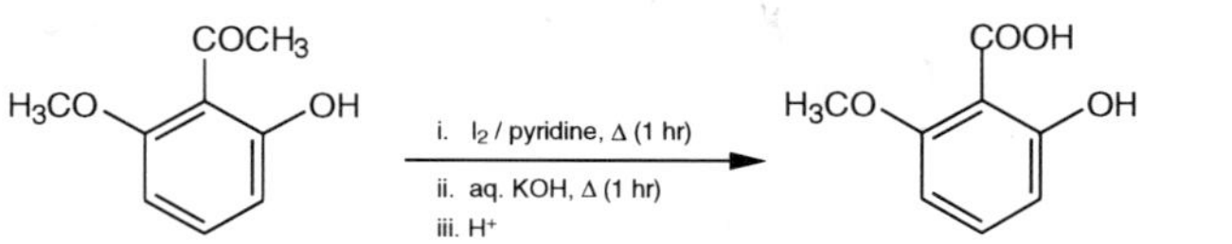

2-Hydroxy-6-methoxyacetophenone → 2-Hydroxy-6-methoxybenzoic acid

A mixture of 2-hydroxy-6-methoxyacetophenone (4.A.17.) (1g, 0.006 mole), iodine (1.5 g, 0.006 mole) and dry pyridine (10 ml) is heated under anhydrous conditions on a steam bath for one hr. The solution is kept at 0° for 24 hr. The separated product is filtered, washed with cold water. It is heated on a steam bath for 1 hr with aqueous potassium hydroxide (60 ml, 2%). The clear, cooled alkaline solution is acidified with hydrochloric acid. The separated product is filtered and purified by dissolving in sodium carbonate solution and acidifying the clear solution. It is crystallised from dilute alcohol as colourless needles. Yield 0.6g (60%). M.p. 136° (lit.[12] m.p. 136-38°).

Note: This is a very convenient method of converting a $-COCH_3$ into $-COOH$.

3.13 3-METHOXY-4-HYDROXYBENZOIC ACID (VANILLIC ACID)

A mixture of potassium hydroxide (56 g, 1.0 mole) and water (8 ml) are placed in a nickel dish set in a deep sand bath and heated by a Bunsen burner. With stirring, the alkali is heated to 120°. The viscous mass is treated with vanillin (20 g, 0.13 mole) in small portions over a period of 10

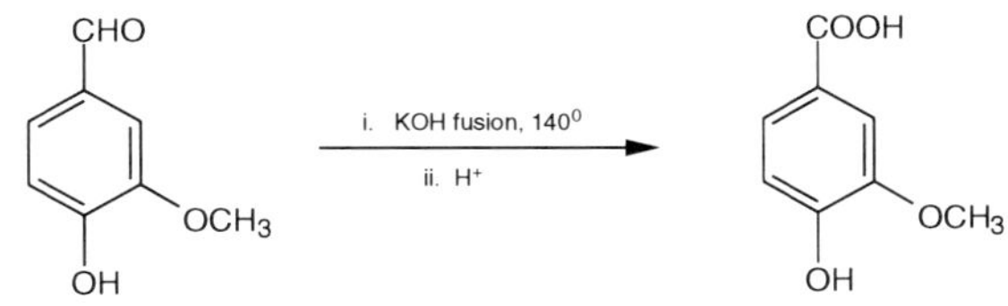

Vanillin Vanillic acid

min. The reaction mixture is stirred after each addition. Considerable evolution of hydrogen takes place after each addition. The temperature rises to 140° and the mixture becomes fluid. It is maintained at this temperature by suitably adjusting the addition of vanillin. After all the vanillin has been added, the mixture becomes pasty and the temperature is raised to 180°, stirring is continued for 10 min more at this temperature. The reaction mixture is cooled and dissolved in water (80 ml) to a red solution. It is acidified with hydrochloric acid. The separated product is filtered and washed with water. It is crystallised from dilute alcohol as colourless needles, Yields 22 g (99.5 %). M.p. 208-209°. (lit.[13,14] m.p. 210°).

3.14 3,4-DIHYDROXYBENZOIC ACID (PROTOCATECHUIC ACID)

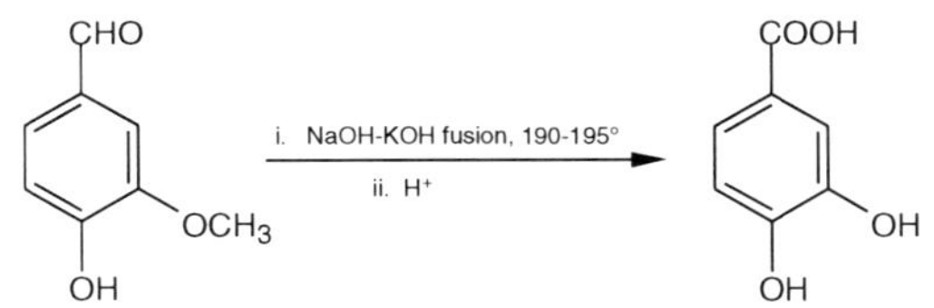

Vanillin Protocatechuic acid

A mixture of sodium hydroxide (80 g, 2 mole), potassium hydroxide (336 g, 6 mole) and water (50 ml) is heated in a nickel dish to 160°. Vanillin (152 g, 1 mole) is added in portions during 3 min at a rate sufficient to maintain the reaction temperature 190-195°. The temperature is then raised to 240-45° with constant stirring. The mixture is heated at this temperature for 5 min and then allowed to cool with stirring. When the mixture has cooled to 150-160°, water (1000 ml) is added and mixture stirred until the fusion mixture dissolves. The solution is transferred to a beaker and sulphur dioxide passed to remove the dark impurities. The solution is acidified with hydrochloric acid (6N, 1.5 litre) and cooled. The separated acid is filtered, washed with water and dried.

Yield 99 g. M.p. 198-199° (lit.[13] m.p. 199-200°). Some more acid (36 g) can be obtained from the mother liquor by ether extraction. Total yield is 135 g (87%).

3.15 METHYL 3,4-DIHYDROXYBENZOATE (METHYL PROTOCATECHUATE)

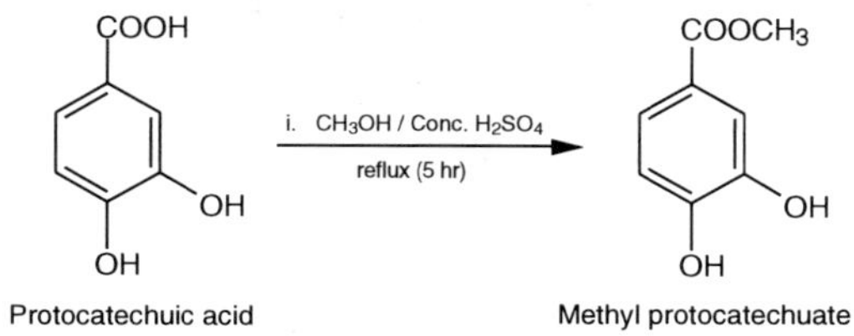

A solution of protocatechuic acid (3.14) (10 g, 0.065 mole) in absolute methanol (25 ml) and concentrated sulphuric acid (3 ml) is refluxed for 5 hr. The methanol is removed under reduced pressure and ice added to the cooled solution. The separated product is filtered, macerated with cold sodium bicarbonate solution and finally washed with cold water. It is crystallized from water as colourless needles. Yield 8 g (73.4%). M.p. 134-135° (lit.[15] m.p. 134-35°).

3.16 3-HYDROXY-4-METHOXYBENZOIC ACID (ISOVANILLIC ACID)

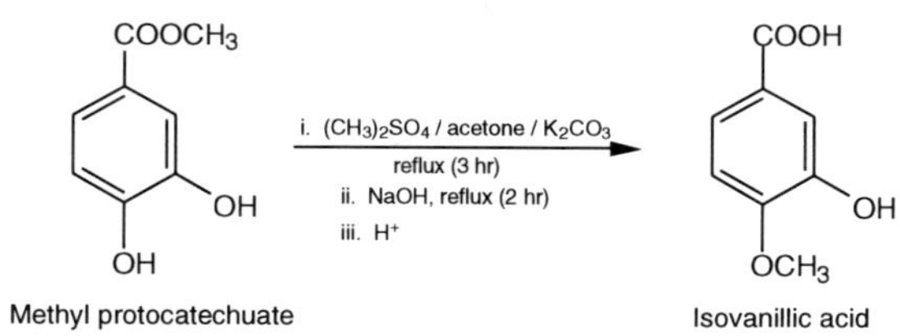

A solution of methyl 3,4-dihydroxybenzoate (methyl protocatechuate) (3.15) (12.2 g, 0.073 mole) in dry acetone (150 ml) is treated with dimethyl sulphate (10 ml, 0.11 mole) and ignited potassium carbonate (25 g). The mixture is refluxed for 3 hr under anhydrous conditions. The acetone is filtered, inorganic salts washed with hot acetone (2 × 40 ml). The combined acetone solution is distilled. The residue is refluxed with sodium hydroxide solution (150 ml, 10%) for 2 hr. The resulting solution is cooled and the clear alkaline solution acidified with hydrochloric acid. The separated isovanillic acid is crystallized from alcohol as colourless prisms.

Yield 10 g (82%), M.p. 254-255° (lit.[16] m.p. 248-249°). (It was obtained earlier[16] by partial demethylation of 3,4-dimethoxybenzoic acid (3.17). The above method is better).

3.17 3,4-DIMETHOXYBENZOIC ACID (VERATRIC ACID)

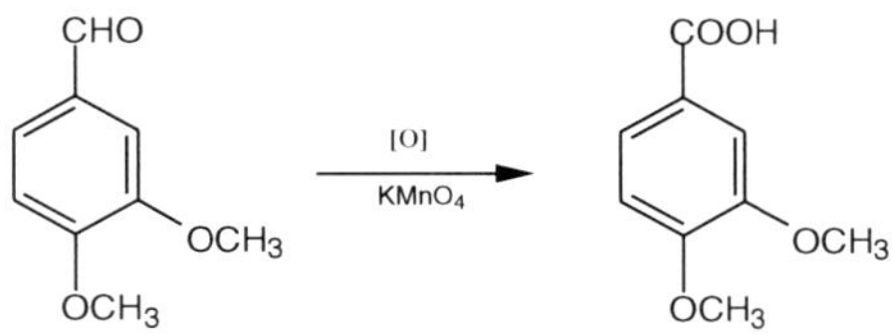

Veratraldehyde Veratric acid

3,4-Dimethoxybenzaldehyde (varatraldehyde) (2.11) (50 g, 0.30 mole) is suspended in water (300 ml) and the suspension warmed to 50-60° (water-bath). A solution of potassium permanganate (71 g, 0.45 mole in 750 ml water) is then added dropwise during 2.5-3 hr to the stirred suspension. The mixture is stirred for 1 hr more. The precipitated manganese dioxide is destroyed by passing a current of sulphur dioxide gas. The separated colourless crystalline product is filtered and dried. Yield 50 g (91.5%). M.p. 179-80° (lit.[17, 18] m.p. 179-80°).

3.18 3,4-DIMETHOXY-6-NITROBENZOIC ACID (6-NITRO VERATRIC ACID)

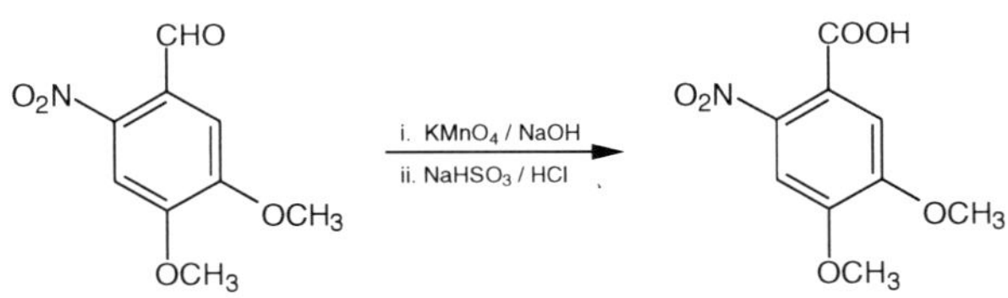

6-Nitro verataldehyde 6-Nitro veratric acid

6-Nitro verataldehyde (2.46) (10 g, 0.047 mole) is suspended in water (200 ml) and sodium hydroxide (5 g) is added. To this stirred suspension is added slowly finely powdered potassium permanganate (10.5 g, 0.066 mole). The solution is stirred continuously until the purple colour of permanganate disappeared (7 hr). The mixture is cooled and sodium bisulphite and hydrochloric acid added to render the solution colourless. The precipitated acid is filtered. It is crystallised from hot water. Yield 4 g (40.2 %). M.p. 190° (lit.[19] m.p. 189-190°).

3.19 3,4-METHYLENEDIOXYBENZOIC ACID (PIPERONYLIC ACID)

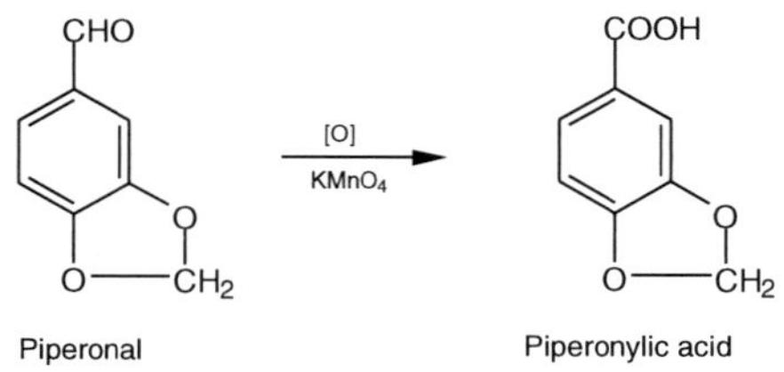

Piperonal (30 g, 0.2 mole) and water (750 ml) are taken in a 3 litre flask. The mixture is heated to 70-80° (steam bath) and to the stirred mixture (vigorous stirring by mechanical stirrer) is added a solution of potassium peremanganate (44.2 g, 0.28 mole) in water (one litre) during 60 min. The solution is stirred for one hr more at 70-80° and sodium hydroxide solution (10%, sufficient to render the mixture alkaline) is added. The hot mixture is filtered, manganese dioxide is washed with hot water (3 × 100 ml). The combined filtrate and washings are cooled. Any unreacted piperonal that separates is filtered and solution is acidified with hydrochloric acid. The separated piperonylic acid is filtered, washed with cold water and dried. It is crystallised from ethyl alcohol. Yield 28 g (84%). M.p. 227-28° (lit.[20] m.p. 227-28°).

3.20 3,5-DIMETHOXYBENZOIC ACID

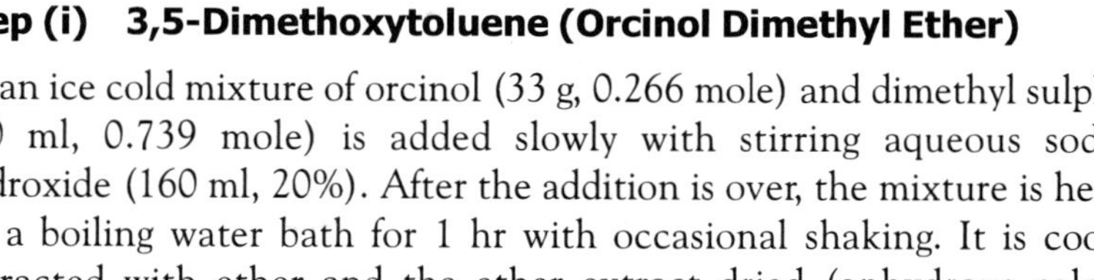

Step (i) 3,5-Dimethoxytoluene (Orcinol Dimethyl Ether)

To an ice cold mixture of orcinol (33 g, 0.266 mole) and dimethyl sulphate (70 ml, 0.739 mole) is added slowly with stirring aqueous sodium hydroxide (160 ml, 20%). After the addition is over, the mixture is heated on a boiling water bath for 1 hr with occasional shaking. It is cooled, extracted with ether and the ether extract dried (anhydrous calcium chloride). The ether is distilled to give 3,5-dimethoxytoluene. Yield 30 g (74.2%). B.p. 240-44° (lit.[21] b.p. 244°).

Step (ii) 3,5-Dimethoxybenzoic Acid

3,5-Dimethoxytoluence (10 g, 0.065 mole) is stirred vigorously with a hot solution of potassium permanganate (20 g, 0.125 mole) in water (500 ml). The mixture is refluxed for 30 min. It is cooled and treated with sodium bisulphite and hydrochloric acid till the solution becomes colourless. The mixture is cooled and the separated product filtered. It is purified by dissolving in sodium bicarbonate solution and acidification of the clear bicarbonate extract. Yield 3.1 g (26.8%). M.p. 181° (lit.[21] m.p. 181°).

3.21 3,5-DIHYDROXYBENZOIC ACID (α-RESORCYLIC ACID)

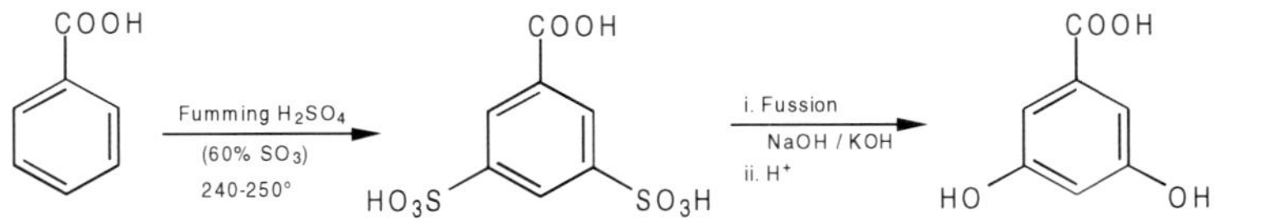

Benzoic acid | Isolated as Ba salt | 3,5-Dihydroxybenzoic acid

Benzoic acid (100 g, 0.87 mole) is added to fuming sulphuric acid [approx. 30% SO_3 made by mixing 500 g of fuming sulphuric acid containing 60% free SO_3 and 500 g of concentrated sulphuric acid (sp. gr. 1.84); 500 ml of this mixture is used] in a Kjeldahl flask and the mixture is heated in an oil bath at 240-250° (bath temperature) in a fume cupboard. The mixture is allowed to stand overnight at room temperature and poured slowly with stirring over crushed ice (1.5 kg). The solution is then neutralised by adding barium carbonate in 50 g lots, stirring occasionally until the gas evolution slackens before each addition (1.2-1.3 kg barium carbonate is required). The mass is filtered (suction) and the barium sulphate is washed with water (5 × 150 ml). The combined filtrate is evaporated nearly to dryness on a steam bath and finally dried at 125-140° (air oven). Yield of crude barium salt is 375 g. The dried powered barium salt is added in 100 g lots to a melt of sodium hydroxide (300 g) and potassium hydroxide (300 g) contained in a copper beaker. The mixture is stirred with a copper stirrer and temperature measured with a thermometer in a copper tube sealed from one end. Each portion is well stirred before the next is added. The temperature is then slowly raised to 250-255°, at this temperature a vigorous reaction occurs with copious evolution of gas. After the reaction has slowed down (25-30 min) the temperature is raised to 290-310° and maintained there for 1 hour and then allowed to come down to 200°. The melt is poured into water (3 litre) (caution). The barium sulphate is filtered by suction and the filtrate acidified with concentrated hydrochloric acid (approx. 1.3 litre). The remaining solution is extracted with ether (3 × 600 ml). The combined ether extracts are concentrated to about 1 litre and dried (anhydrous sodium sulphate). The ether solution is distilled. The residual product is 3,5-dihydroxybenzoic acid. Yield 80 g (65%). It is recrystallized from water. M.p. 228-29° (decomp.). (lit.[22] m.p. 228-29°).

3.22 METHYL 3,5-DIMETHOXYBENZOATE

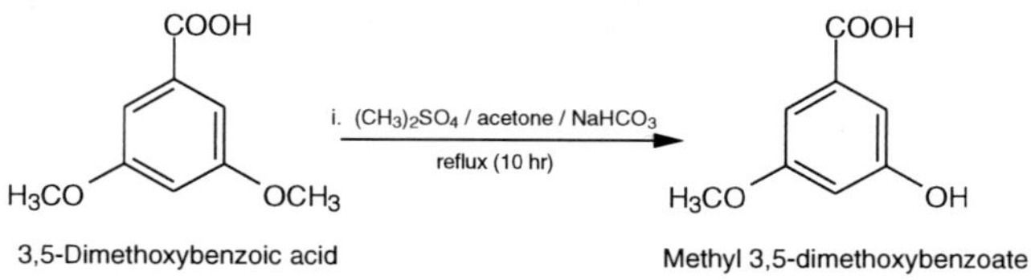

A solution of 3,5-dimethoxybenzoic acid (3.20.) (18 g, 0.099 mole) in dry acetone (200 ml) containing dimethyl sulphate (14.3 ml, 0.15 mole) and anhydrous sodium bicarbonate (20 g) is refluxed for 10 hr under anhydrous conditions. The acetone is filtered and the inorganic residue washed with hot acetone (50 ml). The acetone solution is distilled to give the required ester. Yield 18 g (92.8%). B.p. 296-98° (m.p. 42-43°) (lit.[23] b.p. 298°).

3.23 2,3,4-TRIHYDROXYBENZOIC ACID (PYROGALLOL CARBOXYLIC ACID)

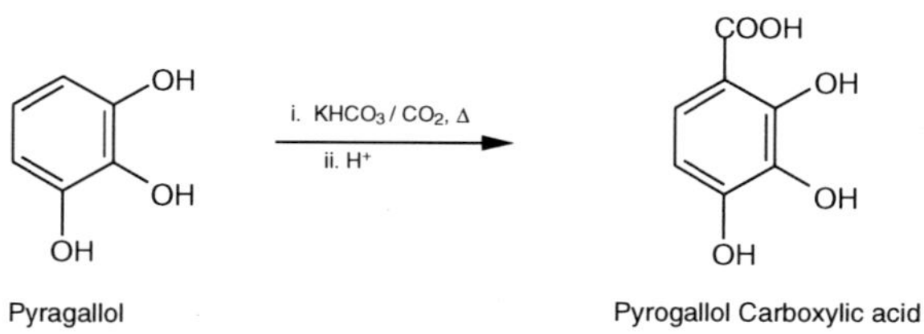

A paste of finely powdered pyrogallol (10 g, 0.079 mole), potassium bicarbonate (20 g, 0.22 mole) and water (5 ml) is heated in an oil bath at 90° under a steam of carbon dioxide. The bath temperature is then raised to 135° during 2 hr and kept at that temperature for 3 hr. The reaction mixture is cooled and dissolved in water (100 ml). It is acidified with hydrocholric acid in cold. The separated product is filtered and washed with cold water. It is crystallised from water (charcoal) as colourless prisms. Yield 7 g (51.8%). M.p. 210° (lit.[24], m.p. 210°).

3.24 METHYL 2,3,4-TRIHYDROXY BENZOATE

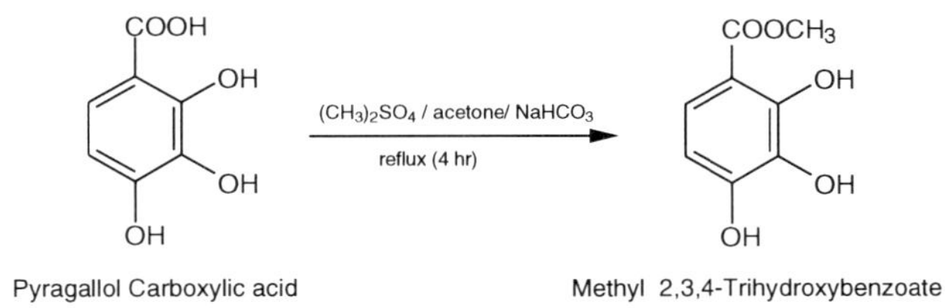

2,3,4-Trihydroxybenzoic acid (3.23) (1 g, 0.0058 mole) in dry acetone (50 ml) is refluxed for 4 hr with dimethyl sulphate (0.64 ml, 0.0067 mole) in presence of anhydrous sodium bicarbonate under anhydrous conditions. The acetone is filtered and inorganic salts washed with hot acetone (1 × 25 ml). The combined acetone solution is distilled and the residue treated with crushed ice. The separated solid is filtered and washed with alcohol. It is crystallised from benzene-petroleum ether as pale yellow plates. Yield 0.8 g (74%). M.p. 151-52° (lit.[25] m.p. 151-52°).

3.25 2-HYDROXY-3,4-DIMETHOXYBENZOIC ACID

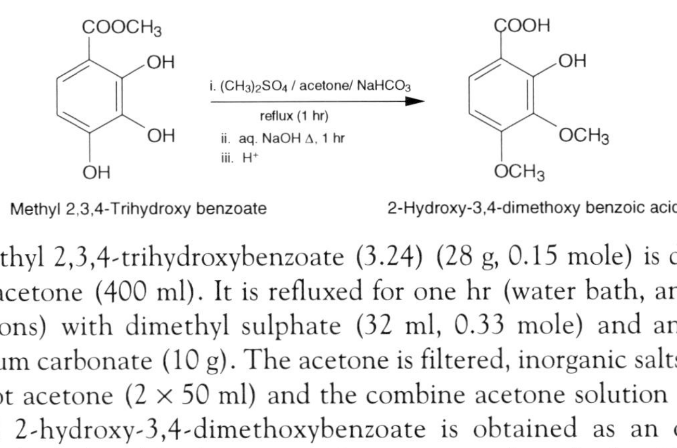

Methyl 2,3,4-trihydroxybenzoate (3.24) (28 g, 0.15 mole) is dissolved in dry acetone (400 ml). It is refluxed for one hr (water bath, anhydrous conditions) with dimethyl sulphate (32 ml, 0.33 mole) and anhydrous potassium carbonate (10 g). The acetone is filtered, inorganic salts washed with hot acetone (2 × 50 ml) and the combine acetone solution distilled. Methyl 2-hydroxy-3,4-dimethoxybenzoate is obtained as an oil. It is directly hydrolysed by heating with aqueous sodium hydroxide, (10% 250 ml) for 1 hr (water bath). The cooled alkaline solution is extracted with ether (100 ml) and the aqueous solution acidified with hydrochloric acid. The separated product is filtered and washed with water. It is crystallised from water as colourless needles. Yield 20 g (66.3%). M.p. 169-70° (lit.[26] m.p. 169-70°).

3.26 3,4,5-TRIMETHOXYBENZOIC ACID (TRI-O-METHYL GALLIC ACID)

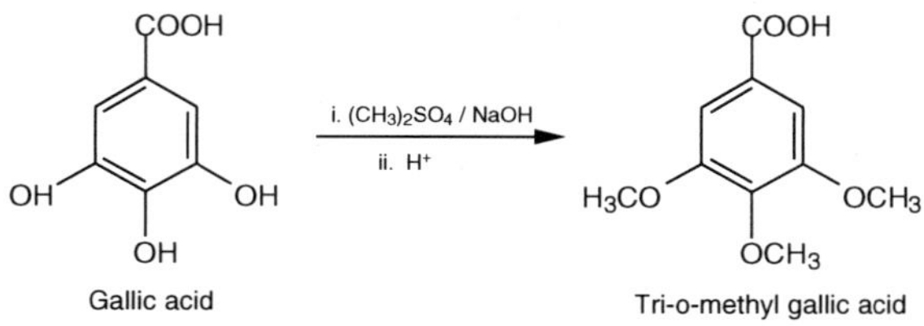

To a cooled solution of sodium hydroxide (40 g, 1 mole) in water (250 ml) in a 500 ml R.B. flask is added gallic acid (25 g, 0.147 mole). The flask is tightly stoppered and the mixture shaken occasionally until a clear solution is obtained. Dimethyl sulphate (44.5 ml, 0.47 mole) is then added and the flask shaken for 20 min with cooling in ice-cold water; the temperature is not allowed to rise above 30-35°. Occasionally the stopper is raised to reduce any pressure. A second portion of dimethyl sulphate (44.5 ml, 0.47 mole) is then added and the shaking continued from 10 min more. During the second addition, the temperature is allowed to rise to 40-45°. The solution is refluxed for 1 hr. Sodium hydroxide solution (10 g in 15 ml water) is added and refluxing continued for 1 hr more. The reaction mixture is cooled and the clear alkaline solution acidified with hydrochloric acid. The solution is cooled and the product filtered. It is crystallised from hot water (charcoal) as colourless needles. Yield 25 g (80.2%). M.p. 166-67° (lit.[27] m.p. 168°).

3.27 METHYL 3,4,5-TRIHYDROXYBENZOATE (METHYL GALLATE)

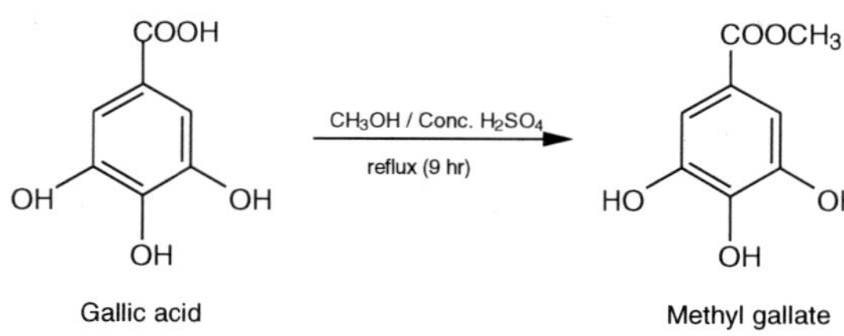

A solution of anhydrous gallic acid (dried at 120°) (25 g, 0.147 mole) in absolute methanol (50 ml) and dry benzene (150 ml) containing concentrated sulphuric acid (2 ml) is refluxed azeotropically (water bath) till the TLC showed the absence of gallic acid (9 hr). The solvent is distilled in vacuo and residue macerated with ice-water (100 ml). The separated solid is filtered and washed with sodium bicarbonate solution (10%, 100 ml) and finally with water. It is crystallized from alcohol as colourless needles. Yield 17g (62.8%). M.p. 155-56°, (lit.[28] m.p. 157°).

3.28 METHYL 3,4,5-TRIMETHOXYBENZOATE

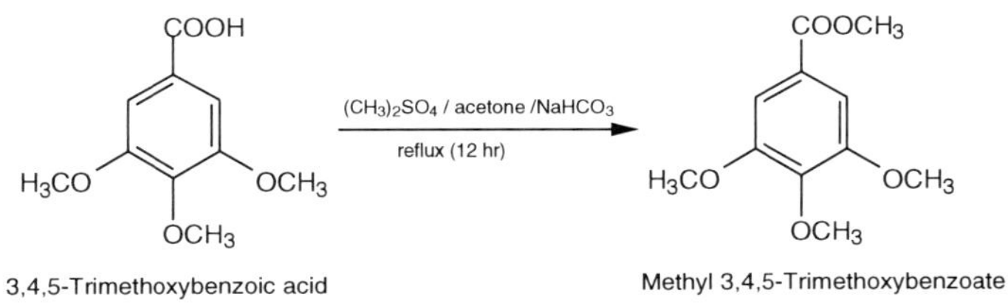

3,4,5-Trimethoxybenzoic acid (3.26) (22.5 g, 0.106 mole) in dry acetone (200 ml) is refluxed with anhydrous sodium bicarbonate (40 g) and dimethyl sulphate (11.25 ml, 0.118 mole) for 12 hr under anhydrous conditions. The solution is filtered and the inorganic salts washed with hot acetone (50 ml). The combined acetone solution is distilled and the residual product macerated with sodium bicarbonate solution (10%). It is filtered and washed with water. The product is crystallised from methanol as colourless needles. Yield 20 g (83.4%). M.p. 80-82° (lit.[27,29] m.p. 82-3°).

3.29 METHYL 3,4-DIMETHOXY-5-HYDROXYBENZOATE[30] (METHYL 3,4 DI-O-METHYLGALLATE)

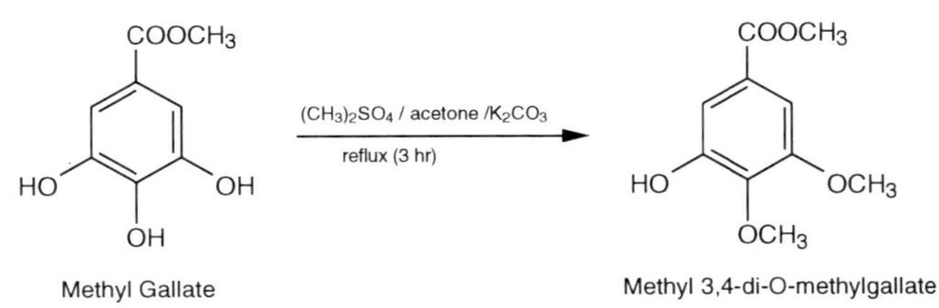

To a solution of methyl 3,4,5-trihydroxybenzoate (3.27) (20 g, 0.108 mole) in dry acetone (150 ml), dimethyl sulphate (21 ml, 0.22 mole) and anhydrous potassium carbonate (30 g) added and the mixture is refluxed from 3 hr (water bath) under anhydrous conditions. The acetone is filtered, inorganic salts washed with hot acetone (50 ml) and the combined acetone solution distilled and the residual product macerated with sodium bicarbonate solution (10%). It is filtered, washed with water and dried. It is purified by chromatography on neutral alumina column (3 × 20 cm) and elution with petroleum ether gave the first fraction of methyl 3,4,5-tri-O-methyl gallate. The second fraction consists of methyl 3,4-di-O-methylgallate. It is crystallised from ethyl acetate-petroleum ether as colourless needles. Yield 15 g (65.2%). M.p. 84°.

3.30 METHYL 3-METHOXY-4,5-DIHYDROXYBENZOATE (METHYL 3-O-METHYLGALLATE)

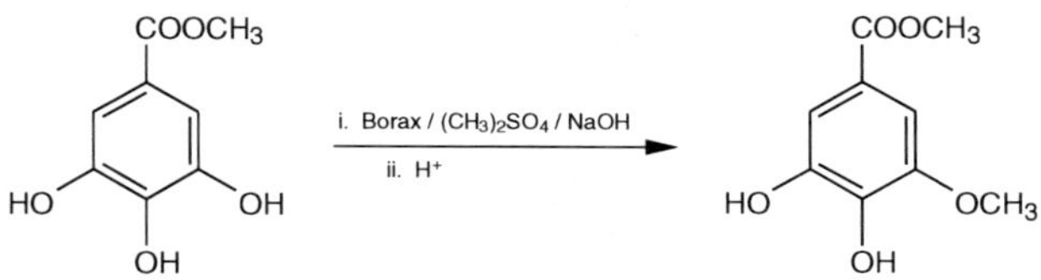

Methyl Gallate   Methyl 3-O-methylgallate

Methyl 3,4,5-trihydroxybenzoate (3.27) (5 g, 0.027 mole) is treated with a warm solution of borax (BDH, AR) (5%, 500 ml) with shaking. Sodium hydroxide solution (5%, 5 ml) is then added to it and the solution is treated with dimethyl sulphate (2.8 ml, 3.78 g, 0.03 mole) added dropwise with constant shaking for 2 hr. During this interval, further quantity of sodium hydroxide solution (5%, 100 ml) is added little by little to ensure that the medium remained alkaline. The reaction mixture is allowed to stand overnight at room temperature and then acidified with hydrochloric acid. It is extracted with ethyl acetate (3 × 75 ml), the extract dried (anhydrous sodium sulphate) and distilled. The residual product is purified by column chromatography over silica gel (2 × 20 cm), methyl 3-methoxy-4,5-dihydroxybenzoate is eluted with ethyl acetate-benzene (5:95). Yield 3 g (55.7 %). M.p. 112° (lit.[31] m.p. 112-13°).

3.31 3,5-DIMETHOXY-4-HYDROXYBENZOIC ACID (SYRINGIC ACID)

3,4,5-Trimethoxybenzoic acid (3.26) (20 g, 0.094 mole) (dried at 110°) is slowly added with vigorous stirring to fuming sulphuric acid (20%, 80 ml) allowing complete solution to take place between successive additions. The solution is maintained at 33-40° until all the trimethoxybenzoic acid has been added. The mixture is stirred for 30 min at 40° and left overnight

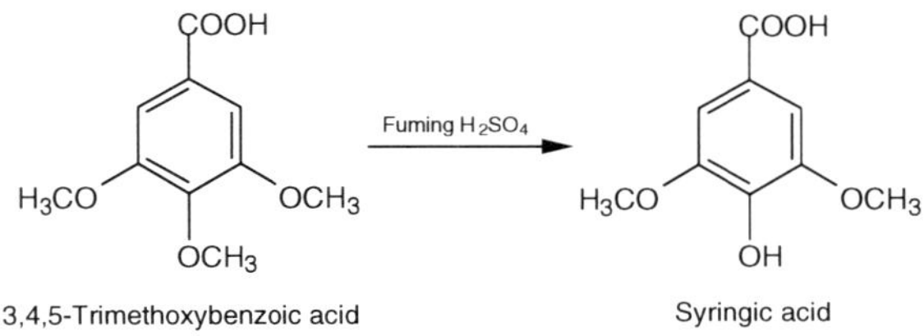

in an ice-box. The viscous mass is poured slowly with stirring, into water (150-170 ml). The precipitate which is first obtained dissolves before the solution reaches its maximum at 120°. After 2 minutes, the syringic acid crystallises out rapidly. The mixture is left for 1 hr at room temperature. The separated product is filtered and washed with cold water. It is crystallised from hot water (charcoal) as colourless needles, yield 13.5 g (72.6%). M.p. 204-205°. (lit.[32] m.p. 206.5°).

3.32 3,4-DIMETHOXY-5-HYDROXYBENZOIC ACID (3,4-DI-O-METHYL GALLIC ACID

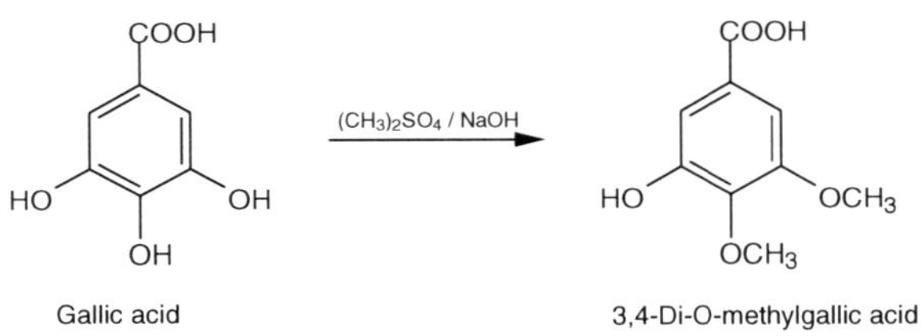

Gallic acid (50 g, 0.29 mole) and water (350 ml) are placed in a 1 litre shaking flask. Into this are added dimethyl sulphate (90 ml, 0.95 mole) and sodium hydroxide solution (43.5 g in 150 ml water) in portions of 36 ml and 60 ml respectively. After each addition the flask is well evacuated and shaken thoroughly for 25 min. If there is any rise in temperature, it should be cooled in cold water. The shaking is continued for 1 hr after the addition is complete. The brownish oily reaction product is taken up in ether. The ether extract is washed successively with potassium bicarbonate solution (10%) (to extract the mixture of methylated gallic acids), sodium carbonate solution 10% (to remove monomethyl gallic acid ester) and finally with sodium hydroxide solution 10% (to remove dimethyl gallic acid ester). By acidification are obtained esters of mono- and di-methyl gallic acids.

To isolate dimethyl gallic acid, the sodium hydroxide extract is acidified with hydrochloric acid. The solution is extracted with ether (3 × 100 ml). It is distilled and the residual oily product hydrolysed by refluxing with sodium hydroxide solution (10%, 150 ml) for 30 min. The cooled solution is acidified and cooled. The separated product is filtered and washed with cold water. It is crystallised from hot water (charcoal) as colourless needles. Yield 29.1 g (50%). M.p. 199-200° (lit.[33,34] m.p. 197-98°).

3.33 2,4,6-TRIHYDROXYBENZOIC ACID (PHLOROGLUCINOL CARBOXYLIC ACID)

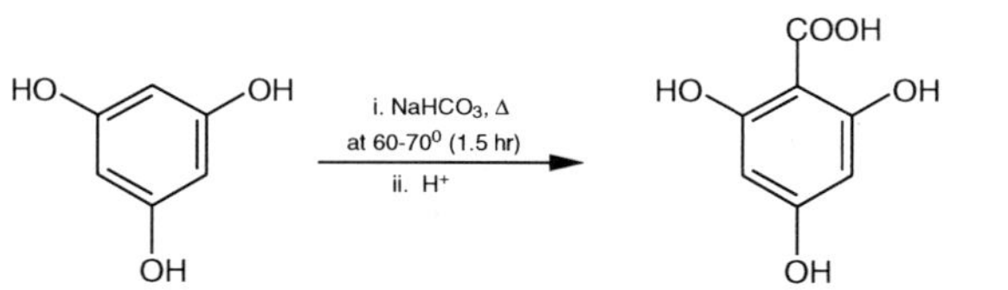

Phloroglucinol   Phloroglucinol carboxylic acid

A solution of phloroglucinol (5 g, 0.039 mole) and sodium bicarbonate (10 g) in water is heated at 60-70° for 1.5 hr. The solution is cooled and acidified. The separated product is filtered and dried in vaccum desiccator. Yield 4 g (65.6%). M.p. 206-207° (lit.[35] m.p. 206°).

3.34 2-HYDROXY-4,6-DIMETHOXYBENZOIC ACID

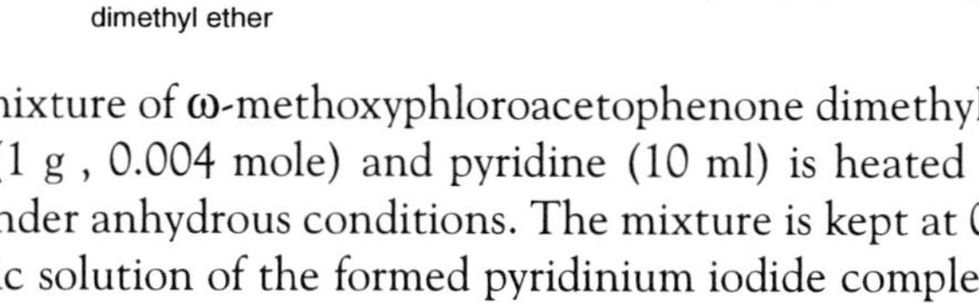

ω-methoxyphloroacetophenone dimethyl ether   2-Hydroxy-4,6-dimethoxybenzoic acid

A mixture of ω-methoxyphloroacetophenone dimethyl ether (4.B.71), iodine (1 g , 0.004 mole) and pyridine (10 ml) is heated for 1 hr (steam bath) under anhydrous conditions. The mixture is kept at 0° for 24 hr. The alcoholic solution of the formed pyridinium iodide complex is taken up in aqueous potassium hydroxide (2%, 60 ml) and heated in a steam bath for 1 hr. It is then acidified with hydrochloric acid in cold. The mixture is extracted with ether, extract washed with water and dried (sodium sulphate). The ether is distilled. The product obtained is purified by dissolving in sodium bicarbonate solution and acidification of the clear solution. It is crystallised from dilute methanol as colourless needles. Yield 0.7 g (80%). M.p. 162-64° (lit.[36] m.p. 160-62°).

Note: 2-Hydroxy-4,6-dimethoxybenzoic acid is also obtained by methylation of phloroglucinol carboxylic acid (3.33) with 3.1 equivalents of dimethyl sulphate in the presence of anhydrous potassium carbonate followed by alkaline hydrolysis of the formed methyl 2-hydroxy-4, 6-dimethoxybenzoate.

3.35 2,5-DIHYDROXY-4-METHOXYBENZOIC ACID

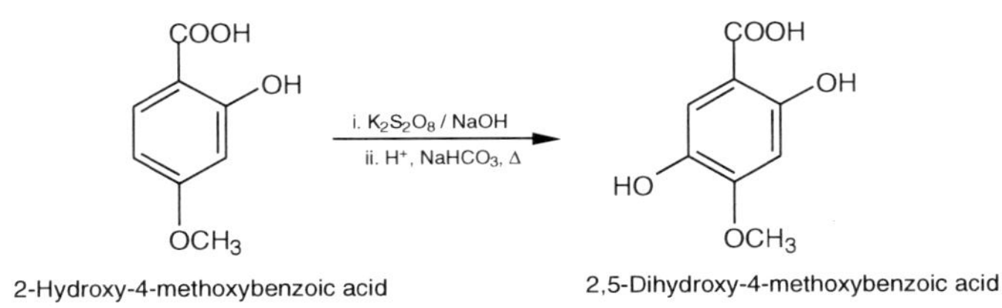

An aqueous solution of potassium persulphate (18 g, 0.066 mole in 200 ml water) is added dropwise during 4 hr to a stirred and cooled solution (15-20°) of 2-hydroxy-4-methoxybenzoic acid (3.8) (10 g, 0.059 mole) in aqueous sodium hydroxide (10%, 100 ml). The reaction mixture is stirred for 2 hr more and left for 48 hr at room temperature. It is cooled and acidified to congo red with dilute hydrochloric acid. The unreacted acid is removed by ether extraction (3 × 100 ml) of the reaction mixture. The remaining aqueous solution is heated for 1 hr on a boiling water bath with sodium bisulphite (10 g) and concentrated hydrochloric acid (60 ml). The solution is cooled and extracted with ether (3 × 100 ml). The ether extract is distilled to give a solid product. It is crystallised from methanol as creamish coloured needles. Yield 3.5 g (31.9%). M.p. 201°. (lit.[37] m.p. 201°).

3.36 2,4,5-TRIMETHOXYBENZOIC ACID

Method (i)

A solution of 2,5-dihydroxy-4-methoxybenzoic acid (3.35) (3.5 g, 0.019 mole) in dry acetone (100 ml) is refluxed for 30 hr under anhydrous conditions with dimethyl sulphate (5.6 ml, 0.059 mole) and anhydrous

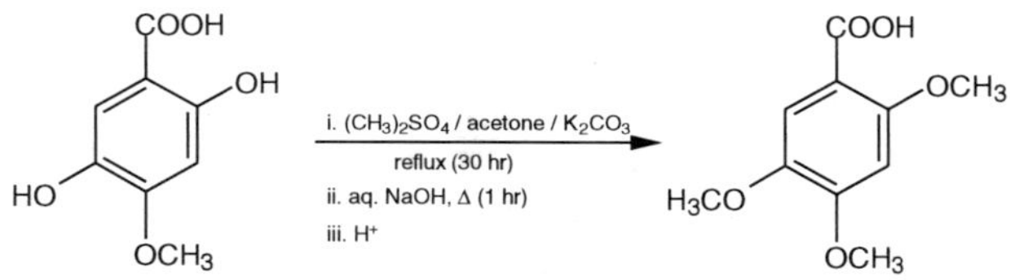

2,5-Dihydroxy-4-methoxybenzoic acid 2,4,5-Trimethoxybenzoic acid

potassium carbonate (15 g). The completion of the methylation is ascertained by negative ferric reaction in an aliquot of the mixture. The mixture is filtered and inorganic salts washed with hot acetone (2 × 25 ml). The combined acetone solution is distilled and the resulting oily product heated for 1 hr with aqueous sodium hydroxide (10%, 50 ml). The cooled, clear alkaline solution is acidified with hydrochloric acid and the product filtered. It is crystallised from ethanol as colourless needles. Yield 3 g (74.4%). M.p. 143-44° (lit.[38] m.p. 142°).

Method (ii)

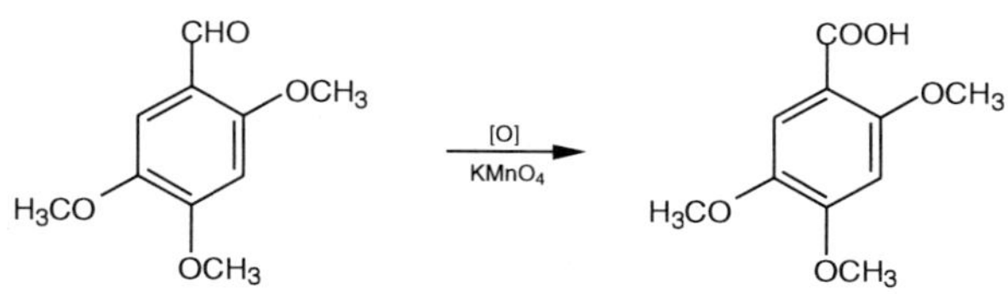

2,4,5-Trimethoxy benzaldehyde 2,4,5-Trimethoxybenzoic acid

A stirred suspension of 2,4,5-trimethoxybenzaldehyde (2.25) (25 g, 0.127 mole) in water (200 ml) is heated at 60-70° and aqueous potassium permanganate (50 g, 0.31 mole in 400 ml water) is added dropwise. The reaction is maintained at 60-70° during addition. The reaction mixture is cooled and sulphur dioxide gas is passed through the solution till the brown colour disappears. The solution is cooled and the separated product filtered and washed with cold water. It is crystallised from alcohol as colourless needles. Yield 18 g (66.6%). M.p. 143-44° (lit.[38] m.p. 144°).

3.37 2,4,5-TRIHYDROXYBENZOIC ACID

2,5-Dihydroxy-4-benzyloxybenzoic acid (3.40) (2.5 g, 0.0095 mole) in ethyl acetate (50 ml) is debenzylated in presence of Pd-C (0.3 g, 10%) with hydrogen gas at room temperature and atomospheric pressure till no more hydrogen is absorbed (3 hr). The reaction mixture is filtered and the

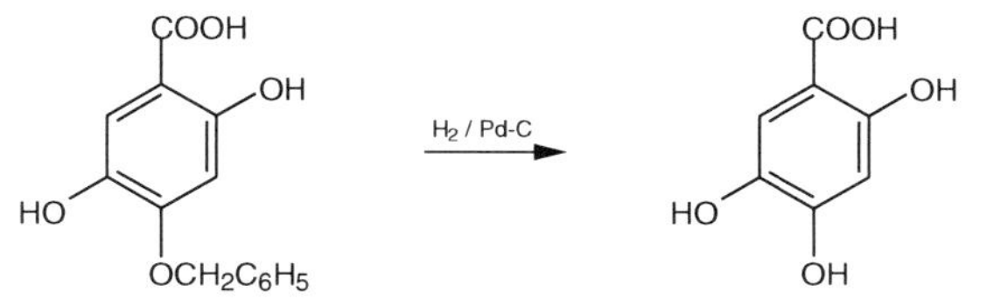

2,5-Dihydroxy-4-benzyloxybenzoic acid   2,4,5-Trihydroxybenzoic acid

catalyst washed with ethyl acetate (2 × 20 ml). The combined ethyl acetate solution is distilled and residual product crystallised from water as colourless needles. Yield 1.5 g (92%) M.p. 216-17° (lit.[39,40] m.p. 217-18°).

3.38 2,6-DIHYDROXY-4-METHYLBENZOIC ACID (p-ORCELLINIC ACID)

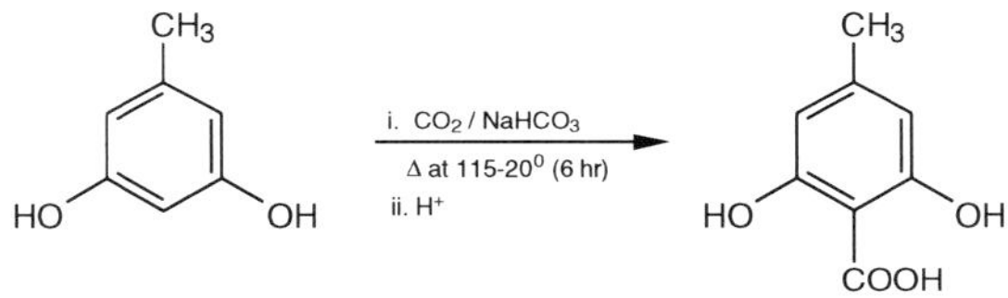

Orcinol   2,6-Dihydroxy-4-methylbenzoic acid

A mixture of orcinol (20 g, 0.16 mole), potassium bicarbonate (40 g) and glycerol (40 ml) is heated in an oil bath at 115-20° for 6 hr in an atmosphere of carbon dioxide. It is found advantageous to stir the reactions for the initial 2 hr. The reaction mixture is cooled, dissolved in water (50 ml) and acidified with hydrochloric acid. The separated product is filtered. It is crystallised from hot water as colourless needles. Yield 13 g (47.9%). M.p. 174-75° (after drying at 100°) (lit.[41,42] m.p. 175-76°).

3.39 2,4-DIHYDROXY-3-METHYLBENZOIC ACID

A solution of 2-methyl resorcinol (1.17) (5 g, 0.04 mole) and potassium bicarbonate (25 g) in water (50 ml) is heated in a steam bath for 4 hr. The solution is refluxed for 30 min while a stream of carbon dioxide gas is passed through the solution. The solution is cooled and acidified with

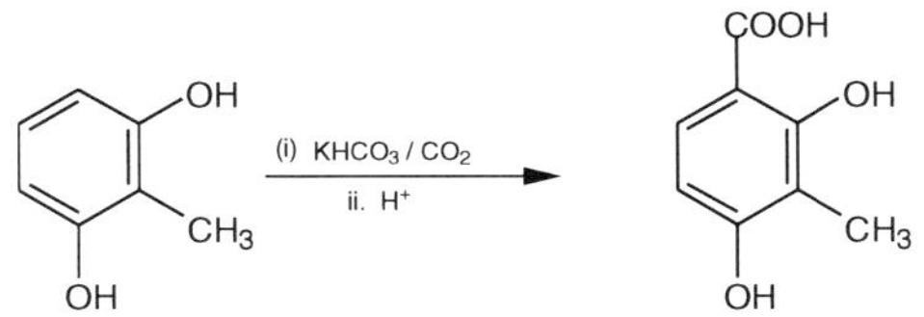

2-Methyl resorcinol   2,4-Dihydroxy-3-methylbenzoic acid

hydrochloric acid. It is cooled and the separated acid filtered. It is crystallised from water as colourless needles. Some more of product is obtained by ether extraction of the mother liquor. Total yield 5 g (73.8%). M.p. 185° (lit.[43] m.p. 185°).

3.40 2,4-DIHYDROXY-5-METHOXYBENZOIC ACID

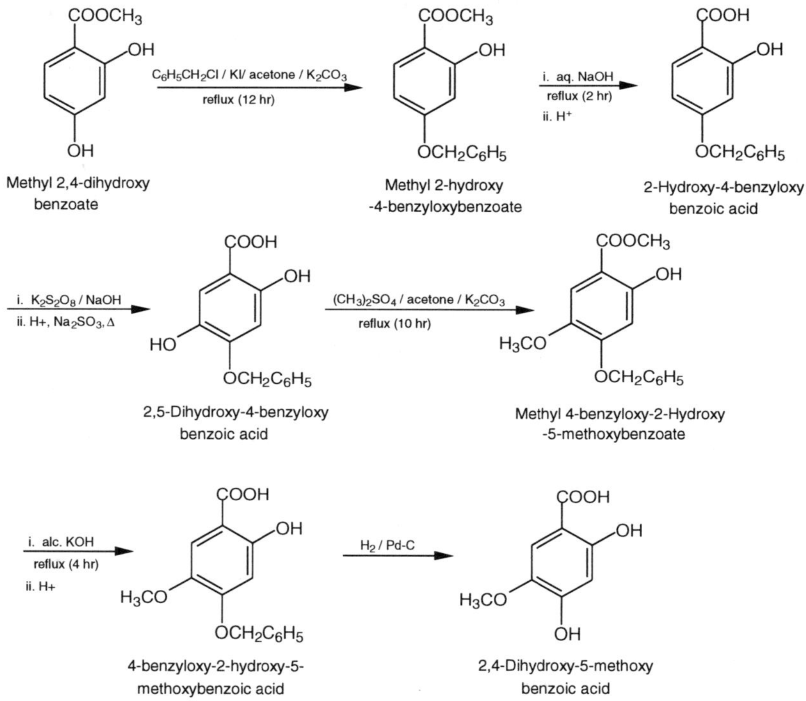

Step (i) Methyl 2-hydroxy-4-benzyloxybenzoate

A solution of methyl 2,4-dihydroxybenzoate (3.7) (28 g, 0.166 Mole) in acetone is treated with dry potassium iodide (22 g, 0.13 mole), benzyl chloride (19 ml, 21g, 0.166 mole) and freshly ignited potassium carbonate (60 g). The reaction mixture is refluxed on a water bath for 12 hr under anhydrous conditions. The mixture is filtered and the inorganic salts washed with hot acetone (2 × 50 ml). The combined acetone solution is distilled and the residual product macerated with ice. It is filtered and cystallised from benzene as creamish needles. Yield 18.2 g (41.8%). M.p. 102-103° (lit.[39] m.p.102-103°).

Step (ii) 2-hydroxy-4-benzyloxybenzoic Acid (β-Resorcyclic acid 4-benzyl Ether)

A solution of the above methyl 2-hydroxy-4-benzyloxybenzoate (5 g, 0.0119 mole) is refluxed for 2 hr with aqueous sodium hydroxide (7%, 100 ml). The cooled solution is acidified with hydrochloric acid. The separated product is filtered and crystallised from dilute alcohol as coloruless long needles. Yield 4 g (84.7%). M.p. 179-80 (lit.[39] m.p. 179-80°).

Step (iii) 2,5-Dihydroxy-4-benzyloxy Benzoic Acid

An aqueous solution of potassium persulphate (8.25 g, 0.03 mole in 120 ml water) is added dropwise during 4 hr to a stirred and cooled solution (10°) of the above 2-hydroxy-4-benzyloxybenzoic acid (7.5 g, 0.03 mole) in aqueous sodium hydroxide (8 g in 75 ml water). The reaction mixture is stirred for 1 hr more and kept overnight at room temperature. It is cooled and acidified to congo red with dilute hydrochloric acid. The mixture is then extracted with ether (3 × 100 ml) to remove the unreacted acid. The remaining aqueous solution is heated with sodium sulphite (5 g) and concentrated hydrochloric acid (40 ml) on a steam bath for 30 min. The reaction mixture is cooled and extracted with ether (4 × 100 ml). Ether is distilled to give a brown solid. It is crystallised from ethyl acetate-petroleum ether as colourless needles. Yield 2.75 g (34.4 %). M.p 202-203° (lit.[39] m.p. 202-203°).

Step (iv) Methyl 4-benzyloxy-2-hydroxy-5-methoxybenzoate

A mixture of the above 2,5-dihydroxy-4-benzyloxybenzoic acid (2.5 g, 0.0095 mole), anhydrous potassium carbonate (10 g), dry dimethyl sulphate (1.95 ml, 0.020 mole) and acetone (75 ml) is refluxed on a water bath for 10 hr under anhydrous conditions. The solution is filtered and the inorganic salts washed with hot acetone (2 × 25 ml). The combined acetone solution is distilled. It is macerated with ice, filtered and crystallised from benzene as colourless needles. Yield 2.1 g (75.9 %) M.p.140-41° (lit.[44] m.p.140-41°).

Step (v) 4-Benzyloxy-2 hydroxy-5-methoxybenzoic Acid

A solution of the above methyl4-benzyloxy-2-hydroxy-5-methoxybenzoate (5 g, 0.017 mole) in alcoholic potassium hydroxide (100 ml, 1:1) is refluxed for 4 hr. The cooled, clear alkaline solution is acidified with hydrochloric acid. The separated product is filtered. It is crystallised from benzene-acetone mixture as light yellow needles. Yield 3.8 g. (80.5%). M.p. 191° (decomp.) (lit.[44] m.p. 191°).

Step (vi) 2,4-Dihydroxy-5-methoxybenzoic Acid

The above acid (3 g, 0.011 mole) in ethyl acetate (100 ml) and Pd-C (0.7 g, 10%) is stirred in an atmosphere of hydrogen at room temperature and atmospheric pressure till no more hydrogen absorbed. The catalyst is filtered and washed with more ethyl acetate (20 ml). The combined extract is distilled. The residual product is crystallised from water as bright yellow needles. Yield 2 g (99.5%). M.p. 201-2° (lit.[44,45] m.p. 201° decompn.)

3.41 2,5-DIHYDROXY-4-METHOXY-3-METHYLBENZOIC ACID

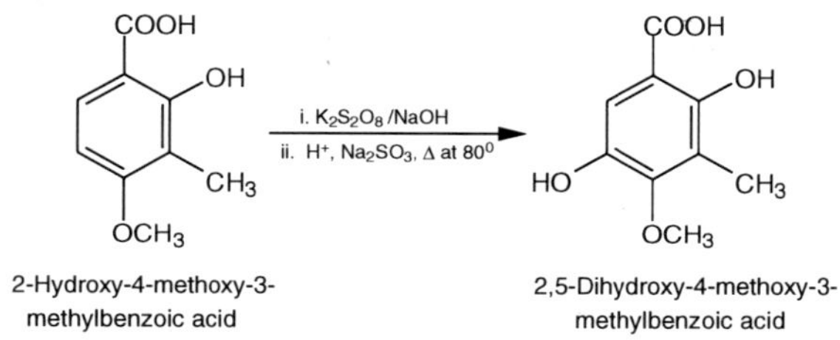

To a cooled solution (10°) of 2-hydroxy-4-methoxy-3-methylbenzoic acid (5 g, 0.027 mole) in aqueous sodium hydroxide (4 g in 40 ml water) is added an aqueous solution of potassium persulphate (8.5 g, 0.031 mole in 150 ml water) with stirring during 4 hr. The mixture is stirred for 1 hr more and left overnight at room temperature. It is cooled and acidified to congo red with dilute hydrochloric acid. It is extracted with ether (3 × 75 ml) to remove the unreacted acid. The remaining aqueous solution is heated with concentrated hydrochloric acid (30 ml) and sodium sulphite (4 g) on a water bath at 80° for 30 min. The solution is cooled and extracted with ether (4 × 100 ml). The ether is distilled and the residual product is crystallised from dilute alcohol as pale brown needles. Yield 0.9 g (16.5%). M.p. 185-86° (lit.[46] m.p. 185-86°).

Note: 2-Hydroxy-4-methoxy-3-methylbenzoic acid is obtained by partial methylation of 2,4-dihydroxy-3-methylbenzoic acid (3.39) with 2.1 equivalent of dimethyl sulphate in dry acetone and ignited potassium carbonate (6 hr refluxing) followed by saponification of the formed methyl 2-hydroxy-4-methoxy-3-methylbenzoate (m.p. 77°, lit.[47] m.p. 77°) with alkali (10%, 2 hr refluxing). Yield 80%. M.p. 218°.

3.42 2-HYDROXY-3,4,5-TRIMETHOXYBENZOIC ACID[48]

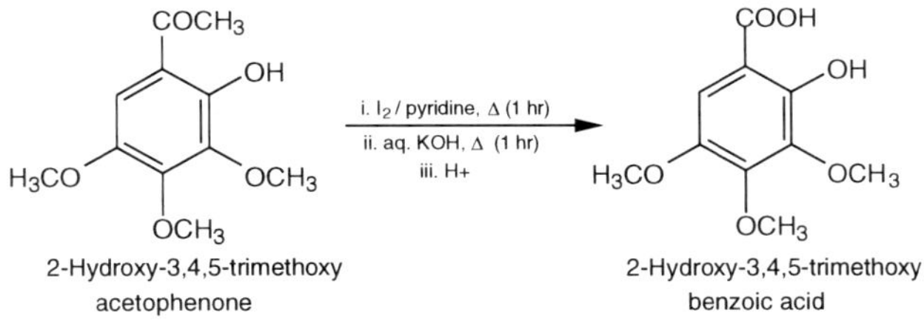

A mixture of 2-hydroxy-3,4,5-trimethoxyacetophenone (4.A.52) (0.67 g, 0.003 mole), iodine (0.8 g, 0.003 mole) and dry pyridine (5 ml) is heated on a steam bath for 1 hr and then kept at 0° for 24 hr. The separated product is filtered, washed with water and heated on a steam bath for 1 hr with aqueous potassium hydroxide (2%, 5 ml). The cooled solution is acidified with hydrochloric acid. It is extracted with ether (4 × 50 ml). The ether extract is dried (sodium sulphate) and distilled. The product obtained is crystallized from ethyl acetate-petroleum ether as colourless needles. Yield 0.5 g (71%, M.p. 190-92° (decompn.) (lit.[48] m.p. 190-92°).

3.43 p-METHOXOPHENYLGLYOXALIC ACID ($p\text{-}OCH_3C_6H_4COCOOH$)

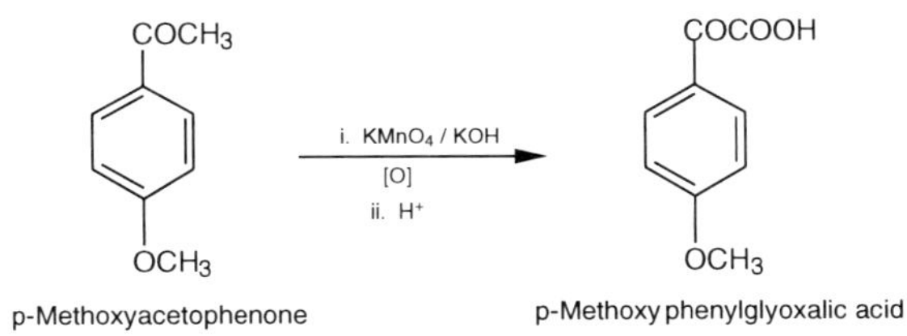

A solution of potassium permanganate (16 g, 0.1 mole) and potassium hydroxide (12 g) in water (250 ml) are added dropwise to a solution of p-methoxyacetophenone (4.A.4) (6 g, 0.055 mole) in pyridine (50 ml) at 15° with constant mechanical stirring during 4 hr. The stirring is continued for 1 hr more and the unreacted potassium permanganate decomposed by addition of sodium bisulphite. The mixture is filtered and the residual manganese dioxide washed with water (2 × 100 ml). The alkaline filtrate is acidified with concentrated hydrochloric acid (100 ml) and the solution left overnight in a refrigerator. The separated anisic acid (about 1 g) is filtered. The filtrate is saturated with sodium chloride and extracted with ether (5 × 100 ml). The ether solution is dried (magnesium sulphate) and distilled. The residual product is crystallised from benzene-petroleum ether as colourless thin rectangular plates. Yield 4 g (39.2%). M.p. 87-88° (lit.[49] m.p. 87-88°).

3.44 β-BENZOYLPROPIONIC ACID ($C_6H_5COCH_2CH_2COOH$)

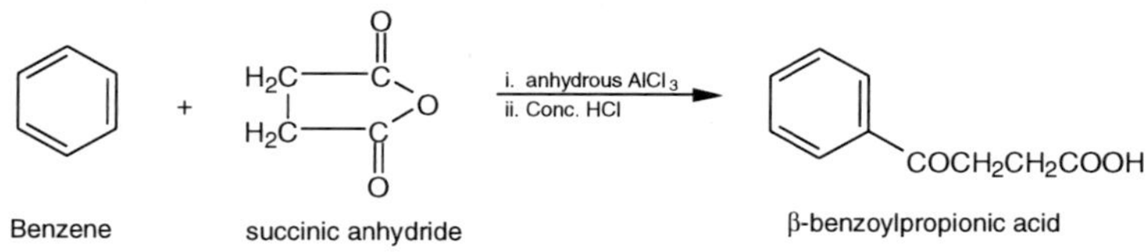

Succinic anhydride (6 g, 0.06 mole) is added to a 500 ml three necked R.B. flask equipped with a mechanical stirrer and a reflux condenser. Pure dry benzene (35 ml) added and the mixture stirred. Powdered anhydrous aluminium chloride (17.6 g, 0.13 mole) is added in one lot. Hydrogen chloride gas is evolved and the mixture becomes hot. It is stirred and heated in a boiling water bath for 30 min. The reaction mixture is cooled to room temperature and water (25 ml) added. Concentrated hydrochloric acid (8.3 ml) is added and benzene removed by steam distillation. The mixture is cooled; the separated oil solidifies. It is filtered, washed with dilute hydrochloric acid (25 ml, 1:3) and cold water (2 × 25 ml). It is purified by dissolving in sodium bicarbonate solution and acidifying the clear filtrate. It is filtered and dried. Yield 10 g (94%). M.p. 115° (lit.[50] m.p. 116°).

3.45 4-PHENYLBUTYRIC ACID ($C_6H_5CH_2CH_2CH_2COOH$)

A mixture of β-benzoylpropionic acid (3.44) (2 g, 0.011 mole), potassium hydroxide (2.13 g, 0.38 mole), and hydrazine hydrate (80%, 1.5 ml, 0.047 mole) in diethylene glycol (15 ml) is refluxed for 1.5 hr. The water is allowed to distill away and the temperature raised to 195° and the mixture is refluxed for 4 hr. The solution is cooled, diluted with water (15 ml) and poured into hydrochloric acid (6N, 9 ml). The separated product is filtered and dried. Yield 1.55 g (75.2%). M.p. 52-3° (lit.[51] m.p. 48-9°).

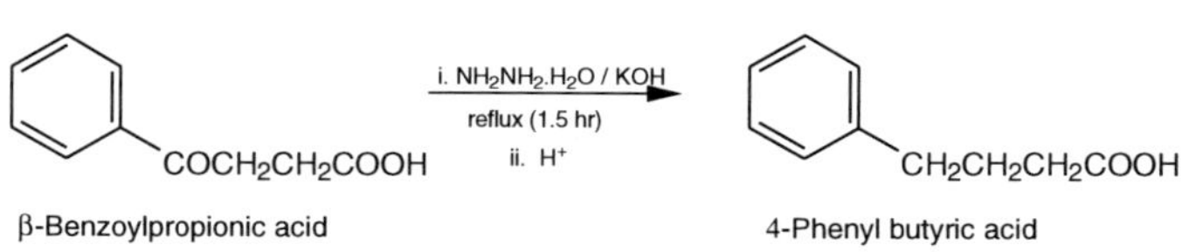

3.46 p-METHOXYPHENYLACETIC ACID

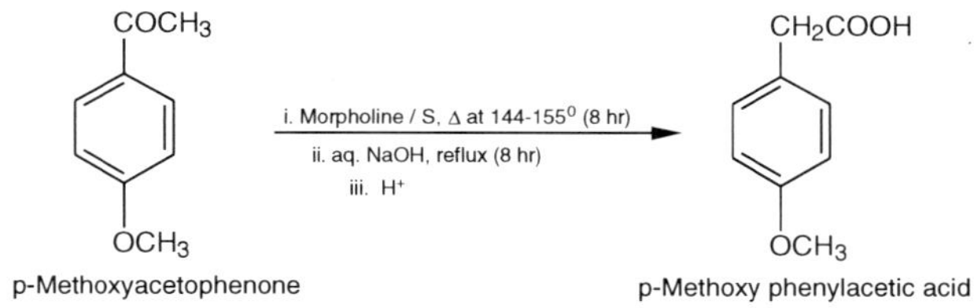

A mixture of p-methoxyacetophenone (4.A.4) (27 g, 0.18 mole), morpholine (24 ml, 24 g, 0.28 mole) and sulphur (9 g, 0.28 mole) is refluxed in an oil bath at 145-155° for 8 hr under anhydrous conditions. The mixture is cooled, extracted with chloroform (5 × 30 ml) and the chloroform solution washed repeatedly with dilute hydrochloric acid till the acid layer becomes faint yellow in colour. The chloroform is distilled and the crude thiomorpholide is refluxed for 8 hr with aqueous sodium hydroxide (80 g in 600 ml water). The clear alkaline solution is acidified with hydrochloric acid and the solution is cooled. The separated product is filtered. It is crystallized from large volume of water as colourless leaflets. Yield 10 g (33.5 %). M.p. 85-6° (lit.[52,53] m.p. 86°).

Note: This is a general method of preparing phenylacetic acids from the corresponding acetophenones. Following phenylacetic acids are similarly obtained:

2,4-Dimethoxyphenylacetic acid[54], m.p. 111-12°

2,5-Dimethoxyphenylacetic acid , m.p. 121-23°

2,4,5-Trimethoxyphenylacetic acid[55], m.p. 81-83°.

3.47 p-NITROPHENYLACETIC ACID

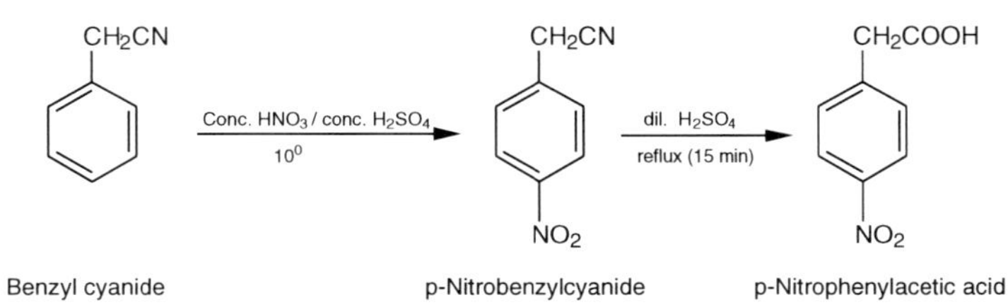

Step (i) p-Nitrobenzylcyanide

Benzyl cyanide (9.95 g, 0.085 mole) is added to previously cooled and stirred mixture of concentrated nitric acid (27.5 ml) and an equal volume of concentrated sulphuric acid at 10°. After the addition, the ice bath is removed and the mixture is stirred for 1hr more. The reaction mixture is then poured over crushed ice. The separated product is filtered and washed with water. It is crystallised from ethanol as yellow flat needles. Yield 7.5 g (54.5%). M.p. 115-116° (lit.[56] m.p. 116-17°).

Step (ii) p-Nitrophenylacetic acid

p-Nitrobenzyl cyanide (5 g, 0.03 mole) is refluxed for 15 min with dilute sulphuric acid (15 ml concentrated acid diluted with 14 ml water). The reaction mixture is diluted with water (30 ml) and cooled (0°). The separated solid is filtered. It is crystallised from water as pale yellow needles. Yield 5.1g (91.4 %). M.p. 151-52° (lit.[57] m.p. 151.5-152°).

3.48 2,4,DINITROPHENYLACETIC ACID

To a cooled solution of phenylacetic acid (14 g, 0.103 mole), a mixture of concentrated sulphuric acid (30 ml) and fuming nitric acid (15 ml, d, 1.5) is added slowly. The temperature of the reaction mixture is kept below 60°. After the addition is over, the mixture is poured into water (500 ml). The separated product is filtered and washed with water. It is crystallised from water as yellow needles. Yield 12 g (51.6%). M.p. 178-80° (decompn.) (lit.[58] m.p. 178-80°).

3.49 CINNAMIC ACID

A mixture of freshly distilled benzaldehyde (9.6 g, 0.1 mole), pure dry acetic anhydride (15 ml) and freshly fused and powdered potassium acetate (5 g) is heated under anhydrous conditions at 155-160° for 1 hr and subsequently at 175-185^0 for 2 hr. The reaction mixture is pourd into

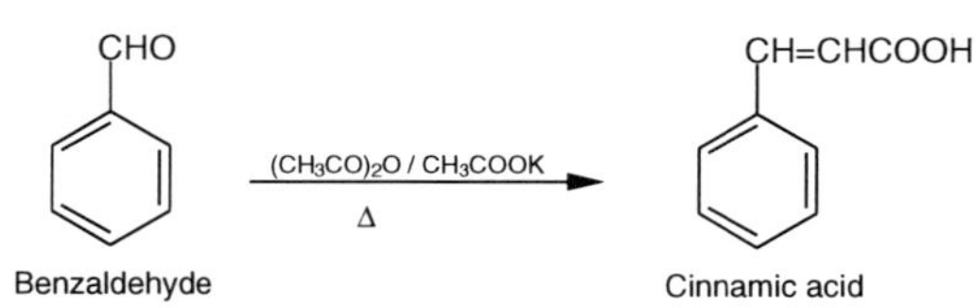

water (50 ml). The solid product is filtered. Cinnamic acid is purified by dissolving the crude product in sodium bicarbonate solution; it is extracted with ether to remove any unreacted benzaldehyde and acidified. The separated product is filtered and crystallised from dilute alcohol as colourless crystals. Yield 12 g (81 %). M.p. 133° (lit.[59] m.p. 133°).

Note: 1. This is a general method of preparing cinnamic acids from the corresponding aldehydes (Perkin Reaction).

2. Use of potassium acetate improves the yield. Moreover sodium acetate can also be used but the reaction is slower and 1-2 hr more heating is required.

3.50 3-PHENYLPROPIONIC ACID (HYDROCINNAMIC ACID)

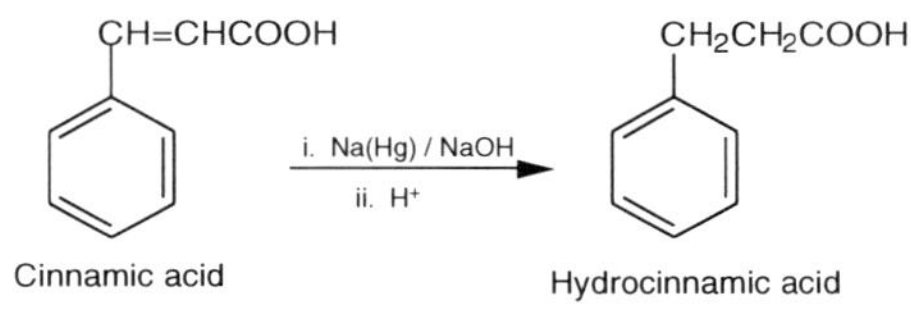

Sodium amalgam (6.A.35 b) (350 g, 0.38 mole of 2.5% amalgam) is gradually added during 1 hr to a stirred solution of cinnamic acid (3.49) (20 g, 0.135 mole) in sodium hydroxide solution (1 M, 73 ml). The reaction is performed in a good fume cupboard. When the evolution of hydrogen ceases, the solution is decanted and the remaining mercury is washed with water. The combined decanted liquid and washings are acidified with dilute hydrochloric acid (1:1). The hydrocinnamic acid separates in the form of oil. It solidifies on cooling. It is filtered, dried and crystallized from petroleum ether as colourless needles. Yield 17 g (85%). M.p. 47-8° (lit.[60] m.p. 46-7°).

3.51 METHYL CINNAMATE

A solution of cinnamic acid (3.49) (29.5 g, 0.2 mole) in absolute methanol (80 ml) containing concentrated sulphuric acid (3 ml) is refluxed for 5 hr

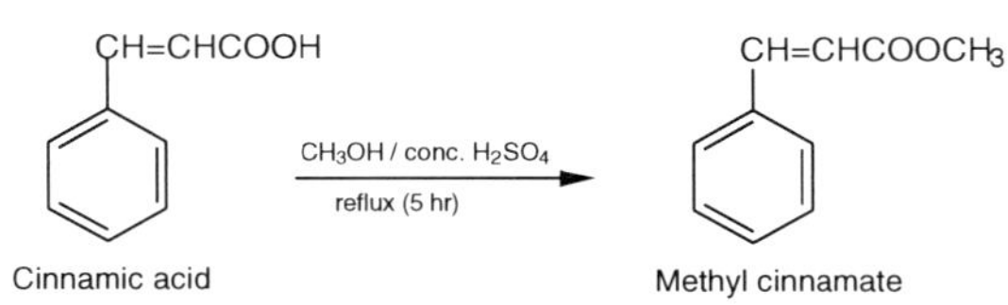

(water bath). The excess methanol is removed by distillation and the residue poured into water (200 ml). The mixture is extracted with ether (2 × 120 ml). The ether extract is washed with water, dried (anhydrous sodium sulphate) and ether removed (rotary evaporator). The residue solidifies on cooling. It is filtered and crystallised from dilute methanol as colourless needles. Yield 29 g (90%). M.p. 36° (lit.[59] m.p. 36°).

3.52 p-HYDROXYCINNAMIC ACID

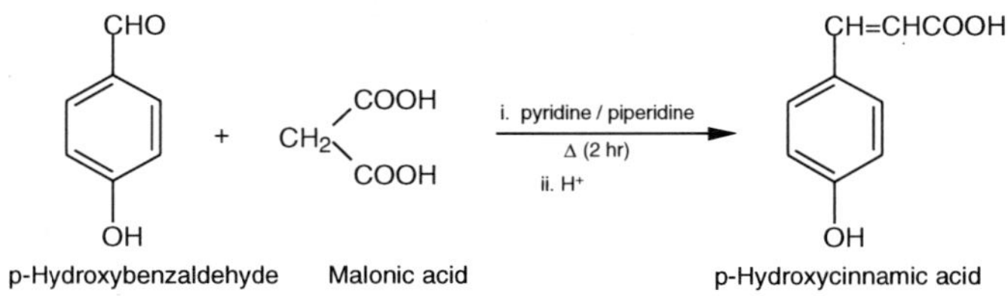

A mixture of p-hydroxybenzaldehyde (10 g, 0.09 mole) and dry malonic acid (25 g, 0.24 mole) in dry pyridine (20 ml) and a few drops of piperidine is heated on a steam bath for 2 hr. The reaction mixture is cooled, acidified with hydrochloric acid and boiled for 5 min. The solution on cooling gave a solid product. It is filtered and crystallised from alcohol as colourless needles. Yield 12 g (77.0%) M.p. 206-7° (lit.[62] m.p. 206°).

Note: This is a general method of making cinnamic acids from the corresponding aldehydes (Knoevengel Reaction).

Following cinnamic acids are similarly prepared.

p-Methoxycinnamic acid, m.p.[63] 174°.

3-Methoxy-4-hydroxycinnamic acid, m.p.[64] 174°.

3,4-Dimethoxycinnamic acid, m.p.[65], 180-81°.

3,5-Dimethoxycinnamic acid, m.p.[66], 174-75°.

References

1. J.F.J. Dippy and R.H. Lewis, J. Chem. Soc., 1937, 1426.
2. Kolbe, J. Prakt. Chem., 1875, 11, 24.
3. W.H. Perkin and F.W. Stoyle, J. Chem. Soc., 1923, 3171.
4. F.E. King, J.H. Gilks and M.W. Partridge, J. Chem. Soc., 1955, 4206.
5. G.A. Edwards et al., J. Chem. Soc., 1925, 127, 197.
6. D.A. Clibbens and H. Nierenstein, J. Chem. Soc., 1915,107, 1494.
7. R. Robinson and R.C. Shah, J. Chem. Soc., 1934, 1491.

8. Dictionary of organic compounds by Chapman and Hall, 5th Edition, D-04247.
9. H. Raistrick and P. Simonart, Biochem. J., 1933, 27, 628.
10. Dictionary of organic compounds by Chapman & Hall, 5th Edition, D-05321.
11. K.S. Pankajmani and T.R. Seshadri, J. Sci. Industrial Research, 1954, 13B, 396
12. F.P. Doyle, K. Haidy, J.H.C. Nayler, M.J. Soulal, E.R. Stove and H.R.J. Waddington, J. Chem. Soc., 1962, 1453.
13. I.A. Pearl, J. Am. Chem. Soc., 1946, 68, 2180.
14. Dictionary of organic compounds by Chapman & Hall, 5th Edition, D-10366.
15. Dictionary of organic compounds by Chapman & Hall, 5th Edition, D-04250.
16. A. Lovely, R. Robinson and S. Sugasawa, J. Chem. Soc., 1930, 817.
17. Perkin and Stoyle, J. Chem. Soc., 1925, 198.
18. W. Wiegrebe, H.M. Stephan and J. Fricke, Helv. Chim. Acta, 1976, 59, 949.
19. C.A. Fetscher and M.T. Bogert, J. Org. Chem., 1939, 4, 71.
20. Fiaig and Mieleck, Ann. 1869, 152, 40.
21. Raistrick et al, Biochem. J., 1941, 35, 1007; Dictionary of organic compounds by Chapmann & Hall, 5th Edition, M-00947.
22. Barth and Senhoefer, Ann., 1871, 159, 222; Graves and Adams, J. Am. Chem. Soc., 1923, 45, 2439.
23. Dictionary of organic compounds by Chapman & Hall, 5th Edition, D-05324.
24. A. Critchlow, R.D. Haworth and P.L. Pausen, J. Chem. Soc., 1951, 1318.
25. Dictionary of organic compounds by Chapman & Hall, 5th Edition, T-03275.
26. Mauthner, J. Pr. Chem., 1914, 89, 304.
27. Overmyer, J. Am. Chem. Soc., 1927, 49, 503.
28. Dictionary of organic compounds by Chapman & Hall, 5th Edition, T-10273.
29. S. Goodwin and B. Witkop, J. Am. Chem. Soc., 1957, 79, 179.
30. V.K. Ahluwalia et al., unpublished work.
31. L. Jurd, J. Am. Chem. Soc., 1959, 81, 4606.
32. M.T. Bogert and B.B. Coyne, J. Am. Chem. Soc., 1929, 51, 569.
33. Spath and Roder, Montash, 1922, 43, 93.

34. Dictionary of organic compounds by Chapman & Hall, 5th Edition, T- 10273.
35. P. Holmes, D.E. White and I.H. Wilson, J. Chem. Soc.(C), 1950, 2810.
36. P.M. DEWICK, Phytochemistry, 1975, 14, 983.
37. S. Rajagopalan, T.R. Seshadri and S. Varadarajan, Proc. Indian Acad. Sci., 1949, 30A, 265.
38. W. Weigrebe, H.M. Stephan and J. Fricke, Helv. Chim. Acta, 1976, 59, 949.
39. D.K. Bhardwaj, S.C. Jain and R. Singh, Indian J. Chem., 1978, 16B, 150.
40. N. Rajagopalan, T.R. Seshadri and S. Varadarajan, Proc. Indian Acad. Sci., 1949, 30A, 265.
41. Julius Kalamar, Eric Steiner, Etienne Charollais, Helv. Chimica Acta, 1974, 57, 2368.
42. Dictionary of organic compounds by Chapman & Hall, 5th Edition, D-04667.
43. Dictionary of organic compounds by Chapman & Hall, 5th Edition, D-04663.
44. D.K. Bhardwaj, R.K. Jain, S.C. Jain and Radhika Singh, Curr. Sci., 1978, 47, 336.
45. L. Paul, Ber., 1906, 39, 2779.
46. J.D. Ramanathan, Ph.D. Thesis, Delhi University, 1961, page 118-119.
47. S. Ramanathan and T.R. Seshadri, Proc. Indian. Acad. Sci., 1937, 6A, 112.
48. V.K. Ahluwalia et al., unpublished work.
49. V.N. Gupta and T.R. Seshadri, Proc. Indian Acad. Sci., 1956, 44A, 223.
50. Kohler and Engelbrecht, J. Am. Chem. Soc., 1919, 41, 768.
51. R.V. Christian, J. Am. Chem. Soc., 1952, 74, 1591.
52. K. Kindler, Arch. Pharm, 1927, 265, 389.
53. K. Kindler, Ber., 1941, 74, 315.
54. F.E. King and K.G. Niell, J. Chem. Soc., 1952, 4754.
55. B.R. Sharma, Ph.D. Thesis, Delhi University, 1965, p.133.
56. S.Gabriel, Ber, 1881, 14, 2341.
57. S. Gabriel, Ber., 1882, 15 , 834.
58. W. Borsche, Ber., 1909, 42, 1313.
59. Dictionary of organic compounds by Chapman & Hall, 5th Edition, p-01516.
60. E. Schwenk and D. Papa, J. Org., 1946, 11, 798.

61. S. Ogawa, Bull. Chem. Soc. Japan, 1927, 2, 25.
62. J. Frederick, J. Dippy and J. E. Page, J. Chem. Soc., 1938, 357.
63. I.A. Pearl and D.L. Beyer, J. Org. Chem., 1951, 16, 216.
64. W.H. Perkin and E. Schiess, J. Chem. Soc., 1904, 85, 164.
65. J. Mukherjee, J. N. Chatterjee and S.C. Sengupta, Indian J. Chemistry, 1975, 13, 859.

4

Ketones

Ketones (aromatic) are prepared by the following methods.

HOUBEN-HOESCH REACTION

Polyhydroxy phenols, phenolic ethers and some reactive hetrocyclic compounds on treatment with nitriles and hydrogen chloride gas in the presence of a Lewis acid (e.g. zinc chloride) in anhydrous ether give the corresponding ketones.

Mechanism

The mechanism of the reaction is not certain. It is believed that the attacking species is the imine hydrochloride. The resulting imine salt on hydrolysis gives the ketone.

$$R{-}C{\equiv}N \;+\; HCl \;+\; ZnCl_2 \longrightarrow R{-}\overset{\oplus}{C}{=}NH.HCl \;+\; [ZnCl_3]^-$$

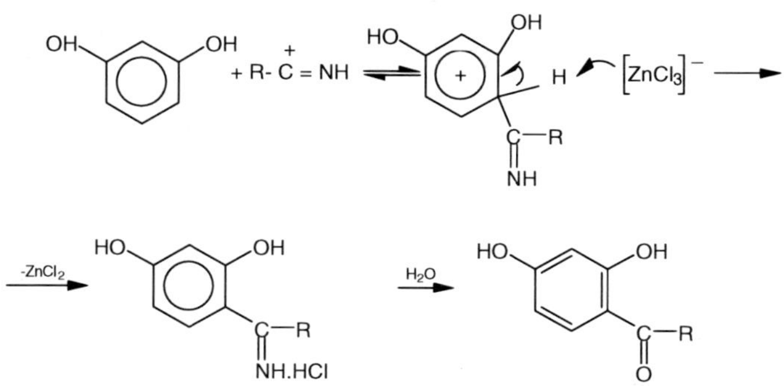

This method is used to synthesize acetophenols and substituted (viz. methoxy, benzyloxy, chloro etc.) acetophenols.

Typical preparation are described viz. 4.A. 25, 27, 28, 44, 56; 4.B. 70, 75, 80, 84, 85, 86, 87, 89, 91, 93, 101; 4.E. 106, 111, 112, 115, 117, 119.

FRIEDEL-CRAFTS REACTION

This reaction involves alkylation or acylation of organic compounds in presence of aluminium chloride or other Lewis acid catalysts.

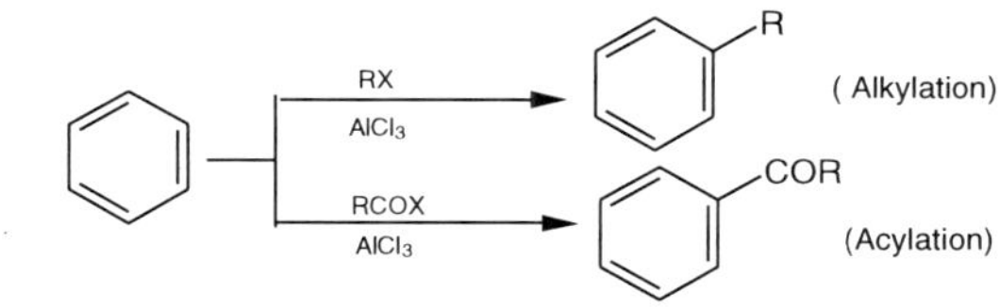

Mechanism

Ketones are obtained by the Friedel Crafts acylation. In this reaction an aromatic compound (methyl ethers) reacts with an acid chloride in the presence of aluminium chloride using anhydrous ether as solvent. The reaction proceeds by formation of an acyl cation, which react with the organic molecule.

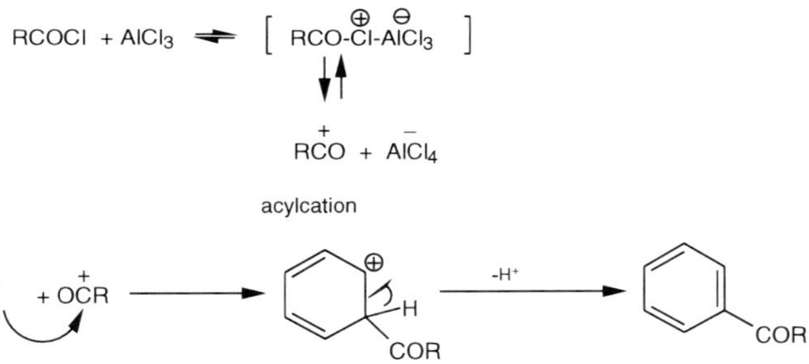

If the reaction is carried out with completely o-methyl ethers, then one of the methoxy group gets demethylated and o-hydroxy ketones are obtained.

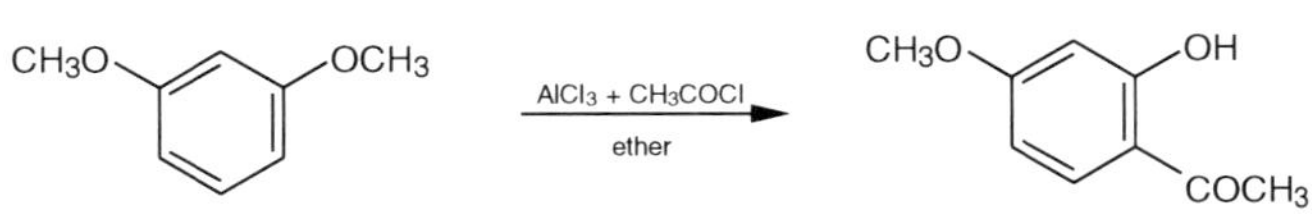

The method is used for the synthesis of o-hydroxy ketones. Typical procedures are described, viz., 4.A. 23, 24, 37, 40, 41, 45, 54, 59, 90; 4.C. 95, 96, 98; 4.D. 102; 4.E. 107, 113, 122, 123.

FRIES MIGRATION

Esters of phenols on heating with anhydrous aluminium chloride undergo rearrangement to give phenolic ketones. This is known as Fries rearrangement. Low temperature favour the formation of p-isomer, while high temperature favour the formation of o-isomer.

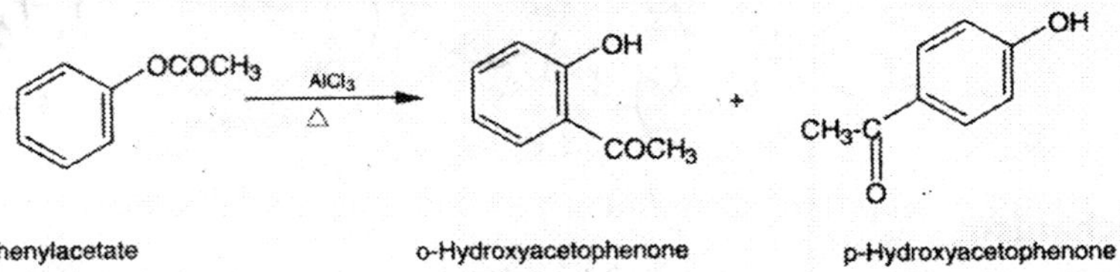

Mechanism

The mechanism of Fries Migration is not certain. There is evidence for an intermolecular mechanism, but recently it has been shown that the rearrangement is at least partly intramolecular. It seems that the mechanism is best regarded as a combination of the mechanism occurring simultaneously.

Intermolecular mechanism

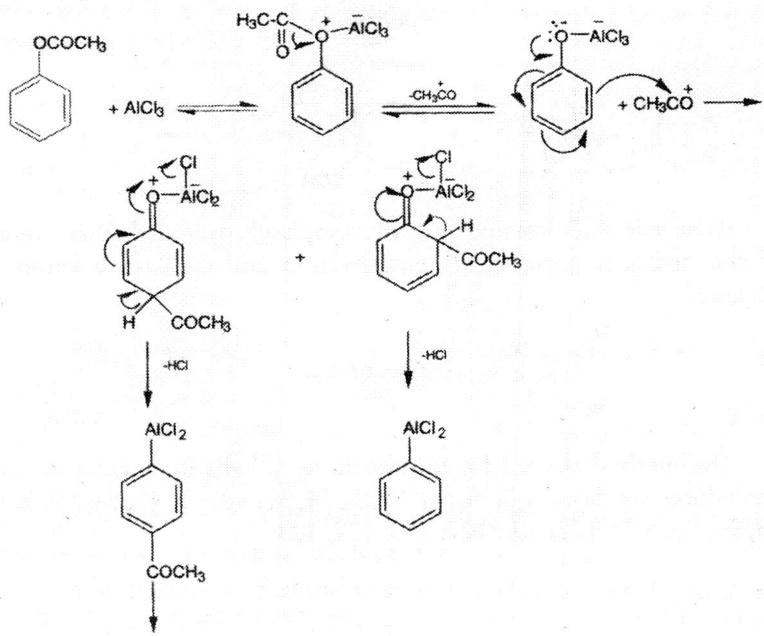

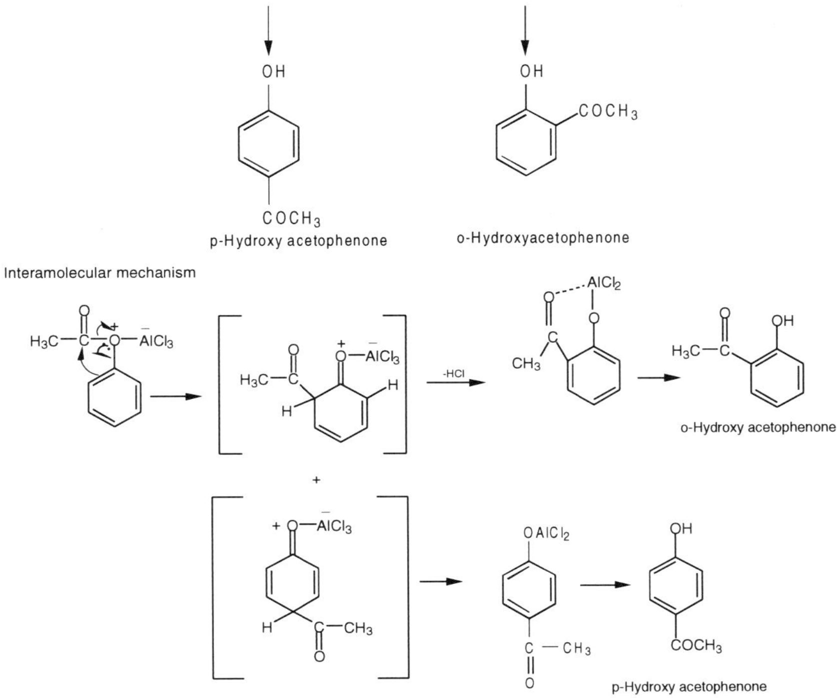

This method is used for the preparation of o-and p-hydroxy-acetophenone, viz. 4.A. 1, 3, 10.

NENCKI REACTION

The ring acylation of phenol with acids in presence of zinc chloride is known as Nencki Reaction.

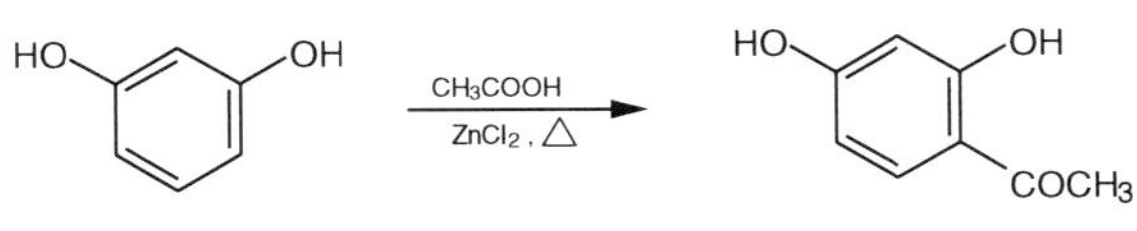

The reaction proceeds by the formation of acylation (CH_3CO) which reacts (as in the case of Friedal Crafts acylation) with phenols to give o-hydroxy ketones. This method is used for the preparation of 4.A. 4, 6, 13a, 34, 35; 4.E. 105, 117.

4. The partial methyl ethers of ketones have also be prepared by methylation or Elbs persulphate oxidation (for details see

Chapter 1. Certain ketones have been prepared by specific procedures. These are discussed in individual cases.

4.A.1 o-HYDROXYACETOPHENONE

Step (i) Phenylacetate

Phenol (26.775 g, 25 ml, 0.28 mole) is dissolved in sodium hydroxide solution (10%, 160 ml) and crushed ice (175 g) added. Acetic anhydride

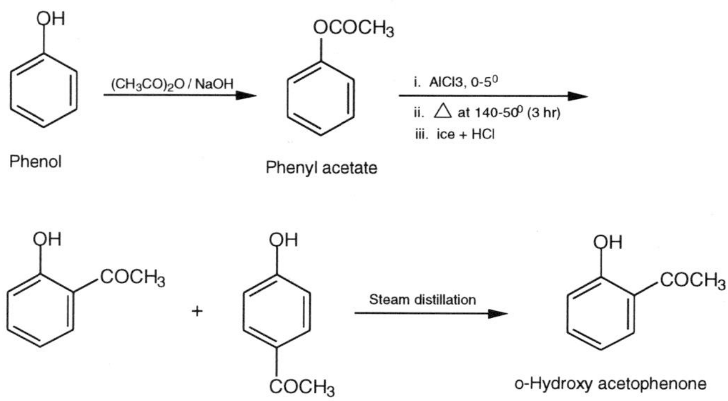

(30 ml, 32.46 g, 0.318 mole) is added and the reaction mixture shaken vigorously for 15 min. The emulsion of phenyl acetate is poured into a separatory funnel and carbon tetrachloride (10 ml) is added to facilitate the separation of two layers. The lower layer of phenyl acetate in carbon tetrachloride is separated, washed with very dilute sodium bicarbonate solution and dried (anhydrous calcium chloride). Carbon tetrachloride is distilled and the fraction boiling at 194-97° is collected. Yield 30 ml, 32.2 g (83.2%).

Step (ii) o-Hydroxyacetophenone

Powdered anhydrous aluminium chloride (56 g, 0.42 mole) is added gradually to phenylacetate (52 g, 48.46 ml, 0.38 mole) in a conical flask at 0-5° (ice-bath). The reaction mixture is heated under anhydrous conditions in an oil bath at 140-50° for 3 hr till the evolution of hydrogen chloride slackens. The cooled reaction mixture is treated with ice and hydrochloric acid (50 ml). The mixture is shaken and allowed to stand for 1 hr. The separated dark oil is extracted with ether and the ether layer dried (sodium sulphate) and distilled giving a mixture of o- and p-hydroxy acetophenones as an oil. o-Hydroxyacetophenone is obtained by steam distillation of the mixture; the steam volatile fraction consist of pure o-hydroxyacetophenone. It is isolated by ether extraction and distillation. Yield 24 g (46%). B.p. 212° (105-106°/20 mm) (lit.[1] b.p. 106°/17 mm). The residue of p-hydroxyacetophenone left after steam distillation solidifies. It is crystallised from alcohol as colourless needles. Yield 24.5 g (47%). M.p. 109° (lit.[2] m.p. 109°).

4.A.2 o-METHOXYACETOPHENONE

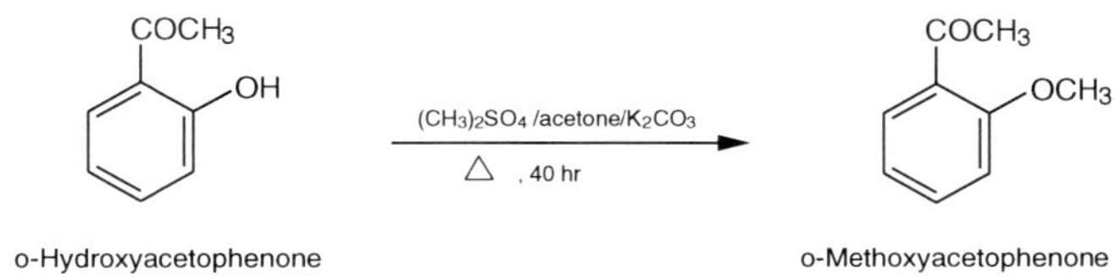

A mixture of o-hydroxyacetophenone (4.A.1) (12 g, 0.088 mole), dimethyl sulphate (9.45 ml, 12.6 g, 0.1 mole) and anhydrous potassium carbonate (25 g) in dry acetone (150 ml) is refluxed under anhydrous conditions for 40 hr. The acetone is filtered and the inorganic salts washed with more of hot acetone (2 × 30 ml). The combined acetone solution is distilled and the residue macerated with ice-water. It is extracted with ether (2 × 100 ml), the extract washed with sodium hydroxide solution (5%, 100 ml), water and dried (sodium sulphate). The ether is distilled to give o-methoxyacetophenonne. Yield 10 g (75.8%). B.p. 228-30° (lit.[1] b.p. 228-30).

4.A.3 p-HYDROXYACETOPHENONE

It is obtained along with o-hydroxyacetophenone during the Fries migration of phenyl acetate (see 4.A.1.). It is separated as given earlier. Yield 45%. M.p. 109°.

4.A.4 p-METHOXYACETOPHENONE

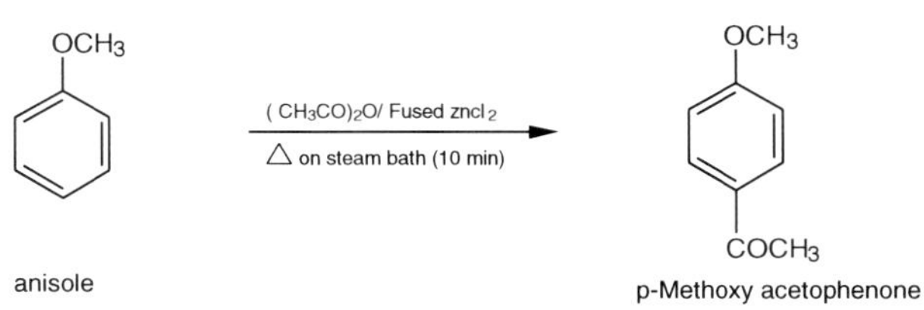

A mixture of anisole (43.2 g, 0.4 mole), acetic anhydride (81.6 g, 75.4 ml, 0.8 mole) and freshly fused zinc chloride (27.2 g, 0.2 mole) is shaken well for 5-10 min. During this time the temperature rises to 50-60° and zinc chloride dissolved. The reaction mixture is heated under anhydrous conditions on a steam bath (80°) for 10 min. The cooled mixture is poured into crushed ice (300 g). The separated oil is extracted with ether. The ether extract is washed with sodium hydroxide solution (5%, 100 ml) and finally with saturated sodium chloride solution (50 ml). The extract is dried (anhydrous sodium sulphate) and distilled. Yield 30 g (50 %). B.p. 110-115°/3.8-4 mm. The oily distillate solidifies on cooling. M.p. 35-37° (lit.[2] m.p. 38-9°).

4.A.5 m-HYDROXYACETOPHENONE

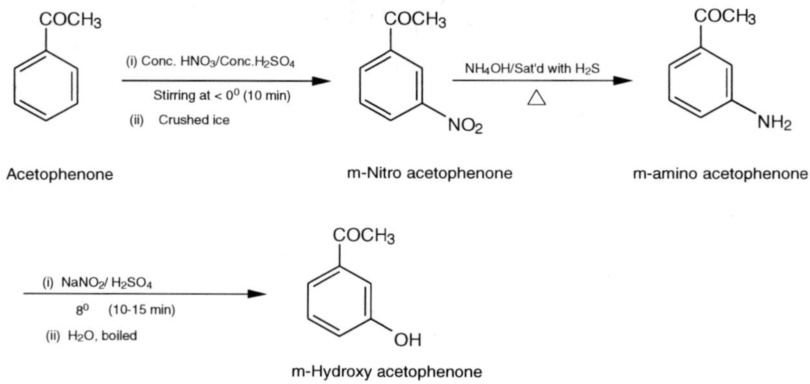

Step (i) m-Nitroacetophenone

Acetophenone (60 g, 58.25 ml, 0.5 mole) is added dropwise during 10-15 min to well cooled (<0°, ice-salt mixture) and stirred concentrated sulphuric acid (150 ml). The temperature during the addition is maintained below 5°. The reaction mixture is cooled to about −7°. Nitrating mixture consisting of nitric acid (40 ml, 0.65 mole, d, 1.42) and concentrated sulphuric acid (60 ml) is cooled to 15-20° and added to the above stirred mixture through a dropping funnel. The temperature of the reaction mixture is kept <0°. After the addition is over, the stirring is continued for 10 min more. The reaction mixture is poured with vigorous stirring into a mixture of crushed ice (750 g) and water (1.5 litre). The separated product is filtered, washed thoroughly with water (till excess acid is removed). The remaining solid is macerated with ice-cold ethyl alcohol to remove any adhering oil and filtered. The nitro compound is dissolved in hot ethyl alcohol (100-120 ml), solution filtered and poured into cold water (1 litre). The separated yellow solid is filtered, washed with cold water and dried. Yield 45 g (55%). It is finally crystallised from alcohol (for this the alcoholic solution has to be cooled in ice-bath and shaken vigorously). M.p. 76-78° (lit.[3,4] m.p. 81°).

Step (ii) m-Aminoacetophenone

A solution of m-nitroacetophenone (40 g, 0.24 mole) in hot ethanol (200 ml) is cooled rapidly. The solution is treated with concentrated ammonium hydroxide (45 ml) and saturated with hydrogen sulphide gas at room temperature. After saturation, the hydrogen sulphide passage is discontinued and the mixture refluxed for 5 hr. Alternate hydrogen sulphide treatment and refluxing is repeated thrice. The solution is diluted with water. The solid product is dissolved in dilute hydrochloric acid and the clear acidic solution is neutralized with ammonium hydroxide. The separated product is filtered and washed with water. It is recrystallised from dilute ethyl alcohol (20%). Yield 26 g (79.5%). M.p. 95° (lit.[5]m.p. 95°).

Step (iii) m-Hydroxyacetophenone

m-Aminoacetophenone (26 g, 0.19 mole) is dissolved in a mixture of sulphuric acid (95%, 60 ml) and water (250 ml) and the solution cooled (8°). A solution of sodium nitrite (13.6 g, 0.197 mole) in water (50 ml) is added dropwise to the above solution is allowed to stand at 8° for 10-15 min and then poured into water (50 ml). The solution is boiled. The separated product is filtered. It is crystallised from hot water (charcoal) as colourless needles. Yield 12 g (46.5%). M.p. 95° (lit.[5] m.p.95°).

4.A.6 2,4-DIHYDROXYACETOPHENONE (β-RESACETOPHENONE)

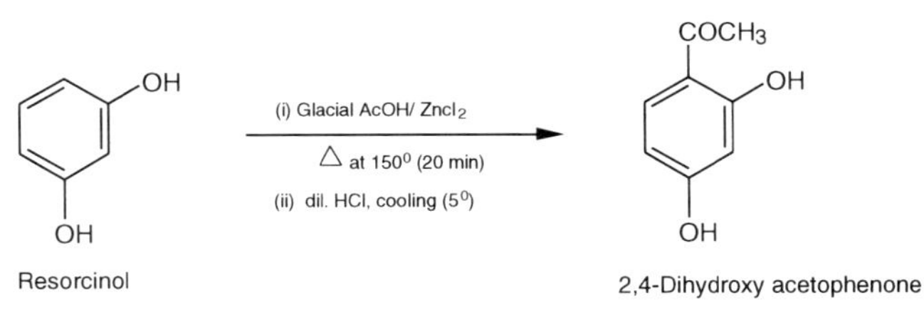

Freshly fused and powdered zinc chloride (33 g, 0.24 mole) is dissolved in glacial acetic acid (32 ml) by heating in a beaker on a sand bath. Dry resorcinol (22 g, 0.2 mole) is added with stirring to the mixture at 140°. The solution is heated until it just begins to boil and kept for 20 min at 150°. Dilute hydrochloric acid (1:1, 100 ml) is added to the mixture and the solution cooled (5°). The separated product is filtered and washed with dilute hydrochloric acid (1:3). It is crystallised from hot water containing little hydrochloric acid. Yield 28 g (92%). M.p. 142-144° (lit.[6] m.p. 133-40°).

4.A.7. 2-HYDROXY-4-METHOXYACETOPHENONE (PEANOL)

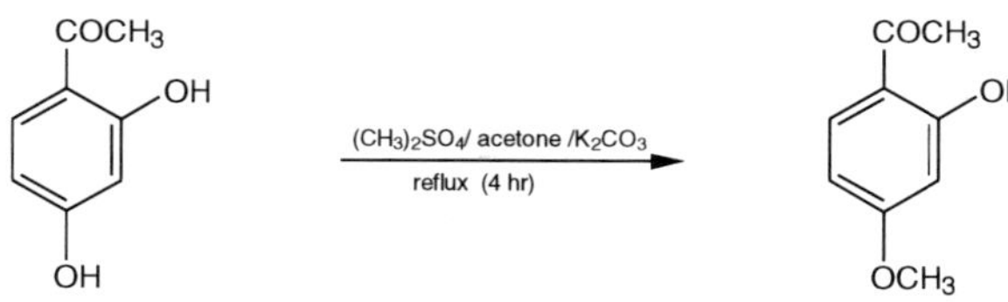

β-Resacetophenone Peanol

A mixture of β-resacetophenone (4.A.6) (15 g, 0.099 mole) in dry acetone (250 ml) is refluxed with dimethyl sulphate (11 ml, 14.66 g, 0.116 mole) and anhydrous potassium carbonate for 4 hr under anhydrous conditions. The inorganic salts are filtered and the residue washed with hot acetone (50 ml). The combined acetone extract is distilled and the oily residue macerated with ice-water. It is extracted with ether, extract washed with water, sodium carbonate solution (5%) and finally extracted with aqueous sodium hydroxide solution (5%).

The clear alkaline solution is acidified with hydrochloric acid and cooled. The separated product is filtered and washed with water. It is crystallised from alcohol as colourless long needles. Yield 15 g (91.5%). M.p. 50-51° (lit.[7] m.p.52-53°).

4.A.8 2,4-DIMETHOXYACETOPHENONE

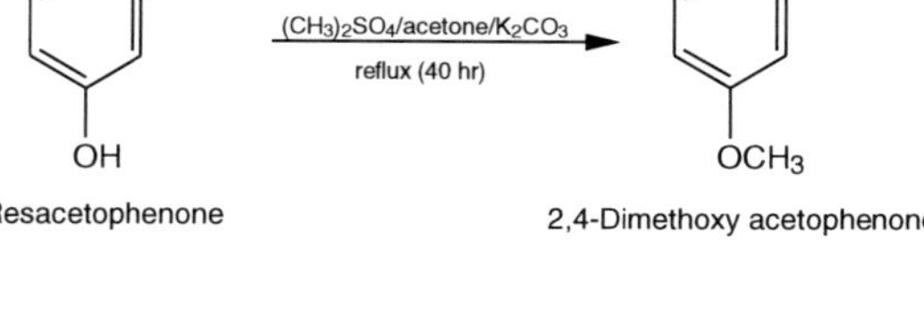

β-Resacetophenone 2,4-Dimethoxy acetophenone

β-Resacetophenone (4.A.6) (10 g, 0.066 mole) is dissolved in dry acetone (150 ml) and refluxed with dimethyl sulphate (18.9 ml, 25.2 g, 0.2 mole) and anhydrous potassium carbonate (50 g) for 40 hr under anhydrous conditions. The acetone solution is filtered and the inorganic salts washed with hot acetone (50 ml). The combined acetone solution is distilled. The oily residue is macerated with ice and cooled. The solid product is filtered and washed with water. It is crystallised from ethyl acetate-petroleum ether as colourless needles. Yield 8 g (67.8%). M.p. 43-44° (lit.[8] m.p. 44°).

4.A.9 2,4-DIHYDROXY-5-NITROACETOPHENONE (5-NITRO-β-RESACETOPHENONE)

β-Resacetophenone (4.A.6) (25 g, 0.164 mole) is added in small portions to cold (ice bath) concentrated nitric acid (200 ml, d, 1.42) with constant stirring. Fresh lot of β-resacetophenone is added only when the previous lot has gone into solution. The mixture is allowed to stand for 1 hr and then poured over crushed ice. The separated product is filtered. It is heated in boiling water (300 ml) and filtered. Treatment with boiling water is done 3-4 times. Finally the yellow product is crystallised from alcohol as light yellow needles. Yield 23 g (71%). M.p. 142° (lit.[9] m.p. 142°).

4.A.10 2,5-DIHYDROXYACETOPHENONE (QUINACETOPHENONE)

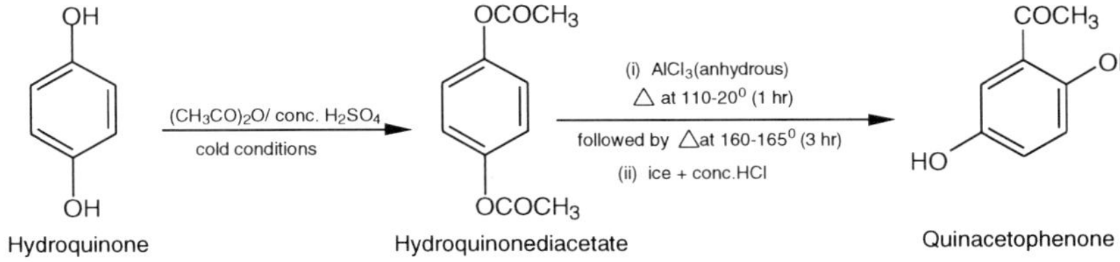

Step (i) Hydroquinone diacetate

To a cooled mixture of hydroquinone (11 g, 0.1 mole) and acetic anhydride (20 ml, 21.64 g, 0.21 mole) is added a drop of concentrated sulphuric acid. The suspension is shaken, the hydroquinone dissolves with the liberation of heat. The clear brown solution thus obtained is poured on to crushed ice. The separated product is filtered and washed with cold water .It is crystallised from dilute alcohol as colourless shining plates. Yield 15 g (77.3%). M.p. 120-21° (lit.[10] m.p. 120-21°).

Step (ii) 2,5-Dihydroxyacetophenone (Quinacetophenone)

An intimate mixture of hydroquinone diacetate (10 g, 0.052 mole) and anhydrous aluminium chloride (23 g, 0.172 mole) is heated at 110-20° for 1 hr and then at 160-65° (oil bath) for 3 hr till the evolution of hydrogen chloride gas ceased. The reaction is carried out under anhydrous conditions. The complex is hydrolysed by adding crushed ice and concentrated hydrochloric acid (20 ml). The solid obtained is filtered and washed with water. It is crystallised from methanol as pale yellow tiny needles. Yield 7 g (89.7%). M.p. 202-203° (lit.[11] m.p. 202-203°).

4.A.11 2-HYDROXY-5-METHOXYACETOPHENONE (QUINACETOPHENONE MONOMETHYL ETHER)

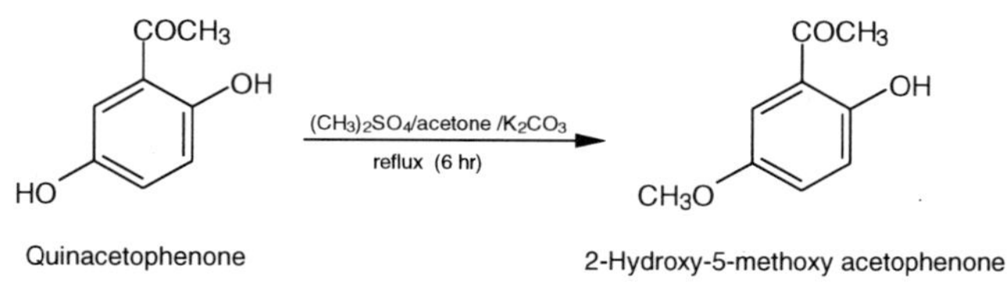

A solution of 2,5-dihydroxyacetophenone (4.A.10) (15 g, 0.099 mole) in dry acetone (150 ml) is refluxed for 6 hr under anhydrous conditions with dimethyl sulphate (11 ml, 14.66 g, 0.116 mole) and ignited potassium carbonate (25 g). The mixture is filtered, inorganic salts washed with hot acetone (2 × 50 ml) and the combined acetone solution distilled. The residue is treated with water and the separated product filtered. It is crystallised from dilute alcohol as colourless small needles. Yield 12 g (73.2%). M.p. 52° (lit.[12] m.p. 49-50°).

4.A.12 2,5-DIMETHOXYACETOPHENONE (QUINACETOPHENONE DIMETHYL ETHER)

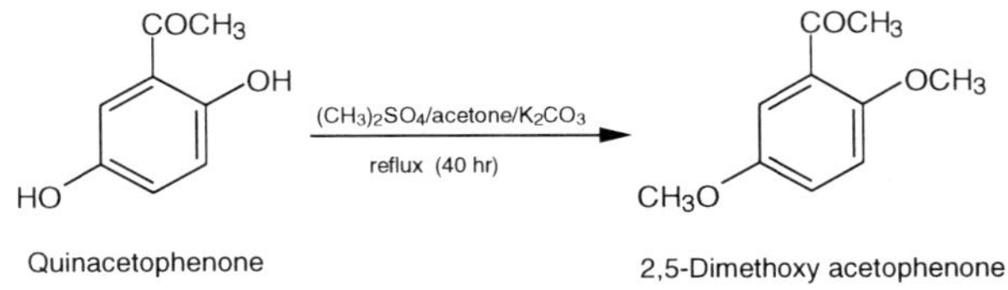

A solution of 2,5-dihydroxyacetophenone (4.A.10) (10 g, 0.065 mole) in dry acetone (150 ml) is refluxed for 40 hr under anhydrous conditions with dimethyl sulphate (18.9 ml, 25.2 g, 0.2 mole) and anhydrous potassium carbonate (25 g). The mixture is filtered and inorganic salts washed with hot acetone (2 × 50 ml). The acetone solution is distilled and the residue macerated with ice-water. The oily product is extracted with ether and extract washed with sodium hydroxide solution (5%) and then with water. The ether solution is dried (anhydrous sodium sulphate) and distilled. 2,5-Dimethoxyacetophenone is obtained as an oil. Yield 9 g (76.3%). B.p. 156-58°/15 mm (lit.[12] b.p. 147-150°/20 mm; lit.[13] b.p. 155-58°/14 mm).

4.A.13 2,3-DIMETHOXYACETOPHENONE

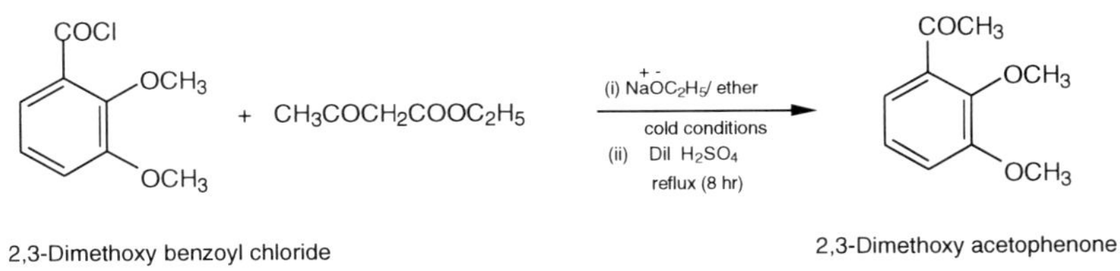

Separate solutions of sodium (2.53 g, 0.11 mole) in absolute alcohol (42 ml) and 2,3-dimethoxybenzoyl chloride[15] (10 g, 0.05 mole) in absolute ether (210 ml) are prepared. Half of the sodium ethoxide solution (21 ml) is mixed with ethylacetoacetate (7.01 g, 6.87 ml, 0.054 mole); to this half of the ethereal solution of 2,3- dimethoxybenzoyl chloride[15] (105 ml) is added and mixture is shaken in cold for 30 min. A mixture of sodium chloride and the sodium salt of dimethoxybenzoyl acetoacetic ester gradually separates out. After 30 min, sodium ethoxide solution (10.05 ml) and the dimethoxybenzoyl chloride solution (50 ml) are added and the mixture shaken for 30 min more. The condensation is completed by the addition of the remaining sodium ethoxide solution (10.05 ml) and the acid chloride solution (50 ml). The reaction mixture is shaken for 30 min and kept overnight in a refrigerator. The separated salt of 2,3-dimethoxybenzoyl acetoacetic ester is filtered and washed with absolute ether (2 × 30 ml). It is hydrolysed by refluxing with excess of dilute sulphuric acid (33%, 200 ml) for 8 hr. The mixture is cooled and the oily product extracted with ether. The ether extract is washed with sodium hydroxide solution (5%) to remove any free acid present and then with water. The ether is then dried (anhydrous magnesium sulphate) and distilled to give the required 2,3-dimethoxyacetophenone as an oil. Yield 4g (44.4%). B.p. 143-44°/14 mm (lit.[14] m.p. 143-44°/14 mm).

4.A.13A 5-ACETYL-2,4-DIHYDROXYACETOPHENONE[16](1,5-DIACETYL-2,4-DIHYDROXYBENZENE)

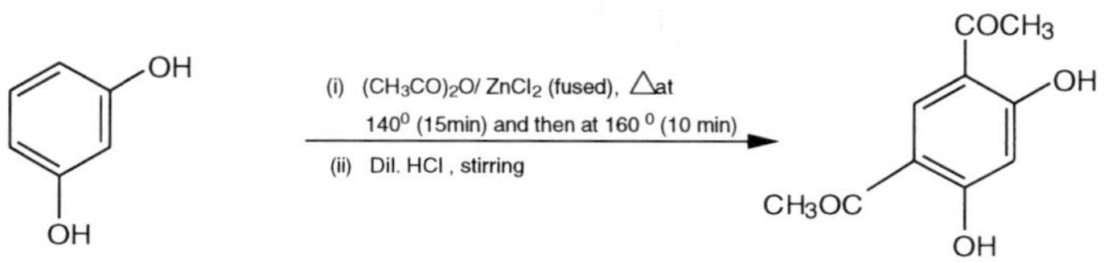

Resorcinol 5-Acetyl-2,4-dihydroxy acetophenone

To a mixture of freshly fused and powdered zinc chloride (10 g, 0.108 mole), and dry acetic anhydride (13 ml, 0.136 mole) contained in a R.B. flask is added dry resorcinol (5.5 g, 0.05 mole) with stirring. The mixture is gently heated to 140° (sand bath) and maintained at this temperature for 15 min. The temperature is raised to 160° and maintained for 10 min. The viscous mixture is allowed to cool to room temperature and treated with dilute hydrochloric acid (15 ml, 1:1) with stirring. After few minutes an orange coloured product separates as an solid product. It is crystallised from methanol as yellow coloured product. M.p. 178-79° (lit.[17] m.p. 178-79°). Yield 7.8 g (75%).

4.A.14 2-HYDROXY-3-METHOXYACETOPHENONE

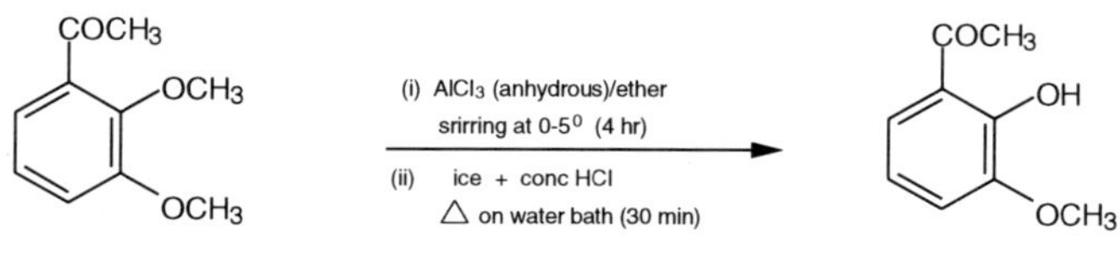

2,3-Dimethoxy acetophenone 2-Hydroxy-3-methoxy acetophenone

A solution of 2,3-dimethoxyacetophenone (4.A.13) (2.5 g, 0.0139 mole) in dry ether (20 ml) is added slowly during 30 min to a cooled solution (0-5°) of anhydrous aluminium chloride (5g, 0.037 mole) in dry ether (40 ml). The mixture is stirred in cold (0-5°) for 4 hr and kept overnight at room temperature. The separated complex is hydrolysed by treatment with crushed ice and concentrated hydrochloric acid (6 ml) followed by heating for 30 min on water bath, the ether is allowed to distil away. The aqueous solution is cooled and extracted with ether.

The extract is extracted with sodium hydroxide solution (10%). The clear cooled alkaline extract is acidified with hydrochloric acid. The separated product is filtered, washed with water and dried. It is crystallised from benzene-petroleum ether as fine colourless needles. Yield 2 g (86.9%). M.p. 55-56° (lit.[18] m.p. 55-56°).

4.A.15 2,3-DIHYDROXYACETOPHENONE

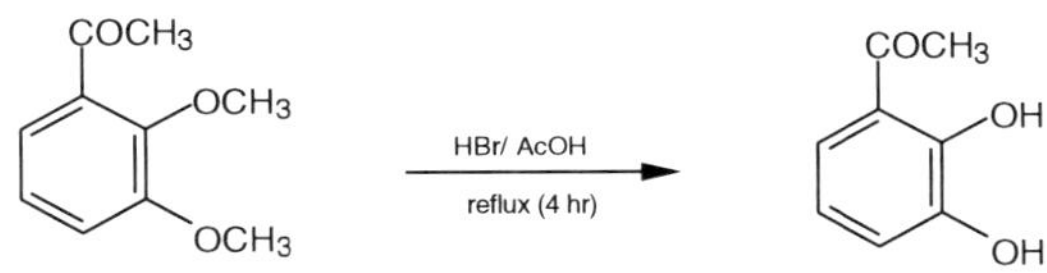

2,3-Dimethoxy acetophenone 2,3-Dihydroxy acetophenone

2,3-Dimethoxyacetophenone (4.A.13) (1 g, 0.0055 mole) in acetic acid (8 ml) is refluxed for 4 hr with hydrobromic acid (8 ml, d, 1.5). The mixture is poured over crushed ice. The separated product is filtered and washed with water. It is crystallised from dilute methanol as pale yellow needles. Yield 0.6 g (71.4%). M.p. 97-8° (lit.[19] m.p. 97-8°)

4.A.16 2,6-DIHYDROXYACETOPHENONE (RESACETOPHENONE)

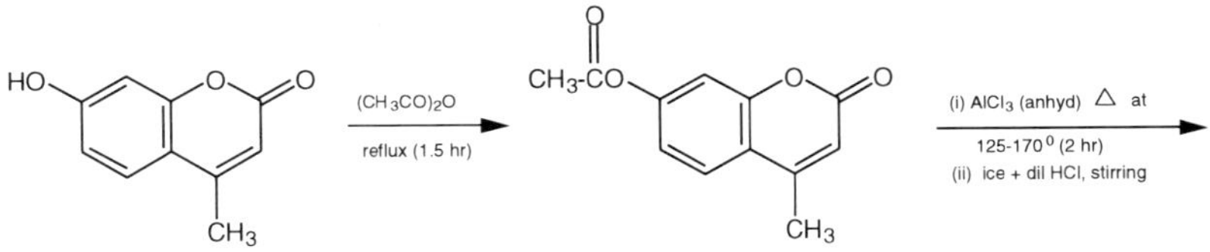

7-Hydroxy-4-methyl coummarin 7-acetoxy-4-methyl coumarin

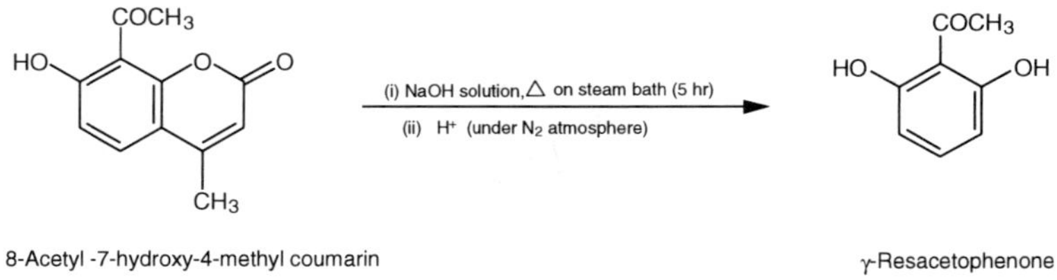

Step (i) 7-Hydroxy-4-methylcoumarin (see 5.A.16).

Step (ii) 7-Acetoxy-4-methylcoumarin

A mixture of 7-hydroxy-4-methylcoumarin (28.2 g, 0.16 mole) and freshly distilled acetic anhydride (57.2 g, 52.87 ml, 0.56 mole) is refluxed for 1.5 hr under anhydrous conditions. While still hot, the solution is poured over crushed ice. The separated product is filtered and washed with water. It is crystallised from alcohol as fibrous needles. Yield 32 g (91.7%). M.p. 150-51° (lit.[20] m.p. 150-51°).

Step (iii) 8-Acetyl-7-hydroxy-4-methylcoumarin

An intimate mixture of 7-acetoxy-4-methylcoumarin (20 g, 0.092 mole) and anhydrous aluminium chloride (45.3 g, 0.34 mole) is heated under anhydrous conditions in an oil bath at 125°. The temperature is raised during 2 hr to 170°. The reaction mixture is treated with crushed ice and dilute hydrochloric acid (1:1, 250 ml), added with stirring. The mixture is left for 2-3 hr so that the complex is fully decomposed. The separated product is filtered and washed with water. It is crystallised from alcohol as long yellow needles. Yield 14.5 g (72.5%). M.p. 162-63° (lit.[21] m.p. 162-63°).

Step (iv) 2,6-Dihydroxyacetophenone (γ-Resacetophenone)

8-Acetyl-7-hydroxy-4-methylcoumarin (14.8 g, 0.068 mole) and distilled water (50 ml) are taken in a 500 ml three necked R. B. flask. A rapid stream of nitrogen gas is bubbled through the suspension to replace all the air from the system. The nitrogen is bubbled slowly and a solution of sodium hydroxide (12.9 g, 0.323 mole) in water (60 ml) is added through a dropping funnel and the mixture is heated for 5 hr on a steam bath. The solution is cooled and acidified with dilute hydrochloric acid (100 ml, 1:3); the nitrogen gas is bubbled throughout the process. The separated solid is filtered, washed with water and dried. It is purified by dissolving in alcohol (100 ml) and charcoalised (15 min heating, steam bath) and water (80 ml) added. The solution is heated for 10 min more and filtered. The clear filtrate on cooling gave 2,6-dihydroxyacetophenone. Some more product is obtained by concentration of the mother liquor. Total yield of purified product is 7.5 g (72.7%). M.p. 154-55° (lit.[21,22] m.p. 155-56°).

4.A.17 2-HYDROXY-6-METHOXYACETOPHENONE(γ-RESACETOPHENONE MONOMETHYL ETHER)

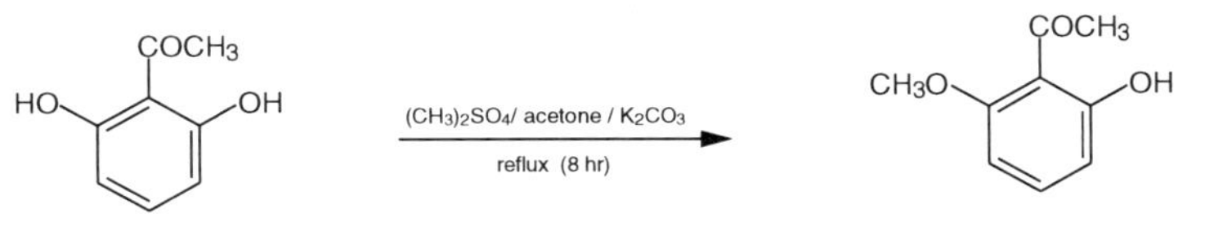

2,6-Dihydroxy acetophenone 2-Hydroxy-6-methoxy acetophenone

A solution of 2,6-dihydroxyacetophenone (4.A.16) (13 g, 0.086 mole) in dry acetone (150 ml) and dimethyl sulphate (9 ml, 12 g, 0.095 mole) is refluxed for 8 hr under anhydrous conditions in presence of ignited potassium carbonate (30 g). The mixture is filtered and the inorganic salts washed with hot acetone (2 × 30 ml). The combined acetone solution is distilled and the residual product macerated with ice-water. The product is filtered and purified by dissolving in sodium hydroxide solution (10%) and acidification of the clear alkaline solution in cold. It is crystallised from methanol as almost colourless fine prisms. Yield 11 g (77.5%). M.p. 59-60° (lit.[23] m.p. 60°).

4.A.18 2,6-DIMETHOXYACETOPHENONE (γ-RESACETOPHENONE DIMETHYL ETHER)

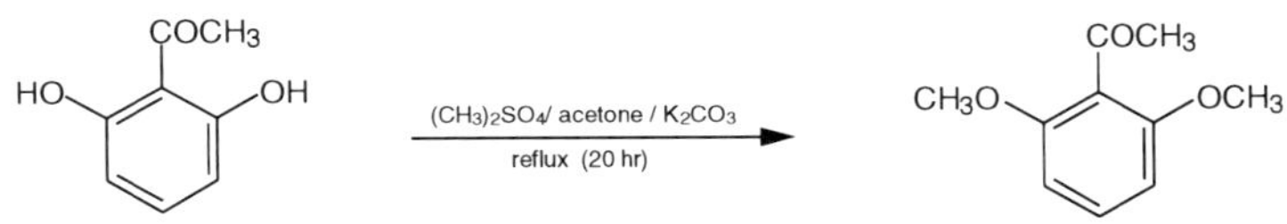

2,6-Dihydroxy acetophenone 2,6-Dimethoxyacetophenone

A solution of 2,6-dihydroxyacetophenone (4.A. 16) (5 g, 0.033 mole) in dry acetone (100 ml) is refluxed with dimethyl sulphate (12.6 g, 9.45 ml, 0.1 mole) in presence of ignited potassium carbonate (30 g) under anhydrous conditions for 20 hr. The mixture is filtered hot, inorganic salts washed with hot acetone (2 × 25 ml) and the combined acetone solution distilled. The residue is treated with crushed ice. The separated product is filtered and washed with water. It is crystallised from dilute methanol as colourless needles. Yield 4.5 g (76.3%). M.p. 72-73° (lit.[24] m.p. 72-73°).

4.A.19 3,4-DIMETHOXYACETOPHENONE

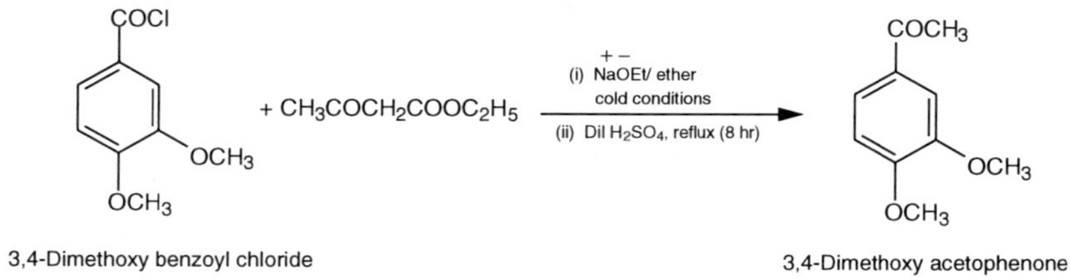

Method (i)[25]

Separate solutions of sodium (2.53 g, 0.11 mole) in absolute alcohol (42 ml) and 3,4-dimethoxybenzoyl chloride (10 g, 0.05 mole) in absolute ether (210 ml) are prepared. Half of the sodium ethoxide solution (21 ml) is mixed with ethylacetoacetate (7.01 g, 6.87 ml, 0.054 mole) and to this half of the ethereal solution of the acid chloride (105 ml) is added. The mixture is shaken in cold for 30 min under anhydrous conditions. To the reaction mixture is added again sodium ethoxide solution (10.05 ml) and the ethereal solution of acid chloride (52.5 ml) and the shaking continue for 30 min. The condensation is completed by adding the remaining sodium ethoxide solution (10.05 ml) and the ethereal solution of acid chloride (52.5 ml). The reaction mixture is shaken again in cold for 30 min and left overnight in a refrigerator. The separated salt is filtered and washed with dry ether. It is refluxed for 8 hr with dilute sulphuric acid (33%, 200 ml). The reaction mixture is cooled and extracted with ether (3×100 ml). The ethereal extract is washed with dilute sodium hydroxide (5%, 2×100 ml) to remove any free acid and finally with water. It is dried (anhydrous sodium sulphate) and distilled to give 3,4-dimethoxyacetophenone. It is crystallised from dilute alcohol as colourless prisms. Yield 4 g (44.4%). M.p. 50° (lit.[26] m.p. 50-1°).

Note: 3,4-Dimethoxybenzoyl chloride required for the above reaction is obtained by refluxing 3,4-dimethoxybenzoic acid (3.17) (10 g, 0.055 mole) with thionyl chloride (7 ml, 11.4 g, 0.096 mole) for 30 min (water bath). The excess thionyl choride is removed under reduced pressure and the remaining acid chloride is used as such.

Method (ii)

A mixture of 3,4-dihydroxyacetophenone (4.A.22) (20 g, 0.13 mole) in dry acetone (150 ml), dimethyl sulphate (28 ml, 37.3 g, 0.296 mole) and ignited potassium carobonate (50 g) is refluxed for 8 hr under anhydrous

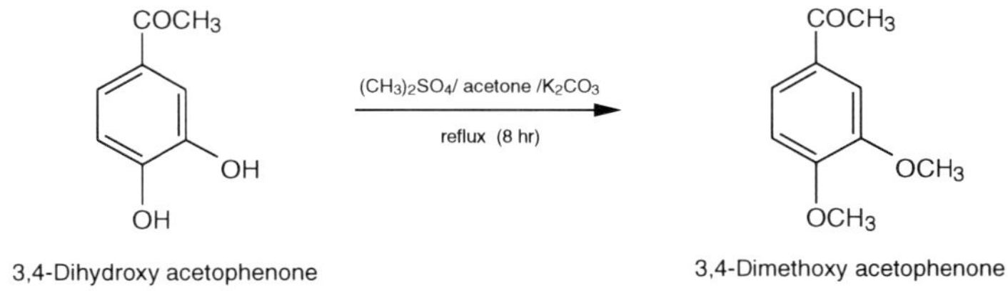

conditions. The reaction mixture is filtered and the residual inorganic salts washed with more acetone (2 × 30 ml). The combined acetone solution is distilled and the oily residue macerated with ice. The separated product is filtered and washed with water. It is crystallised from dilute alcohol as colourless prisms. Yield 16 g (67.8%). M.p. 48° (lit.[27] m.p. 48°).

4.A.20 3-HYDROXY-4-METHOXYACETOPHENONE AND

4.A.21 4-HYDROXY-3-METHOXYACETOPHENONE

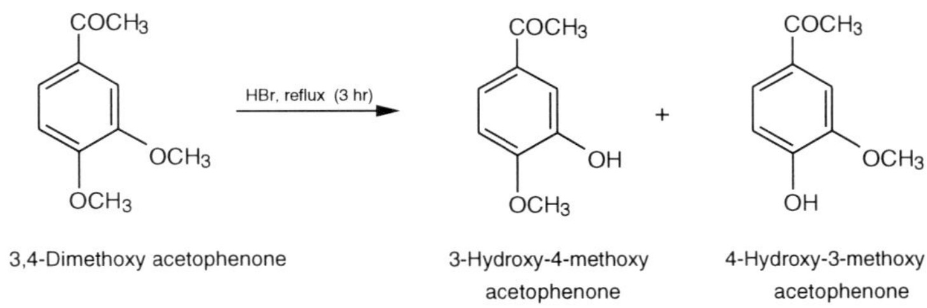

A mixture of 3,4-dimethoxyacetophenone (4.A.19) (5g, 0.028 mole) and hydrobromic acid (43%, 50 ml) is refluxed for 3 hr and poured over crushed ice. The dark tarry matter is extracted with chloroform (4 × 50 ml). The solvent is distilled and the oily residue dissolved in aqueous potassium hydroxide (5%, 100 ml) and freed from any neutral substance by extraction with ether. The clear alkaline solution is acidified with hydrochloric acid and the solution extracted with chloroform. The extract is dried (sodium sulphate) and distilled to give a mixture of the two ketones (TLC). It is separated by chromatography over silica gel column

(2.5 × 25 cm). Elution with benzene-petroleum ether (1:1) gives 4-hydroxy-3-methoxyacetophenone. Yield 0.5 g (10.8%). M.p. 115° (lit.[28,29] m.p. 115°). Further elution of the column with petroleum ether-benzene (1:3) gave 3-hydroxy-4-methoxyacetophenone. Yield 0.2 g (4.3%). M.p. 91° (lit.[30] m.p. 91°).

4.A.22 3,4-DIHYDROXYACETOPHENONE

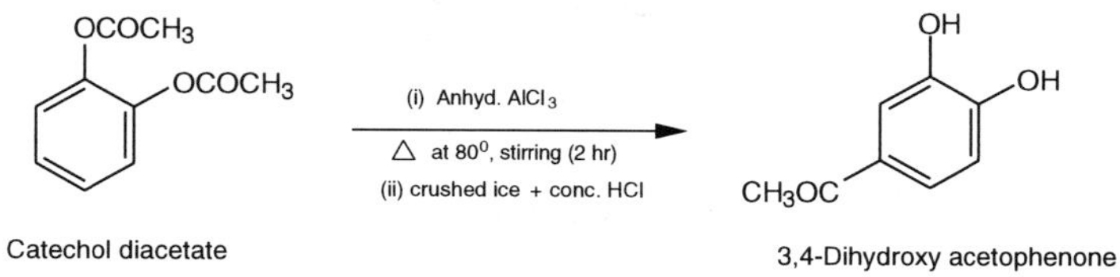

A solution of catechol diacetate (40 g, 0.206 mole) in dry chlorobenzene (80 ml) is added to a well stirred mixture of powdered anhydrous aluminum chloride (80 g, 0.6 mole) in chlorobenzene (100 ml) at such a rate that the evolution of hydrogen chloride does not become too vigorous. The reaction is done under anhydrous conditions and is exothermic. The thickened reaction mixture is then heated with stirring for 2 hr at 80°, cooled to room temperature and poured into crushed ice (400 g) and concentrated hydrochloric acid (50 ml). After decomposition of the complex (2-3 hr at room temperature), the solid product is filtered and washed with chloroform, water and dried. It is crystallised from benzene as pale yellow crystals. Yield 25 g (79.8%). M.p. 115° (lit.[31,32] m.p. 114°).

4.A.23 2,5-DICHLOROACETOPHENONE

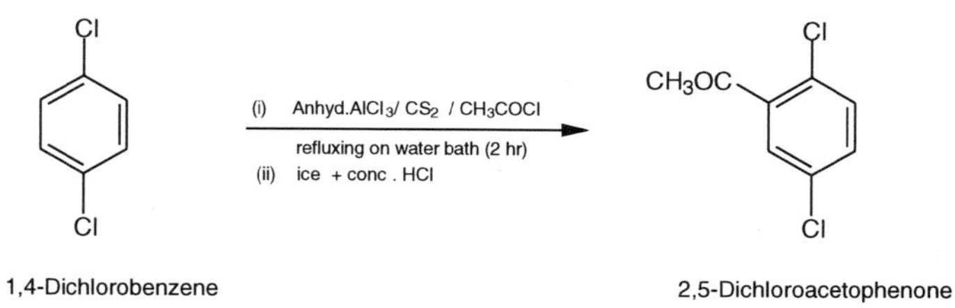

A mixture of 1,4-dichlorobenzene (53 g, 0.36 mole), dry carbon disulphide (200 ml) and powdered anhydrous aluminium chloride (150 g, 1.12 mole) is heated gently on a water bath under anhydrous conditions. Acetyl chloride (47.5 ml, 52.44 g, 0.668 mole) is added dropwise; gentle refluxing is maintained during addition and refluxing continued for 2 hr more. Carbon disulphide is distilled off and the reaction mixture treated with crushed ice (200 g) and concentrated hydrochloric acid (150 ml). The mixture is left at room temperature for 2-3 hr and extracted with ether (4 × 100 ml). The ether extract is washed with aqueous sodium hydroxide (10%), water and dried (magnesium sulphate) and the solvent is removed by distillation. The crude 2,5-dichloroacetophenone thus obtained is purified by chromatography over silica gel column (2 × 40 cm) and elution with petroleum ether. Yield 50 g (73.4%). B.p. 245-50° (lit.[33] b.p. 245-50°).

4.A.24 2,5-DIBROMOACETOPHENONE

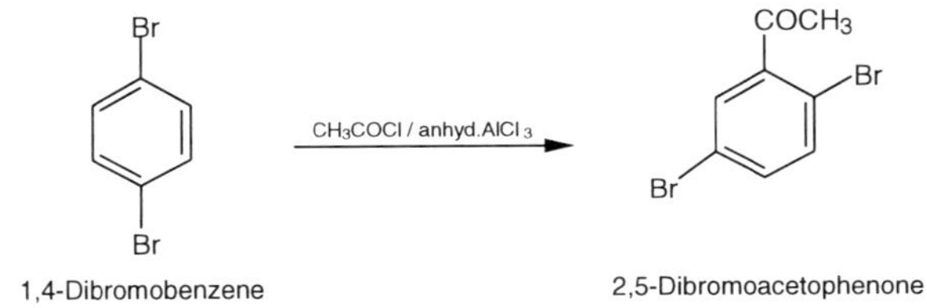

It is prepared as in the case of 2,5-dichloroacetophenone (4.A.23) using 1,4-dibromobenzene (47 g, 0.2 mole), acetyl chloride (28.4 ml, 31.4 g, 0.4 mole) and anhydrous aluminium chloride (80 g, 0.6 mole). The reaction is worked up to give 2,5-dibromacetopheneone as an oil. It is purified by chromatography over silica gel column and elution with petroleum ether. Yield 44.5 g (80%). M.p. 40-41° (lit.[34] m.p. 41°).

4.A.25 2,4-DIHYDROXY-3-METHYLACETOPHENONE[35]

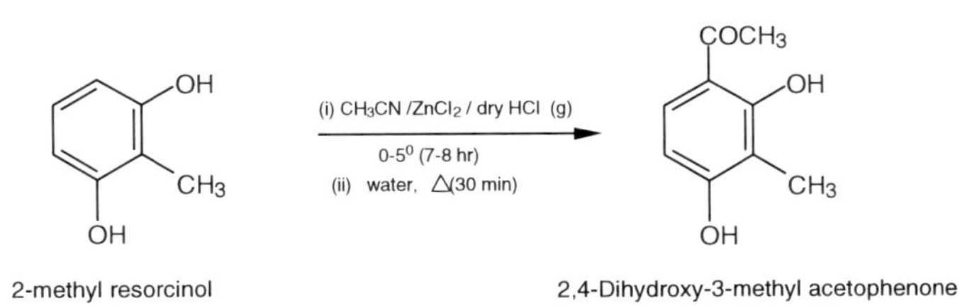

A rapid stream of dry hydrogen chloride gas is passed through a solution of 2-methyl resorcinol (1.17) (10 g, 0.08 mole), acetonitrile (9 ml, 7.07 g, 0.17 mole) and fused zinc chloride (2 g, 0.0147 mole) in dry ether (100 ml) at 0-5° for 7-8 hr with occasional shaking. The reaction is done under anhydrous conditions and the reaction mixture left in a refrigerator for 24 hr. The clear ether solution is decanted and the ketimine hydrochloride washed with dry ether (2 × 30 ml). It is dissolved in water (200 ml) and the solution is heated on a water bath for 30 min and cooled. The separated product is filtered and washed with cold water. It is crystallised from dilute alcohol as colourless shining needles. Yield 8 g (59.8%). M.p. 155-56° (lit.[36] m.p. 157-58°).

4.A.26 2-HYDROXY-4-METHOXY-3-METHYLACETOPHENONE[35]

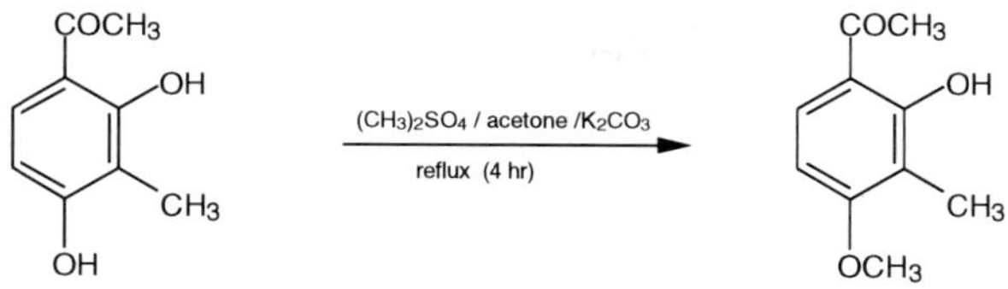

2,4-Dihydroxy-3-methylacetophenone 2-Hydroxy-4-methoxy-3 methylacetophenone

A solution of 2,4-dihydroxy-3-methylacetophenone (4.A.25) (2 g, 0.012 mole) in dry acetone (70 ml) is refluxed for 4 hr under anhydrous conditions, with dimethyl sulphate (1.3 ml, 1.73 g, 0.0137 mole) in the presence of anhydrous potassium carbonate (5 g). The reaction mixture is filtered and the inorganic salts washed with more hot acetone (2 × 30 ml). The combined acetone solution is distilled and the oily residue macerated with ice. The separated product is filtered, washed with water and dried. It is crystallised from petroleum ether as pink shining needles. Yield 2 g (92.6%). M.p. 82-83°.

4.A.27 2,4-DIHYDROXY-5-METHYLACETOPHENONE

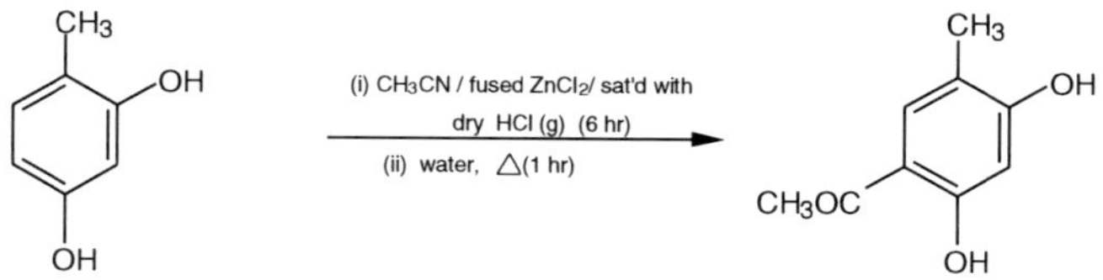

2,4-Dihydroxy toluene 2,4-Dihydroxy-5-methyl acetophenone

To a solution of 2,4-dihydroxy toluene (4-methyl resorcinol) (1.18) (6 g, 0.048 mole) in dry ether (30 ml) is added acetonitrile (5 ml, 3.93 g, 0.096 mole) and freshly fused and powdered zinc chloride (2 g, 0.0146 mole). The solution is cooled (0-5°) and saturated for 6 hr with dry hydrogen chloride gas under anhydrous conditions. The reaction mixture is left overnight at room temperature. The clear ether solution is decanted and the solid product washed with more dry ether (2 × 25 ml). The residual product is dissolved in water (25 ml) and the solution heated for 1 hr on a boiling water bath. The solution is cooled, the separated product is filtered and washed with water. It is crystallised from dilute alcohol as colourless crystals. Yield 5 g (62.3%). M.p. 178° (lit.[36a] m.p. 178°).

4.A.28 2,4-DIHYDROXY-6-METHYLACETOPHENONE (β-ORCACETOPHENONE)

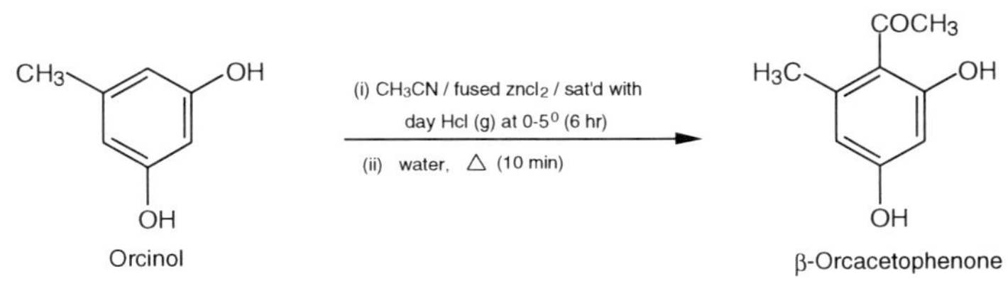

A solution of dry orcinol (6 g, 0.048 mole) and acetonitrile (5 ml, 3.93g, 0.096 mole) in dry ether (50 ml) containing fused zinc chloride (2 g, 0.0146 mole) is saturated with dry hydrogen chloride gas at 0-5° under anhydrous conditions. After 4 hr, yellowish ketimine hydrochloride starts separating, the flow of hydrogen chloride gas is continued for 2 hr more. The mixture is left overnight in a refrigerator. The ether solution is decanted and the solid product washed with dry ether (2 × 20 ml). It is dossolved in water (20 ml) and heated in a boiling water bath for 10 min. The solution is then neutralised with dilute ammonia and cooled. The separated product is filtered, washed with water and dried. It is crystallised from benzene as colourless needles. Yield 5 g (62.3%). M.p. 159-60° (lit.[37] m.p. 159-60°).

4.A.29 2-HYDROXY-4-METHOXY-6-ETHYLACETOPHENONE (β-ORCACETOPHENONE 4-0-METHYL ETHER)

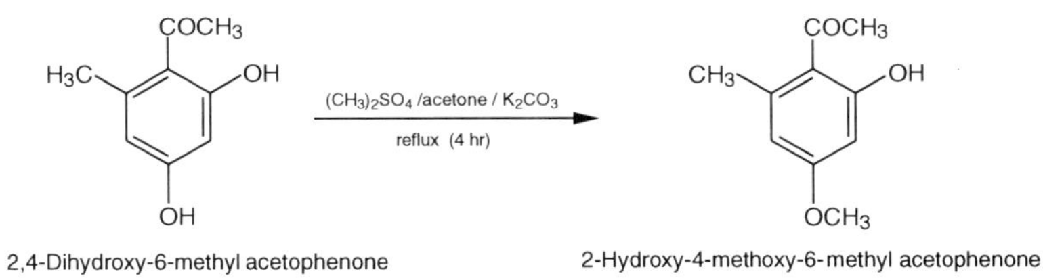

A solution of 2,4-dihydroxy-6-methylacetophenone (4.A.28) (5 g, 0.03 mole) in anhydrous acetone (80 ml) is refluxed for 4 hr under anhydrous conditions with dimethyl sulphate (3.2 ml, 4.27 g, 0.034 mole) and ignited potassium carbonate (20 g). The mixture is filtered and the inorganic salts washed with hot acetone (2 × 30 ml). The combined acetone solution is distilled, residue macerated with ice and extracted with ether (2 × 50 ml). The ether extract is washed with sodium carbonate solution (5%) and then extracted with sodium hydroxide solution (5%). The clear alkaline extract is acidified and the solid product filtered, washed with water and dried. It is crystallised from ethyl acetate as long colourless needles. Yield 4 g (74%). M.p. 79° (lit.[38,39] m.p. 78-78.5°).

4.A.30 2,4-DIMETHOXY-6-METHYLACETOPHENONE (β-ORCACETOPHENONE DIMETHYL ETHER)

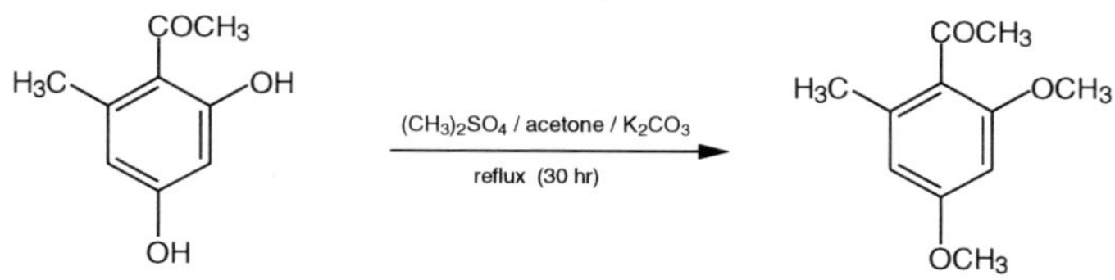

2,4-Dihydroxy-6-methylacetophenone 2,4-Dimethoxy-6-methyl acetophenone

A solution of 2,4-dihydroxy-6-methylacetophenone (4.A.28) (2 g, 0.012 mole) in dry acetone (50 ml) is refluxed for 30 hr under anhydrous conditions with dimethyl sulphate (3.5 ml, 4.66 g, 0.37 mole) and ignited potassium carbonate (10 g). The reaction mixture is filtered and the inorganic salts washed with hot acetone (2 × 25 ml). The combined acetone solution is distilled and the residue macerated with crushed ice. The solid product is filtered, washed with sodium hydroxide solution (5%) and with water and dried. It is crystallised from ether–petroleum ether as colourless needles, yield 1.8 g (78.3%). M.p. 48° (lit.[40] m.p. 48°).

4.A.31 2,6-DIHYDROXY-4-METHYLACETOPHENONE (γ-ORCACETOPHENONE)

Step (i) Orcinol diacetate

Orcinol (5 g, 0.04 mole) is refluxed with acetic anhydride (7.6 ml, 8.22 g, 0.08 mole) and pyridine (5 drops) for 2 hr at 135-140°. The cooled reaction mixture is poured over crushed ice and allowed to stand for 2-3 hr. It is extracted with ether (3 × 50 ml), ether extract washed with sodium bicarbonate solution (5%) and with water. It is dried (anhydrous magnesium sulphate) and distilled. Orcinol diacetate is obtained as an oil. Yield 5g (59.6%). B.p. 240°/10 mm (lit.[41] b.p. 240°/10 mm).

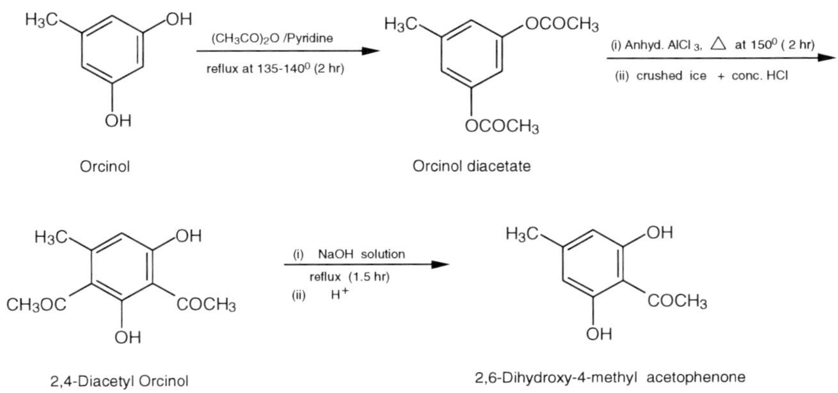

Step (ii) 2,4- Diacetylorcinol

An intimate mixture of orcinol diacetate (5 g, 0.024 mole) and powdered anhydrous aluminum chloride (12 g, 0.09 mole) is heated at 150° for 2 hr. The reaction mixture is cooled and treated with crushed ice (100 g) and concentrated hydrochloric acid (20 ml). The mixture is allowed to stand for 2-3 hr at room temperature. The separated product is filtered and washed with water. It is crystallised from dilute acetic acid as pale yellow needles. Yield 3 g (60%). m.p. 96° (lit.[41] m.p. 96°).

Step (iii) 2-6-Dihydroxy-4-methylacetophenone (γ-orcacetophenone)

2,4-Diacetylorcinol (1 g, 0.0048 mole) is refluxed with sodium hydroxide solution (25 ml, 16%) for 1.5 hr. The clear brownish alkaline solution is cooled and acidified with hydrochloric acid. The separated product is filtered and washed with water. It is crystallised from benzene as light yellow shining needles. Yield 0.70 g (87.7%). M.p146-47° (lit.[41] m.p. 146°).

4.A.32 2-HYDROXY-6-METHOXYACETOPHENONE (γ-ORCACETOPHENONE-6-0-METHYL ETHER)

A solution of 2,6- dihydroxy-4-methylacetophenone (4.A.31) (5 g, 0.03 mole) in dry acetone (60 ml) is refluxed with dimethyl sulphate (3.25 ml, 4.33 g, 0.34 mole) and anhydrous potassium carbonate (15 g) for 6 hr

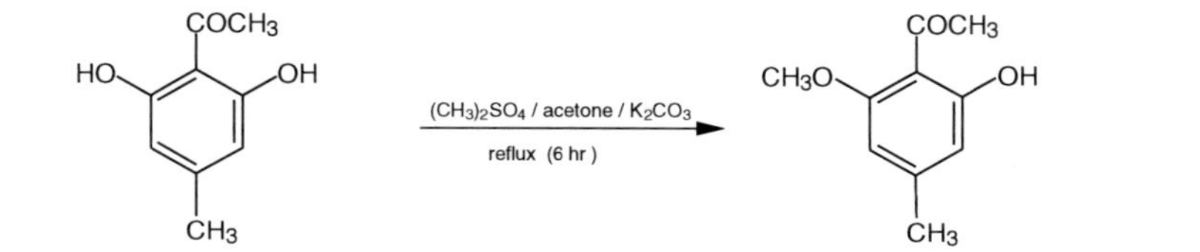

2,6-Dihydroxy-4-methyl acetophenone 2-Hydroxy-6-methoxy-4-methyl acetophenone

under anhydrous condition. The mixture is filtered and the inorganic residue washed with hot acetone (2 × 25 ml). The combined acetone solution is distilled and the residue macerated with crushed ice. It is extracted with ether (2 × 50 ml). The ether extract is washed with sodium carbonate solution (5%) and then with sodium hydroxide solution (5%). The clear alkaline solution is acidified in cold. The separated product is filtered, washed with water and dried. It is crystallised from hexane as colourless needless. Yield 4 g (73.8%). M.p. 80-81° (lit.[39] m.p. 80-81°).

4.A.33 2,6-DIMETHOXY-4-METHYLACETOPHENONE (γ-ORCACETOPHENONE DIMETHYL ETHER)

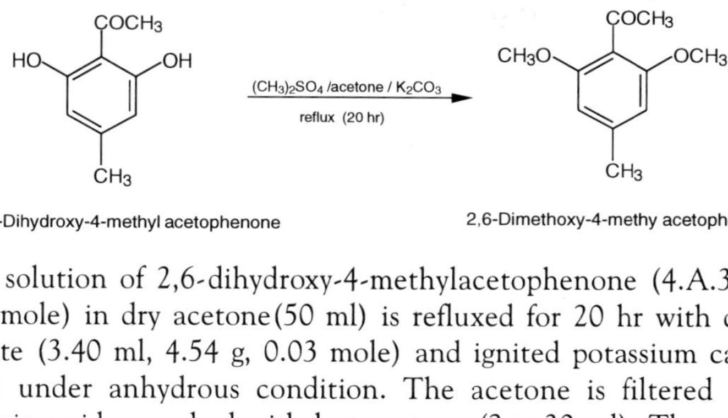

2,6-Dihydroxy-4-methyl acetophenone 2,6-Dimethoxy-4-methy acetophenone

A solution of 2,6-dihydroxy-4-methylacetophenone (4.A.31) (2 g, 0.012 mole) in dry acetone(50 ml) is refluxed for 20 hr with dimethyl sulphate (3.40 ml, 4.54 g, 0.03 mole) and ignited potassium carbonate (15 g) under anhydrous condition. The acetone is filtered and the inorganic residue washed with hot acetone (2 × 30 ml). The combined acetone solution is distilled and the residue macerated with crushed ice. The solid product is filtered, washed with water and dried. It is crystallised from ethyl acetate petroleum ether as colourless needles. Yield 1.7 g (73%). M.p. 85-86° (lit.[42] m.p. 89°).

4.A.34 5-ACETYL-2,4-DIHYDROXY-3-METHYLACETOPHENONE[43] (1,5-DIACETYL-2,4- DIHYDROXY-3-METHYLBENZENE)

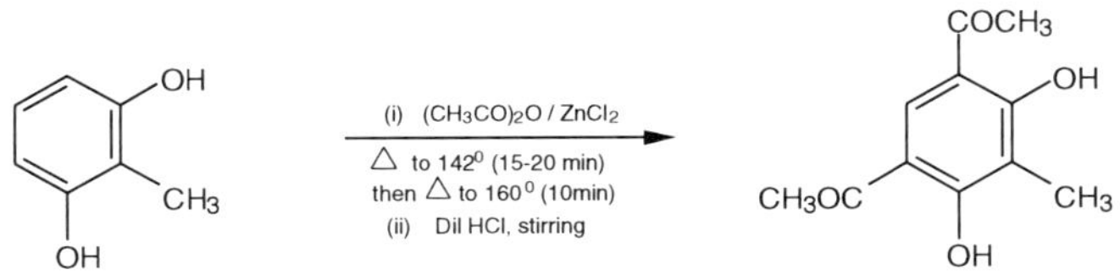

1,3-Dihydroxy-2-methyl benzene — 5-Acetyl-2,4-dihydroxy-3-methyl acetophenone

To a mixture of freshly fused and powdered zinc chloride (10 g, 0.073 mole) in dry acetic anhydride (13 ml, 14.09 g, 0.138 mole) contained in a R.B. flask is added 1,3-dihydroxy-2-methyl benzene (orcinol) (5 g, 0.04 mole) with stirring. The mixture is gently heated to 142° (sand bath) and maintained at this temperature for 15-20 min. The temperature is raised to 160° and maintained for 10 min. The viscous mixture is allowed to cool to room temperature and treated with dilute hydrochloric acid (8 ml, 1:1) with stirring. After few minutes an orange coloured product separates as an oil (7.5 g) which is found to be a mixture of two products (TLC). It is extracted with ether, extract washed with water, dried (sodium sulphate)and distilled. The oily product is subjected to column chromatography on silica gel (2 × 30 cm). It is eluted with benzene-petroleum ether (1:9). This gave a minor product (1.85 g, 27.8%), 2,4-dihydroxy-3-methyl acetophenone (β-orcacetophenone), m.p. 156°. Subsequently, the column is eluted with benzene- petroleum ether (3:7); this gives the required 1,5-diacetyl-2,4-dihydroxy-3-methylbenzene (4.7g, 56%) as colurless needles. M.p 146-47°.

4.A.35 2,3,4-TRITHYDROXYACETOPHENONE (GALLACETOPHENONE)

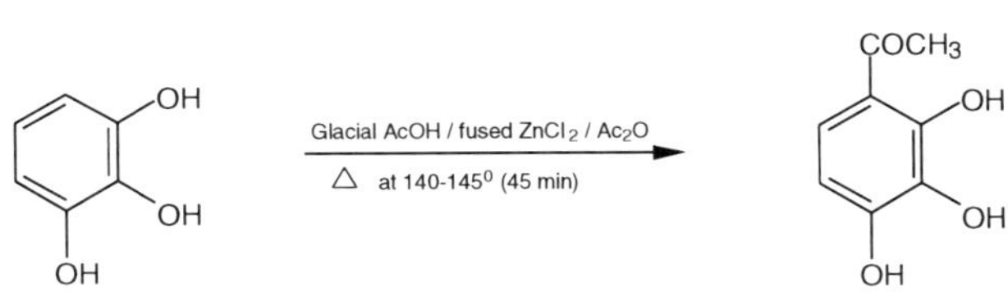

Pyrogallol — Gallacetophenone

Freshly fused and powdered zinc chloride (28 g, 0.205 mole) is dissolved in glacial acetic acid (38 ml, 0.66 mole) by heating at 135-140° (oil bath). To the clear solution is added acetic anhydride (40 ml, 43.3 g, 0.424 mole) followed by the addition of pyrogallol (50.5 g, 0.4 mole) in one lot. The mixture is heated at 140-45° (oil bath) for 45 min with frequent stirring. Excess acetic anhydride is removed in vaccuo and the residual reaction mixture treated with ice water. The mixture is stirred and cooled. The separated product is filtered and washed with cold water. It is crystallised from hot water (about 500 ml) saturated with sulphur dioxide. Gallacetophenone separates as straw coloured product. Yield 36 g (53.5%). M.p 171-72° (lit.[44,45] m.p. 169-171°).

4.A.36 2,3-DIHYDROXY-4-METHOXYACETOPHENONE (GALLACETOPHENONE-4-0-METHYL ETHER)

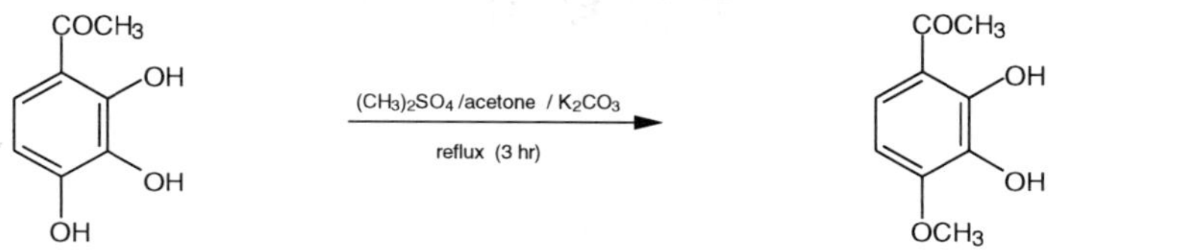

Gallacetophenone 2,3-Dihydroxy-4-methoxy acetophenone

A solution of gallacetophenone (4.A.35) (1 g, 0.006 mole) in dry acetone (50 ml)and ignited potassium carbonate(3.0 g) is refluxed for 3 hr under anhydrous conditions with dimethyl sulphate (0.6 ml, 0.8 g, 0.0063 mole). The mixture is filtered and the inorganic salts washed with hot acetone (2 × 20 ml). The combined acetone solution is distilled and residual oily product macerated with crushed ice. It is extracted with ether and the ether extract extracted with a saturated solution of borax. The borax solution is treated with hydrochloric acid and extracted with ether. The ether is distilled. The product obtained is crystallised from alcohol as colourless needles. Yield 0.4 g (37.0%). M.p. 127-28° (lit.[46,47] m.p. 127-28°).

4.A.37 2-HYDROXY-3,4-DIMETHOXYACETOPHENONE (GALLACETOPHENONE DIMETHYL ETHER)

Method (i)

A solution of gallacetophenone (4.A.35) (20 g, 0.12 mole) is dissolved in dry acetone (200 ml) and refluxed for 8 hr under anhydrous conditions

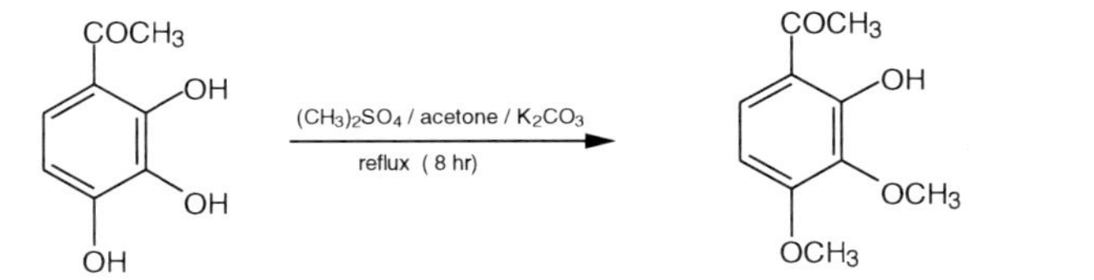

Gallacetophenone 2-Hydroxy-3,4-dimethoxy acetophenone

with dimethyl sulphate (26.8 ml, 35.7 g, 0.28 mole) and ignited potassium carbonate (70 g). The mixture is filtered in hot and the inorganic salts washed with hot acetone (2 × 50 ml). The combined acetone is distilled and the residue macerated with crushed ice. The separated product is filtered and washed with water. It is crystallised from methanol as colourless needles. Yield 18 g (77.3%). M.p. 77° (lit.[48,49] m.p. 77°).

Method (ii)

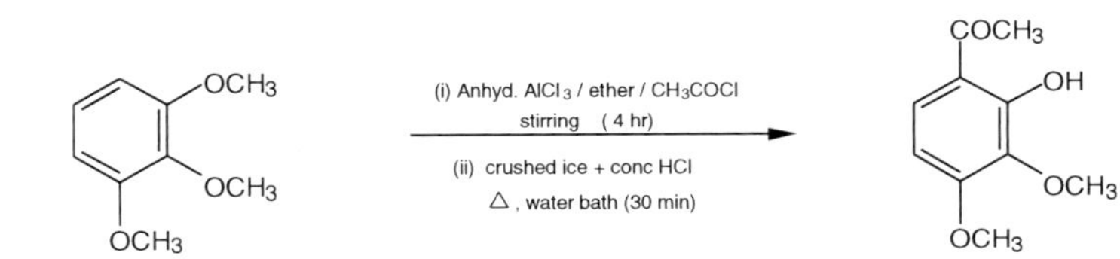

1,2,3-Trimethoxybenzene 2-Hydroxy-3,4-dimethoxy acetophenone

A solution of 1,2,3-trimethoxy benzene (pyrogallol trimethyl ether) (1.9) (6.4 g, 0.038 mole) in absolute ether (20 ml) is added to a cold solution of anhydrous aluminum chloride (15 g, 0.112 mole) in absolute ether (50 ml). The cooled (0-5°) and stirred solution is treated with acetyl chloride (6 ml, 6.62 g, 0.084 mole), added dropwise during 1 hr. The mixture is stirred for 4 hr more. The whole reaction is done under anhydrous condition. The mixture is left overnight at room temperature and treated in cold with crushed ice (40 g) and concentrated hydrochloric acid (15 ml). It is heated for 30 min on a water bath, the ether is allowed to distil away. The mixture while hot, is extracted with benzene (4 × 30 ml). The benzene extract is washed with sodium carbonate (5%) and then extracted with sodium hydroxide solution (10%). The clear alkaline solution is acidified in cold with hydrochloric acid. The separated product is filtered, washed with water and dried. It is crystallised from benzene-petroleum ether as colourless fine prisms. Yield 6.5 g (87%). M.p.77° (lit.[49] m.p. 77°).

4.A.38 2,3,4-TRIMETHOXYACETOPHENONE (GALLACETOPHENONE TRIMETHYL ETHER)

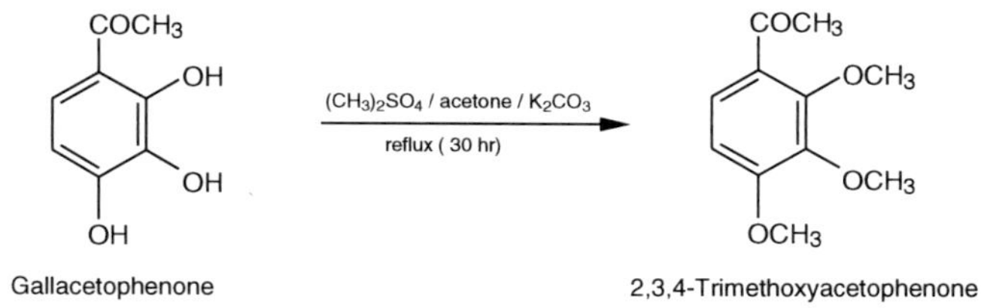

A solution of gallacetophenone (4.A.35.) (12.5 g, 0.074 mole) in dry acetone (150 ml) is refluxed for 30 hr with dimethyl sulphate (35 ml, 46.65 g, 0.37 mole) and ignited potassium carbonate (80 g) under anhydrous conditions. The acetone is filtered and the inorganic salts washed with more hot acetone (2 × 30 ml). The combined acetone solution is distilled, oily residue macerated with crushed ice and extracted with ether (2 × 50 ml). The ether solution is washed with water, dried (anhydrous sodium sulphate) and distilled. Yield 11g (70.5%). B.p. 174°/19 mm (lit.[50] b.p. 174°/19 mm).

4.A.39 2,3,6-TRIMETHOXYACETOPHENONE

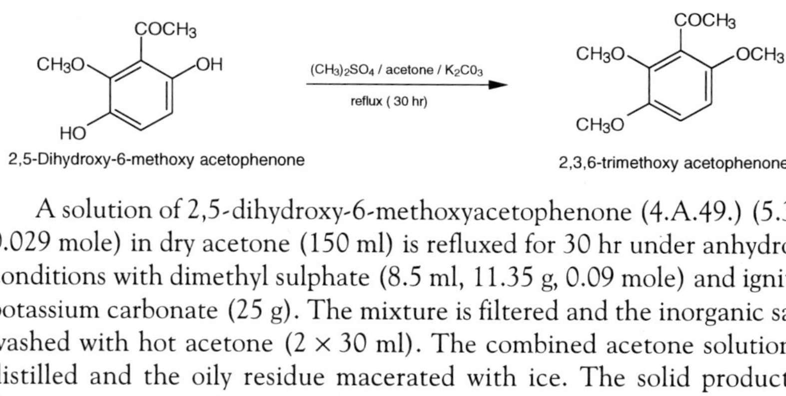

A solution of 2,5-dihydroxy-6-methoxyacetophenone (4.A.49.) (5.3 g, 0.029 mole) in dry acetone (150 ml) is refluxed for 30 hr under anhydrous conditions with dimethyl sulphate (8.5 ml, 11.35 g, 0.09 mole) and ignited potassium carbonate (25 g). The mixture is filtered and the inorganic salts washed with hot acetone (2 × 30 ml). The combined acetone solution is distilled and the oily residue macerated with ice. The solid product is filtered and washed with cold water and dried. It is crystallised from petroleum ether as colourless needles. Yield 4.8 g (78.4%). M.p. 41-42°. (lit.[51] m.p 41.5°).

4.A.40 2-HYDROXY-3,6-DIMETHOXYACETOPHENONE

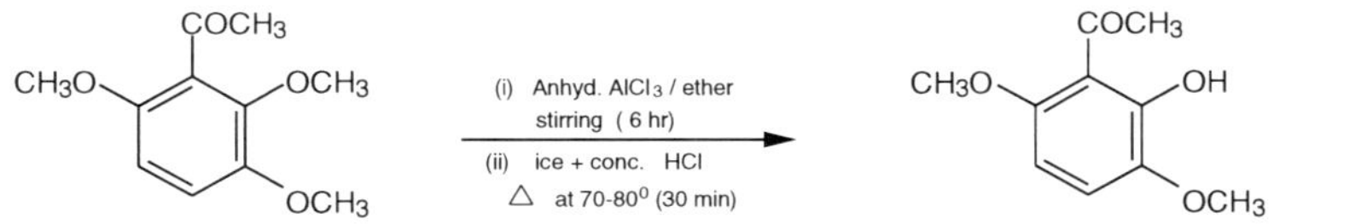

2,3,6-Trimethoxy acetophenone 2-Hydroxy-3,6-dimethoxy acetophenone

To a stirred and cooled solution (0-5°) of anhydrous aluminium chloride (11.4 g, 0.085 mole) in dry ether (50 ml) is added slowly a solution of 2,3,6-trimethoxyacetohphenone (4.A.39.) (6 g, 0.028 mole) in dry ether (50 ml). To this stirred solution is added freshly distilled acetyl chloride (2.5 ml, 2.76 g, 0.035 mole). The stirring is continued for 6 hr (the reaction is done under anhydrous conditions) and the mixture allowed to stand for 48 hr at room temperature with occasional shaking. The aluminium chloride complex is decomposed with ice (150 g) and concentrated hydrochloric acid (12 ml). The mixture is heated for 30 min at 70-80° (water bath). It is cooled and extracted with ether (3 × 50 ml). The ethereal solution is extracted with sodium hydroxide solution (5%, 4 × 25 ml). The clear alkaline solution is acidified with dilute hydrochloric acid and extracted with ether. The ether extract is washed with water, dried (anhydrous sodium sulphate) and distilled. The oily product solidifies on cooling. It is filtered, washed with water and dried. It is crystallised from petroleum ether as lemon yellow plates. Yield 4.8 g (85.7%). M.p. 62-63° (lit.[52] m.p. 61°).

4.A.41 2-HYDROXY-4,5-DIMETHOXYACETOPHENONE

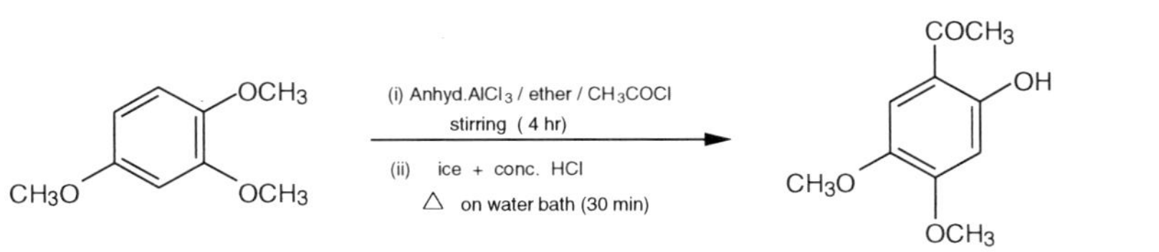

1,2,4-Trimethoxy benzene 2-Hydroxy-4,5-dimethoxy acetophenone

To cooled (0-5°) and stirred solution of anhydrous aluminium chloride (20 g, 0.15 mole) in dry ether (50 ml) is added a solution of 1,2,4-trimethoxybenzene (1.22) (8.4 g, 0.05 mole) in dry ether (20 ml). To this stirred solution is added freshly distilled acetyl chloride (5 ml, 5.52 g, 0.07 mole) in dry ether (20 ml) dropwise during 1 hr under anhydrous conditions. The stirring is continued for 4 hr more. During this time a greenish aluminium complex separates initially as an oil which finally solidifies. The mixture is left overnight at room temperature and treated with crushed ice (150 g) and concentrated hydrochloric acid (20 ml). The mixture is heated on a water bath for 30 min, the ether is allowed to distil away. While hot, the mixture is extracted with benzene (2 × 75 ml) and the benzene layer washed with dilute hydrochloric acid and then extracted with sodium hydroxide solution (10%, 2 × 50 ml). The cooled and clear alkaline solution is acidified with dilute hydrochloric acid. The separated product is filtered and washed with water. It is crystallised from alcohol as colourless needles. Yield 8 g (81.6%). M.p.110-11° (lit.[53] m.p. 112°).

4.A.42 2,4,5-TRIMETHOXYACETOPHENONE

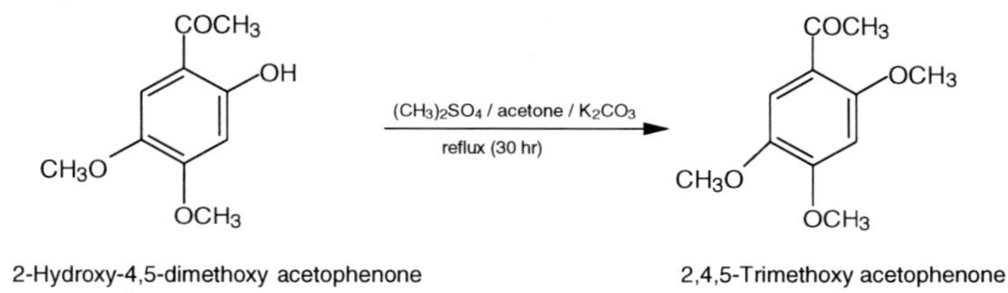

2-Hydroxy-4,5-dimethoxyacetophenone (4.A.41) (5 g, 0.0255 mole) in dry acetone (100 ml) is refluxed for 30 hr under anhydrous conditions with dimethyl sulphate (2.8 ml, 3.78 g, 0.03 mole) and ignited potassium carbonate (20 g). The mixture is filtered and the inorganic residue washed with hot acetone (2 × 20 ml). The combined acetone solution is distilled and the oily residue macerated with crushed ice. The solid product is filtered, washed with water and dried. It is crystallised from benzene as colourless needles. Yield 4.5 g (84.1%). M.p. 97-9° (lit.[54] m.p. 102°).

4.A.43 2,5-DIHYDROXY-4-METHOXYACETOPHENONE

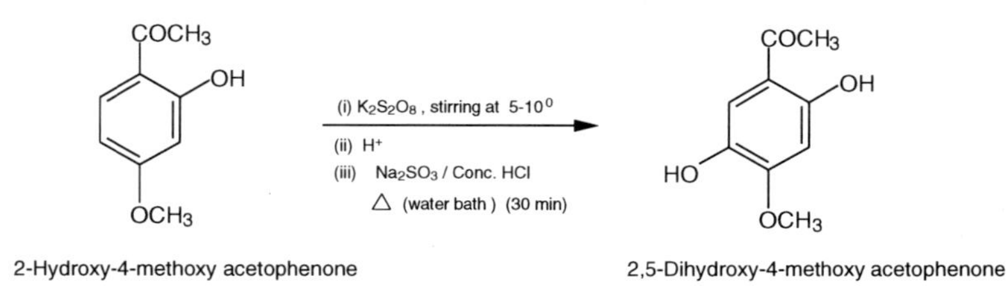

A solution of potassium persulphate (7.5 g 0.0277 mole) in 250 ml water, is added dropwise during 2 hr to a stirred and cooled solution (5-10°) of 2-hydroxy-4-methoxyacetophenone (4.A.7) (4 g, 0.024 mole) in aqueous potassium hydroxide (7.5 g in 60 ml water). The reaction mixture is stirred for 3 hr more and left overnight at room temperature. The cooled mixture is acidified to congo red with dilute hydrochloric acid (1:1) and the separated unreacted ketone is removed by filtration. The filtrate is extracted with ether (2 × 40 ml) to remove completely the starting ketone. It is heated for 30 min in a boiling water bath with sodium sulphite (4 g) and concentrated hydrochloric acid (30 ml). The mixture is cooled and the separated product filtered and washed with cold water. It is crystallised from hot water as pale brown elongated prisms. Yield 1.5 g (34.2%). M.p. 164-66° (lit.[55] m.p. 164-66°).

On partial methylation with 1.1equivalent of dimethyl sulphate in anhydrous acetone in the presence of ignited potassium carbonate (4 hr refluxing) and usual working up is obtained the partial methyl ether, viz. 2-hydroxy-4,5-dimethoxyacetophenone, m.p. 110-11°, identical with the product described earlier (4.A.41).

4.A.44 2,4,6-TRIHYDROXYACETOPHENONE (PHLOROACETOPHENONE)

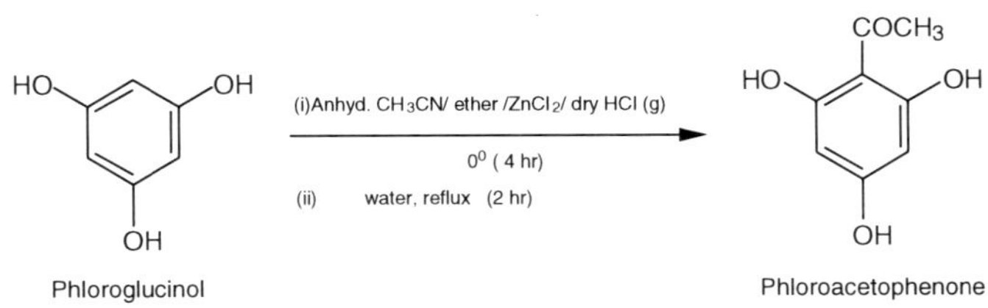

To a solution of dry phloroglucinol (20 g, 0.159 mole) and anhydrous acetonitrile (10 g, 12.72, ml 0.244 mole) in dry ether (80 ml) is added freshly fused and powdered zinc chloride (4 g, 0.029 mole). The mixture is cooled to 0° and a rapid stream of dry hydrogen chloride gas is passed for 4 hr. The mixture is left overnight in ice-chest, the ether solution decanted and the solid ketimine hydrochloride washed with dry ether (2 × 30 ml). The solid product is refluxed for 2 hr with water (200 ml). A small amount decolourising carbon is added to the solution, boiled for 5 min more and filtered under suction while hot. The solution is cooled. The separated product is filtered and dried at 120°. Yield 22 g (82.5%). M.p. 219° (lit.[56,57] m.p. 219°).

4.A.45 2-HYDROXY-4,6-DIMETHOXYACETOPHENONE (PHLOROACETOPHENONE DIMETHYL ETHER)

Method (i)

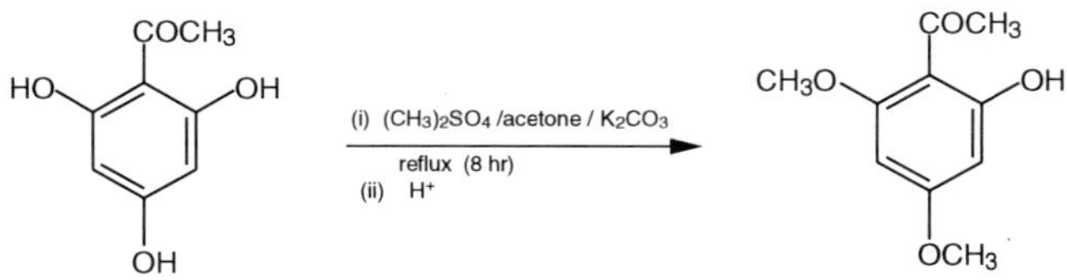

Phloroacetophenone 2-Hydroxy-4, 6-dimethoxyacetophenone

A solution of dry phloroacetophenone (4.A.44) (4.2 g, 0.025 mole) in dry acetone (75 ml) is refluxed for 8 hr under anhydrous conditions with dimethyl sulphate (6.5 ml, 8.7 g, 0.069 mole) and ignited potassium carbonate (15 g) under anhydrous condition. The mixture is filtered and the inorganic salts washed with hot acetone (2 × 30 ml). The combined acetone solution is distilled and the oily residue macerated with crushed ice. It is extracted with ether (2 × 50 ml) and the ethereal solution extracted with sodium hydroxide solution (5%, 2 × 50 ml). The clear aqueous alkaline solution is acidified in cold with dilute hydrochloric acid. The separated product is filtered and washed with water. It is crystallised from alcohol as colourless needles. Yield 3.5 g(71.4%). M.p. 81-2°. (lit.[49]m.p. 81-2°).

Method (ii)[16]

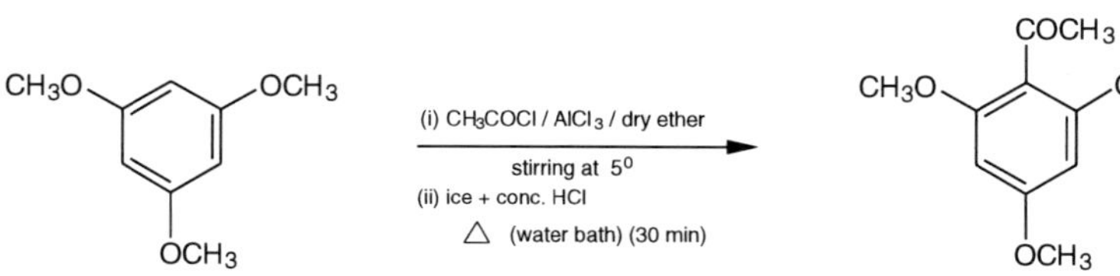

Phloroglucinol trimethyl ether 2-Hydroxy-4,6-dimethoxy acetophenone

To a cooled (5°) and stirred solution of anhydrous aluminium chloride (9 g, 0.067 mole) in dry ether (50 ml) is added slowly a solution of phloroglucinol trimethyl ether (1.14.) (5 g, 0.03 mole) under anhydrous conditions. To this mixture, freshly distilled acetyl chloride (2.5 ml, 2.76 g, 0.035 mole) in dry ether (10 ml) is added dropwise during 1 hr and stirring continued for 4 hr more. The mixture is left for 48 hr at room temperature.

The separated oily aluminium chloride complex is decomposed with ice (100 g) and concentrated hydrochloric acid (20 ml) and the mixture heated in a water bath for 30 min; the ether is allowed to distil away. The cooled solution is extracted with ether (3 × 50 ml). The ether extract is extracted with dilute alkali (5%, 3 × 50 ml). The clear alkaline solution is acidified with hydrochloric acid. The separated product is filtered and washed with water. It is crystallised from alcohol as colourless needles. Yield 5 g (85.8%). M.p. 81-82°.

4.A.46 2,4,6-TRIMETHOXYACETOPHENONE (PHLOROACETOPHENONE TRIMETHYL ETHER)

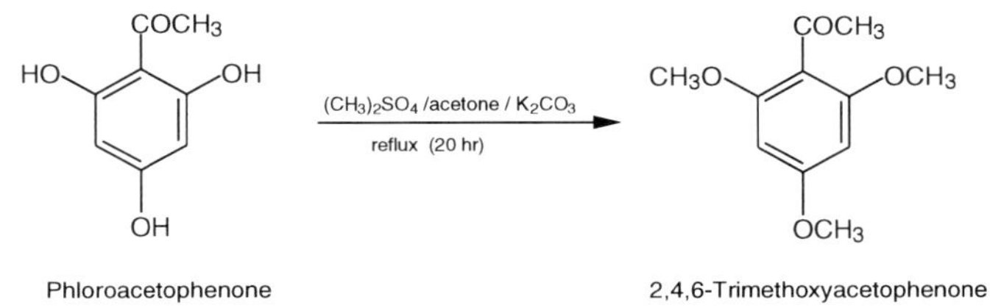

Phloroacetophenone (4.A.44) (5 g, 0.03 mole) is dissolved in dry acetone (100 ml) and refluxed with excess of dimethyl sulphate (15 ml, 20 g, 0.159 mole) in presence of ignited potassium carbonate (35 g) for 20 hr under anhydrous conditions. The hot mixture is filtered and the inorganic residue washed with hot acetone (2 × 50 ml). The combined acetone solution is distilled and the residue macerated with crushed ice. The separated product is filtered and washed with water. It is crystallised from alcohol as colourless needles. Yield 4.5 g (72%). M.p.102-103°. (lit.[58] m.p. 103°).

4.A.47 2,4-DIHYDROXY-6-METHOXYACETOPHENONE (PHLOROACETOPHENONE-6-0-METHYL ETHER)

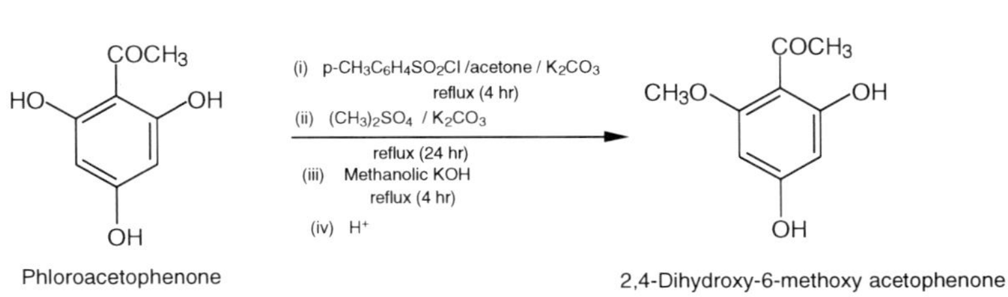

Phloroacetophenone (4.A.44) (16.8 g, 0.1 mole) is refluxed for 4 hr under anhydrous condition with p-toluenesulphonyl choloride (38.2 g, 0.2 mole) in dry acetone (400 ml) in presence of anhydrous potassium carbonate (60 g). To this mixture is added dimethyl sulphate (13 ml, 17.3 g, 0.137 mole) and anhydrous potassium carbonate (30 g) and refluxing is continued for 24 hr. The reaction mixture is filtered and the inorganic residue washed with hot acetone (2 × 50 ml). The combined acetone solution is distilled and the syrupy product is refluxed for 4 hr in methanolic potassium hydroxide (5%, 350 ml). The methanol is distilled under reduced pressure, the residue diluted with water and acidified in cold with hydrochloric acid. The separated solid product is filtered. It is purified by dissolving in sodium hydroxide solution (5%) and acidification of the clear alkaline solution. Yield 9 g (49.5%). M.p. 203-204° (lit.[59] m.p. 203-204°).

4.A.48 2,6-DIHYDROXY-4-METHOXYACETOPHENONE (PHLOROACETOPHENONE-4-0- METHYL ETHER)

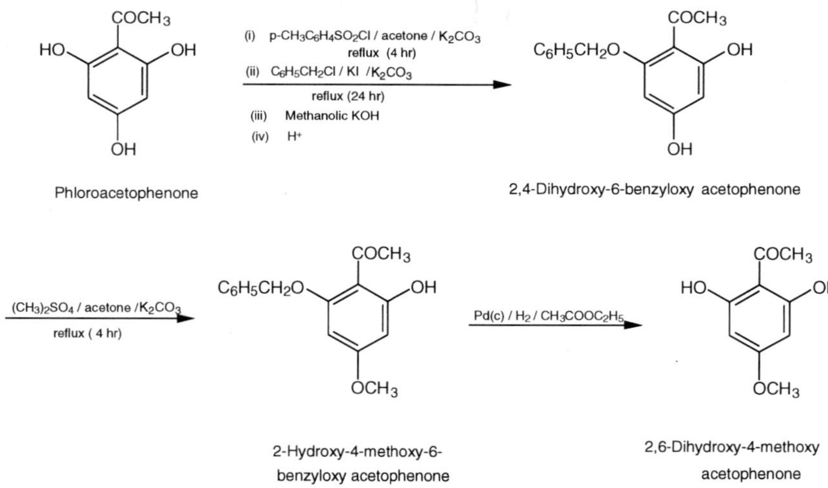

Step (i) 2,4-Dihydroxy-6-benzyloxyacetophenone[60]

Phloroacetophenone (4.A.44.) (8.4 g, 0.05 mole) is refluxed with p-toluenesulphonyl chloride (19.5 g, 0.102 mole) in acetone (200 ml) in the presence of ignited potassium carbonate (30 g) for 4 hr under anhydrous condition. To the reaction mixture is added benzyl chloride (8 ml, 8.8 g,

0.0695 mole), potassium iodide (12 g, 0.072 mole) and anhydrous potassium carbonate (15 g) and refluxing is continued for 24 hr more. The reaction mixture is filtered and the inorganic salts washed with hot acetone (2 × 50 ml). The combined acetone solution is distilled and the residual syrupy product is refluxed for 4 hr with methanolic potassium hydroxide (5%, 175 ml). The methanol is removed in vacuo, the residue diluted with water and acidified in cold with hydrochloric acid. The separated product is filtered. It is purified by dissolving in aqueous sodium hydroxide (5%) and acidifying clear alkaline solution. The product is filtered and washed with water. It is crystallised from methanol as colourless shining needles. Yield 6 g (46.5%). M.p. 225-26°.

Step (ii) 2-Hydroxy-4-methoxy-6-benzyloxyacetophenone[60]

To a solution of 2,4-dihydroxy-6-benzyloxyacetophenone (5 g, 0.019 mole) in dry acetone (100 ml) is added dimethyl sulphate (1.89 ml, 2.5 g, 0.02 mole) and ignited potassium carbonate (10 g). The mixture is refluxed for 4 hr under anhydrous conditions. It is filtered and the inorganic salts washed with hot acetone (2 × 20 ml). The combined acetone solution is distilled. The residue is macerated with crushed ice, filtered, washed with water and dried. It is crystallised from benzene-petroleum ether as cream coloured needles. Yield 4 g (75.9%). M.p. 110-12°.

Step (iii) 2,6-Dihydroxy-4-methoxyacetophenone[61]

Palladised charcoal (5%, 0.5 g) in ethyl acetate (30 ml) is saturated with hydrogen gas with stirring. A solution of 2-hydroxy-4-methoxy-6-benzyloxyacetophenone (5 g, 0.018 mole) in ethyl acetate (40 ml) is then added and the hydrogenation continued at atmospheric pressure till the absorption of hydrogen ceases. The reaction mixture is filtered and the solvent is distilled to give the required product. Yield 2.5 g (75.8%). M.p. 148-49°.

4.A.49 2,5-DIHYDROXY-6-METHOXYACETOPHENONE

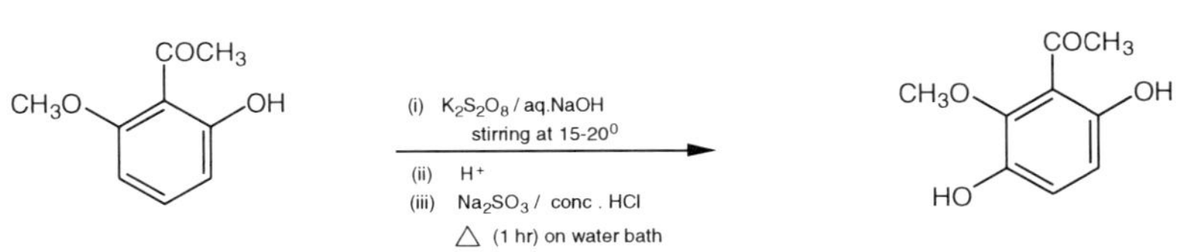

2-Hydroxy-6-methoxy acetophenone

2,5-Dihydroxy-6-methoxyacetophenone

To a solution of 2-hydroxy-6-methoxyacetophenone (4.A.17) (12.5 g, 0.075 mole) in aqueous sodium hydroxide (10%, 150 ml) is added dropwise during 4 hr a solution of potassium persulphate (20.5 g, 0.075 mole in 400 ml water), the temperature during addition is maintained at 15-20°. The mixture is stirred for 1 hr more and allowed to stand overnight at room temperature. The solution is cooled and acidified to congo red with hydrochloric acid. The separated starting ketone is filtered and the filtrate extracted with ether (2 × 100 ml). The remaining solution is heated for 1hr in a water bath with sodium sulphite (3 g) and concentrated hydrochloric acid (150 ml). The solution is cooled and the separated product filtered, washed with water and dried. It is crystallised from petroleum ether as long yellow prisms. Yield 3 g (21.9%). M.p. 90-91° (lit.[23] m.p. 90°). Some more of the product (1g) is obtained by ether extraction of the aqueous mother liquor.

4.A.50 2-HYDROXY-5,6-DIMETHOXYACETOPHENONE

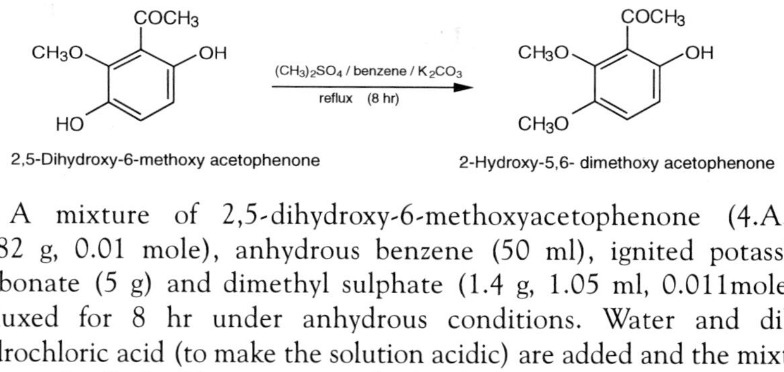

2,5-Dihydroxy-6-methoxy acetophenone — 2-Hydroxy-5,6- dimethoxy acetophenone

A mixture of 2,5-dihydroxy-6-methoxyacetophenone (4.A.49) (1.82 g, 0.01 mole), anhydrous benzene (50 ml), ignited potassium carbonate (5 g) and dimethyl sulphate (1.4 g, 1.05 ml, 0.011mole) is refluxed for 8 hr under anhydrous conditions. Water and dilute hydrochloric acid (to make the solution acidic) are added and the mixture steam distilled. The steam volatile fraction gave 2-hydroxy-5,6-dimethoxyacetophenone as bright yellow oil. Yield 1.3 g (66.3%). B.p. 162-63°/22 mm (lit.[23] b.p. 162-63°/22 mm).

4.A.51 2,5-DIHYDROXY-3,4-DIMETHOXYACETOPHENONE

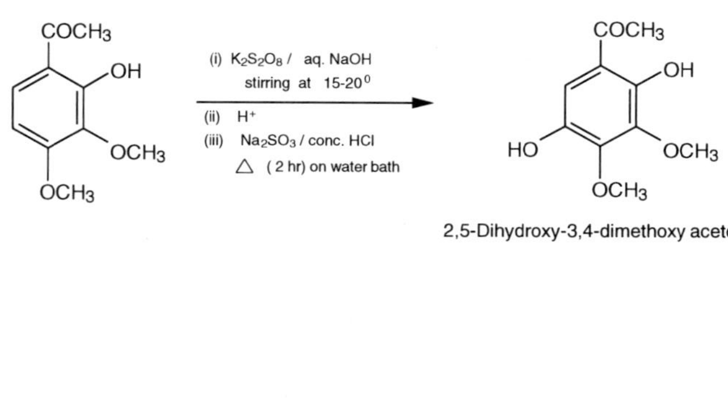

2,5-Dihydroxy-3,4-dimethoxy acetophenone

To a stirred solution of 2-hydroxy-3,4-dimethoxyacetophenone (4.A.37) (10 g, 0.051 mole) in aqueous sodium hydroxide(15%, 100 ml) is added dropwise during 4 hr, a solution of potassium persulphate (20 g, 0.074 mole in 400 ml water); the temperature during addition is maintained at 15-20°. After keeping the reaction mixture for 48 hr at room temperature the cooled solution is acidified to congo red with hydrochloric acid. The separated starting compound is filtered and the filtrate extracted with ether (2 × 100 ml). The remaining aqueous solution is heated for 2 hr on boiling water bath with sodium sulphite (3 g) and concentrated hydrochloric acid (60 ml). The hot solution is charcoalised and filtered. The solution is cooled and the separated product filtered. Yield 3 g (27.8%) m.p. 118° (lit.[62] m.p. 119-21°). Some more of the required product (0.5 g) is obtained by ether extraction of the aqueous mother liquor.

4.A.52 2-HYDROXY-3,4,5-TETRIMETHOXYACETOPHENONE

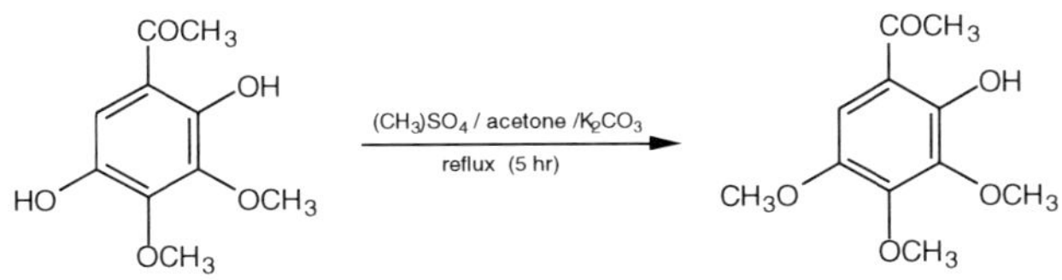

2,5-Dihydroxy-3,4-dimethoxy acetophenone 2-Hydroxy-3,4,5-trimethoxy acetophenone

A solution of 2,5-dihydroxy-3,4-dimethoxyacetophenone (4.A.51) (1 g, 0.0047 mole) in dry acetone (50 ml) is refluxed for 5 hr with dimethyl sulphate (0.5 ml, 0.67 g, 0.0053 mole) and ignited potassium carbonate (5 g) under anhydrous conditions. The reaction mixture is filtered and the inorganic salts washed with hot acetone (2 × 15 ml). The combined acetone solution is distilled and the residue macerated with crushed ice. The separated product is filtered and purified by dissolving in sodium hydroxide solution (5%) and acidifying the clear alkaline solution. The product obtained is filtered, washed with water and dried. It is crystallised from petroleum ether as colourless needles. Yield 0.8 g (74.8%). M.p. 85-86° (lit.[62,63] m.p .86°).

4.A.53 2,3,4,5-TETRAMETHOXYACETOPHENONE

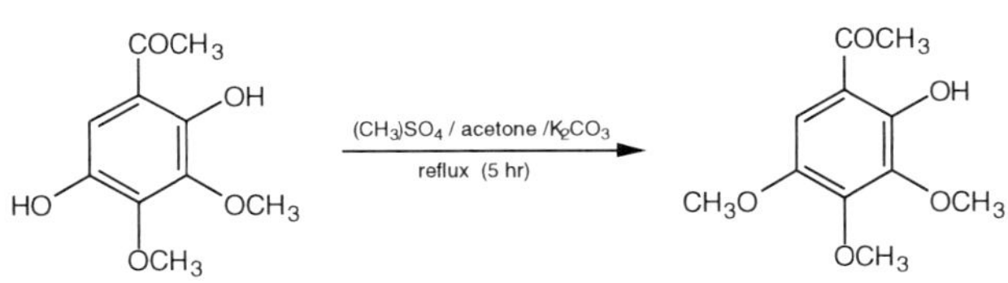

2,5-Dihydroxy-3,4-dimethoxy acetophenone 2-Hydroxy-3,4,5-trimethoxy acetophenone

2,5-Dihydroxy-3,4-dimethoxyacetophenone (4.A.51) (3.5 g, 0.0165 mole) in dry acetone (75 ml) is refluxed for 24 hr under anhydrous conditions with dimethyl sulphate (5 ml, 6.67 g, 0.053 mole) and ignited potassium carbonate (15 g). The acetone is filtered and the inorganic salts washed with hot acetone (2 × 15 ml). The combined acetone solution is distilled and the oily residue macerated with crushed ice. It is extracted with ether, the extract dried (anhydrous sodium sulphate) and distilled. Yield 3 g (75.8%). B.p. 102°/15 mm (lit.[64] b.p. 102°/15 mm) (lit.[6] m.p. 24-5°).

4.A.54 2-HYDROXY-3,4,6-TRIMETHOXYACETOPHENONE

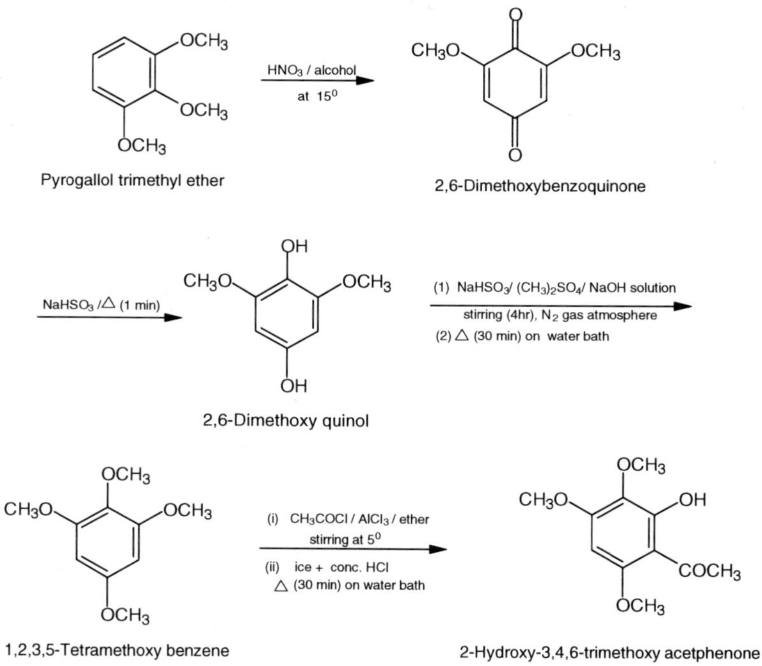

Step (i) 2,6-Dimethoxybenzoquinone

A solution of pyrogallol trimethyl ether (1.9) (31 g, 0.184 mole) in alcohol (155 ml) is cooled to 15° and treated with nitric acid (155 ml, d, 1.2). The temperature of the reaction is raised to 35° (water bath) when a vigorous reaction sets in. The temperature is controlled at this stage so that it remains below 50°. After the reaction subsides, the reaction mixture is left at room temperature for 4 hr with occasional shaking. The yellow crystals of the compounds are filtered and treated with benzene (50 ml) so as to remove the nitro compound formed during oxidation. The remaining yellow needles are pure 2,6-dimethoxybenzoquinone. Yield 26 g (83.9%). M.p. 252-53° (lit.[62] m.p. 252-53°).

Step (ii) 2,6-Dimethoxyquinol

To a suspension of 2,6-dimethoxybenzoquinone (30 g, 0.178 mole) in boiling water (300 ml) is added sodium hydrosulphite (60 g, 0.344 mole) and the mixture heated for 1 min. The resulting deep red solution is cooled rapidly and kept aside for 3 hr. The separated product is filtered, washed with a little ice-cold water. It is crystallised from hot water as colourless solid. Yield 25 g (82.4%). M.p. 166-67° (lit.[62] m.p. 166-67°).

Step (iii) 1,2,3,5,-Tetramethoxybenzene

To a continuously stirred mixture of 2,6-dimethoxyquinol (11 g, 0.065 mole) in alcohol (26 ml), is added sodium hydrosulphite (1 g). Dimethyl sulphate (20 ml, 26.66 g, 0.21 mole) is added in an atmosphere of hydrogen gas. Sodium hydroxide solution (16 g in 36 ml water) is then added dropwise during 4 hr to the stirred mixture and the reaction mixture kept at 15-20°. The mixture is heated on a water bath for 30 min with frequent shaking. It is diluted with water (100 ml) and cooled. The solid obtained is filtered, washed with water and dried. Yield 11 g (85.9%). M.p. 38-40° (lit.[62] m.p. 38-40°).

Step (iv) 2-Hydroxy-3,4,6-trimethoxyacetophenone

To a well cooled and stirred solution of anhydrous aluminium chloride (50 g, 0.375 mole) in dry ether (200 ml) is added 1,2,3,5-tetramethoxybenzene (50 g, 0.25 mole) in dry ether (50 ml). To the above stirred mixture at 5° is added freshly distilled acetyl chloride (25 ml, 27.6 g, 0.35 mole) in dry ether (40 ml) dropwise during 1 hr and the stirring is continued for 6 hr. During this time a greenish aluminium chloride complex is separated initially as an oil but it finally solidifies. The reaction mixture is left overnight at room temperature and is treated cautiously with ice (100 g) and concentrated hydrochloric acid (50 ml). The complex is hydrolysed by heating on a water bath for 30 min, ether is allowed to distil away. While hot, the aqueous solution is extracted with benzene (200 ml). The benzene layer is washed with dilute hydrochloric acid and then extracted with sodium hydroxide solution (10%). The clear alkaline solution is acidified with hydrochloric acid. The separated product is filtered and washed with water. It is crystallised from alcohol as light fawn coloured rectangular prisms. Yield 45 g (78.9%). M.p. 104-5° (lit.[62] m.p. 104-105°).

4.A.55 2,3,4,6-TETRAMETHOXYACETOPHENONE

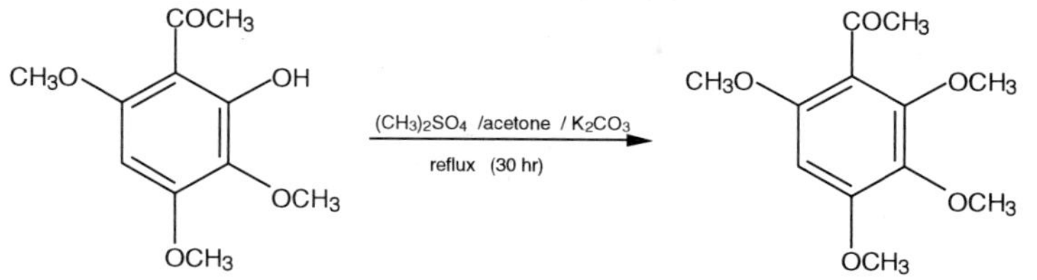

2-Hydroxy-3,4,6-trimethoxy acetophenone — 2,3,4,6-Tetramethoxy acetophenone

2-Hydroxy-3,4,5-trimethoxyacetophenone (4.A.52) (5 g, 0.022 mole) in dry acetone (75 ml) is refluxed for 30 hr with dimethyl sulphate (3.5 g, 2.63 ml, 0.028 mole) and ignited potassium carbonate (10 g). The mixture is filtered and the inorganic salts washed with hot acetone (2 × 15 ml). The combined acetone solution is distilled and the oily product macerated with crushed ice. The mixture is extracted with ether (2 × 50 ml). The ether extract is washed with sodium hydroxide solution (5%), water, dried (sodium sulphate) and distilled. Yield 4 g (75.5%). B.p. 310°. The product solidifies, m.p. 53-4° (lit.[65,66] m.p. 53-4°, b.p. 310°).

4.A.56 2,4-DIHYDROXY-3,6-DIMETHOXYACETOPHENONE

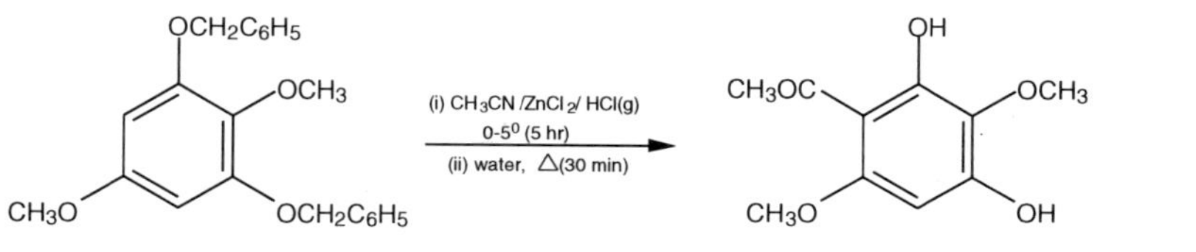

1,3-Dibenzyloxy-2,5-dimethoxy benzene — 2,4-Dihydroxy-3,6-dimethoxy acetophenone

To 1,3-dibenzyloxy-2,5-dimethoxybenzene (4.B.80) (3.5 g, 0.01 mole) in dry ether (50 ml) is added freshly distilled acetonitrile (1.5 g, 1.91 ml, 0.0365 mole) and freshly fused and powdered zinc chloride (1.5 g, 0.0109 mole). The mixture is cooled (0-5°) and a rapid stream of dry hydrogen chloride gas is passed under anhydrous condition for 5 hr. The mixture is left for 48 hr at room temperature. The clear ether is decanted and the solid ketimine hydrochloride washed with anhydrous ether (2 × 30 ml). The residual product is dissolved in water (30 ml) and heated on a water bath for 30 min. The solution on cooling gave a solid product. It is filtered and washed with cold water. It is crystallised from hot water as colourless long needles. Yield 1.8 g (85%). M.p. 128° (lit.[67] m.p. 129°).

4.A.57 2,5-DIHYDROXY-4,6-DIMETHOXYACETOPHENONE

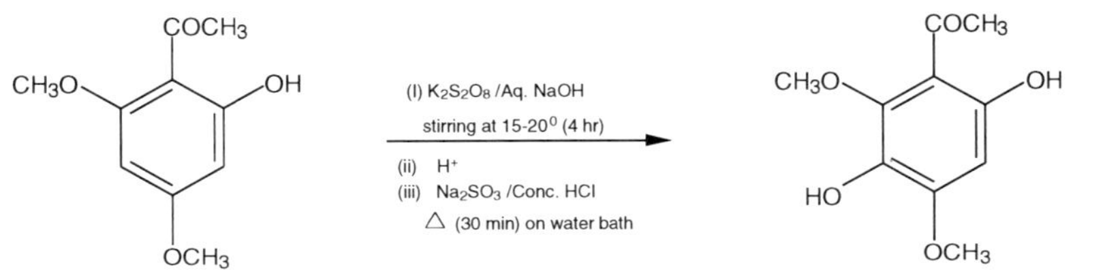

2-Hydroxy-4,6-dimethoxy acetophenone 2,5-Dihydroxy-4-6-dimethoxy acetophenone

A solution of potassium persulphate (11 g, 0.0405 mole) in water (200 ml) is added dropwise during 4 hr to a cooled (15-20°) and stirred solution of 2-hydroxy-4, 6-dimethoxyacetophenone (4.A.45) (7.5 g, 0.038 mole) in aqueous sodium hydroxide (10 g in 100 ml water). The reaction mixture is stirred for 1 hr more and allowed to stand for 24 hr at room temperature. The dark reddish brown reaction mixture is cooled and acidified to congo red with hydrochloric acid. The separated unreacted ketone is filtered and the filtrate extracted with ether (2 × 50 ml). The clear aqueous filtrate is treated with sodium sulphite (3 g) and animal charcoal (5-6 g), mixture stirred for 5-10 min and filtered again. The filtrate is treated with concentrated hydrochloric acid (80 ml) and heated for 30 min on a boiling water bath. The solution is cooled and the separated product is filtered and washed with water. It is crystallised from aqueous alcohol as rhombic yellow prisms. Yield 2.5 g (30.8%). M.p. 163-64° (lit.[49] m.p. 164-65°). Some more product about (1 g) is obtained by ether extraction of the combined mother liquor.

4.A.58 2,5-DIHYDROXY-3,4,6-TRIMETHOXYACETOPHENONE

To a stirred solution of 2-hydroxy-3,4,6-trimethoxyacetophenone (4.A.54) (5.65 g, 0.025 mole) in aqueous sodium hydroxide solution (10 g in 70 ml water) at 15-20° is added dropwise a solution of potassium persulphate

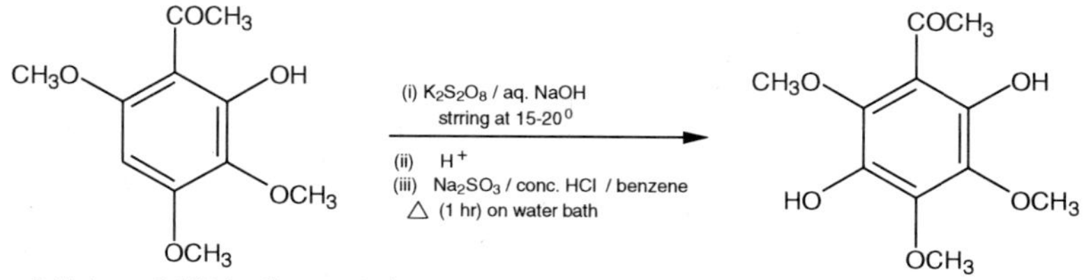

2-Hydroxy- 3,4,6-trimethoxyacetophenone   2,5-Dihydroxy- 3,4,6-trimethoxy acetophenone

(8.8 g, 0.0325 mole in 160 ml water) during 4 hr. The stirring is continued for 2 hr more and the reaction mixture allowed to stand overnight at room temperature. It is cooled and acidified to congo red with dilute hydrochloric acid. The separated unreacted starting ketone is filtered and the filtrate is extracted with ether (2 × 50 ml). The remaining aqueous solution is heated for 1 hr in a boiling water bath with benzene (150 ml), sodium sulphite (2.0 g) and concentrated hydrochloric acid (75 ml). The organic layer is separated, dried (anhydrous sodium sulphate) and distilled. The residual product is crystallised from hot water as yellow needles. Yield 1.6 g (26.4%). M.p. 116-17° (lit.[48] m.p. 116-17°).

4.A.59 2-HYDROXY-4,5,6-TRIMETHOXYACETOPHENONE

Method (i)

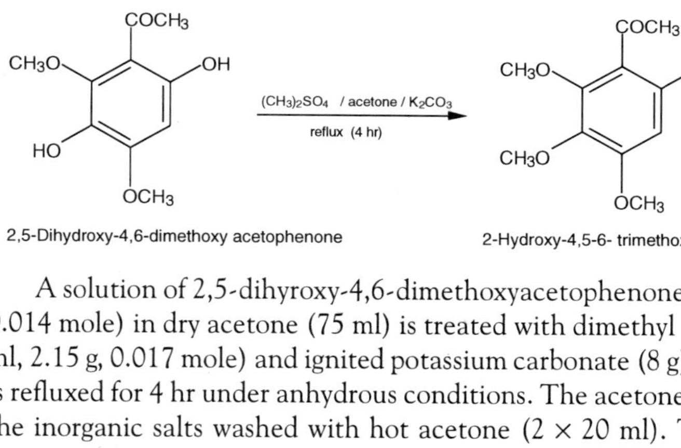

2,5-Dihydroxy-4,6-dimethoxy acetophenone   2-Hydroxy-4,5-6- trimethoxy acetophenone

A solution of 2,5-dihyroxy-4,6-dimethoxyacetophenone (4.A.57) (3 g, 0.014 mole) in dry acetone (75 ml) is treated with dimethyl sulphate (1.61 ml, 2.15 g, 0.017 mole) and ignited potassium carbonate (8 g). The mixture is refluxed for 4 hr under anhydrous conditions. The acetone is filtered and the inorganic salts washed with hot acetone (2 × 20 ml). The combined acetone solution is distilled and the residue macerated with crushed ice. The separated product is filtered and purified by dissolving in aqueous sodium hydroxide (5%) and acidifying the clear alkaline solution. The required product is obtained by ether extraction as pale yellow oil. Yield 2.5 g (78.2%). It solidifies on cooling. M.p. 46-47° (lit.[49,68] oil).

Method (ii)

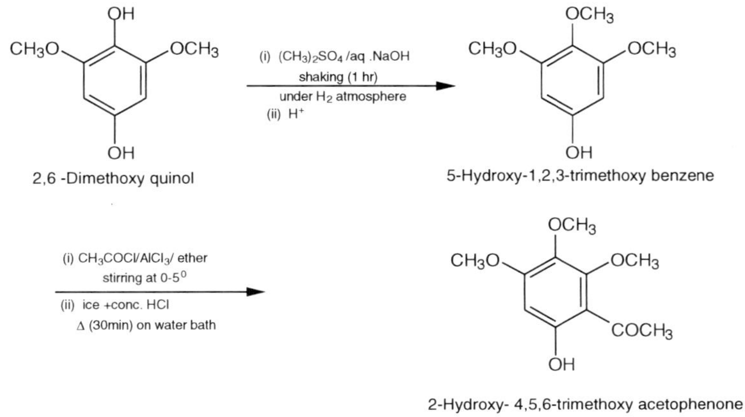

Step (i) 5-Hydroxy-1,2,3-Trimethoxybenzene (antiarol)

A solution of sodium hydroxide (20 g in 140 ml water) is added to 2,6-dimethoxyquinol (4.A.54) (34 g, 0.2 mole) in a R.B. flask from which air has been excluded by a current of hydrogen. Dimethyl sulphate (32 g, 24 ml, 0.25 mole) is then added in one lot and shaking continued for 1 hr. The mixture is extracted with ether and the ether solution extracted with sodium hydroxide solution (5%). The clear alkaline extract is acidified in cold with hydrochloric acid. The separated product is filtered. It is crystallised from hot water as colourless needles. Yield 20 g (54.3%). M.p. 146° (lit.[68] m.p. 148°).

Step (ii) 2-Hydroxy-4,5,6-trimethoxyacetophenone

A solution of the above 5-hydroxy-1,2,3-trimethoxybenzene (antiarol) (3 g, 0.016 mole) is added in one lot to a cooled (0-5°) solution of anhydrous aluminium chloride (5 g, 0.0375 mole) in dry ether (50 ml). To the above vigorously stirred and cooled mixture is added under anhydrous conditions, freshly distilled acetyl chloride (3 ml, 3.3 g, 0.042 mole). The acetyl chloride is added during 1 hr and the mixture stirred in cold for 6 hr. more. The reaction mixture is left overnight at room temperature. It is treated with crushed ice (20 g) and concentrated hydrochloric acid (5 ml) and the solution heated in a water bath for 30 min, the ether is allowed to distil away. The cooled solution is extracted with ether (2 × 50 ml). The ether solution is washed with water, aqueous sodium carbonate solution (5%) and extracted with sodium hydroxide solution (10%, 3 × 20 ml). The clear alkaline extract is acidified with dilute hydrochloric acid. The oily product is extracted with ether (2 × 50 ml), ether extract washed with water, dried (anhydrous sodium sulphate) and distilled. The oily product is purified by column chromatography over silica gel (2 × 30 cm) and eluted with benzene petroleum ether (1:4). The required product is obtained by distillation of eluate. Yield 1.2 g (32.6%). M.p. 46-47° (lit.[49,68] oil).

4.A.60 2-HYDROXY-3,4,5,6-TETRAMETHOXYACETOPHENONE

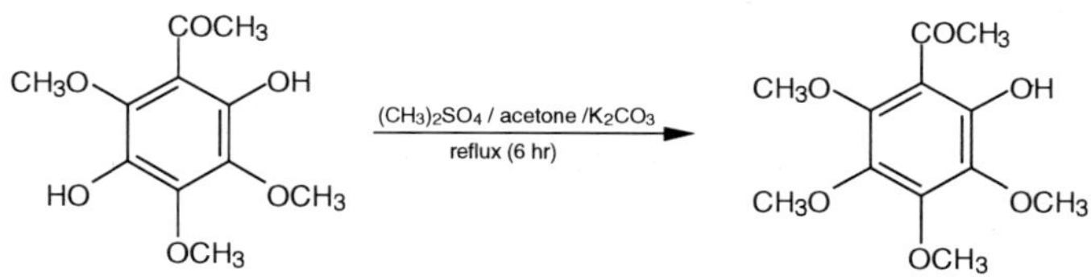

2,5-Dihydroxy- 3,4,6-trimethoxy acetophenone 2-Hydroxy -3,4,5,6-tetramethoxy acetophenone

A solution of 2,5-dihydroxy-3,4,6-trimethoxyacetophenone (4.A.58) (2 g, 0.008 mole) in dry acetone (50 ml) is refluxed for 6 hr under anhydrous condition with dimethyl sulphate (0.88 ml, 1.17 g, 0.009 mole) in presence of ignited potassium carbonate (5 g). The acetone is filtered and inorganic salts washed with hot acetone (2 × 15 ml). The combined acetone solution is distilled and the residue macerated with ice. It is extracted with ether. Ether extract is washed with water, dried (anhydrous sodium sulphate) and distilled. Yield 1.8 g (85.7%). B.p.148°/14 mm (lit.[69] b.p. 148°/14 mm).

4.A.61 2,3,4,5,6-PENTAMETHOXYACETOPHENONE

A solution of 2,5-dihydroxy-3,4,6-trimethoxyacetophenone (4.A.58) (3.5 g, 0.0145 mole) in dry acetone (100 ml) is refluxed for 30 hr under anhydrous conditions with dimethyl sulphate (4.5 ml, 6 g, 0.0475 mole) and ignited potassium carbonate (15 g). The acetone is filtered and the

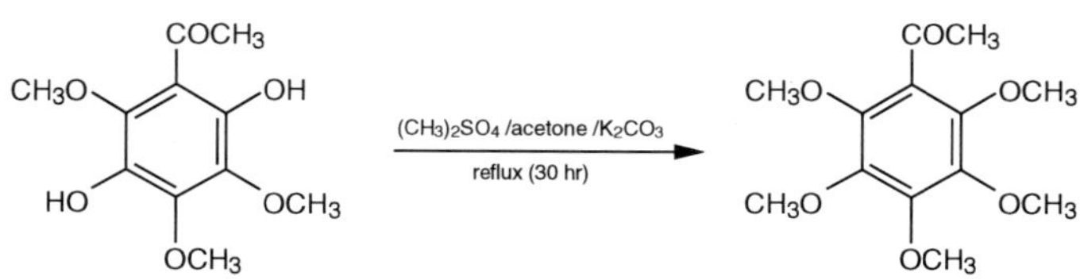

2,5-Dihydroxy-3,4,6-trimethoxyacetophenone 2,3,4,5,6-Pentametoxyacetophenone

inorganic salts washed with hot acetone (2 × 25 ml). The combined acetone solution is distilled and the oily residue macerated with crushed ice. The separated product is filtered, washed with water and dried. It is crystallised from petroleum ether as colourless fine needles. Yield 3.2 g (82%). M.p. 43° (lit.[70] m.p. 43°).

4.A.62 2-HYDROXY-3-METHYL-4,6-DIMETHOXYACETOPHENONE (3-C-METHYL-2-HYDROXY-4,6-DIMETHOXY ACETOPHNONE) (C-METHYLPHLOROACETOPHENONE DIMETHYL ETHER)

Method (i)

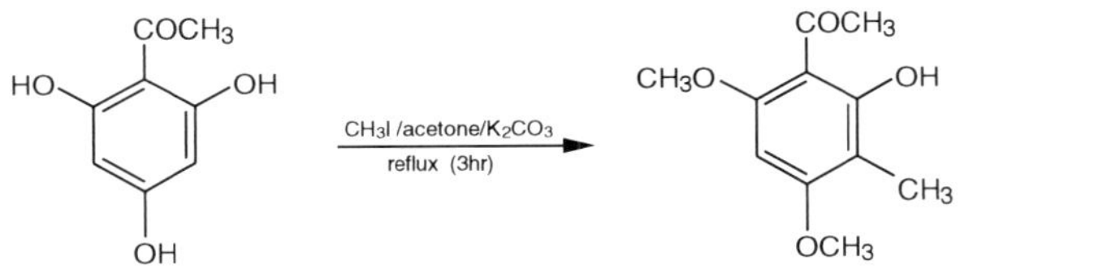

Phloroacetophenone (4.A.44) (3 g, 0.0178 mole) is dissolved in dry acetone (100 ml) and refluxed with methyl iodide (10 ml, 22.8 g, 0.16 mole) and ignited potassium carbonate (15 g) for 3 hr under anhydrous condition. Acetone and excess methyl iodide are distilled off and water added to dissolve the inorganic salts. The undissolved product is filtered and washed with water. It is crystallised thrice from alcohol when 2-hydroxy-3-methyl-4,6-dimethoxyacetophenone is obtained as pale yellow rectangular prisms. Yield 0.7 g (18.66%). M.p. 142-43°. (lit.[71] m.p. 142-43°). The ethanolic mother liquor on concentration and cooling gave 2-hydroxy-4,6-dimethoxyacetophenone (4.A.45) m.p. 82-83°.

Method (ii)

Step (i) 2-Hydroxy-4,6-dimethoxytoluene

A solution of 2-hydroxy-4,6-dimethoxybenzaldehyde (2.16) (2.5 g, 0.0135 mole) in alcohol (30 ml) is added to zinc amalgam dust and [prepared from 10 g, 0.153 mole zinc dust and 6 g, 0.022 mole mercuric chloride], followed by the addition of concentrated hydrochloric acid (5 ml). The mixture is refluxed for 30 min, filtered while hot and the residue washed with hot

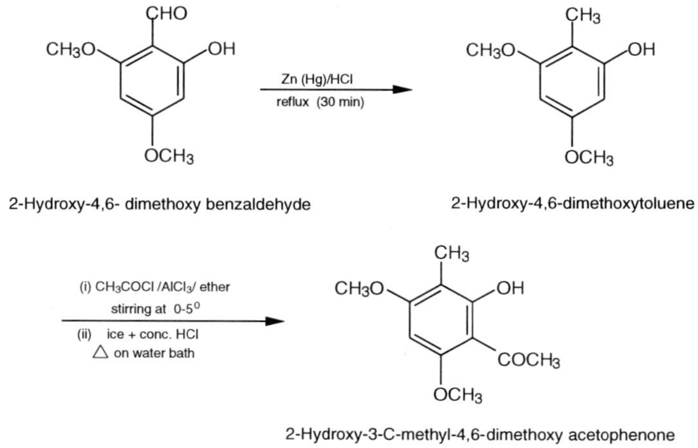

alcohol (20 ml). The combined alcoholic solution is distilled, the residue diluted with water and extracted with ether (2 × 50 ml). The ether extract is washed with water, dried (sodium sulphate) and distilled. The residual product is passed through a column of silica gel (2 × 20 cm) and eluted with petroleum ether-benzene (4:1). The required compound is obtained by concentration of the eluate as feathery coloured needles. Yield 2 g (86.9%). M.p. 60° (lit.[72] m.p. 60°).

Step (ii) 3-C-Methyl-2-hydroxy-4,6-dimethoxyacetophenone

Freshly distilled acetyl chloride (2 ml, 22 g, 0.028 mole) is added dropwise during 15 min to a stirred and cooled (0°) solution of 2-hydroxy-4,6-dimethoxytoluene (1 g, 0.0059 mole) and anhydrous aluminium chloride (3.2 g, 0.024 mole) in dry ether (60 ml). The mixture is stirred in cold for 1 hr more and left in ice-chest for 72 hr. The complex is decomposed with ice-hydrochloric acid and the mixture extracted with ether. The ether solution is washed with sodium bicarbonate solution (5%), water, dried (anhydrous sodium sulphate) and distilled. The residue is chromatographed over sillica gel column (2 × 20 cm) and eluted with petroleum ether-benzene (3:1) to give the required product. It is crystallised from methanol as almost colourless needles. Yield 0.7 g (56%). M.p. 140-2° (lit.[71] m.p. 141-42°).

4.A.63 2-HYDROXY-3,4-DIMETHOXY-6-METHYLACETOPHENONE

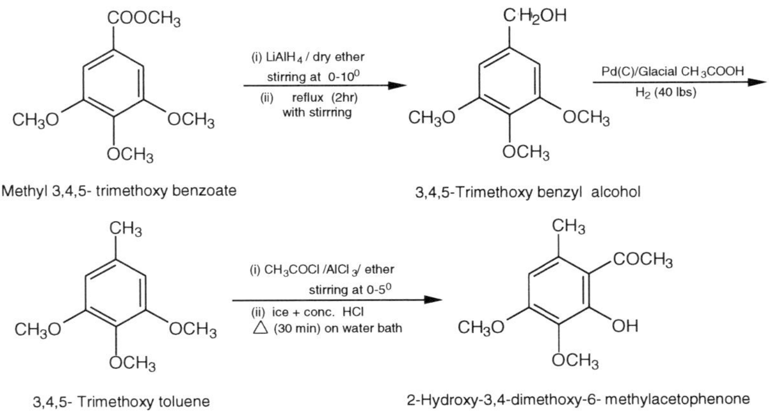

Step (i) 3,4,5-Trimethoxybenzyl Alcohol

A solution of methyl 3,4,5-trimethoxybenzoate (3.28) (22.6 g, 0.1 mole) in dry ether (200 ml) is added dropwise during 30 min to a stirred slurry of lithium aluminum hydride (4.6 g, 0.12 mole) in dry ether (100 ml) at 0-10°. The mixture is stirred under reflux for 2 hr. Excess of lithium aluminum hydride is decomposed with cold water. The ether is decanted and the residue treated with cold dilute sulphuric acid (10%, 25 ml). The gummy solid thus obtained is extracted with ether (2 × 50 ml), extract washed with water, dried (anhydrous sodium sulphate) and distilled to give the required product. Yield 14.7 g (74.2%). B.p. 135-37°/0.25 mm (lit.[73] b.p. 135-37°/ 0.25 mm).

Step (ii) 3,4,5-Trimethoxytoluene

A mixture of 3,4,5-trimethoxybenzyl alcohol (12 g, 0.06 mole), Pd/C (10%, 5 g) and glacial acetic acid (100 ml) is hydrogenated at a pressure of 40 lb. The mixture is shaken for 15 hr until no more hydrogen is absorbed. The catalyst is removed by filtration and the residue washed with more acetic acid (20 ml). The combined acetic acid solution is concentrated in vacuo at 50° to about 30 ml. It is diluted with water (150 ml) and extracted with ether. The ether extract is washed with sodium bicarbonate solution (5%) and then with water. It is dried (anhydrous sodium sulphate) and distilled to give the required product. Yield 10.3 g (93.4%). B.p. 95°/1 mm. (lit.[74] b.p. 95°/1 mm).

Step (iii) 2-Hydroxy-3,4-dimethoxy-6-methylacetophenone

Anhydrous aluminum chloride (3.5 g, 0.026 mole) is dissolved in dry ether (20 ml) with cooling and a solution of 3,4,5-trimethoxytoluene (3.5 g, 0.019 mole) in dry ether (20 ml) added under anhydrous condition. The mixture is vigorously stirred, cooled in ice-bath and treated with freshly distilled acetyl chloride (2 ml, 2.21 g, 0.028 mole) in dry ether (10 ml) which is added dropwise during 1 hr. The stirring is continued for 6 hr. During this time the yellow aluminium chloride complex which separates as oil finally solidifies. The mixture is left overnight and treated cautiously with crushed ice (50 g) and concentrated hydrochloric acid (5 ml). The mixture is heated on a water bath for 30 min, the ether is allowed to distil away. While hot, the mixture is extracted with benzene (2 × 50 ml). The benzene solution is washed with dilute hydrochloric acid and extracted with aqueous sodium hydroxide (5%). The clear alkaline solution is acidified with dilute hydrochloric acid. The separated product is filtered and washed with water. It is crystallised from methanol as colourless fawn coloured rectangular prisms. Yield 3.2 g (79.2). m.p. 90-91° (lit.[75] m.p.92°).

4.A.64 2,4,6-TRIHYDROXY-3-METHYLACETOPHENONE (3-C-METHYLPHLOROACETOPHENONE)

Method (i)

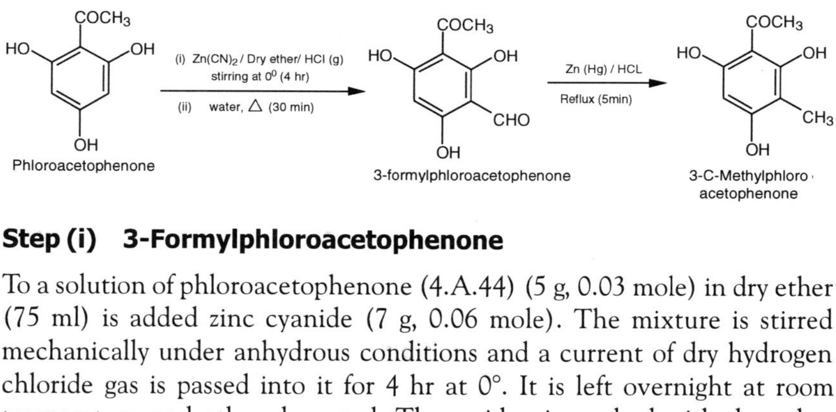

Step (i) 3-Formylphloroacetophenone

To a solution of phloroacetophenone (4.A.44) (5 g, 0.03 mole) in dry ether (75 ml) is added zinc cyanide (7 g, 0.06 mole). The mixture is stirred mechanically under anhydrous conditions and a current of dry hydrogen chloride gas is passed into it for 4 hr at 0°. It is left overnight at room temperature and ether decanted. The residue is washed with dry ether (2 × 25 ml). It is dissolved in water (30 ml), the solution is heated for 30 min in a boiling water bath and the solution cooled. The separated product is filtered, washed and dried. It is crystallised from xylene as colourless tiny needles. Yield 2.5 g (43%). M.p. 182-83° (Lit.[76,77] m.p. 182-83°).

Step (ii) 3C-Methylphloroacetophenone

The above formylacetophenone (9 .0 g, 0.046 mole) is refluxed gently for 5 min with methanol (75 ml), hydrochloric acid (50 ml), water (25 ml) and amalgamated zinc (prepared from 40 g, 0.612 mole zinc dust and 24 g, 0.088 mole mercuric chloride). It is filtered and diluted with water and cooled. The separated product is filtered and washed with cold water. It is crystallised from hot water as almost colourless needles. Yield 5 g (59.9%). M.p. 211-12° (lit.[76,78] m.p. 210-11°).

Method (ii)

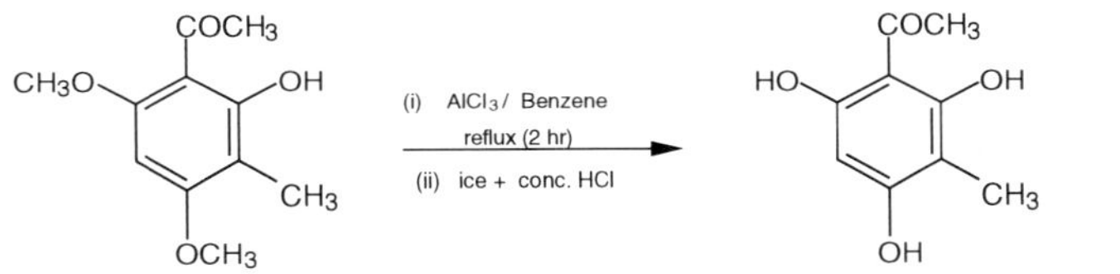

2- Hydroxy -3-methyl-4,6-dimethoxy acetophenone

3-C-Methylphloroacetophenone

Anhydrous aluminium chloride (6 g, 0.045 mole) is added to a solution of 2-hydroxy-3-methyl-4,6-dimethoxyacetophenone (4.A.62) (2 g, 0.01 mole) in dry benzene (20 ml). The mixture is refluxed for 2 hr under anhydrous conditions. The benzene is then distilled off and the cooled reaction mixture is treated with ice (50 g) and concentrated hydrochloric acid (10 ml). After 2 hr, the almost colourless product is filtered. It is purified by dissolving in aqueous sodium carbonate solution (10%, 100 ml) and acidifying the clear sodium carbonate extract with hydrochloric acid. It is filtered, washed with cold water and crystallised from hot water as clusters of colourless needles. Yield 1.5 g (86.5%). M.p. 211-12° (lit.[76,78] m.p. 210-11°).

4.B.65 ω-METHOXYACETOPHENONE ($C_6H_5COCH_2OCH_3$)

Step (i) Diazomethylphenyl Ketone ($C_6H_5COCHN_2$)

Benzoyl chloride (14 g, 11.56 ml, 0.1 mole) in ether (200 ml) is added dropwise with stirring to a cooled (–15°) ethereal solution of diazomethane

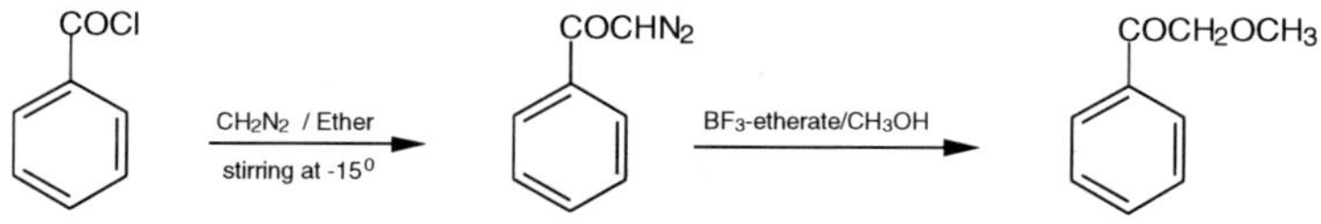

Diazomethyl phenyl ketone ω-Methoxy Acetophenone

(6.A.17) (from 5 g of nitrosomethyl urea). After keeping overnight, the solution is concentrated in vacuo and the oily product directly used for the next stage.

Step (ii) ω-Methoxyacetophenone

The above diazoketone (10 g, 0.068 mole) is dissolved in methanol at room temperature and boron trifluoride etherate (0.5 ml, 0.58 g, 0.004 mole) added. A brisk effervescence persists for about 30 min. Methanol is removed in vacuo and the remaining solution is diluted with ether. The ether solution is washed with sodium bicarbonate solution (5%, 200 ml), water and dried (anhydrous sodium sulphate). Evaporation of the solvent gives ω-methoxyacetophenone. Yield 4.5 g (43.8%). b.p. 226-28° (lit.[79,80] b.p. 228-30°).

4.B.66 2,4-DIHYDROXY-ω-METHOXYACETOPHENONE (ω-METHOXYRESACETOPHENONE)

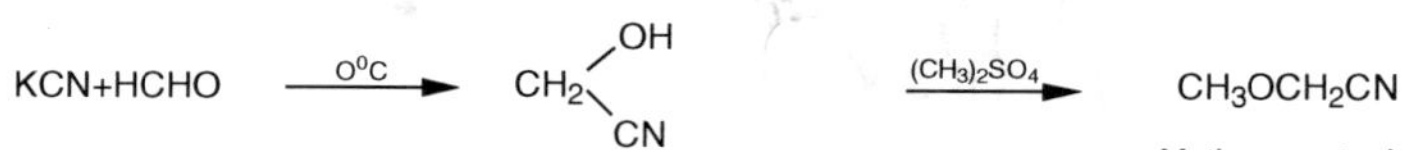

Methoxyacetonitrile

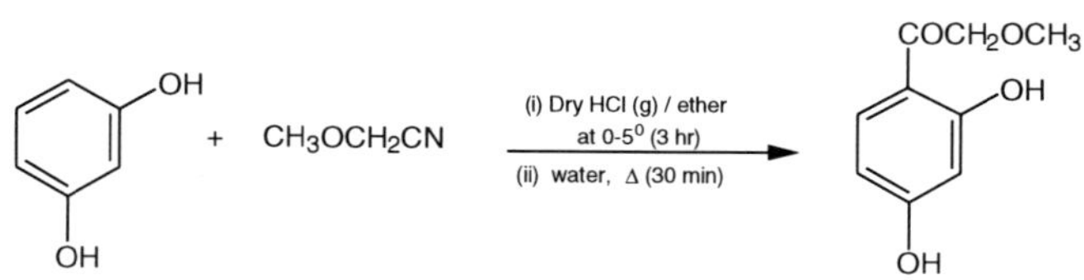

Resorcinol ω-Methoxyresacetophenone

Step (i) Methoxycetonitrile

An ice cold (0°) solution of potassium cyanide (26 g, 0.4 mole) in water (75 ml) is treated with an aqueous solution of formaldehyde (40%, 35 ml, 14 g, 0.466 mole formaldehyde and 35 ml water), which is added in small portions with shaking. Care is taken to see that the temperature of the mixture does not rise appreciably above 0°. After the addition of formalin is over (30 min), the mixture is left in ice-bath for 1 hr with shaking at intervals. It is treated with dimethyl sulphate (50 ml, 66.56 g, 0.528 mole), added little by little with constant shaking. During the addition of dimethyl sulphate the temperature of reaction mixture is kept as low as possible. The reaction mixture is then removed from the ice-bath and allowed to attain room temperature all the while the mixture is gently stirred. The reaction sets in vigorously in the neighborhood of 25° as is evident from the sudden rise of temperature. By the judicious use of ice bath, the temperature is adjusted to be not more than 35°. The completion of the reaction is indicated by the total disappearance of the dimethyl sulphate which formed a heavy layer at the bottom during the initial stage. The upper layer of methoxyacetonitrile formed as a colourless smelling oil is separated and the aqueous layer extracted with ether (2 × 100 ml). The combined organic layer is dried (anhydrous sodium sulphate) and ether removed by distillation. The remaining oily product is distilled and fraction between 118-28° is collected. This fraction is redistilled over phosphorus pentaoxide (5 g) using a small fractionating column (0.5 × 6") and pure methoxyacetonitrile is collected between 118-22° (lit.[81] b.p. 120-22°). Yield 10 g (35.2%).

Step (ii) ω-Methoxyresacetophenone

A solution of anhydrous resorcinol (9 g, 0.08 mole) in dry ether (75 ml) is treated with methoxyacetonitrile (9 ml, 8.6 g, 0.12 mole) and solution is cooled (0-5°) in ice salt mixture. A current of dry hydrogen chloride gas is passed through the cooled solution for 3 hr under anhydrous conditions. The flask is well stoppered and left in a refrigerator for 2 days. The ether is decanted and the solid ketimine hydrochloride washed with dry ether (2 × 25 ml). The ketimine hydrochloride is dissolved in minimum amount of water and the solution is heated for 30 min on a water bath (80°). The solution is cooled and the separated product filtered. It is crystallised from hot water as pink shining plates. Yield 9g (60.4%). M.p. 136-37° (lit.[82] m.p. 136°).

4.B.67 2-HYDROXY-ω,4-DIMETHOXYACETOPHENONE (ω-METHOXYRESACETOPHENONE-4-0-METHYL ETHER)

ω-Methoxyresacetophenone (4.B.66) (5 g, 0.027 mole) in dry acetone (150 ml) is refluxed for 4 hr with dimethyl sulphate (2.8 ml, 3.73 g, 0.03 mole) in the presence of freshly ignited potassium carbonate (10 g)

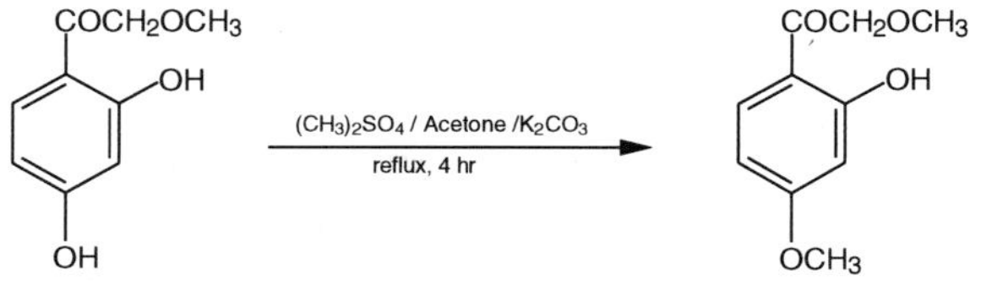

ω -Methoxyresacetophenone — 2- Hydroxy-ω,4-dimethoxy acetophenone

under anhydrous conditions. The acetone solution is filtered and the inorganic residue washed with hot acetone (2 × 30 ml). The combined acetone solution is distilled and the product macerated with crushed ice. It is filtered, washed with water, dried and crystallised from benzene-petroleum ether mixture as colourless prisms. Yield 4.5 g (83.6%). M.p. 67-69° (lit.[83] m.p. 65-66°).

4.B.68 ω,2,4-TRIHYDROXYACETOPHENONE (FISETOL)

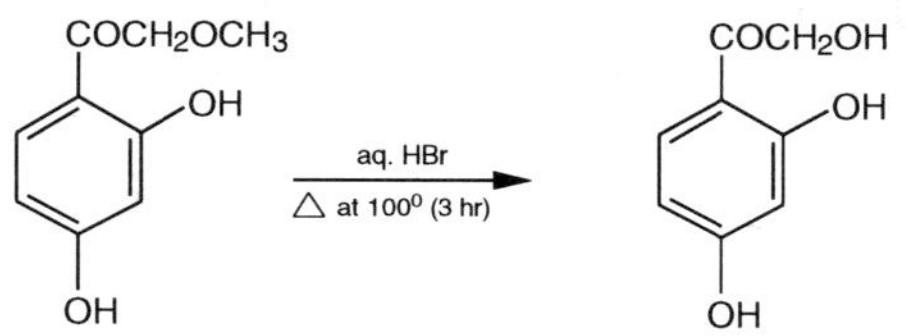

ω -Methoxyresacetophenone — ω,2,4-Trihydroxy acetophenone

ω-Methoxyresacetophenone (4.B.66) (1 g, 0.0055 mole) is heated with aqueous hydrobromic acid (40%, 20 ml) for 3 hr at 100°. It is filtered through cotton wool to remove resinous matter and the solution extracted with ether (3 × 30 ml). The ether solution is washed with water and distilled. The product obtained is crystallised from hot water containing few drops of hydrochloric acid as colourless needles. Yield 0.15 g (16.3%). M.p. 187-88° (lit.[84] m.p. 189°).

4.B.69 ω,2-DIHYDROXY-4-METHOXYACETOPHENONE

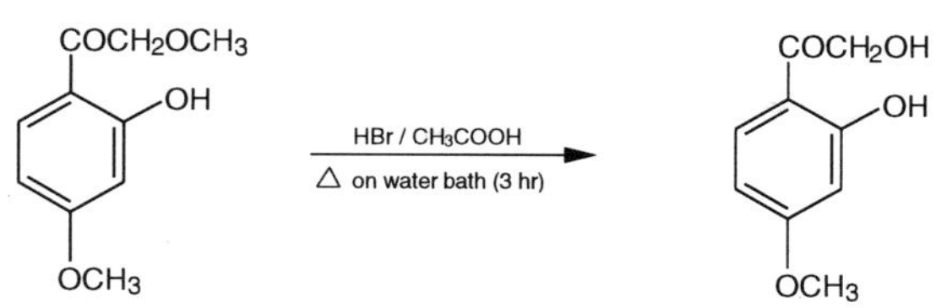

2-Hydroxy-ω,4-dimethoxyacetophenone — ω,2-Dihydroxy-4-methoxyacetophenone

2-Hydroxy-ω,4-dimethoxyacetophenone (4.B.67) (2 g, 0.01 mole) is dissolved in acetic acid (15 ml) and hydrobromic acid (45%, 40 ml) added to it. The solution is heated on a boiling water bath for 3 hr and poured over crushed ice. The mixture is extracted with ether (3 × 30 ml), extract washed with sodium bicarbonate solution (5%), water and then distilled. The product obtained is crystallised from dilute alcohol as colourless long flat needles. Yield 0.4 g (21.5%). M.p. 127° (lit.[85] m.p. 127°).

4.B.70 2,4,6-TRIHYDROXY-ω-METHOXYACETOPHENONE (ω-METHOXYPHLOROACETOPHENONE)

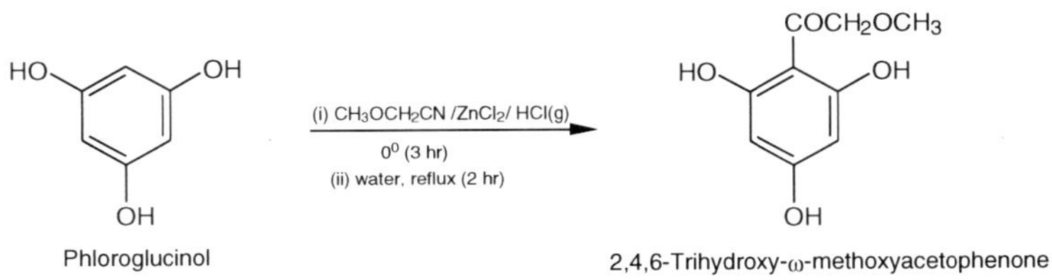

To a solution of dry phloroglucinol (10 g, 0.079 mole) in dry ether (80 ml) is added freshly powdered zinc chloride (1.5 g, 0.011 mole) and methoxyacetonitrile (4.B.66) (10 ml, 9.56 g, 0.134 mole). The mixture is cooled (0°, ice salt mixture) and is saturated with dry hydrogen chloride gas for 3 hr under anhydrous conditions. In about 30 min, the solution becomes turbid and the ketimine hydrochloride separates as a crust. The current of hydrogen chloride gas is continued for 1 hr more after saturation and the reaction mixture is left in ice-chest for 24 hr. The clear ether is decanted and solid washed with dry ether (2 × 30 ml). The pale yellow ketimine hydrochloride is dissolved in water (400 ml) and hydrolysed by refluxing for 2 hr. The solution is cooled the separated product filtered and washed with cold water. It is crystallised from hot water. Yield 11 g (70.1%). M.p. 191-92° (after drying at 110-120-° for 3 hr) (lit.[86] m.p. 191-92°).

4.B.71 2-HYDROXY-ω,4,6-TRIMETHOXYACETOPHENONE (ω-METHOXYPHLOROACETOPHENONE-4,6-DIMETHYL ETHER)

ω-Methoxyphloroacetophenone (4.B.70) (5 g, 0.025 mole) in dry acetone (150 ml) is refluxed for 4 hr under anhydrous conditions with dimethyl

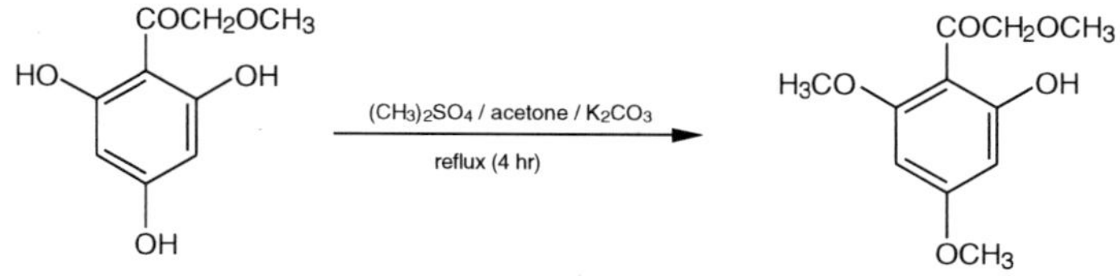

ω-methoxyphloroacetophenone 2-Hydroxy-ω,4,6-trimethoxyacetophenone

sulphate (6.15 ml, 8.2 g, 0.0625 mole) in presence of ignited potassium carbonate (15 g). The solution is filtered and the inorganic salts washed with hot acetone (2 × 30 ml). The combined acetone solution is filtered and ice added to the residual product. The solid product is filtered and washed with water. It is crystallised from aqueous methanol as colourless needles. Yield 4 g (70.2%). M.p. 102-104° (lit.[87] m.p. 102-104°).

4.B.72 2,4-DIHYDROXY-ω,6-DIMETHOXYACETOPHENONE[88] (ω-METHOXYPHLOROACETOPHENONE-6-0-METHYL ETHER)

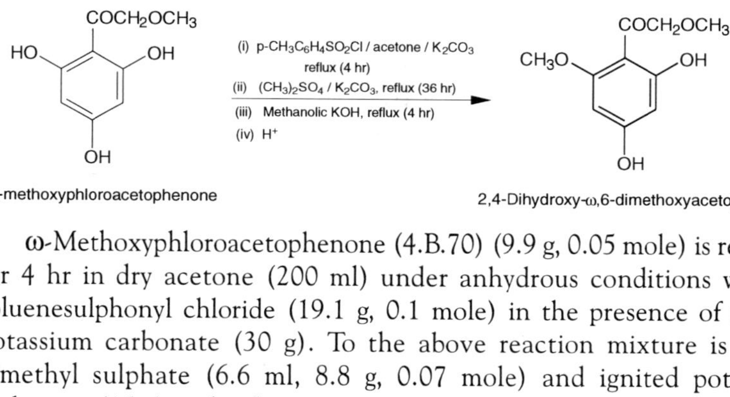

ω-methoxyphloroacetophenone 2,4-Dihydroxy-ω,6-dimethoxyacetophenone

ω-Methoxyphloroacetophenone (4.B.70) (9.9 g, 0.05 mole) is refluxed for 4 hr in dry acetone (200 ml) under anhydrous conditions with p-toluenesulphonyl chloride (19.1 g, 0.1 mole) in the presence of ignited potassium carbonate (30 g). To the above reaction mixture is added dimethyl sulphate (6.6 ml, 8.8 g, 0.07 mole) and ignited potassium carbonate (15 g) and refluxing is continued for 36 hr more. The reaction mixture is filtered in hot, inorganic residue washed with hot acetone (2 × 50 ml). The combined acetone solution is distilled and the residual syrup is refluxed for 4 hr with methanolic potassium hydroxide (5%, 200 ml). The methanol is removed in vacuo, the clear alkaline solution is diluted with water and acidified with hydrochloric acid. The separated product is filtered and washed with water. It is purified by dissolving in sodium carbonate solution (5%) followed by acidification of the clear alkaline solution. The solid product obtained is filtered and washed with water. It is crystallised from methanol as colourless prisms. Yield 4.5 g (42.5%). M.p. 190-92°.

4.B.73 ω,2-DIHYDROXY-4,6-DIMETHOXYACETOPHENONE

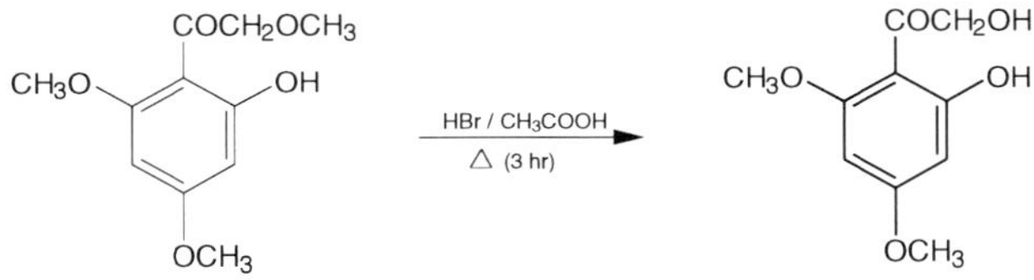

2-Hydroxy-ω,4,6-trimethoxyacetophenone ω, 2-Dihydroxy-4,6-dimethoxyacetophenone

2-Hydroxy-ω,4,6-trimethoxyacetophenone (4.B.71) (2.5 g, 0.11 mole) is dissolved in acetic acid (25 ml) and hydrobromic acid (40%, 25 ml) added. The solution is heated for 3 hr on a boiling water bath. It is poured over ice (300 g). The separated product is filtered and washed with water. It is crystallised twice from methanol as colourless long rhombohedral tablets. Yield 0.75 g (31.9%). M.p. 139-40°. (lit.[85] m.p. 139-40°).

4.B.74 ω-HYDROXY-2,4,6-TRIMETHOXYACETOPHENONE

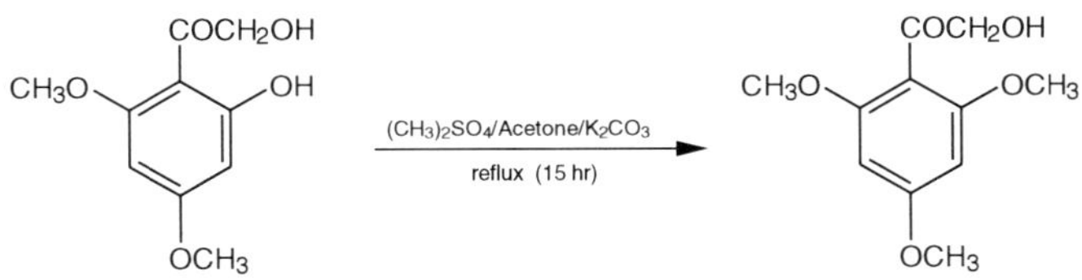

ω-2-Dihydroxy-4,6-dimethoxy acetophenone ω-Hydroxy-2,4,6-trihydroxy acetophenone

ω-2-Dihydroxy-4,6-dimethoxyacetophenone (4.B.73) (2 g,0.0094 mole) in dry acetone (50 ml) is refluxed for 15 hr under anhydrous conditions with dimethyl sulphate (3 ml, 4 g, 0.03 mole) and ignited potassium carbonate (10 g). The reaction mixture is filtered and the inorganic salts washed with hot acetone (2 × 10 ml). The combined acetone solution is distilled and the residue macerated with crushed ice. The solid product is filtered and washed with water. It is crystallised from dilute alcohol as colourless polyhedral prisms. Yield 1.2 g (56.3%). m.p. 185-86° (lit.[85] m.p. 185-86°).

4.B.75 2,4-DIHYDROXY-6-METHYL-ω-METHOXYACETOPHENONE[89] (ω-METHOXY-ORCACETOPHENONE)

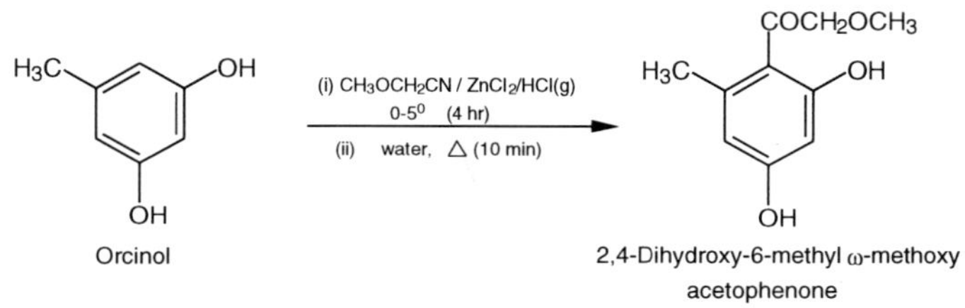

A solution of orcinol (6 g, 0.048 mole) and methoxyacetonitrile (4.B.66) (6 ml, 5.74 g, 0.08 mole) in dry ether (30 ml) containing freshly fused zinc chloride (2 g, 0.147 mole) is saturated with dry hydrogen chloride gas at 0-5° for 4 hr under anhydrous conditions. An yellowish ketimine hydrochloride starts separating, the flow of gas is continued for 2 hr more and the reaction mixture left overnight in a refrigerator. The clear ether solution is decanted off and the solid product washed with dry ether (2 × 20 ml). It is dissolved in water (20 ml) and the solution heated for 10 min in boiling water bath. The solution is neutralised with dilute ammonia and the solution cooled. The separated product is filtered. It is crystallised from hot water as colourless needles. Yield 5 g (52.7%). M.p. 182-83°.

4.B.76 2,5-DIHYDROXY-ω,4-DIMETHOXYACETOPHENONE

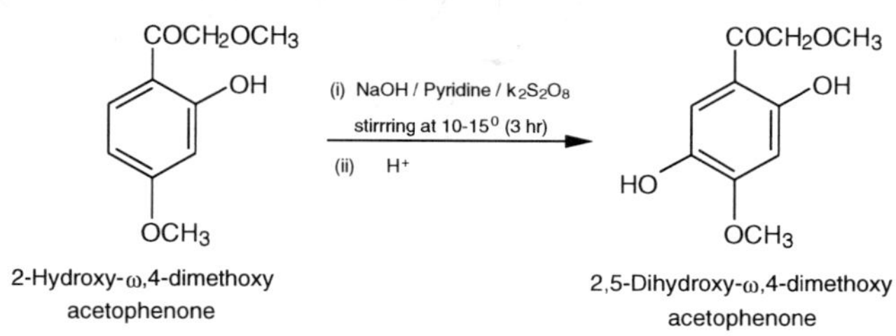

2-Hydroxy-ω,4-dimethoxyacetophenone (4.B.67) (10 g, 0.051 mole) is suspended in aqueous sodium hydroxide solution (10 g in 120 ml water) and the separated sodium salt is brought into solution by the addition of small amount of pyridine (15 ml). A solution of potassium persulphate (25 g, 0.092 mole) in water (200 ml) is added dropwise during 3 hr to the above stirred solution at 10-15°. The stirring is continued for 1 hr more and the reaction mixture left overnight. The reaction mixture is cooled, acidified to congo red with hydrochloric acid and the separated starting compound is filtered. The filtrate is extracted with ether (2 × 50 ml) and the clear solution is heated for 30 min on water bath at 80° with sodium bisulphite (5 g) and concentrated hydrochloric acid (75 ml). The mixture is cooled and the separated product filtered. It is crystallised from hot water as colourless broad plates. Yield 2 g (18.5%). M.p. 145-46° (lit.[90] m.p. 145-46°).

4.B.77 2-HYDROXY-ω,4,5-TRIMETHOXYACETOPHENONE

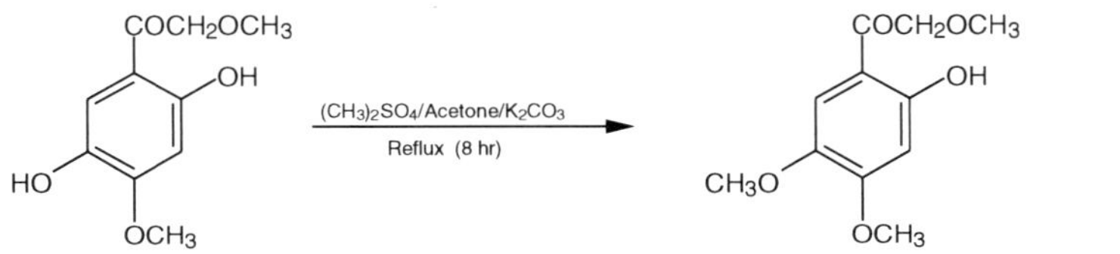

2,5-Dihydroxy-ω,4-dimethoxy acetophenone   2-Hydroxy-ω,4,5-trimethoxy acetophenone

A solution of 2,5-dihydroxy-ω,4-dimethoxyacetophenone (4.B.76) (1.6 g, 0.0075 mole) in dry acetone (75 ml) is refluxed for 8 hr (water-bath) under anhydrous conditions with dimethyl sulphate (1 ml, 1.3 g, 0.01 mole) and freshly ignited potassium carbonate (5 g). The acetone is filtered and the inorganic salts washed with hot acetone (2 × 25 ml). The combined acetone solution is distilled and the residue macerated with crushed ice. The separated product is filtered and purified by dissolving in sodium hydroxide solution and acidifying the clear alkaline solution with hydrochloric acid. The separated product is filtered and crystallised from alcohol as colourless rectangular plates. Yield 1 g (58.8%). M.p. 90-92° (lit.[90] m.p. 90-91°). The alkali insoluble product is filtered and crystallised from alcohol as colourless needles of ω, 2,4,5-tetramethoxyacetophenone. Yield 0.5 g (27.7%). M.p. 131-32° (lit.[90] m.p. 131-32°).

4.B.78 2,5-DIHYDROXY-ω,4,6-TRIMETHOXYACETOPHENONE

To a mechanically stirred solution of 2-hydroxy-ω,4,6-trimethoxyacetophenone (4.B.71) (5 g, 0.022 mole) in aqueous sodium hydroxide (5% ,100 ml) kept at 15-20° is added during 3 hr an aqueous solution of potassium persulphate (7 g, 0.026 mole in 100 ml water). The mixture is stirred for 1 hr more and left for 24 hr at room temperature. It is cooled acidified to

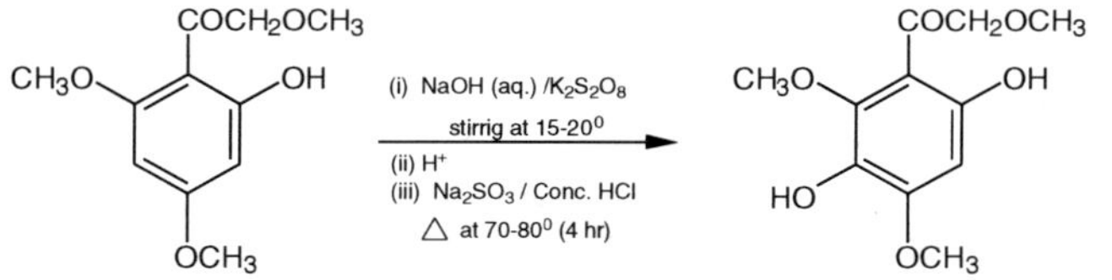

2-Hydroxy-ω,4,6-trimethoxy acetophenone — 2,5-Dihydoxy-ω,4,6-trimethoxy acetophenone

congo red with dilute hydrochloric acid and is extracted repeatedly with ether (4 × 50 ml) to remove the unreacted ketone. The clear aqueous solution is heated on a water bath at 70-80° for 4 hr with sodium sulphite (5 g) and concentrated hydrochloric acid (50 ml). The solution is cooled, the separated product filtered and washed with water. It is crystallised from alcohol as yellow needles. Yield 1.2 g (22.4%). M.p. 136-37° (lit. [91] m.p. 135-36°). A little more of the product (0.7 g) is obtained by ether extraction of the mother liquor.

4.B.79 2-HYDROXY-ω,4,5,6-TETRAMETHOXYACETOPHENONE

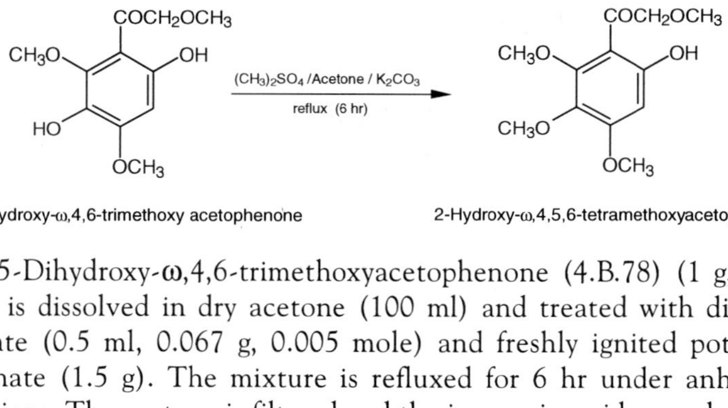

2,5-Dihydroxy-ω,4,6-trimethoxy acetophenone — 2-Hydroxy-ω,4,5,6-tetramethoxyacetophenone

2,5-Dihydroxy-ω,4,6-trimethoxyacetophenone (4.B.78) (1 g, 0.004 mole) is dissolved in dry acetone (100 ml) and treated with dimethyl sulphate (0.5 ml, 0.067 g, 0.005 mole) and freshly ignited potassium carbonate (1.5 g). The mixture is refluxed for 6 hr under anhydrous conditions. The acetone is filtered and the inorganic residue washed with hot acetone (30 ml). The combined acetone solution is distilled and the residue macerated with crushed ice. The solid product is filtered and purified by dissolving in sodium hydroxide solution (10%) and acidifying the clear alkaline solution. Finally it is crystallised from alcohol as colourless transparent prisms. Yield 0.8 g (75.5%). M.p. 77-8° (lit.[91] m.p. 77-8°).

4.B.80 2,4- DIHYDROXY-ω,3,6-TRIMETHOXYACETOPHENONE

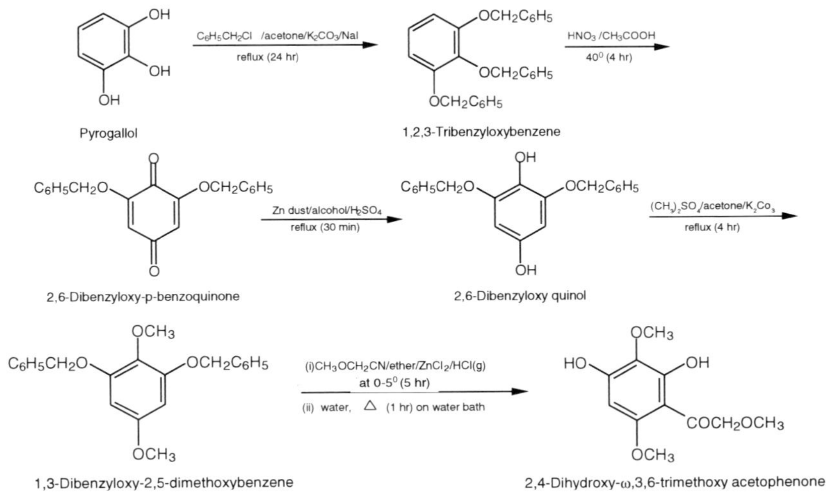

Step (i) 1,2,3-Tribenzyloxybenzene (Pyrogallol Tribenzyl Ether)

Ignited potassium carbonate (225 g) is added to a stirred solution of pyrogallol (100 g, 0.079 mole) in dry acetone (1l) containing anhydrous sodium iodide (125 g, 0.083 mole) and benzyl chloride (305 g, 277.3 ml, 2.41 mole). The mixture is refluxed for 24 hr with occasional shaking under anhydrous condition; a further addition of ignited potassium carbonate (225 g) is made after the first six hr. The mixture is filtered and the inorganic salts washed with hot acetone (4 × 150 ml). The combined acetone solution is distilled. Water (50 ml) is added to the residue and is subjected to steam distillation to remove excess of benzyl chloride. The residue solidifies on cooling. It is filtered, washed with water and dried. Yield 260 g (82.7%). M.p. 65° (lit.[92] m.p. 65°). It is used as such for the next step.

Step (ii) 2,6-Dibenzyloxy-p-benzouinone

Pyrogallol tribenzyl ether (40 g, 0.1 mole) is dissolved in acetic acid (400 ml) and the solution kept at 40°. Nitric acid (20 ml, d, 1.19) is added in one lot to the above solution. After keeping the reaction mixture for 4 hr at 40°, the precipitated solid consisting mainly of 5-nitro-1,2,3-tribenzyloxy-benzene is filtered. To the deep red filtrate, a further quantity of nitric acid

(20 ml, d, 1.19) is added. There is a gradual separation of the quinone. The reaction mixture is left overnight at room temperature. The yellow precipitate of the quinone is filtered and washed with a small quantity of acetic acid (20 ml) and finally with water. The quinone (24 g) consisting of mostly the required quinone and some nitro compound is suspended in carbon tetrachloride (300 ml) and the mixture boiled. The hot solution is filtered quickly taking care not to allow the pores of the funnel to clog. The residue is washed with a further volume of boiling carbon tetrachloride (3 × 150 ml). By this procedure the nitro compound is selectively removed by carbon tetrachloride leaving behind the quinone in a pure state. Yield 18 g (55.7%). M.p. 201-202° (lit.[92] m.p. 201-202°).

Step (iii) 2,6-Dibenzyloxyquinol

Sulphuric acid (25%, 60 ml) is added to a gently boiling mechanically stirred mixture of 2,6-dibenzyloxy-p-benzoquinone (40 g, 0.125 mole), alcohol (400 ml) and zinc dust (80 g) taken in a three necked R.B. flask. The mixture is refluxed for 30 min during which time the yellow quinone disappears. The brown liquid is filtered from zinc salts and washed with hot alcohol (2 × 40 ml). To the combined alcoholic solution is added sodium bisulphite (10 g) and dilute sulphuric acid (10%, 100 ml). The mixture is diluted with water (1.5 l) and left overnight. The separated quinol is filtered, washed with cold water (3 × 250 ml) and dried in vacuo over concentrated sulphric acid. Yield 35 g (86.9%). M.p. 116-17° (lit.[92] m.p. 116-17°).

Step (iv) 1,3-Dibenzyloxy-2,5-dimethoxybenzene

2,6-Dibenzyloxyquinol (20 g, 0.062 mole) in dry acetone (200 ml) is refluxed for 4 hr under anhydrous conditions with dimethyl sulphate (27 g, 0.21 mole, added in four lots) and ignited potassium carbonate (50 g). The reaction mixture is filtered and the inorganic salts washed with hot acetone (2 × 30 ml). The combine acetone solution is distilled and the residue macerated with crushed ice. The solid product is filtered and washed with water. It is crystallised from alcohol as colourless crystals. Yield 21 g (96.6%). M.p. 82-3° (lit.[92] m.p. 82-3°).

Step (v) 2,4-Dihydroxy-ω,3,6-trimethoxyacetophenone

1,3-Dibenzyloxy–2,5-dimethoxybenzene (7 g, 0.02 mole) is dissolved in anhydrous ether (100 ml). Methoxy acetonitrile (9.B.66) (3.5 ml, 3.35 g,

0.047 mole) and freshly fused and powdered zinc chloride (4.5 g, 0.033 mole) added to it. A stready stream of dry hydrogen chloride gas is bubbled through the solution for 5 hr, the mixture is cooled throughout in ice-salt bath (0-5°). The mixture is left overnight in an ice-chest. The ether solution is decanted and the ketimine hydrochloride washed with dry ether (2 × 50 ml). It is dissolved in water (20 ml) and again extracted with ether to remove completely any unreacted starting compound. The clear aqueous solution is heated on a boiling water bath for 1 hr and then cooled. The separated product is filtered. It is crystallised from hot water (charcoal) as colourless compound. Yield 2.5 g (51.7%). M.p. 150-51° (lit.[93] 150-51°).

4.B.81 2-HYDROXY-ω,3,4,6-TETRAMETHOXYACETOPHENONE

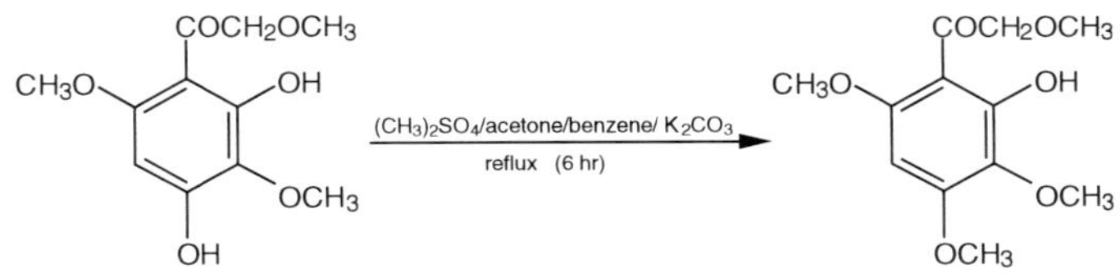

2,4-Dihydroxy-ω,3,6-trimethoxy acetophenone 2-Hydroxy-ω,3,4,6-tetramethoxy acetophenone

2,4-Dihydroxy-ω,3,6-trimethoxyacetophenone (4.B.80) (2 g, 0.008 mole) in dry acetone (20 ml) and dry benzene (40 ml) is refluxed for 6 hr under anhydrous condition with dimethyl sulphate(1 ml, 1.3 g, 0.01 mole) and ignited potassium carbonate (5 g). The mixture is filtered and inorganic salts washed with hot acetone (2 × 20 ml). The combined filtrate is distilled and the residue macerated with crushed ice. The solid product is filtered. It is purified by dissolving in sodium hydroxide solution (5%) and acidifying the clear alkaline extract. It is crystallised from dilute alcohol as long colourless needles. Yield 0.8 g (37.7%). M.p. 114-15° (lit. [94] m.p. 115-16°).

4.B.82 2,5-DIHYDROXY-ω,3,4,6-TETRAMETHOXYACETOPHENONE

2-Hydroxy-ω,3,4,6-tetramethoxyacetophenone (4.B.81) (1.5 g, 0.0059 mole) is dissolved in pyridine (2 ml) and aqueous sodium hydroxide (2 g in 20 ml water) added. A solution of potassium persulphate (2.4 g, 0.0089 mole in 35 ml water) is then added dropwise during 4 hr to the above continuously stirred solution at 15-20°. The stirring is continued for 2 hr

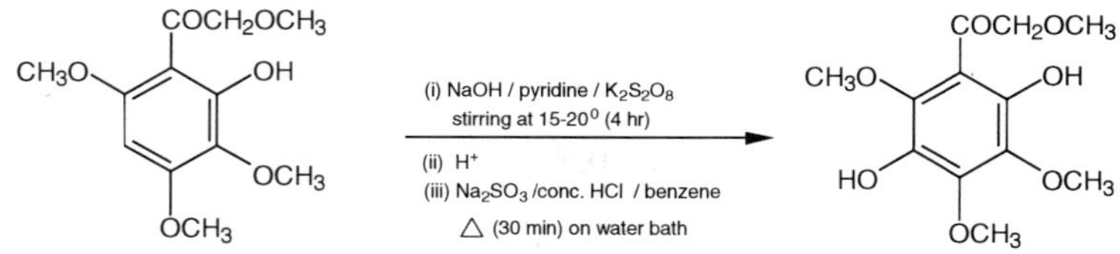

2-Hydroxy-ω,3,4,6-tetramethoxy acetophenone — 2,5-Dihydroxy-ω,3,4,6-tetramethoxy acetophenone

more and the deep red solution allowed to stand for 24 hr at room temperature. The reaction mixture is cooled and acidified to congo red with dilute hydrochloric acid. The separated unreacted ketone is filtered and the filtrate extracted with ether (2 × 25 ml). The aqueous solution is heated in a boiling water bath for 30 min with sodium sulphite (1 g), concentrated hydrochloric acid (5 ml) and benzene (50 ml). The benzene layer is separated and the aqueous layer extracted with ether (3 × 50 ml). The combined benzene and ether extracts are dried (sodium sulphate) and the solvent distilled. The residual product is chromatographed over silica gel column (2 × 20 cm). The required product is eluted with petroleum ether- ethyl acetate (9:1). Yield 0.5 g (31.5%). M.p. 102-103°. (lit.[95] m.p. 102-103°).

4.B.83 2-HYDROXY-ω-3,4,5,6-PENTAMETHOXYACETOPHENONE

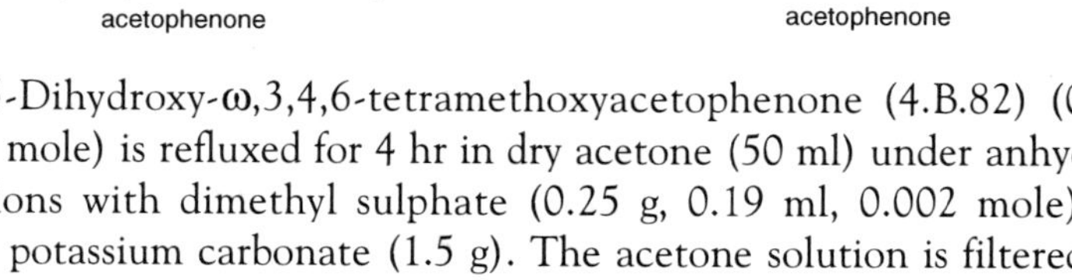

2,5-Dihydroxy-ω,3,4,6-tetramethoxy acetophenone — 2-Hydroxy-ω,3,4,5,6-pentamethoxy acetophenone

2,5-Dihydroxy-ω,3,4,6-tetramethoxyacetophenone (4.B.82) (0.5 g, 0.0018 mole) is refluxed for 4 hr in dry acetone (50 ml) under anhydrous conditions with dimethyl sulphate (0.25 g, 0.19 ml, 0.002 mole) and ignited potassium carbonate (1.5 g). The acetone solution is filtered and the inorganic salts washed with hot acetone (2 × 20 ml). The combined acetone solution is distilled and residue macerated with crushed ice. The oily product is extracted with ether (2 × 30 ml), ether extract dried (sodium sulphate) and distilled. The residual oily product is purified by column chromatography over silica gel (2 × 25 cm). The required product is obtained as yellow oil by elution with petroleum ether-ethyl acetate (19:1). Yield 0. 4g (75.7%) (lit.[96] oil).

4.B.84 2,4,6-TRIHYDROXY-ω-ETHOXYACETOPHENONE (ω-ETHOXYPHLORO-ACETOPHENONE)

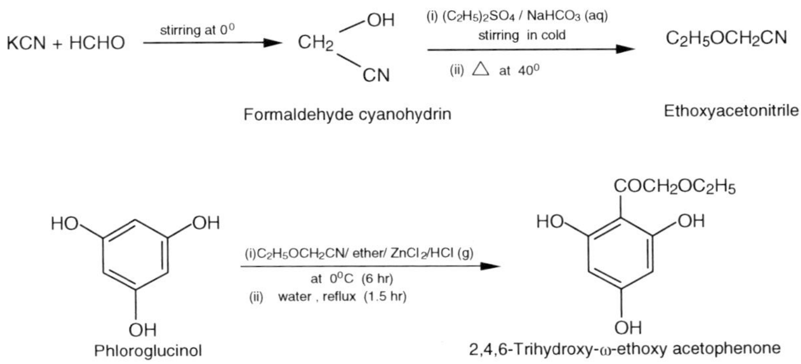

Step (i) Ethoxyacetonitrile

Powdered potassium cyanide (26 g, 0.4 mole) is dissolved in water (75 ml) and the solution cooled to –5° (ice-salt mixture). Formaldehyde (40%, 35 ml, 14 g, 0.46 mole) dilute with water (35 ml) is then added little by little with stirring during 45 min. The temperature is not allowed to rise above 0° throughout the reaction. After the addition is over, the mixture is left in ice-chest for 1 hr with occasional shaking. Diethyl sulphate (64.74 g, 55 ml, 0.42 mole) previously washed with cold water, sodium bicarbonate solution (5%) and dried (over anhydrous calcium chloride) is added in 5-6 lots to the cooled solution of formaldehyde cyanohydrin. The mixture is vigorously stirred with a mechanical stirrer for 20 min. There is no visible reaction in cold. The mixture is slowly heated to 40° in a water bath. The heavy layer of diethyl sulphate at the bottom disappears in 10 min and the lighter reddish layer of ethoxyacetonitrile appears at the top. It is already separated (separatory funnel), diluted with ether (100 ml), the ether extract dried (sodium sulphate) and distilled. The residual product is distilled in vacuo and the fraction distilling <80°/15 mm is collected (lit.[97] b.p.< 80°/15 mm). Yield (7g (20.6%).

Step (ii) ω-Ethoxyphloroacetophenone

Anhydrous phloroglucinol (5 g, 0.04 mole) is dissolved in dry ether (100 ml) containing ethoxyacetonitrile (8 g, 0.094 mole) and fused zinc chloride (5 g, 0.037 mole) added. A stream of dry hydrogen chloride gas is passed through the reaction mixture for 6 hr cooled to 0° (ice-salt bath) (with occasional shaking) under anhydrous conditions. The reaction mixture is kept in ice-chest for 48 hr. The ether is decanted and the yellow ketimine hydrochloride washed with dry ether (2 × 25 ml). The pale yellow solid ketimine hydrochloride is dissolved in water (100 ml), solution refluxed for 1.5 hr and cooled. The separated product is filtered, washed with cold water and dried. It is crystallised from benzene-petroleum ether as yellow cyrstalline substance. Yield 3 g (35.7%). M.p. 198-200^0 (lit.[98] m.p. 198-200°). Some more of the product (1.5 g, 17.8%) is obtained by ether extraction of the aqueous mother liquor with ethyl acetate, the extract is dried, concentrated and passed through column of silica gel and elution with benzene-petroleum ether (75:25).

4.B.85 2,4-DIHYDROXY-ω-BENZOYLOXYACETOPHENONE (ω-BENZOYLOXY RESACTOPHENONE)

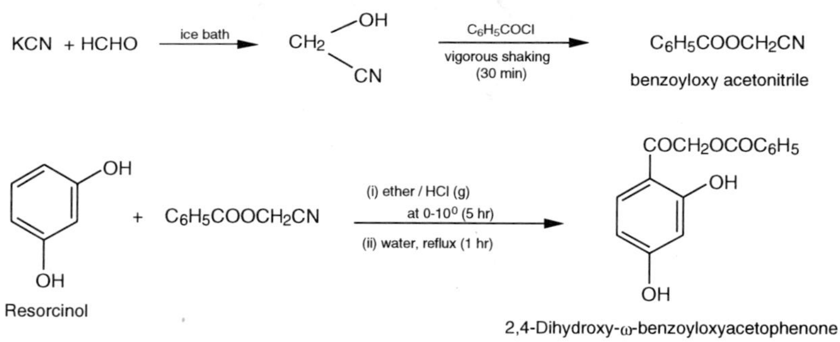

Step (i) Benzoyloxyacetonitrile

To an aqueous solution of potassium cyanide (26 g, 0.4 mole) in water (50 ml) is added formalin (40%, 35 ml, 14 g, 0.466 mole) in small lots keeping the reaction mixture cooled in ice bath. After the addition of formalin is complete, benzoyl chloride(56.2 g, 68 ml, 0.4 mole) is added in one lot and the solution vigorously shaken for 30 min with occasional cooling in ice-bath. Benzoyloxy acetonitrile separates as a pleasant smelling liquid as the upper layer. It is extracted with ether (1 × 100 ml), ether extract dried over anhydrous sodium sulphate and solvent removed. The remaining oily compound is distilled and the fraction boiling at 255-260° is collected. It solidifies on cooling. M.p. 27-8°. Yield 29 g (45%) (lit.[99] m.p. 27-8°, b.p. 255-60°).

Step (ii) 2,4-Dihydroxy-ω-benzoyloxyacetophenone (ω-Acetoxyresactophenone)

An ice cold solution of benzyloxyacetonitrile (10.9 g, 0.068 mole) and anhydrous resorcinol (5.5 g, 0.05 mole) in dry ether (100 ml) is saturated with dry hydrogen chloride gas for 5 hr at 0-10° under anhydrous conditions. The reaction mixture is left overnight at room temperature. The ether is decanted and the solid product washed with dry ether (2 × 30 ml). The solid product is dissolved in water (100 ml) and refluxed for 1 hr. The separated product is filtered and washed with water. It is crystallised from alcohol as colourless crystals. Yield 10 g (77.5%). M.p. 202-203° (lit.[100] m.p. 202-203°).

4.B.86 2,4,6-TRIHYDROXY-ω-BENZOYLOXYACETOPHENONE (ω-BENZOYLOXYPHLOROACETOPHENONE)

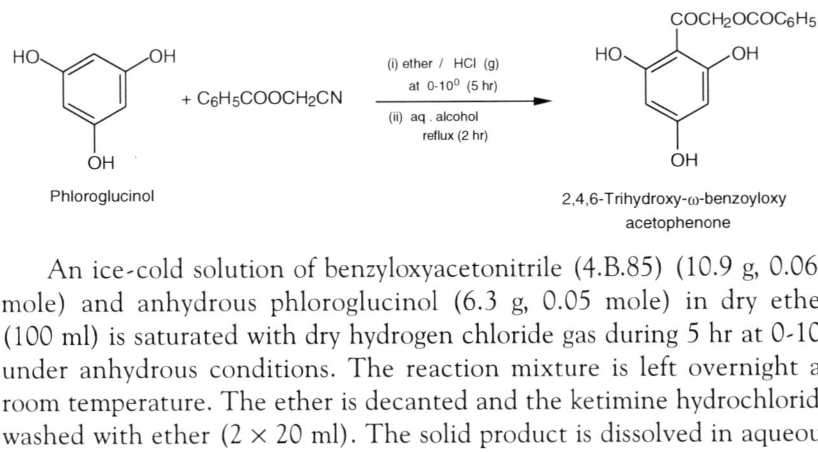

An ice-cold solution of benzyloxyacetonitrile (4.B.85) (10.9 g, 0.068 mole) and anhydrous phloroglucinol (6.3 g, 0.05 mole) in dry ether (100 ml) is saturated with dry hydrogen chloride gas during 5 hr at 0-10° under anhydrous conditions. The reaction mixture is left overnight at room temperature. The ether is decanted and the ketimine hydrochloride washed with ether (2 × 20 ml). The solid product is dissolved in aqueous alcohol (50%, 200 ml) and refluxed for 2 hr. Most of the alcohol is removed by distillation and the solution cooled. The separated product is filtered. It is crystallised from dilute alcohol as pale yellow leaflets. Yield 8.8 g (61%). M.p. 231-33° (lit.[100] m.p. 231-33°).

4.B.87 2,4-DIHYDROXY-ω-ACETOXYACETOPHENONE(ω-ACETOXYRESACETOPHENONE)

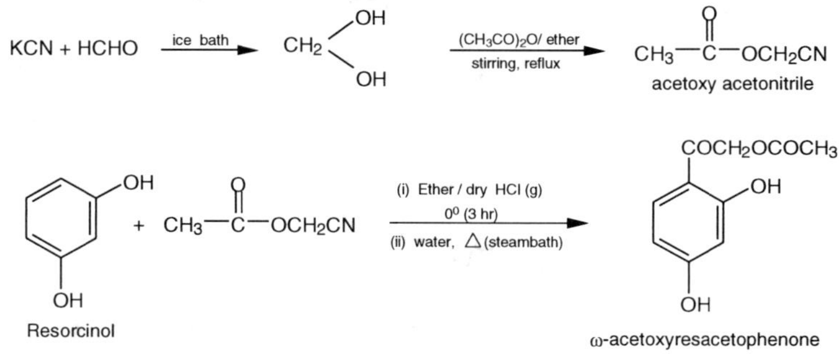

Step (i) Acetoxyacetonitrile

Formalin (40%, 60 ml, 24 g, 0.8 mole) is slowly added to an ice-cold solution. (<15°) of potassium cyanide (56 g, 0.86 mole) in water (125 ml). A solution of acetic anhydride (132 g, 122 ml, 1.29 mole) and dry ether (200 ml) is gradually added with vigorous stirring. The ether refluxed and the success of the operation depends on the efficiency of stirring. The ethereal layer is separated and the aqueous portion extracted with ether (3 × 100 ml). The combined ether solution is dried over anhydrous sodium sulphate and distilled when acetoxyacetonitrile is obtained. Yield 25 g (29.4%). B.p. 175°. (lit.[101] b.p. 175°).

Step (ii) ω-Acetoxyresacetophenone

A mixture of dry resorcinol (5.5 g, 0.05 mole), acetoxyacetonitrile (5 g, 0.05 mole) in dry ether (50 ml) is saturated at 0^0 under anhydrous conditions with hydrogen chloride gas for 3 hr. The mixture is left for 24 hr overnight in ice-chest. The ether is decanted and solid product washed with ether (2 × 30 ml). The remaining ketimine hydrochloride is completely freed from hydrochloric acid in vacuum desiccator over potassium hydroxide. The salt is mixed with ten times its weight of water and gently heated on a steam bath until it dissolves; simultaneously a clear brown oil separates. It solidifies quickly and is filtered. It is crystallised from water as long needles. Yield 7 g (66.7%). M.p. 164° (lit.[102] m.p. 164.5°).

4.B.88 2-HYDROXY-4-METHOXY-ω-NITROACETOPHENONE

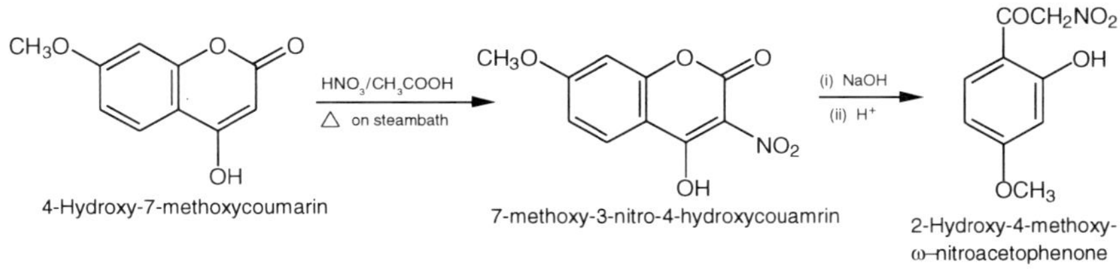

Step (i) 7-Methoxy-3-Nitro-4-Hydroxycoumarin

To a suspension of 4-hydroxy-7-methoxycoumarin (5.A.10) (5 g, 0.026 mole) in glacial acetic acid (15 ml) is added slowly a solution of concentrated nitric acid (2.20 ml) in glacial acetic acid (2 ml). The mixture is placed on a steam bath and the temperature is raised to 80° at which a vigorous reaction starts. The reaction mixture is cooled in ice-cold water to keep the reaction mixture under control. After completion of the reaction, crushed ice is added. The separated product is filtered and washed with water. It is crystallised from methanol as pale yellow needles. Yield 3.5 g (56.7%). M.p. 185° (lit.[103] m.p. 185°).

Step (ii) 2-Hydroxy-4-Methoxy-ω-Nitroacetophenone

The above 7-methoxy-3-nitro-4-hydroxycoumarin (3 g, 0.0126 mole) is dissolved in sodium hydroxide solution (5%, 75 ml) and the reaction mixture is kept at room temperature for 24 hr. Insoluble material is filtered and the clear filtrate acidified with hydrochloric acid. The separated product is filtered and washed with water. It is crystallised from methanol as pale yellow needles. Yield 1.5 g (56.2%). M.p. 123-25° (lit.[103] m.p.123-25°).

4.B.89 2,4-DIHYDROXY-ω-CHLOROACETOPHENONE (ω-CHLORORESACETOPHENONE)

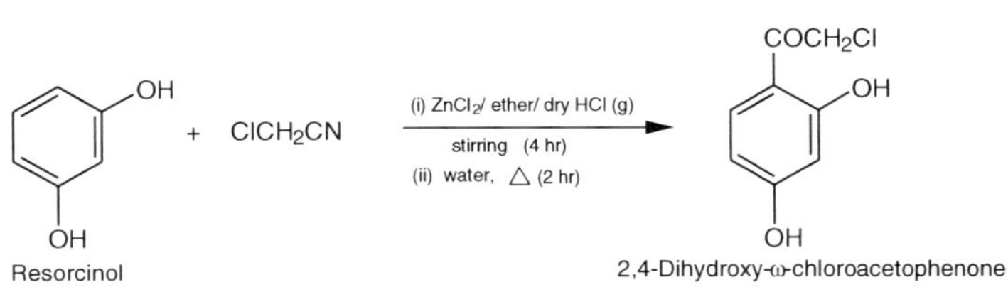

Dry hydrogen chloride gas is passed for 4 hr under anhydrous conditions into a stirred mixture of dry resorcinol (20 g, 0.18 mole), chloroacetonitrile (6.A.12) (14 g, 11.7 ml, 0.185 mole), fused zinc chloride (2 g, 0.0147 mole) and dry ether (150 ml). The reaction mixture is left at room temperature for 18 hr. The clear ether solution is decanted and the residual product washed with dry ether (2 × 30 ml). It is dissolved in water (150 ml) and heated at 100° for 2 hr and the solution cooled. The separated product is filtered and washed with cold water. It is crystallised from hot water as long needles. Yield 18 g (53.1%). M.p. 131° (lit.[104] m.p. 131°).

4.B.90 2-HYDROXY-4-METHOXY-ω-CHLOROACETOPHENONE (ω-CHLORORESACETOPHENONE-4-O-METHYL ETHER)

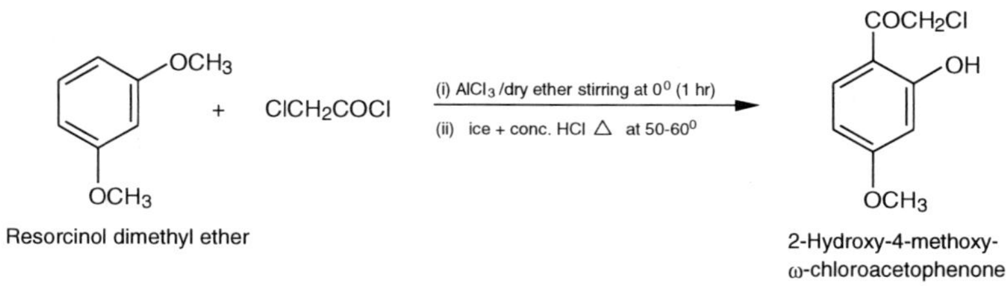

Resorcinol dimethyl ether (1.4) (18 g, 0.13 mole) is added to a solution of anhydrous aluminium chloride (65 g, 0.49 mole) in dry ether (400 ml). To this solution, chloroacetyl chloride (15 ml, 21.27 g, 0.188 mole) in dry ether (50 ml) is added dropwise at 0°. The reaction mixture is dried for 1 hr at 0° and left for 24 hr at room temperature. The ether is distilled off and to the residue is added ice (100 g) and concentrated hydrochloric acid (30 ml). The solution is warmed (50-60°) and cooled. The separated product is filtered and washed with water. It is crystallised from alcohol as colourless needles. Yield 15 g (57.3%), m.p. 116° (lit.[105] m.p. 116°).

4.B.91 2,4,6-TRIHYDROXY-ω-CHLOROACETOPHENONE (ω-CHLOROPHLOROACETOPHENONE)

Dry hydrogen chloride gas is passed for 4 hr under anhydrous conditions into a stirred mixture of phloroglucinol (20 g, 0.159 mole), chloroacetonitrile (6.A.12) (12.5 g. 10.5 ml, 0.166 mole) and fused zinc chloride (6 g, 0.044 mole) in dry ether (150 ml) at 0°. The mixture is left

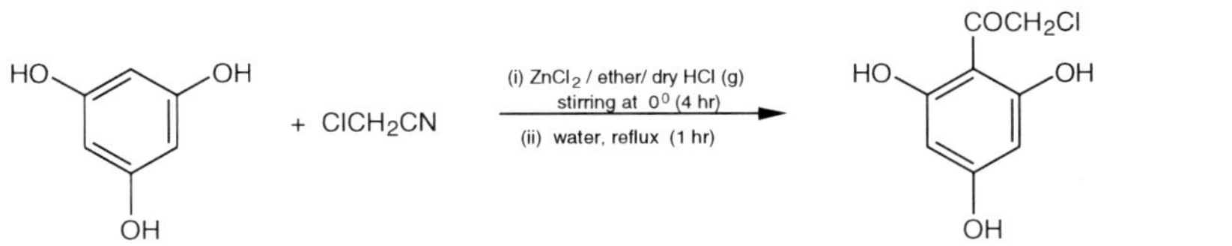

Phloroglucinol — 2,4,6-Trihydroxy-ω-chloroacetophenone

overnight, the ether decanted and the residual solid washed with dry ether (2 × 30 ml). It is dissolved in water (180 ml) and the solution is refluxed for 1 hr. The product obtained on cooling is filtered. It is crystallised from hot water as pink coloured needles. Yield 20 g (62.3%). M.p. 190-91° (lit.[106] m.p. 188-91°).

4.B.92 2-HYDROXY-4,6-DIMETHOXY-ω-CHLOROACETOPHENONE (ω-CHLORO-PHLOROACETOPHENONE DIMETHYL ETHER)

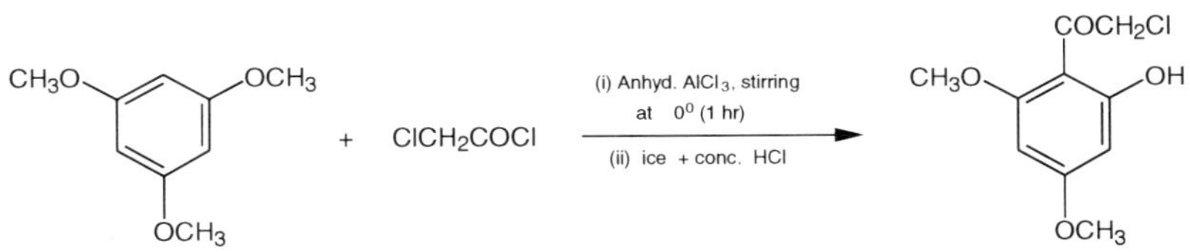

Phloroglucinol trimethyl ether — 2-Hydroxy-4,6-dimethoxy-ω-chloroacetophenone

Powdered anhydrous aluminum chloride (22 g, 0.165 mole) is slowly added under anhydrous conditions to a cooled solution of phloroglucinol trimethyl ether (1.14) (5 g, 0.03 mole) in ether (100 ml). A solution of chloroacetyl chloride (2.6 ml, 3.7 g, 0.033 mole) in dry ether (20 ml) is added dropwise to the cooled solution at 0° (ice-salt mixture). The mixture is stirred for 1 hr more and ether distilled. The residue is treated with ice (50 g) and concentrated hydrochloric acid (20 ml). The mixture is kept for 1 hr at room temperature and the separated product filtered and washed with water. It is crystallised from ethanol as colourless needles. Yield 4 g (58.3%). M.p. 140-42° (lit.[107] m.p. 140-42°).

4.C.93 2-4-DIHYDROXYPROPIOPHENONE (RESPROPIOPHENONE)

An ice cold solution (0-5°, ice-salt mixture) of dry resorcinol (10 g, 0.09 mole) and propionitrile (6.2 ml, 4.96 g, 0.09 mole) in dry ether (50 ml) containing fused zinc chloride (2 g, 0.0147 mole) is saturated with a rapid

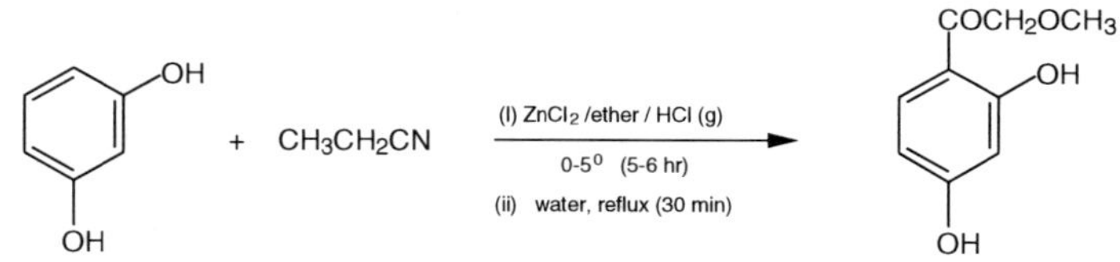

Resorcinol 2,4-Dihydroxypropiophenone

stream of dry hydrogen chloride gas for 5-6 under anhydrous conditions. The mixture is allowed to stand overnight in a refrigerator. The ether is decanted, the solid residue washed with dry ether (2 × 30 ml) and dissolved in water (100 ml). The solution is refluxed for 30 min and cooled. The separated product is filtered, washed with water and dried. It is crystallised from benzene as colourless needles. Yield 10 g (66.2%). M.p. 102-3° (lit.[108] m.p. 96°).

4.C.94 2-HYDROXY-4-METHOXYPROPIOPHENONE

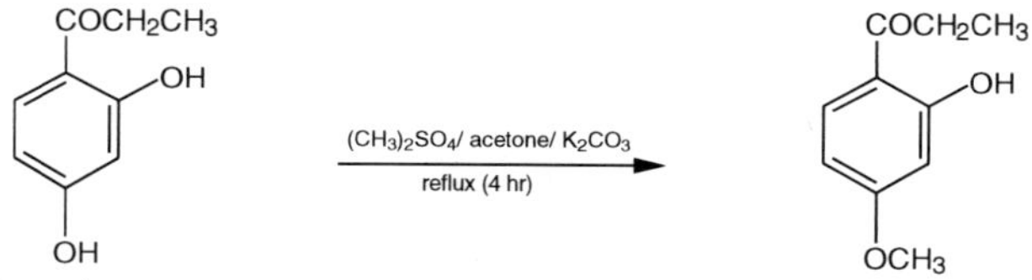

Respropiopnenone 2-Hydroxy-4-methox propiophenone

A solution of respropiophenone (4.C.93.) (5 g, 0.03 mole) in dry acetone (120 ml) is refluxed for 4 hr under anhydrous conditions with dimethyl sulphate (3.2 ml, 4.27 g, 0.034 mole) in presence of ignited potassium carbonate (10 g). The acetone is filtered and the inorganic salts washed with hot acetone (2 × 25 ml). The combined acetone solution is distilled and residue macerated with crushed ice. The solid product is filtered, washed with water and dried. It is crystallised from benzene-petroleum ether as colourless plates. Yield 4.5 g (83%). M.p. 57° (lit.[109] m.p. 58°).

4.C.95 2-HYDROXY-3-DIMETHOXYPROPIOPHENONE[110]

Anhydrous powdered aluminium chloride (15 g, 0.112 mole) is dissolved in anhydrous ether (50 ml) with cooling in ice-chest. Pyrogallol trimethyl ether (1.9) (6 g, 0.036 mole) in dry ether (30 ml) is added to the above solution. The mixture is vigorously stirred, cooled (5°) and treated

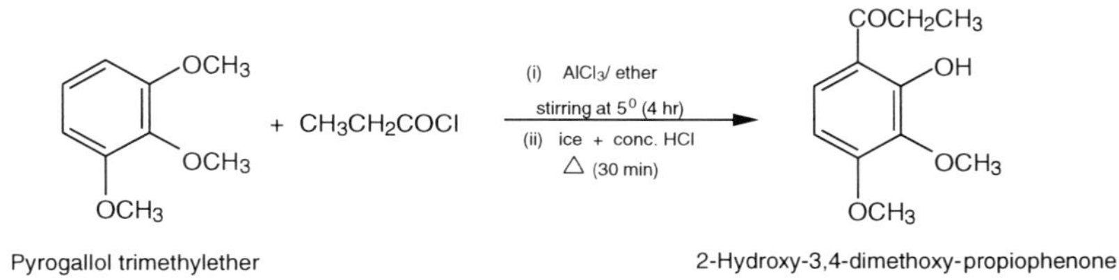

dropwise under anhydrous conditions with a solution of propionyl chloride[111] (3.15 ml, 3.35 g, 0.036 mole) in ether (30 ml) during 1 hr. The mixture is stirred for 4 hr more and allowed to stand overnight at room temperature. The mixture is treated with crushed ice (40 g) and concentrated hydrochloric acid (15 ml) and heated on a water bath for 30 min, ether being allowed to distil away. While hot, the resultant mixture is extracted with benzene (4 × 25 ml). The benzene extract is washed with aqueous sodium carbonate (5%) and then extracted with excess of aqueous sodium hydroxide solution (10%). The clear alkaline solution is acidified in cold. The separated product is filtered, washed with water and dried. It is crystallised from benzene–petroleum ether as pale yellow prisms. Yield 4.2 g (56%). M.p. 93-94°.

4.C.96 2,4,6-TRIHYDROXYPROPIOPHENONE (PHLOROPROPIOPHENONE)

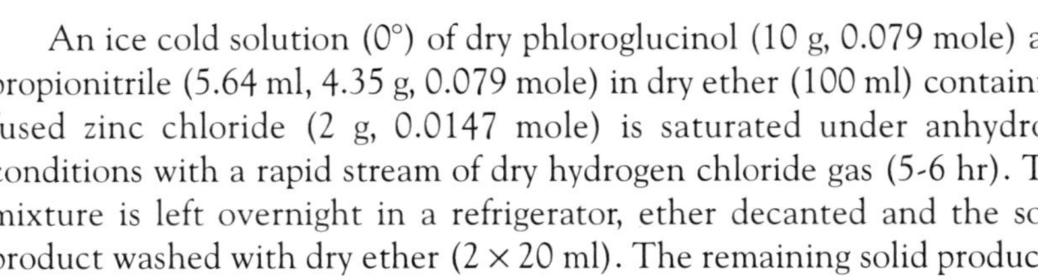

An ice cold solution (0°) of dry phloroglucinol (10 g, 0.079 mole) and propionitrile (5.64 ml, 4.35 g, 0.079 mole) in dry ether (100 ml) containing fused zinc chloride (2 g, 0.0147 mole) is saturated under anhydrous conditions with a rapid stream of dry hydrogen chloride gas (5-6 hr). The mixture is left overnight in a refrigerator, ether decanted and the solid product washed with dry ether (2 × 20 ml). The remaining solid product is dissolved in water (100 ml) and the solution refluxed for 30 min and cooled. The separated product is filtered and washed with water and dried. It is crystallised from benzene as colourless needles. Yield 12 g (83.1%). M.p. 174-75° (lit.[112] m.p. 174-75°).

4.C.97 2-HYDROXY-4,6-DIMETHOXYPROPIOPHENONE

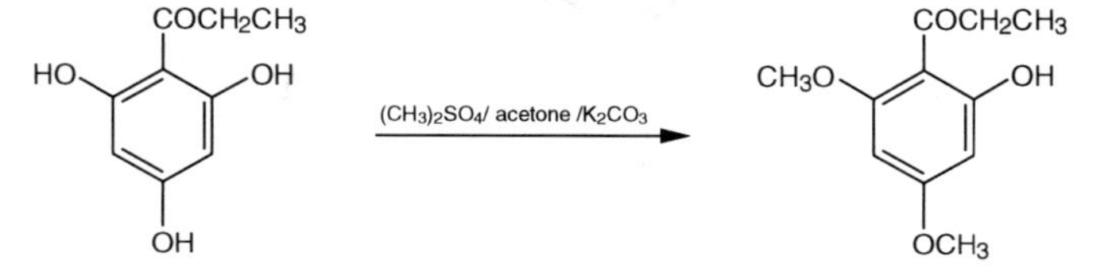

2,4,6-Trihydroxypropiophenone

2-Hydroxy-4,6-dimethoxy propiophenone

A solution of 2,4,6-trihydroxypropiophenone (4.C.96) (5.5 g, 0.03 mole) in dry acetone (150 ml) is refluxed for 4 hr under anhydrous condition with dimethyl sulphate (6.6 ml, 8.8 g, 0.07 mole) and ignited potassium carbonate (15 g). The inorganic salts are filtered and washed with hot acetone (2 × 20 ml). The combined acetone solution is distilled and the residue macerated with ice. The separated product is filtered, washed with water and dried. It is crystallised from benzene-petroleum ether as colourless needles. Yield 4.5 g (71%). M.p. 110-12° (lit.[112] m.p. 111°).

4.C.98 2,4-DIHYDROXY-6-METHYLPROPIOPHENONE

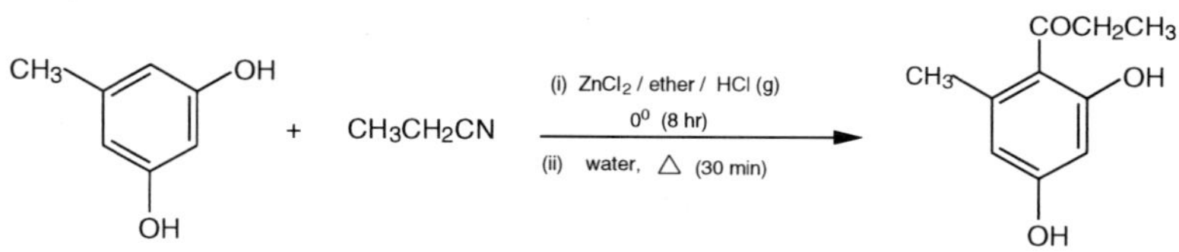

Oricinol

2,4-Dihydroxy-6-methyl propiophenone

A solution of dry orcinol (6 g, 0.048 mole) fused zinc chloride (2 g, 0.0147 mole) and propionitrile (3 g, 3.9 ml, 0.054 mole) in dry ether (50 ml) is saturated at 0° (ice- salt mixture) with dry hydrogen chloride gas for 8 hr under anhydrous conditions. The mixture is left for 12 hr at room temperature. The ether is decanted and oily ketimine hydrochloride washed with dry ether (2 × 20 ml). It is dissolved in water (50 ml) and the solution heated for 30 min in a boiling water bath. The solution is neutralised with dilute ammonia and cooled. The separated product is filtered, washed with water and dried. It is crystallised from benzene from benzene as colourless needles. Yield 5g (57.4%). M.p. 128° (lit.[113] m.p. 128°).

4.C.99 2-HYDROXY-4-METHOXY-6-METHYLPROPIOPHENONE[114]

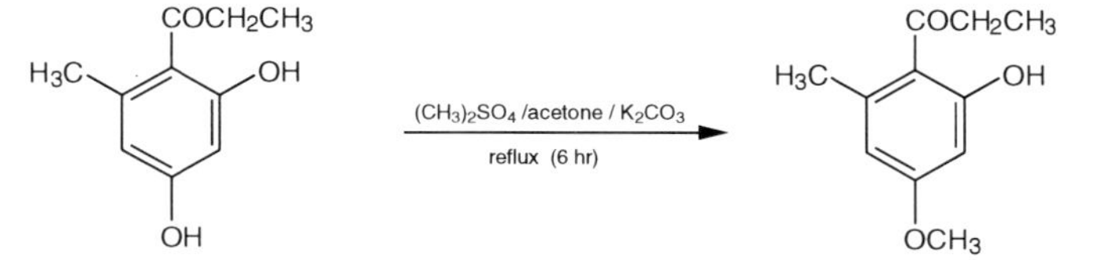

2,4-Dihydroxy-6-methyl propiophenone   2-Hydroxy-4-methoxy-6-methyl propiophenone

A solution of 2,4 dihydroxy-6-methylpropiophenone (4.C.98.) (2 g, 0.011 mole) in dry acetone (50 ml) is refluxed for 6 hr under anhydrous condition with dimethyl sulphate (1.3 ml, 1.7 g, 0.0137 mole) and ignited potassium carbonate (5 g). The mixture is filtered and the inorganic salts washed with hot acetone (2 × 20 ml). The combined acetone solution is distilled and the residue macerated with crushed ice (50 g). The separated product is filtered. It is purified by dissolving in aqueous sodium hydroxide solution and acidifying the clear alkaline solution in cold. It is crystallised from ethyl acetate as glistening needles. Yield 1.4 g (65%). M.p. 84-5°.

4.C.100 2,4-DIMETHOXY-6-METHYLPROPIOPHENONE

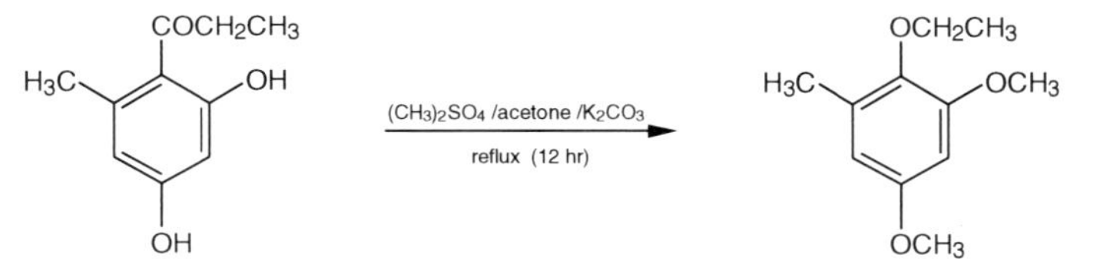

2,4-Dihydroxy-6-methyl propiophenone   2,4-Dimethoxy-6-methyl propiophenone

A solution of 2,4-dihydroxy-6-methylpropiophenone (4.C.98) (2 g, 0.011 mole) in dry acetone (60 ml) is refluxed for 12 hr under anhydrous condition with dimethyl sulphate (2.7 ml, 3.6 g, 0.0285 mole) and ignited potassium carbonate (10 g). The mixture is filtered, and the residue washed with hot acetone (2 × 20 ml). The combined acetone solution is distilled and the oily product macerated with crushed ice. The solid product is filtered, washed with sodium hydroxide solution (5%), water and dried. It is crystallised from ether-petroleum ether as colourless shining needles. Yield 1.8 g (78.3%). M.p. 84-6° (lit.[115] m.p. 84-6°).

4.D.101 2-HYDROXY-4-METHOXYBUTYROPHENONE[110]

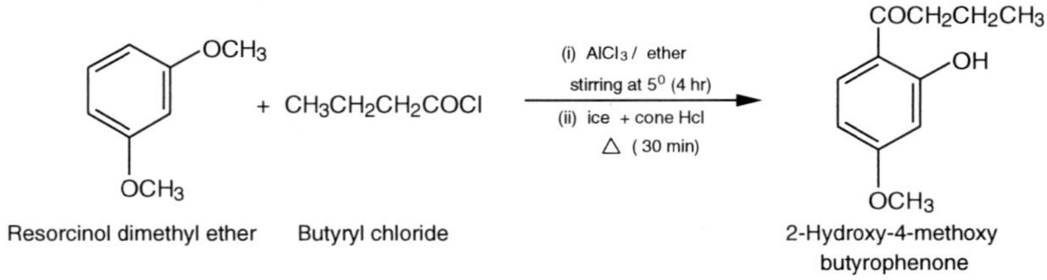

Anhydrous aluminium chloride (15 g, 0.112 mole) is dissolved in absolute ether (50 ml) with cooling in ice-chest and then resorcinol dimethyl ether (1.4.) (10 g, 0.072 mole) in anhydrous ether (20 ml) added. The mixture is vigorously stirred, cooled (5°) and treated dropwise during 30 min with a solution of butyryl chloride[116] (8.5 ml, 8.7 g, 0.082 mole) in ether (30 ml) under anhydrous conditions. The stirring is continued for 4 hr and the reaction mixture left overnight at room temperature. The mixture is treated with ice (40 g) and concentrated hydrochloric acid (15 ml) and heated on a water bath for 30 min, ether being allowed to distil away. While hot the mixture is extracted with benzene (4 × 25 ml). The benzene layer is washed with aqueous sodium carbonate (5%) and extracted with excess of aqueous sodium hydroxide solution (10%). The clear alkaline solution is acidified with hydrochloric acid. The separated oil is extracted with ether, extract washed with water, dried (anhydrous sodium sulphate) and distilled to give the required product as oil. Yield 10.4 g, (74%).

4.D.102 2-HYDROXY-3,4-DIMETHOXYBUTYROPHENONE

Anhydrous aluminium chloride (15 g, 0.0112 mole) is dissolved in anhydrous ether (50 ml) with cooling in ice chest and then pyrogallol trimethyl ether (1.9.) (10 g, 0.06 mole) in anhydrous ether (20 ml) added.

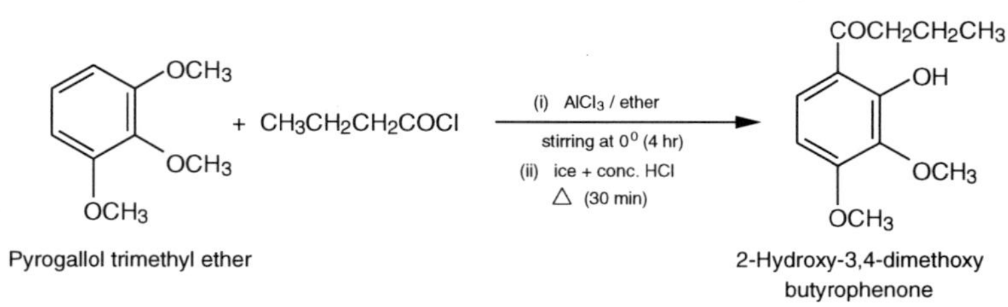

To the above mixture which is vigorously stirred and cooled is added dropwise during 1 hr under anhydrous conditions a solution of butyryl chloride[116] (7 ml, 7.2 g, 0.067 mole) in dry ether (20 ml) and the stirring is continued for 4 hr more. The reaction mixture is left overnight at room temperature. Crushed ice (50 g) and concentrated hydrochloric acid (15 ml) are added and the mixture heated on a water bath for 30 min, ether being allowed to distil away. While hot the mixture is extracted with benzene (4 × 25 ml). The benzene extract is washed with aqueous sodium carbonate solution (5%) and extracted with excess aqueous sodium hydroixde solution (10%). The clear alkaline solution is acidified with hydrochloric acid. The separated product is filtered, washed with water and dried. It is crystallised from benzene-petroleum ether as pale yellow plates. Yield 6 g (45.1%). M.p. 53-4°.

4.D.103 2-HYDROXY-4,6-DIMETHOXYBUTYROPHENONE[110]

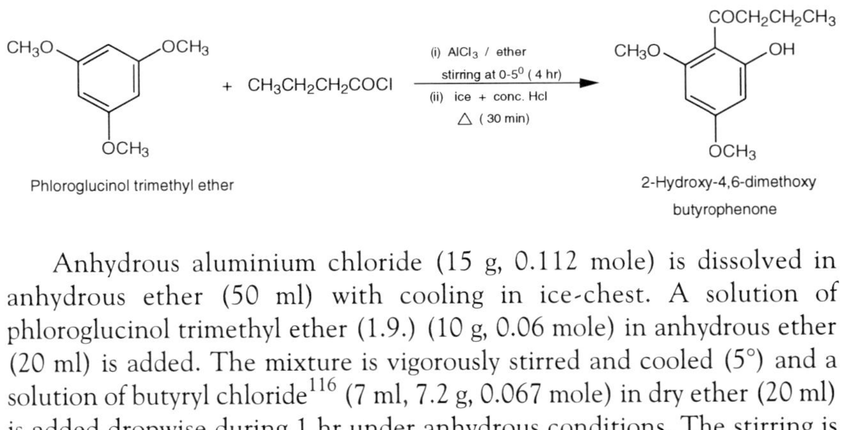

Anhydrous aluminium chloride (15 g, 0.112 mole) is dissolved in anhydrous ether (50 ml) with cooling in ice-chest. A solution of phloroglucinol trimethyl ether (1.9.) (10 g, 0.06 mole) in anhydrous ether (20 ml) is added. The mixture is vigorously stirred and cooled (5°) and a solution of butyryl chloride[116] (7 ml, 7.2 g, 0.067 mole) in dry ether (20 ml) is added dropwise during 1 hr under anhydrous conditions. The stirring is continued for 4 hr more and the reaction mixture left overnight at room temperature. Crushed ice (40 g) and concentrated hydrochloric acid (15 ml) are added and the mixture heated on a water-bath for 30 min, ether being allowed to distil away. The hot mixture is extracted with benzene (4 × 25 ml). The bnezene extract is washed with sodium carbonate solution (5%) and extracted with excess of sodium hydroxide solution (10%). The separated oily product is extracted with ether, ether extract washed with water, dried (anhydrous magnesium sulphate) and distilled. The solid product is crystallised from benzene-petroleum ether as yellow flakes. Yield 9 g (67.7%). M.p. 70-1°.

4.D.104 2,4-DIHYDROXY-ω-PHENYLBUTYROPHENONE

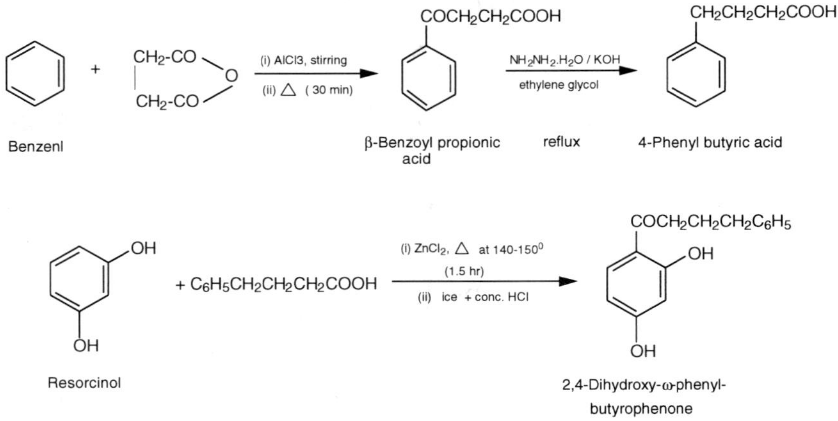

Step (i) β-Benzoylpropionic Acid, $C_6C_5COCH_2CH_2CO_2H$

Succinic anhydride (6 g, 0.66 mole) is added to three necked flask (500 ml capacity) equipped with a mechanical stirrer and a reflux condenser. Dry benzene (55 ml) is added to succinic anhydride and mixture stirred. Powdered anhydrous aluminium chloride (17.6 g, 0.0132 mole) is then added. Hydrogen chloride gas evolves and the mixture become hot. The stirring is continued and the mixture is heated for 30 min in boiling water bath. It is cooled to room temperature and water (25 ml) slowly added. Concentrated hydrochloric acid (83 ml) is then added and the benzene removed by steam distillation. The mixture is cooled when the separated oily product solidifies. It is filtered, washed with cold dilute hydrochloric acid (1:4, 25 ml) and cold water (25 ml). It is purified by dissolving in sodium bicarbonate solution (5%, 150 ml), filtering the insoluble material and acidifying the clear filtrate. The separated product washed with water and dried. Yield 10 g (94%). M.p. 114-15°.

Step (ii) 4-Phenylbutyric Acid, $C_6H_5CH_2CH_2CH_2COOH$

A mixture of β-benzoyl propionic acid (2 g, 0.11 mole), potassium hydroxide (2.14 g, 0.04 mole), hydrazine hydrate (80%, 1.5 ml, 1.2 g, 0.037 mole) in ethylene glycol (15 ml) is refluxed for 1.5 hr using a Dean-Stark apparatus. The water formed is removed. The temperature is allowed to rise to 195^0 and the refluxing is continued for 4 hr. The solution is cooled, diluted with water (15 ml) and poured into hydrchloric acid (6N, 9 ml). The separated product is filtered, washed with water and dried. Yield 1.55 g (84.2%). M.p. 52-3° (lit.[117] m.p.52°).

Step (iii) 2,4-Dihydroxy-ω-phenylbutyrophenone

A mixture of dry resorcinol (1.67 g, 0.015 mole), 4-phenylbutyric acid (2.7 g, 0.016 mole) and freshly fused zinc chloride (2.4 g, 0.0176 mole) is heated for 1.5 hr at 140-150° (oil bath) under anhydrous conditions. The mixture is allowed to cool to room temperature and treated with crushed ice (50 g) and concentrated hydrochloric acid (15 ml). The separated product is filtered, washed with water and dried. It is purified by column chromatography over sillica gel (2 × 30 cm). Initial elution with petroleum ether-benzene (60:40) gave the unreacted acid (1 g, m.p. 52-3°). Subsequent fractions gave the required ketone using petroleum ether-benzene (20:80) as eluant. Yield 1.17 g(30.2%). M.p. 145-46° (lit.[118] m.p. 145-46°).

4.E.105 p-HYDROXYBENZOPHENONE

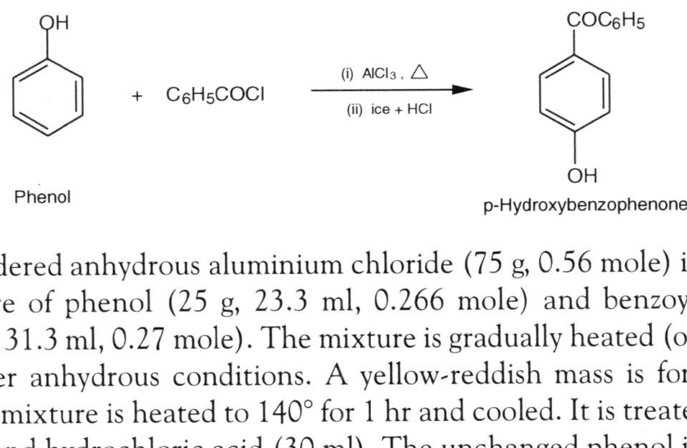

Powdered anhydrous aluminium chloride (75 g, 0.56 mole) is added to a mixture of phenol (25 g, 23.3 ml, 0.266 mole) and benzoyl chloride (37.95 g, 31.3 ml, 0.27 mole). The mixture is gradually heated (oil-bath) to 75° under anhydrous conditions. A yellow-reddish mass is formed. The reaction mixture is heated to 140° for 1 hr and cooled. It is treated with ice (200 g) and hydrochloric acid (30 ml). The unchanged phenol is removed by steam distillation. The residual solid product (obtained after cooling) is filtered and dissolved in sodium hydroxide solution (10%, 200 ml) to separate the hydroxybenzophenone from phenyl benzoate. The clear alkaline solution is acidified in cold. The separated product is filtered and washed with water and dried. It is crystallised from benzene as colourless plates. Yield 20 g (37.9%). M.p. 134° (lit.[119] m.p. 134-35°).

4.E.106 2,4-DIHYDROXYBENZOPHENONE (RESBENZOPHENONE)

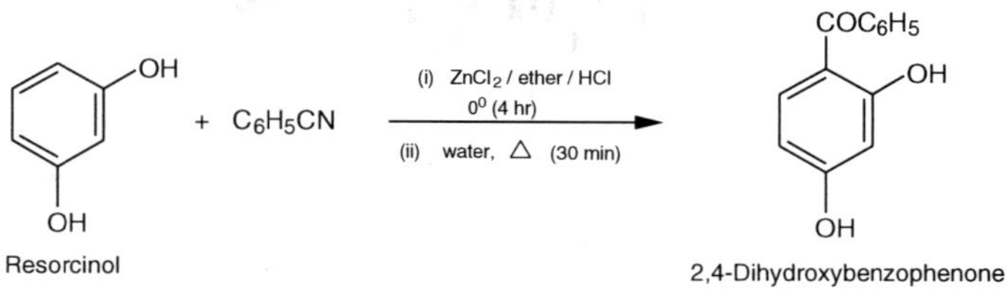

A rapid stream of dry hydrogen chloride gas is passed through a solution of dry resorcinol (20 g, 0.18 mole) and benzonitrile (20 ml, 20.2 g, 0.196 mole) in dry ether (80 ml) in presence of freshly fused zinc chloride (6 g, 0.044 mole) at 0^0 for 4 hr under anhydrous conditions. The reaction mixture is left overnight in refrigerator. The ether solution is decanted and the solid ketimine hydrochloride washed with dry ether (2 × 20 ml). It is dissolved in water (200 ml) and the solution heated in a boiling water for 30 min. The solution is cooled, the separated product filtered and washed with water. It is crystallised from hot water as colourless needles. Yield 20 g (51.4%). M.p. 144° (lit.[120,121] m.p. 143-44°).

4.E.107 2-HYDROXY-4-METHOXYBENZOPHENONE

Method (i)

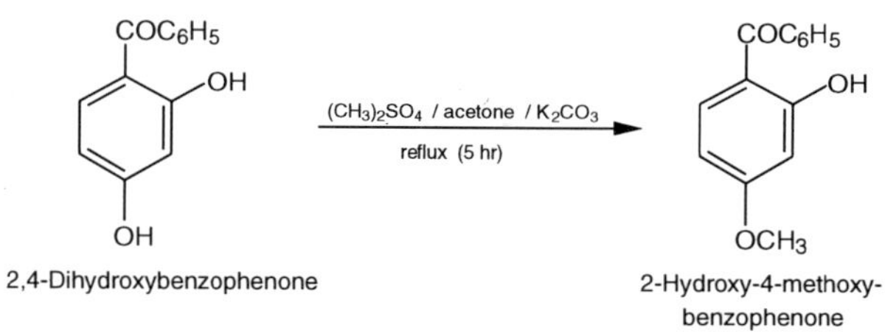

A mixture of 2,4-dihydroxybenzophenone (4.E.106) (5 g, 0.023 mole) and dimethyl sulphate (2.5 ml, 3.33 g, 0.026 mole) in dry acetone (60 ml) containing ignited potassium carbonate (10 g) is refluxed for 5 hr under anhydrous conditions. The mixture is filtered and the inorganic salts washed with hot acetone (2 × 20 ml). The combined acetone solution is distilled and the residue macerated with crushed ice. The oily product is extracted with ether (2 × 30 ml), ether extract washed with sodium carbonate solution, water and dried (sodium sulphate). Ether is distilled and the product is crystallised from dilute alcohol as yellow needles. Yield 4 g (75%). M.p. 66° (lit.[122] m.p. 66°).

Method (ii)

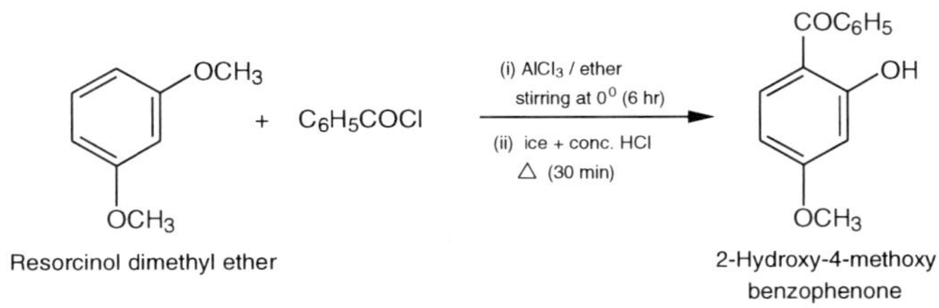

Anhydrous aluminium chloride (20 g, 0.15 mole) is dissolved in anhydrous ether (200 ml) with cooling and resorcinol dimethyl ether (6.4 ml, 14.8 g, 0.107 mole) added. The mixture is vigorously stirred and cooled to 0° (ice-bath) and freshly distilled benzoyl chloride (15 ml, 18.2 g, 0.129 mole) is added dropwise during 1 hr under anhydrous conditions. The mixture is stirred for 6 hr more. An oily red aluminium chloride complex separates. The reaction mixture is kept for 48 hr at room temperature and treated with crushed ice (100 g) and concentrated hydrochloride acid (25 ml). The mixture is heated on a water bath for 30 min, the ether being allowed to distil away. The cooled mixture is extracted with ether (2 × 100 ml). The ether extract is washed with sodium carbonate solution (5%) and then extracted with sodium hydroxide solution (10%, 200 ml). The cooled, clear alkaline solution is acidified with hydrochloric acid and the separated product is filtered and washed with cold water. It is crystallised from dilute alcohol as light yellow needles. Yield 18 g (73.8%). M.p. 66° (lit.[122,123] m.p. 66°).

4.E.108 2,6 DIHYDROXYBENZOPHENONE

Step (i) 7-Benzoyloxy-4-phenylcomarin

7-Hydroxy-4-phenylcoumarin (5.A.18) (9.3 g, 0.039 mole) is heated with benzoyl chloride (6 ml, 7.27 g, 0.052 mole) for 45 min at 185-90° under anhydrous conditions. The reaction mixture is cooled to room temperature and poured over crushed ice. The separated product is filtered, washed with sodium carbonate solution (10%) and finally with water. It is crystallised from alcohol as colourless needles. Yield 11 g (82.3%). M.p. 137° (lit.[124] m.p. 137°).

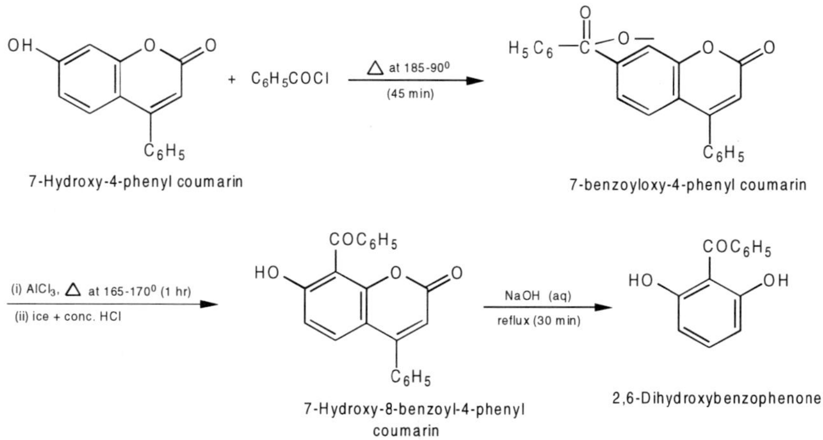

Step (ii) 7-Hydroxy-8-benzoyl-4-phenylcoumarin

An intimate mixture of the above 7-benzoyloxy-4-phenylcoumarin (20 g, 0.058 mole) and anhydrous aluminium chloride (50 g, 0.37 mole) is heated for 1 hr at 165-70° (oil-bath) under anhydrous conditions. The reaction mixture is treated with crushed ice (100 g) and hydrochloric acid (30 ml) and allowed to stand for 2 hr at room temperature with occasional shaking. The separated product is filtered. It is purified by dissolving in sodium hydroxide solution (1N, 70 ml) and acidifying the clear alkaline solution. The product obtained is filtered and washed with water. It is crystallised from glacial acetic acid as flat microscopic needles. Yield 19.3 g (96.5%). M.p. 225° (lit. [124] m.p. 225°).

Step (iii) 2,6-Dihydroxybenzophenone

The above 7-hydroxy-8-benzoyl-4-phenylcoumarin (1 g, 0.0029 mole) is refluxed in sodium hydroxide solution (2N, 7.4 ml) for 30 min on a sand bath. The alkaline solution is steam distilled. The remaining alkaline solution is cooled and acidified with hydrochloric acid. The separated product is filtered and washed with cold water. It is crystallised from hot water as wooly needles. Yield 0.3 g (48%). M.p. 135° (lit.[125] m.p. 135°).

4.E.109 2,5-DIHYDROXY-4-METHOXYBENZOPHENONE

To a stirred and cooled solution (10-15°) of 2-hydroxy-4-methoxybenzophenone (4.E.107) (3 g, 0.013 mole) in aqueous potassium hydroxide

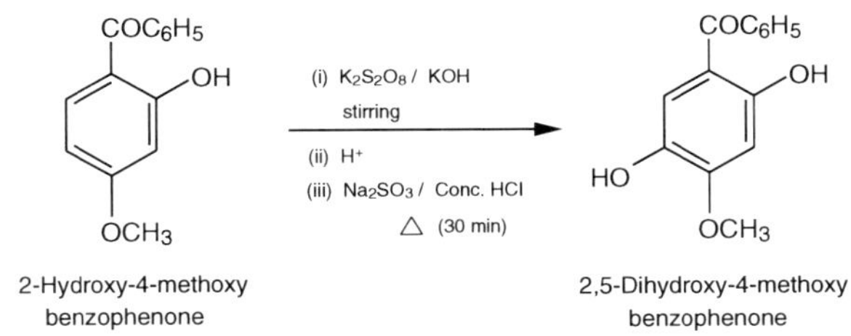

(4 g in 30 ml water) is added dropwise during 3 hr a solution of potassium persulphate (4 g, 0.0148 mole in 100 ml water). The reaction mixture is stirred for 1 hr more and left for 24 hr at room temperature. The mixture is cooled and acidified to congo red with hydrochloric acid (1:1). The separated unreacted ketone is filtered and the aqueous solution extracted with ether (2 × 50 ml). The remaining aqueous solution is heated for 30 min on a water bath with sodium sulphite (4 g) and concentrated hydrochloric acid (20 ml). The solution is cooled and the separated product is filtered. It is crystallised from alcohol as deep yellow thin plate. Yield 0.7 g (21.7%). M.p. 188° (lit.[126] m.p. 188°).

4.E.110 2-HYDROXY-4,5-DIMETHOXYBENZOPHENONE

Method (i)

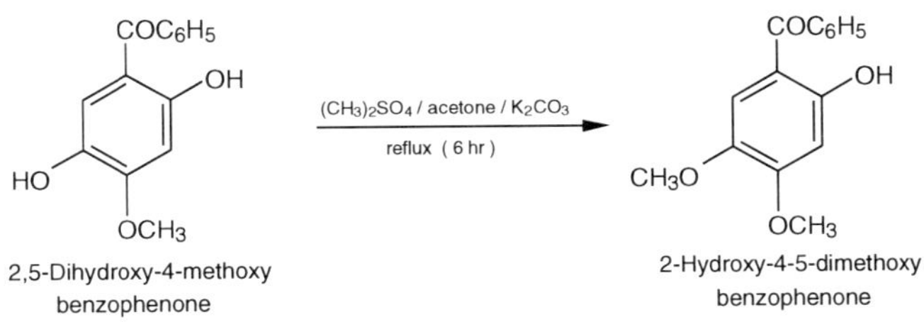

A solution of 2,5-dihydroxy-4-methoxybenzophenone (4.E.109) (5 g, 0.02 mole) in acetone (100 ml) is refluxed for 6 hr under anhydrous conditions with dimethyl sulphate (2.1 ml, 2.8 g, 0.022 mole) in presence of ignited potassium carbonate (8 g). The acetone is filtered and the inorganic salts washed with hot acetone (2 × 25 ml). The combined acetone solution is distilled and the residue macerated with crushed ice (50 g). The separated product is filtered and washed with water. It is crystallised from methanol as almost colourless needles. Yield 4.2 g (79.5%). M.p. 109-110° (lit.[127,128] m.p. 106-107°).

Method (ii)[16]

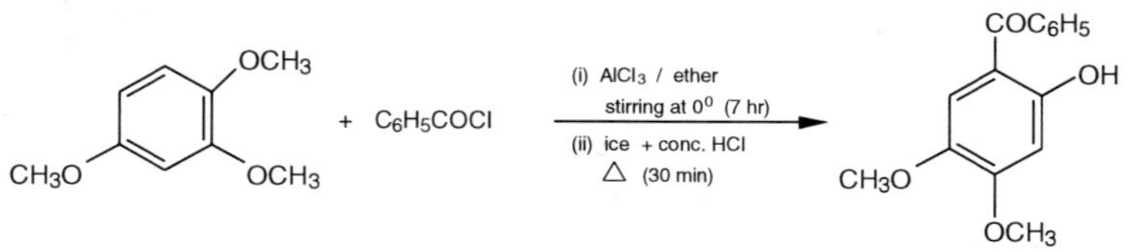

1,2,4-Trimethoxy benzene

2-Hydroxy-4,5-dimethoxy benzophenone

Anhydrous aluminium chloride (20 g, 0.15 mole) is dissolved in anhydrous ether (175 ml) with cooling (0°) and 1,2,4-trimethoxybenzene (1.22) (17.9 g, 0.107 mole) added. The mixture is vigorously stirred and cooled to 0° (ice-bath) and freshly distilled benzoyl chloride (15 ml, 18.2 g, 0.129 mole) is added during 1 hr under anhydrous conditions. The mixture is stirred for 6 hr more and kept for 48 hr at room temperature. Crushed ice (100 g) and concentrated hydrochloric acid (25 ml) are added and the mixture is heated on a water bath for 30 min, the ether is being allowed to distil away. The cooled mixture is extracted with ether (2 × 100 ml). The ether extract is washed with sodium carbonate solution (5%) and then extract with sodium hydroxide solution (10%, 200 ml). The cooled, clear alkaline solution is acidified with hydrochloric acid. The solid product is filtered and washed with cold water. It is crystallised from dilute alcohol as light yellow needles. Yield 20.4 g (74.2%). M.p. 109-110° (lit.[128] m.p. 106-107°).

4.E.111 2,4-DIHYDROXY-3-METHYLBENZOPHENONE

A solution of 2-methyl resorcinol (1.17) (12 g, 0.097 mole) and benzonitrile (10 ml, 10.1 g, 0.098 mole) in dry ether (75 ml) containing freshly fused zinc chloride (4 g, 0.029 mole) is saturated with dry hydrogen chloride gas at 0-5° for 6 hr under anhydrous conditions. The reaction mixture is left overnight in a refrigerator. The ether solution is decanted

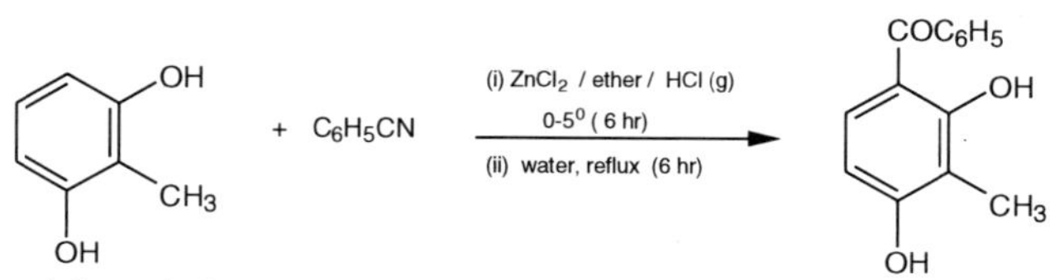

2-methyl resorcinol

2,4-Dihydroxy-3-methyl benzophenone

and the remaining product washed with dry ether (2 × 25 ml). It is dissolved in water (50 ml) and the solution refluxed for 1 hr. The solution is cooled the separated product filtered and washed with water. It is crystallised from methanol as light yellow needles. Yield 12 g (54.3%). M.p. 177-78° (lit.[129] m.p. 177-78°).

4.E.112 2,4,6-TRIHYDROXYBENZOPHENONE (PHLOROBENZOPHENONE)

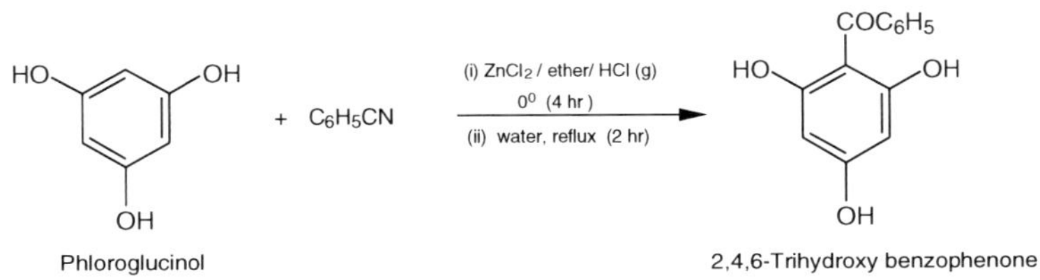

A rapid system of dry hydrogen chloride gas is passed into a solution of anhydrous phloroglucinol (20 g, 0.159 mole) and benzonitrile (20 ml, 20.2 g, 0.196 mole) in dry ether (100 ml) in presence of freshly fused zinc chloride (4 g, 0.029 mole) at 0° for 4 hr under anhydrous conditions. The reaction mixture is left for 24 hr in a refrigerator. The ether solution is decanted and the residual ketimine hydrochloride washed with dry ether (2 × 20 ml). The remaining product is dissolved in water (200 ml) and the solution refluxed for 2 hr. The product separated on cooling is filtered and washed with cold water. It is crystallised from hot water as pale yellow crystal. Yield 22 g (66.3%). M.p. 165-66° (lit.[130] m.p. 165-66°).

4.E.113 2-HYDROXY-4,6-DIMETHOXYBENZOPHENONE

Method (i)

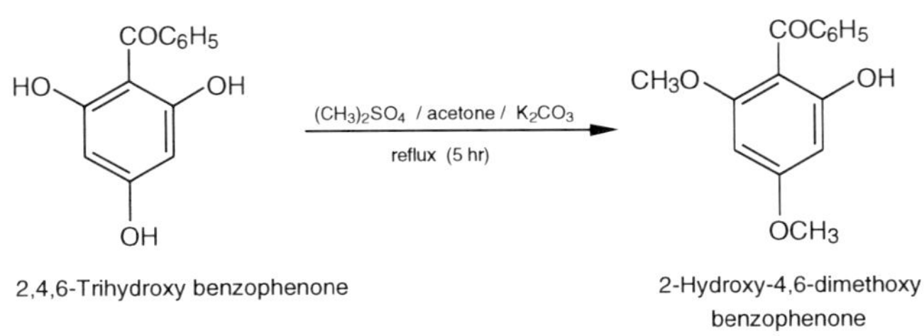

A solution of 2,4,6-trihydroxybenzophenone (4.E.112) (5 g, 0.022 mole) and dimethyl sulphate (4.64 ml, 6.18 g, 0.049 mole) in dry acetone (100 ml) is refluxed for 5 hr with ignited potassium carbonate (10 g) under anhydrous conditions. The reaction mixture is filtered and the inorganic salts washed with hot acetone (2 × 30 ml). The combined acetone solution is distilled and the residual oily product macerated with crushed ice. It is extracted with ether (2 × 50 ml), ether extract washed with sodium carbonate solution (5%) and extracted with sodium hydroxide solution (5%, 150 ml). The clear alkaline solution is acidified with hydrochloric acid. The separated product is filtered and washed with water. It is crystallised from ethanol as light yellow needles. Yield 5 g (89.3%). M.p. 99-100° (lit.[131] m.p. 95-6°).

Method (ii)[132]

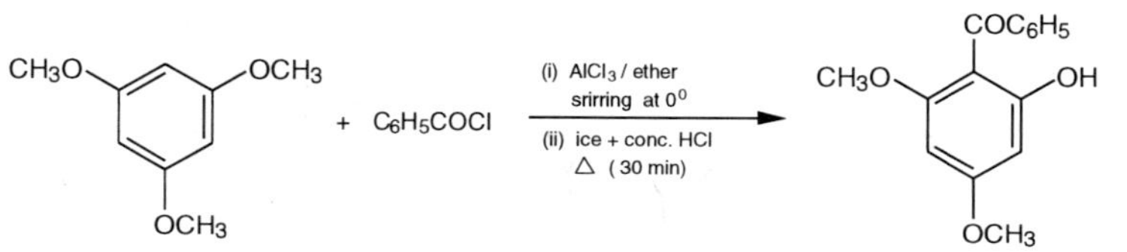

Phloroglucinol trimethyl ether

2-Hydroxy-4,6-dimethoxy benzophenone

To a cooled solution of anhydrous aluminum chloride (9 g, 0.067 mole) is added a solution of phloroglucinol trimethyl ether (1.14) (5 g, 0.03 mole) in dry ether (40 ml). The mixture is cooled (0°) and stirred vigorously. A solution of distilled benzoyl chloride (4.5 ml, 5.45 g, 0.039 mole) in dry ether (20 ml) is added dropwise during 1 hr. The stirring is continued for 6 hr more and the mixture left at room temperature for 48 hr. Ice (100 g) and concentrated hydrochloric acid (15 ml) are added to the reaction mixture. It is heated in a water bath for 30 min, ether being allowed to distil away. The hot reaction mixture is extracted with benzene (2 × 100 ml). The benzene extract is washed with sodium carbonate solution (5%) and extracted with sodium hydroxide solution (5%, 150 ml). The cooled, clear alkaline solution is acidified with hydrochloric acid. The separated product is filtered and washed with water. It is crystallised from alcohol as colourless shining needles. Yield 4.5 g (58.6%). M.p. 99-100° (lit.[131] m.p. 95-6°).

4.E.114 2,4-DIHYDROXY-6-METHOXYBENZOPHENONE[133]

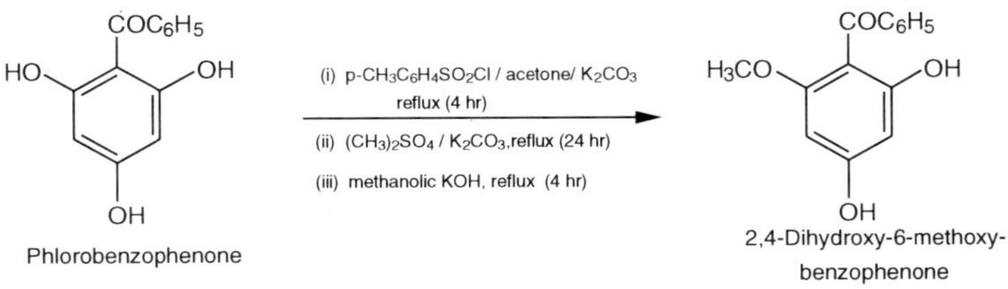

Phlorobenzophenone (4.E.112) (10 g, 0.043 mole) is refluxed with p-toluenesulphonyl chloride (17.7 g, 0.093 mole) in dry acetone (300 ml) in the presence of ignited potassium carbonate (30 g) for 4 hr under anhydrous conditions. Dimethyl sulphate (7 ml, 9.3 g, 0.074 mole) and more of ignited potassium carbonate (15 g) are then added to the reaction mixture and the refluxing is continued for 24 hr more. The acetone is filtered and inorganic salts washed with hot acetone (2 × 30 ml). The combined acetone solution is distilled and the remaining syrupy mass is refluxed for 4 hr with metahnolic potassium hydroxide solution (5%, 350 ml). The methanol is removed in vacuo and the residue is diluted with water (75 ml). The clear alkaline solution is acidified in cold with hydrochloric acid. The separated product is filtered, washed with water and dried. It is crystallised from benzene-petroleum ether as cream yellow needles. Yield 7 g (66%). M.p. 162-3° (lit.[134] m.p. 162°).

4.E.115 2,4-DIHYDROXY-6-METHYLBENZOPHENONE[132]

A solution of dry orcinol (6 g, 0.048 mole) and benzonitrile (6 ml, 6.06 g, 0.059 mole) in dry ether (50 ml) containing freshly fused zinc chloride (3 g, 0.022 mole) is saturated at 0-5° for 6 hr with dry hydrogen chloride gas under anhydrous conditions. After 6 hr, the brownish oily ketimine hydrochloride started separating, the flow of gas is continued for 2 hr more

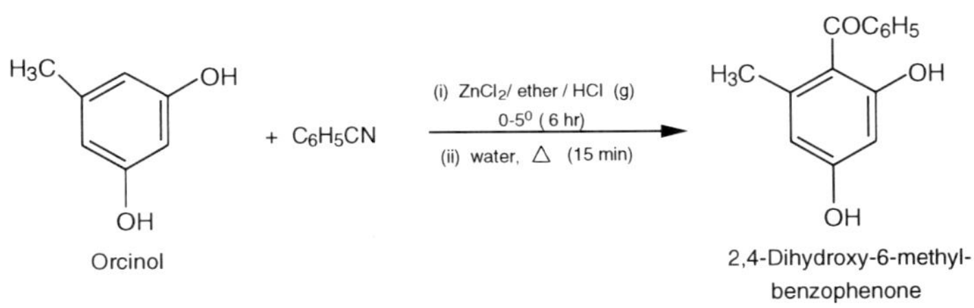

and the reaction mixture left overnight in a refrigerator. The ether is decanted and the solid product washed with dry ether (2 × 20 ml). It is dissolved in water (20 ml) and the solution heated for 15 min, in a boiling water bath. The solution is neutralized with dilute ammonia and the resulting solution cooled. The separated oily product is extract with sodium hydroxide solution (5%). The clear alkaline solution is acidified in cold with hydrochloric acid. The separated yellow product is filtered, washed with water and dried. It is purified by column chromatography over silica gel (2 × 30 cm). Elution with chloroform gave 2,6-dihydroxy-4-methylbenzophenone as yellow solid. It is crystallised from benzene as shining yellow fluffy flakes. Yield 0.5 g (4.5%). M.p. 153-54°. Further elution of the column with methanol -chloroform (1:99) gave the required 2,4-dihdroxy-6-methyl benzophenone. Yield 4.2 g (38%). It is crystallised from benzene as yellow shining needles. M.p 145-46°.

4.E.116 2-HYDROXY-4-METHOXY-6-METHYLBENZOPHENONE[132]

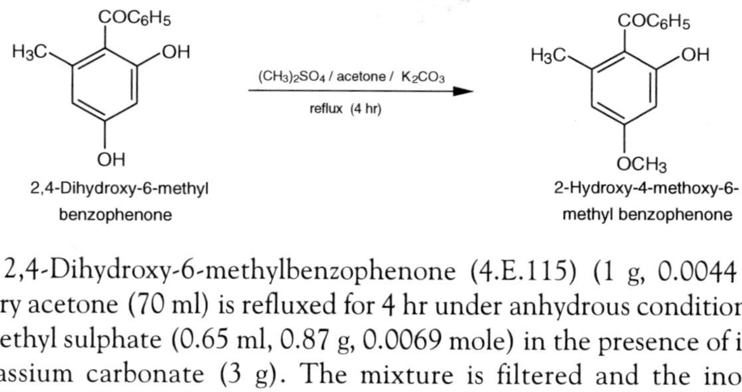

2,4-Dihydroxy-6-methyl benzophenone → 2-Hydroxy-4-methoxy-6-methyl benzophenone

2,4-Dihydroxy-6-methylbenzophenone (4.E.115) (1 g, 0.0044 mole) in dry acetone (70 ml) is refluxed for 4 hr under anhydrous conditions with dimethyl sulphate (0.65 ml, 0.87 g, 0.0069 mole) in the presence of ignited potassium carbonate (3 g). The mixture is filtered and the inorganic residue washed with hot acetone (2 × 20 ml). The combined acetone solution is distilled and the residue macerated with crushed ice. The separated product is filtered and washed with water. It is crystallised from methanol as colourless needles. Yield 0.8 g (75.5%). M.p. 93-4°.

4.F.117 2,4-DIHYDROXYPHENYLBENZYL KETONE (2,4-DIHYDROXYDEOXYBENZOIN)

Method (i)

A continuous stream of dry hydrogen chloride gas is passed through a cooled (0-5°) solution of dry resorcinol (10 g, 0.09 mole) and freshly

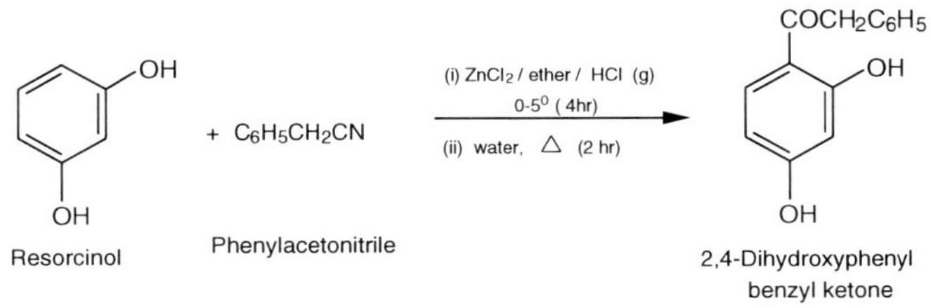

distilled phenylacetonitrile (11 ml, 11.17 g, 0.095 mole) in dry ether (100 ml) containing freshly fused zinc chloride (4 g, 0.029 mole) for 4 hr under anhydrous conditions. The reaction mixture is left for 24 hr in ice-chest. The ether solution is decanted and the solid ketimine hydrochloride washed with dry ether (2×20 ml). It is dissolved in water (100 ml) and the solution heated for 2 hr on a water bath. The solution is cooled and the separated solid product filtered and washed with water. It is crystallised from dilute alcohol as buff coloured needles. Yield 8 g (38.6%). M.p. 114-15° (lit.[135,136] m.p. 114-15°).

Method (ii)

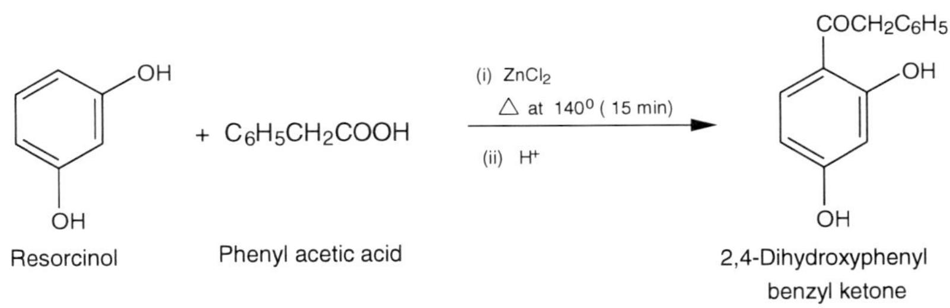

Phenyl acetic acid (30 g, 0.22 mole) is heated on a sand bath in beaker and to the molten mass is added finely powdered fused zinc chloride (25 g, 0.18 mole). After most of the zinc chloride goes into solution at 110°, dry resorcinol (20 g, 0.18 mole) is added and the temperature raised slowly to 140°. The reaction mixture is kept at this temperature for 15 min, cooled and hydrochloric acid (1:1, 100 ml) added. The solid product obtained is filtered, washed with aqueous sodium bicarbonate solution (400 ml), dilute hydrochloric acid and finally with water. It is crystallised from ethyl acetate-petroleum ether as buff coloured needles. Yield 30 g (72.3%). M.p. 116° (lit[135,136] m.p. 114-15°).

4.F.118 2-HYDROXY-4-METHOXYPHENYLBENZYL KETONE

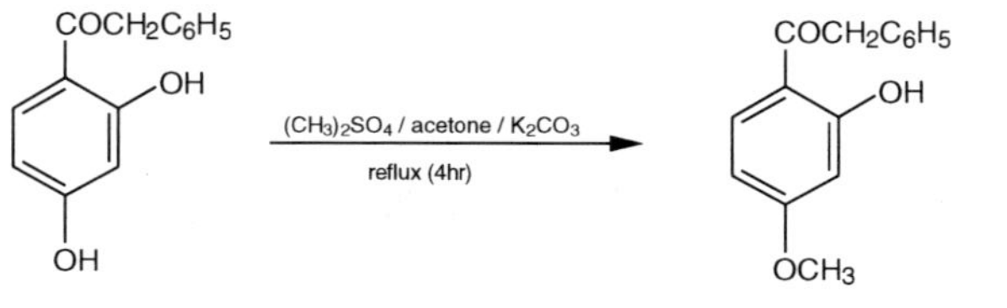

2-Hydroxy-4-methoxyphenyl benzyl ketone

A solution of 2,4-dihydroxyphenyl benzyl ketone (4.F.117) (5 g, 0.022 mole) in dry acetone (150 ml) is refluxed for 4 hr under anhydrous condition with dimethyl sulphate (2.2 ml, 2.9 g, 0.23 mole). The acetone is filtered and the residue washed with hot acetone (2 × 20 ml). The combined acetone solution is distilled and the residual product macerated with crushed ice (70 g). The separated product is filtered and washed with water. It is crystallised from alcohol as colourless prisms. Yield 4.5 g (84.9%). M.p. 87-8° (lit.[136] m.p. 88°).

4.F.119 2,4,6-TRIHYDROXYPHENYL BENZYL KETONE

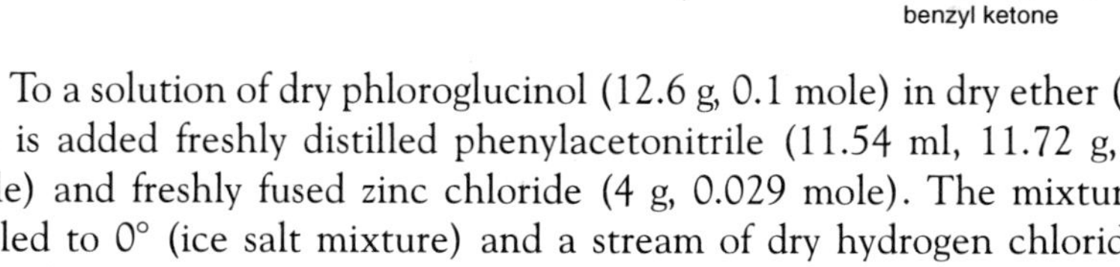

Phloroglucinol

2,4,6-Trihydroxyphenyl benzyl ketone

To a solution of dry phloroglucinol (12.6 g, 0.1 mole) in dry ether (150 ml) is added freshly distilled phenylacetonitrile (11.54 ml, 11.72 g, 0.1 mole) and freshly fused zinc chloride (4 g, 0.029 mole). The mixture is cooled to 0° (ice salt mixture) and a stream of dry hydrogen chloride is passed for 4 hr under anhydrous conditions. The reaction mixture is left for 24 hr in ice-chest. The ether solution is decanted and the solid product washed with dry ether (2 × 25 ml). It is dissolved in water (100 ml) and the solution is heated for 2 hr on water bath. The solution is cooled and the solid product is filtered and washed with cold water. It is crystallised from hot water as long colourless needles. Yield 12 g (49%). M.p. 162-63° (anhydrous sample) (lit.[135,136] m.p. 162°).

4.F.120 2-HYDROXY-4,6-DIMETHOXYPHENYL BENZYL KETONE

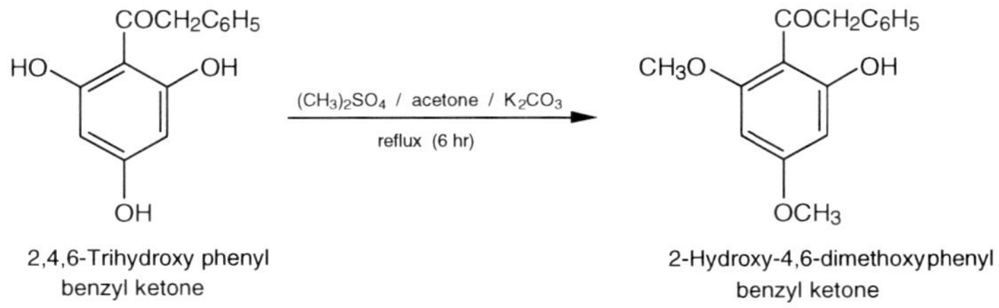

A solution of 2,4,6-trihydroxyphenyl benzyl ketone (4.F.119) (12 g, 0.049 mole) in dry acetone (250 ml) is refluxed for 6 hr under anhydrous conditions with dimethyl sulphate (10 ml, 13.33 g, 0.106 mole) in presence of ignited potassium carbonate (25 g). The mixture is filtered and the inorganic salts washed with hot acetone (2 × 20 ml). The combined acetone solution is distilled and the residue macerated with crushed ice (100 g). The separated product is filtered and washed with water. It is crystallised from methanol as colourless needles. Yield 10 g (74.6%). M.p. 115-16° (lit.[136] m.p. 117°).

4.F.121 2,4-DIHYDROXY-6-METHOXYPHENYL BENZYL KETONE

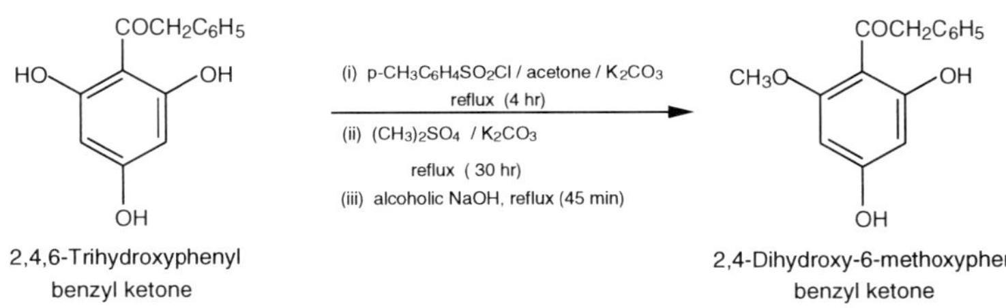

To a solution of 2,4,6-trihydroxyphenyl benzyl ketone (4.F.119) (2.5 g, 0.01 mole) in dry acetone (100 ml) is added p-toluenesulphonyl chloride (4.3 g, 0.023 mole) and ignited potassium carbonate (20 g). The mixture is refluxed for 4 hr under anhydrous conditions. To the reaction mixture is added dimethyl sulphate (1 ml, 1.3 g, 0.011 mole) and ignited potassium carbonate (5 g) and refluxing is continued for 30 hr more. The acetone is filtered and the inorganic residue washed with hot acetone (2 × 25 ml). The combined acetone solution is distilled and the residual product is refluxed for 45 min with alcoholic sodium hydroxide (10 g in 50 ml water and 50 ml alcohol). The alcohol is removed in vacuo and water (100 ml) added. The cooled solution is acidified with hydrochloric acid. The separated product is filtered and washed with water. It is purified by dissolving in sodium carbonate solution (5%) and acidifying the clear solution. The separated product is filtered. It is crystallised from benzene-petroleum ether as small light brown needles. Yield 1 g (37.9%). M.p. 145-46° (lit.[137] m.p. 145-46°).

4.G.122 2,4,6-TRIHYDROXYISOBUTYROPHENONE (PHLOROISOBUTYROPHENONE)

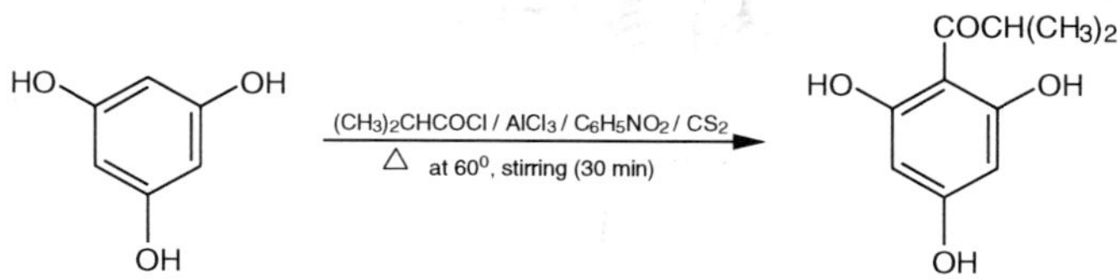

Phloroglucinol

2,4,6-Trihydroxy isobutyrophenone

A mixture of anhydrous phloroglucinol (7.5 g, 0.06 mole), anhydrous powdered aluminium chloride (24 g, 0.28 mole) dry carbon disulphide (30 ml) is stirred for 30 min in a three necked R.B.flask (250 ml capacity) fitted with a mechanical stirrer, a dropping funnel and a reflux condenser protected by a calcium chloride drying tube. Dry nitrobenzone (25 ml) is added and the mixture is stirred for 30 min more. The reaction mixture is heated in a water bath at 60° and a solution of isobutyryl chloride[138] (6.5 g, 0.06 mole) in nitrobenzene (10 ml) is added dropwise during 30 min keeping the temperature of the water bath at 60°. The mixture is stirred for 30 min and then poured over crushed ice. Rochelle salt (sodium potassium tartarate) (213 g, 0.75 mole) is added and the solution neutralised with aqueous sodium hydroxide (40%). The nitrobenzene and carbon disulphide are removed by vigorous steam distillation (20 min). The mixture is cooled (overnight, refrigerator) and the separated product is filtered. Yield 4.7 g (40%). It is crystallised twice from hot water. m.p.78-80°. The anhydrous sample is obtained by drying in a vacuum desiccator over phosphorus pentoxide for 2-3 days. It melts at 138-40° (lit.[139] m.p. 70° (hydrated), 138° (anhydrous).

4.G.123 2,4,6-TRIHYDROXYISOVALEROPHENONE (PHLOROISOVALEROPHENONE)

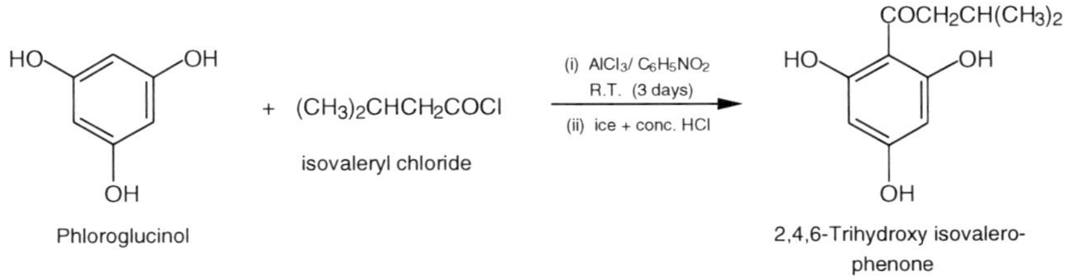

To a solution of anhydrous phloroglucinol (16 g, 0.127 mole) and anhydrous powdered aluminum chloride (20 g, 0.15 mole) in dry nitrobenzene (200 ml) is added isovaleryl chloride[140] (15.67 g, 15.85 ml, 0.13 ole) slowly with shaking. The mixture is left for 3 days at room temperature with occasional shaking. It is poured over crushed ice (200 g) and hydrochloric acid (200 ml) with stirring. The mixture is allowed to stand for 2-3 hr at room temperature and extracted with ether. The ether extract is washed with sodium bicarbonate solution (5%) and distilled. The remaining solution is steam distilled to remove nitrobenzene. The aqueous solution is cooled and the separated product filtered. It is crystallised from hot water to give buff coloured needles. Yield 9 g (33.8%). M.p. 145° (anhydrous sample) (lit.[141] m.p. 145°).

4.G.124 BIS(2,4-DIHYDROXY-3-ACETOPHENYL) METHANE R-CH_2-R (R=2,4-DIHYDROXY-3-ACETYLPHENOL)

β-Resacetophenone (4.A.6) (5 g, 0.033 mole) is dissolved in ethanolic sodium ethoxide (4%, 50 ml) and methylene iodide (2 ml, 6.65 g, 0.025 mole) added. The mixture is left for 18 hr at room temperature. It is heated for 1 hr at 60-70° and the solvent is distilled off. Water (100 ml) is added to the residue and the mixture extracted with ether (2 × 50 ml). The clear

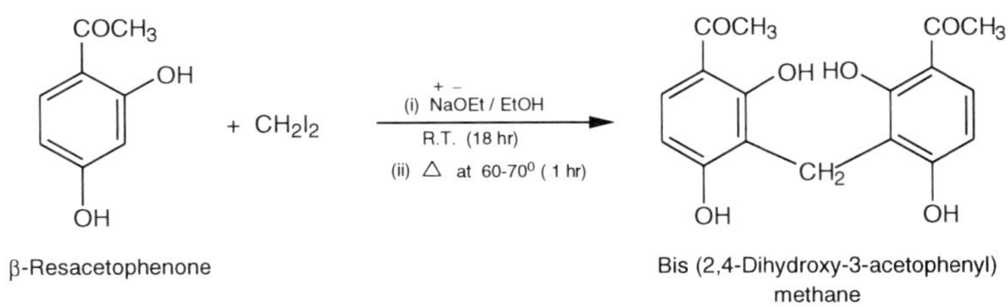

alkaline solution is cooled and acidified with hydrochloric acid. The separated product is filtered and washed with water. It is crystallised from alcohol as colourless small prisms. Yield 1 g (9.6%). M.p. 204-205° (lit.[142] m.p. 204-205°).

4.G.125 5,5-DIMETHYL-1,3-CYCLOHEXANEDIONE (DIMEDONE)

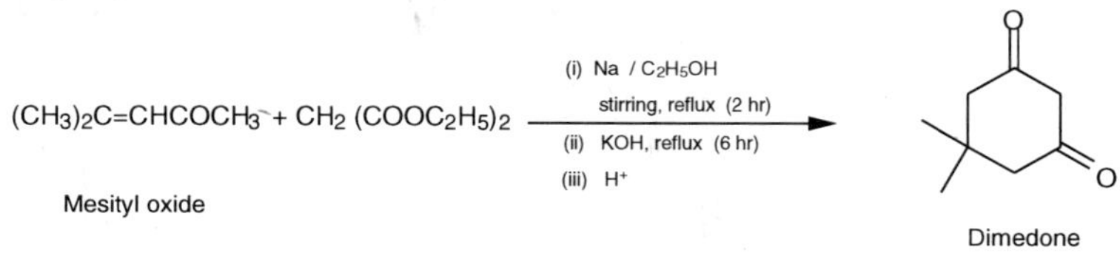

Absolute alcohol (200 ml) is placed in 1 litre three necked R. B. flask fitted with a mercury sealed stirrer, a dropping funnel and a reflux condenser fitted with a calcium chloride drying tube. Sodium (12.5 g, 0.54 mole) is added through the condenser in small amount so that the solution is kept at refluxing temperature. After all the sodium has dissolved, ethyl malonate (85 g, 80.57 ml, 0.53 mole) is added followed by addition of pure mesityl oxide (50 g, 58.3 ml, 0.51 mole) through the dropping funnel. The solution is constantly stirred and refluxed for 2 hr and then aqueous potassium hydroxide (62.5 g, 1.1 mole in 280 ml water) is added, mixture stirred and refluxed for 6 hr (water bath). The hot mixture is acidified to litmus with dilute hydrochloric acid (1:2 approx. 280 ml is required). The excess alcohol is distilled off (water bath), the remaining mixture is treated with decolorizing charcoal and filtered. The solution is again acidified to litmus with dilute hydrochloric acid (approx. 75 ml) boiled and charcoalised. The hot clear solution is made distinctly acidic to methyl orange with more dilute hydrochloric acid (30-50 ml), boiled for few a minutes and cooled. The separated dimedone is filtered, washed with cold water. It is crystallised from acetone. Yield 60 g, (85%). M.p. 147-48° (lit.[143] m.p. 147-48°).

References

1. Dictionary of organic compounds by Chapman and Hall 5th edition, H-01169
2. Dictionary of organic compounds by Chapman and Hall, 5th Edition, H-01171

3. Dictionary of organic compounds by Chapman and Hall, 5th Edition, N-00667
4. Barkenbus and Clements, J. Am. Chem. Soc., 1870, 3, 886.
5. E.G. Cobb, Proc. S. Dakola Acad. Sci., 1945, 25, 64; Chem. Abstr., 1946, 7180.
6. Robinson and Shah, J. Chem. Soc, 1934,1491.
7. Dictionary of organic compounds by Champman & Hall, 5th Edition, D-04142.
8. R. Huisgen, G. Seidl and I. Wimmes, Justus Liebigs Ann. Chem., 1964, 677, 21.
9. E.O. Robert and Clifts S. Hamilton, J. Am. Chem. Soc., 1937, 59, 642
10. Ciusa and Sollazo, Chem. Zentr., 1943, 114, II, 615; Heile, Ann, 1906, 350, 344.
11. Rosenmund and Lohfert, Ber., 1928, 61, 2606.
12. A.T. Shulgin and D.C. Dyer, J. Med. Chem. 1975, 18, 1201.
13. Dictionary of organic compounds by Chapman & Hall, 5th Edition, D-04143
14. V.K. Ahluwalia, D.S. Gupta and T.R. Seshadri, Proc. Indian Acad. Sci., 1953, 38A, 480.
15. 2,3-Dimethoxybenzoyl chloride is prepared by refluxing 2,3-dimethoxybenzoic acid (III.5) with 1.2 equivalents of thionyl chloride (water-bath, 30 min). The product obtained after removal of excess thionyl chloride is crystallised from petroleum ether as colourless stout needles. M.p. 54-5°.
16. V.K. Ahluwalia et al., Unpublished results.
17. A.S.R. Anjaneyulu, U.V. Malavadham, Y. Vekateswarlu and A.V. Rama Prasad, Indian J. Chem., 1987,823
18. Devendra Kumar, Ph.D. Thesis, Delhi University, 1976, p.105.
19. Dictionary of organic compounds by Chapman & Hall, 5th Edition, D-04141
20. Dictionary of organic compounds by Chapman & Hall, 5th Edition D-04144.
21. Baker, J. Chem. Soc., 1934,1953.
22. A. Bobik, G.M. Holder and A.J. Rayan, J. Med. Chem., 1977,20, 1194.
23. W. Baker, J.Chem.Soc., 1939, 1956.
24. A. Bobik, G.M.Holder and A.J. Rayan , J. Med. Chem., 1977, 20, 1194
25. Kanchan Kapoor , Ph.D. Thesis, Delhi University, 1981, p, 93.
26. W. Ploeg, Red Trav. Chim., Pays-Bas Belg, 1926, 45, 392

27. H. Stephen and C. Weizmann, J. Chem. Soc., 1914, 105, 1051.
28. V.K. Ahluwalia, Kanchan Kapoor and Saroj Manchanda, Indian J. Chem., 1982, 21B, 186.
29. I. Reichstein, Helv. Chim. Acta, 1927, 10, 392.
30. W. Schneider and E. Kroft, Ber, 1922,55B, 1892.
31. Moed, Dijk and Niewind, Recuel, 1958, 77, 278.
32. Dictionary of organic compounds by Chapman & Hall, 5th Edition, D-04145
33. K.C. Kshatriya , N.S. Shodhan and K.S. Nargund, J. Indian Chem. Soc., 1947, 24, 373
34. C.S. Gibson and B. Levin, J. Chem. Soc., 1931, 2395.
35. S. Rangaswamy and T.R. Seshadri, Proc. Indian Acad. Sci., 1938, 8A, 214.
36. F. Kallay and G. Janzso, Tetrahedron Lett., 1978, 1443.
36a. Dictionary of organic compounds by Chapman & Hall, 5th Edition, D-04617
37. K. Hoesch, Chem. Ber., 1915, 48, 1122.
38. Dictionary of organic compounds by Chapman & Hall, 5th Edition, D-04618.
39. T. Suga, H. Hitrata and F. Walls, J. Sci. Hiroshima Univ., 1974, 38, 327.
40. J. Tambor, Ber., 1908, 41, 793.
41. P.R. Saraiya and R.C. Shah, Proc. Indian Acad. Sci., 1950, 31A, 220.
42. Dictionary of organic compounds by Chapman & Hall, 5th Edition, D-04621.
43. V.K. Ahluwalia, Neelu Kalia and Shashi Bala, Indian J. Chem., 1986, 25B, 663.
44. Nencki and Sieber, J. Prakt. Chem., 1881, 23, 151, 538; Crabtree and Robinson, J. Chem., Soc., 1922, 121, 1038.
45. P. Price and S.S. Israelstam, J. Org. Chem., 1964, 29, 2800.
46. S.K. Aggarwal, S.K. Grover and T. R. Seshadri, Indian J. Chem., 1972, 10, 911.
47. I. Das, N. Narsimhachari and T.R. Seshadri, Proc. Indian Acad. Sci., 1952, 37A, 603.
48. W. Baker, J. Chem. Soc., 1941, 662.
49. V.D.N. Shastri and T.R. Sheshadri, Proc. Indian Acad. Sci., 1946, 23A, 262.
50. M. Blumberg and B. Kostanecki , Ber., 1903, 36, 2191
51. W. Baker, J. Chem. Soc., 1939, 956.

52. Dictionary of organic compounds by Chapman & Hall, 5th Edition, T-03239.
53. L.E. Smith and H.L. Haller, J. Am. Soc. Chem., 1934, 56, 237.
54. Dictionary at organic compounds by Chapman & Hall, 5th Edition. T-03240
55. K.R. Laumas, S. Neelkantan and T.R. Seshadri, Proc. Indian Acad. Sci., 1957, 46A, 343
56. R. Robinson and K. Venkataraman, J. Chem. Soc., 1926,2347.
57. Dictionary of organic compounds by Chapman & Hall, 5th Edition, T-70248.
58. R. Huisgen, G.seidl and I. Wimmer, Justus Liebigs Ann Chem., 1964, 677, 21.
59. B.S. Bajwa, Pyare Lal and T.R. Seshadri, Indian J. Chem., 1971,9,17.
60. V.K. Ahluwalia, N. Rani and R. Gupta, Nat. Acad. Sci Letters, 1978, 1(10), 369.
61. Anjula Jain, Ph.D. Thesis, Delhi University, 1983, p.141.
62. W. Baker, J. Chem. Soc., 1941,665
63. G. Bergelline and A. Olverio, Ber., 1942, 75B, 2083;
64. P. Roland Hul, A. Jubler, J. Andulux and A. Resplande, Bull. Soc. Chim. France, 1970, 10, 3617.
65. P. F. Wiley, J. Am. Chem. Soc., 1954, 74, 4329
66. Dictionary of organic compounds by Chapman & Hall, 5th Edition, T-01170.
67. V.D.N Sastri and T.R. Seshadri, Proc. Indian Acad. Sci., 1946, 24A, 248.
68. E. Chapman, A. G. Perkin and R. Robinson, J. Chem. Soc., 1927, 3015.
69. Z. Horri, J. Pharm. Soc. Japan, 1940, 60, 614; Chem. Abstr. 1941, 35, 7964.
70. W. Baker, J. Chem. Soc., 1941,665
71. F.H. Curd and A. Robertson, J. Chem. Soc., 1933, 437.
72. A.C. Jain, P. Lal and T.R. Seshadri, Tetrahedron, 1969, 25, 283
73. M.U. Tsao, J. Am. Chem. Soc., 1951, 73, 5495.
74. S. Goodwin and B. Witkop, J. Am. Chem. Soc., 1957, 79,179.
75. W. Waker and H. Raistrick , J. Chem. Soc., 1941,671
76. A. Robertson and W. B. Whally, J. Chem. Soc., 1951,3355
77. H.A. Shah and R.C. Shah, J. Chem. Soc., 1939,949
78. S.K. Mukerjee, T.R. Seshadri and S. Varadarajan, Proc. Indian Acad. Sci., 1953, 37A, 140.

79. R.S.Mc. Credri, E. Ritchie and W.C. Taylor, Abstr. J. Chem., 1969, 22, 1011.
80. Dictionary of organic compounds by Chapman & Hall, 5th Edition, H-01168.
81. W.K. Slater and H. Stephen, J. Chem ,Soc., 1920.117, 312.
82. Dictionary of organic compounds by Chapman & Hall, 5th Edition, T-03235
83. J. Chopin, P. Durual and M. Chandensen, Bull. Soc. Chim. Fr., 1965, 4, 3572
84. M. Nierenstein, D.G. Wang and J.C. Warr, J. Am. Chem. Soc., 1924, 46, 2551
85. S.K. Grover, V.N. Gupta, A.C. Jain and T.R. Seshadri, J. Sci. Industrial Res., 1960, 19B, 258.
86. R. Robinson and K. Venkataraman, Chem. Soc., 1929, 63.
87. R.L. Shriner and C.J. Hull, J. Org. Chem., 1945, 10, 228.
88. V.K. Ahluwalia and Chandra Prakash , Indian J. Chem., 1976, 14B,586.
89. V.K. Ahhluwlia, Chandra Prakash and R.P. Singh, Chemistry and Industry, 1980, 464.
90. L.R. Row and T.R. Seshadri , Proc. Indian Acad. Sci., 1945, 21A, 155
91. L.R. Row and T.R. Seshadri, Proc. Indian Acad. Sci., 1946, 23A, 23.
92. W. Baker, R. Nodzu and R. Robinson, J. Chem. Soc., 1929, 74.
93. P.R. Rao, P.S. Rao and T.R. Seshadri, Proc. Indian Acad. Sci., 1944, 19A, 88.
94. V.V.S Murti, L.R. Rao and T.R. Seshadri, Proc. Indian Acad. Sci., 1946, 24A, 233.
95. P.R. Raw and T.R. Seshadri, Proc. Indian Acad. Sci., 1945, 22A, 157.
96. W. Karrer, Helv. Chim. Acta, 1934, 17, 1560.
97. L.R. Rao and T.R. Thiruvendadam, Curr. Sci.,1947,16 ,379.
98. Meera Ahuja, Ph.D. Thesis, Delhi University , 1974, p.150.
99. Aloy and Rabaut, Bull. Soc. Chim., 1913, 13, 457.
100. T. Heap and R. Robinson, J. Chem. Soc., 1926, 2336.
101. Chavan and Robinson, J. Chem. Soc., 1933, 369.
102. Charlesworth, Chavan and Robinson, J. Chem. Soc., 1933, 371.
103. K. Paradhasaradhi, Ph.D. Thesis, Delhi University, 1954, p. 153.
104. A. Sonn, Ber., 1917, 50, 1267.
105. Von Auwers and Pohl, Ann., 1914, 405, 264.
106. R.L. Shiner and F. Grosser, J. Am. Chem. Soc., 1942, 382.
107. P. Friedlaender and L.C. Schnell, Ber., 1897, 30, 2153.

108. C.M. Brewster and J.C. Harris, J. Am. Chem. Soc., 1930, 52, 4869.
109. Dictionary of oragnic compounds by Chapman & Hall, 5th Edition, D-04994.
110. V.K. Ahluwalia, Chandra Prakash and Ravinder Singh Jolly, Gazette, Chem. Italina, 1979, 109, 643.
111. Propinol chloride is obtained by heating propionic acid and thionyl chloride (equimolar quantities on a water bath; excess thionylchloride, if any, is removed (water bath)and the resultant acid chloride used as such.
112. F.W. Canter, F.H. Curd and A. Robertson, J. Chem. Soc., 1931, 1245.
113. P.L. Trivedi, S.M. Sethna and R.C. Shah, J. Univ. Bombay, 1942, 11, 144.
114. V.K.Ahluwalia, Devendra Kumar, Nimmi Rani and Sunita, Indian J. Chem., 1977, 15B, 329.
115. S.M. Sethna and R.C. Shah, J. Indian Chem. Soc., 1940, 17, 211. Devendra Kumar, Ph.D. Thesis, Delhi University, 1976, p.41.
116. Butyryl chloride is obtained by the reaction of equimolar quantities of butyric acid and thionyl chloride (heating water bath), excess thionyl chloride, if any, is removed by heating (water bath) and the resultant acid chloride used as such.
117. Dictionary of organic compounds by Chapman and Hall, 5th Edition, P-01040.
118. Samta Goyal, Ph.D. Thesis, Delhi University, 1992, p.103.
119. R. Phadke and R.C. Shah, J. Ind .Chem. Soc., 1950, 27, 349.
120. R.D. Desai and M. Ekhlas, Proc. Indian Acad. Sci., 1938, 8A, 575.
121. P. Price and S.S. Israelstam, J. Org. Chem., 1964, 29, 2800.
122. B.Konig and St.V. kostaecki, Ber., 1906, 39, 4029
123. K.R. Laumas, Ph.D. Thesis, Delhi University, 1955, p.47.
124. Limaye and Munje, Rasaynam, 1937, 1, 84.
125. A. Bobik, G.M. Holder and A.J. Rayan, J. Med. Chem. 1977, 20, 1194.
126. H.C. Barnes, W.D. Ollis, I.O. Sutherland, O.R. Gottleib and M.T. Magalhales, Tetrahedron, 1965, 21, 2707.
127. K.R. Laumas, Ph.D. Thesis, Delhi University, 1955, p. 48.
128. Bergellini and Martgiani, Atti, R. Acad. Lineei., 1911, 20, 183.
129. A. Robertson, R.B. Water and E. T. Jones, J. Chem. Soc., 1932, 1681.
130. Hoosch, Ber., 1915, 48, 1131.
131. J. Pelonsky, Bull. Soc. Chim. Fr., 1955, 541; Chem. Abstr., 1955, 49, 12495i.
132. V.K. Ahluwalia, Devendra Kumar and M.C. Gupta, Indian J. Chem., 1978, 16B, 574.

133. V.K. Ahluwalia, M. Khanna and R.P. Singh, Indian J. Chem., 1981, 20B, 990

134. P. Karrer, Helv. Chim. Acta, 1919, 2, 486.

135. E. Chapman and H. Stephen, J. Chem. Soc., 1923,404.

136. G.G. Badcock, G.W.K. Cavill, A. Robertson and W.B. Whalley, J. Chem. Soc., 1950, 2961.

137. V.R. Lalitha and N.V.S. Rao, Proc. Indian Acad. Sci., 1973, 77, 157.

138. Isobutyryl chloride is prepared by adding slowly 1.1 equivalent of distilled thionyl chloride dropwise to 1 equivalent of isobutyric acid. The mixture is heated under anhydrous conditions for 30 min the mixture is distilled and the isobutyryl chloride is collected at 90-93°. Yield 70%.

139. M. Lounasamaa, C.J. Widen and A. Huhtikangas, Acta Chem. Scand. Ser. B, 1974, 28, 1209.

140. Isovaleryl chloride is obtained by adding dropwise phosphorus trichloride (8.25 ml) to isovaleric acid (21 ml) under anhydrous conditions. The mixture is refluxed for 30 min. The solution is cooled and upper layer of acid chloride is separated from the syrupy H_3PO_3. The upper layer is distilled to give isovaleryl chloride. Yield 19 ml. B.p. 114-15°.

141. T.S. Kenny, A. Robertson and S.W. George, J. Chem. Soc., 1939, 1601.

5

Heterocyclic Compounds

COUMARINS

Coumarins are prepared by the following methods.

PECHMANN CONDENSATION

Condensation of phenols with malic acid or β-ketoesters in the presence of conc. sulphuric acid or other condensing agent give the corresponding coumarin. The method is illustrated by the condensation of resorcinol with ethyl acetoacetate.

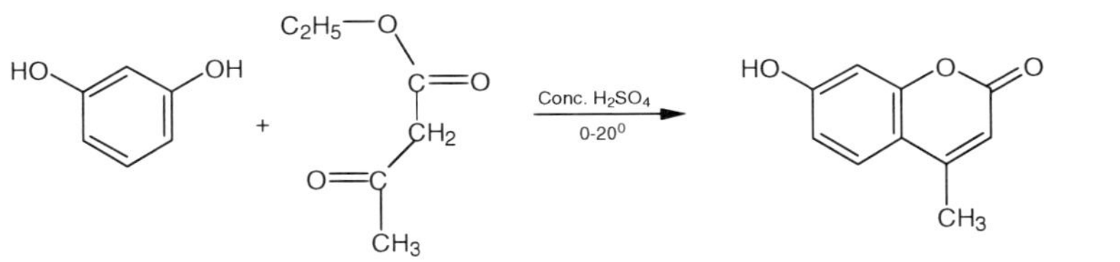

Resorcinol Ethylacetoacetate 4-Methyl-7- hydroxycoumarin

Use of ethyl benzoyl acetate or 2-methylacetoacetate (in place of ethyl acetoacetate) give the corresponding 4-phenyl or 3,4-dimethyl coumarin. Using this method following coumarins have been obtained, 5.A 16, 18, 20 and 28.

Simple coumarins, i.e coumarins without any substituents in the pyrone ring are prepared by the condensation of phenols with malic acid in the presence of concentrated sulphuric acid.

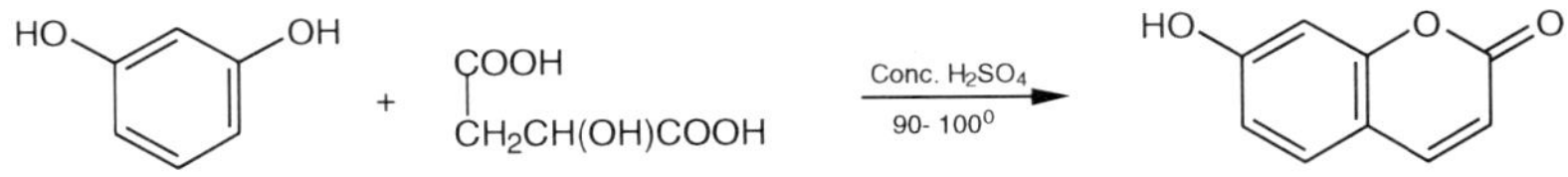

Resorcinol Malic acid 7-Hydroxycoumarin

Using this procedure following coumarins have been prepared 5.A. 1, 5 and 6.

Mechanism

In case of reaction with β-ketoesters, the reaction proceeds by the formation of β-hydroxyester, which cyclises and dehydrates to yield the coumarin.

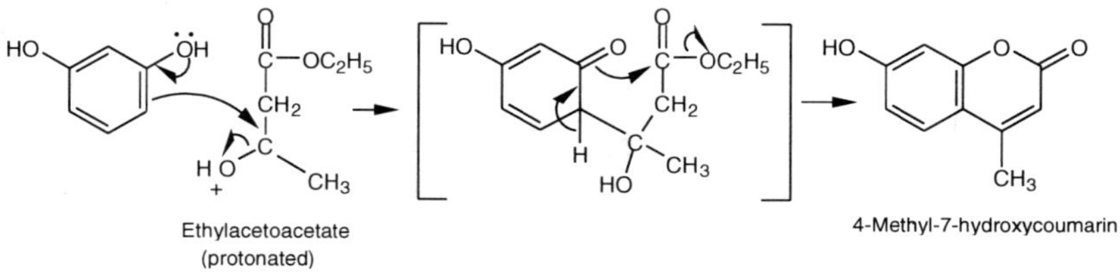

If malic acid is used then malanoaldehyde is first formed which reacts with phenols to give the corresponding coumarins.

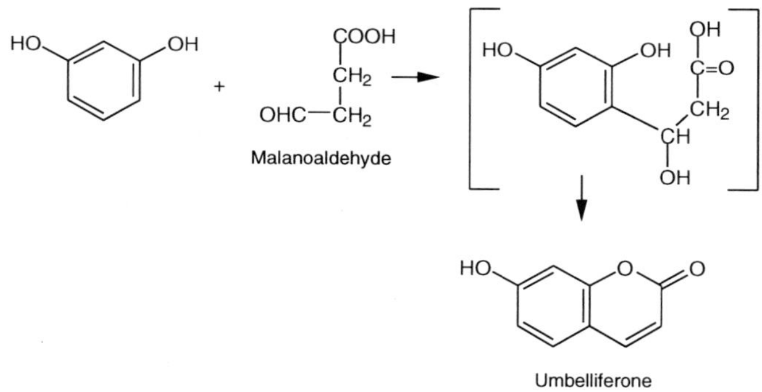

PERKIN REACTION

It has already been mentioned in connection of the preparation of α, β-unsaturated carboxylic acids (page 73) that aromatic aldehydes on heating with aliphatic acid anhydride (e.g., acetic anhydride) in the presence of sodium or potassium salts of the corresponding anhydride give the corresponding cinnamic acid. The mechanism of this reaction (known as Perkin reaction) has also been discussed (page 73). In this reaction if an o-hydroxybenzaldehyde is used, the corresponding o- hydroxycinnamic acid formed cyclises to give the coumarin.

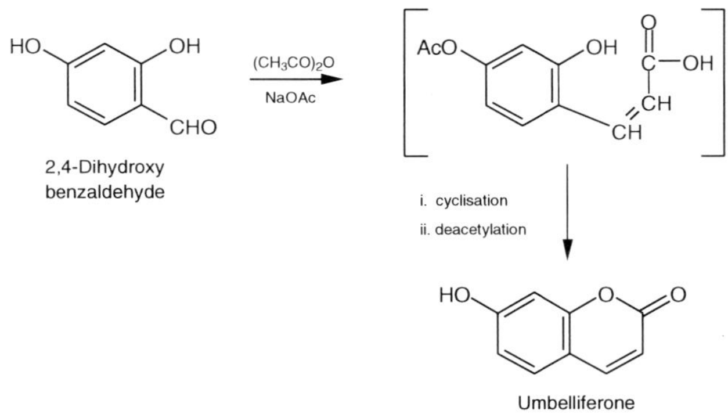

Using Perkin's method following coumarins have been prepared: 5.A. 3, 15, 19 and 22.

NUCLEAR OXIDATION

This method is used for the introduction of a hydroxy group in a preformed coumarin. In this method the coumarin is warmed with aqueous alkali when o-hydroxycinnamic acid is formed. This is subjected to nuclear oxidation with sodium persulphate and the formed product on acidification gives the 6-hydroxycoumarin. Thus

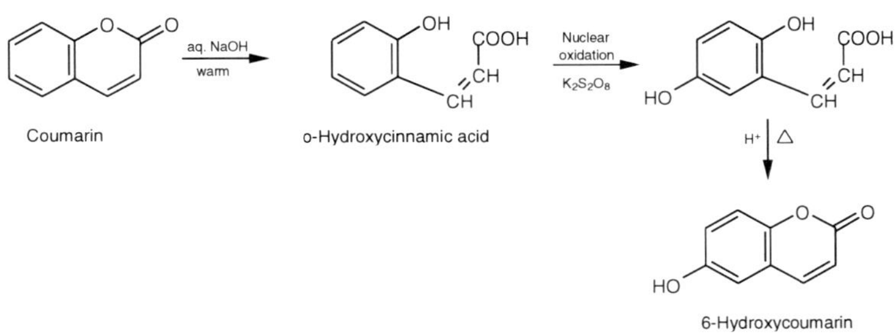

Using this method 6-hydroxycoumarin (5.A. 2) has been obtained.

The mechanism of persulphate oxidation has already been discussed (page 3).

THE OXAZOLONE METHOD

This method is used for the synthesis of 3-hydroxycoumarins. In this method, an o-hydroxybenzaldehyde on heating with hippuric acid in the presence of acetic anhydride and fused sodium acetate give the corresponding oxazolone. The oxazolone on refluxing with aqueous alkali followed by acidification give the corresponding 3-hydroxycoumarin.

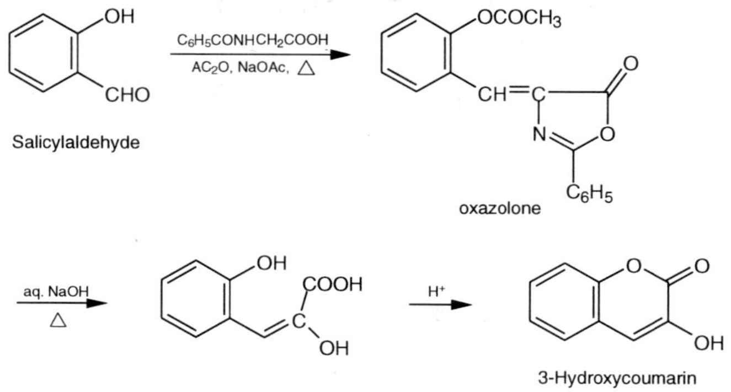

Using this method, 3-hydroxycoumarin (5.A.7) and 3-hydroxy-7-methoxycoumarin (5.A.8) have been prepared.

4-Hydroxycoumarins

These are prepared either from phenols or o-hydroxyacetophenones.

(a) From Phenols

(i) Hoesch reaction of a phenol with ethylcyanoacetate followed by hydrolysis of the intermediate ketimide give the corresponding 4-hydroxycoumarin

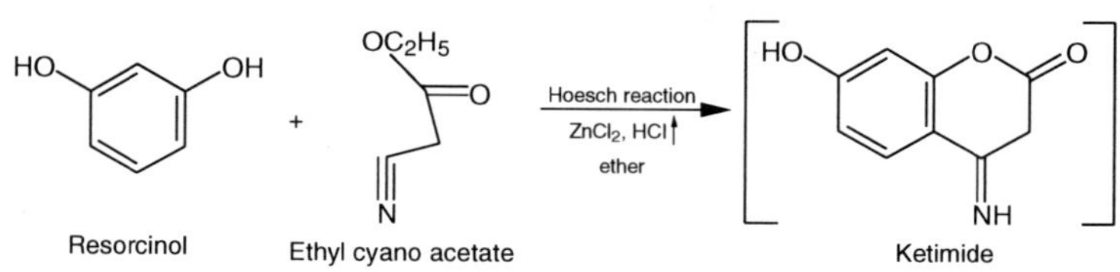

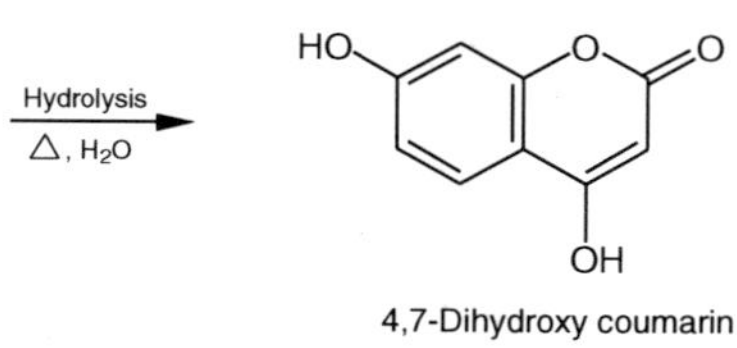

Using this method 4,7-dihydroxycoumarin (5.A.11) has been prepared. The mechanism of Hoesch reaction has been discussed earlier. (page 108).

(ii) Reaction of phenol with malonic acid in the presence of phosphorus oxochloride and fused zinc chloride gives the corresponding 4-hydroxycoumarin.

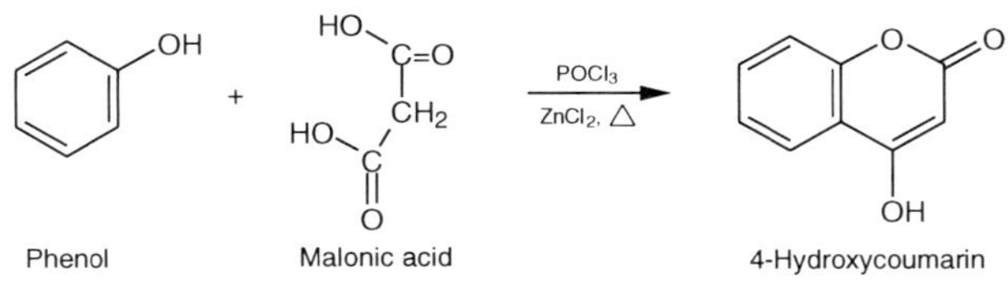

Using this method 4-hydroxycoumarin (5.A.9) and 4,5,7-trihydroxycoumarin (5.A.13) have been prepared.

(b) From Ketones

(i) Reaction of an o-hydroxyacetophenone with diethylcarbonate in the presence of pulverised sodium or sodium ethoxide gives the corresponding 4-hydroxycoumarin.

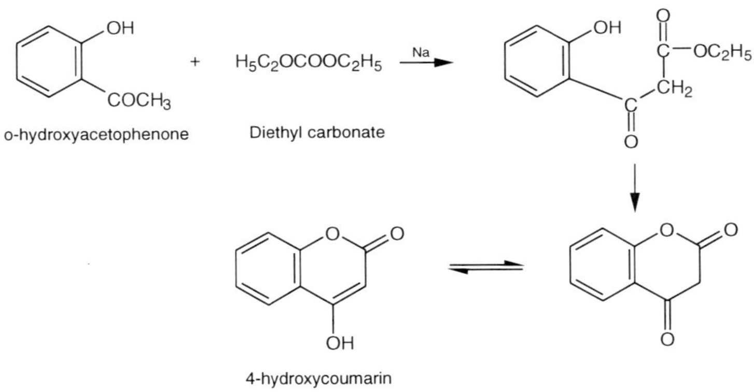

In the above method all hydroxyl groups except the o-hydroxy group must be protected. Using this method following coumarins have been prepared: 5.A. 9, 10, 14, 24, 25 and 26.

5.A.1 7-HYDROXYCOUMARIN (UMBELLIFERONE)

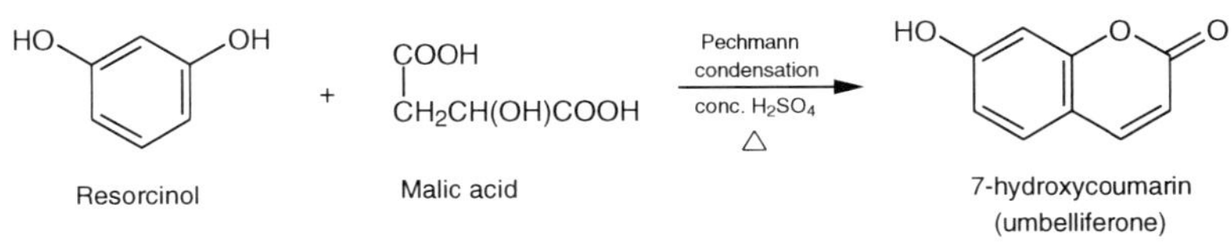

A mixture of resorcinol (10 g, 0.09 mole), malic acid (12 g, 0.09 mole) and concentrated sulphuric acid (25 ml) is heated at 120° (oil bath) under anhydrous conditions till effervescence ceased (2 hr). The mixture is poured over crushed ice and left overnight. The separated solid is filtered and dried. It is purified by its acetylation with acetic anhydride (30 ml) and pyridine (1 ml) (room temperature, 24 hr). The acetate obtained by pouring the mixture over crushed ice is filtered, washed with water and dried. It is passed through a column of silica gel (2 × 30 cm) and eluted with benzene–petroleum ether (1:1). The acetate is deacetylated with concentrated sulphuric acid (20 ml, 0° for 2 hr). On pouring over crushed ice, umbelliferone separates out as colourless solid. It is filtered and washed with water. It is crystallised from dilute alcohol as colourless rhombic prisms. Yield 6 g (40.8%). M.p. 228° (lit.[1] m.p. 227-28).

5.A.2 6-HYDROXYCOUMARIN

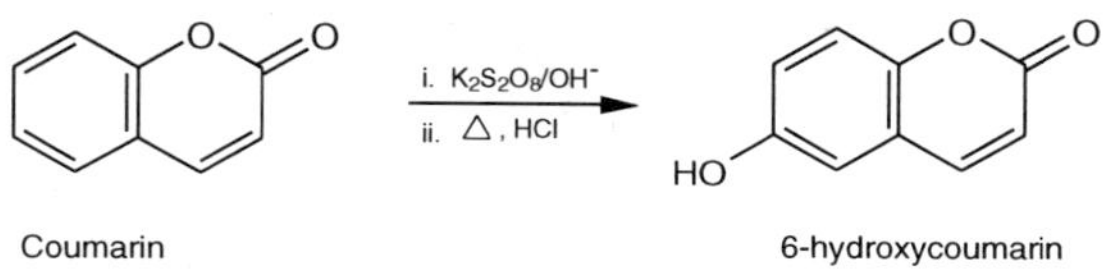

Coumarin (12 g, 0.082 mole) is dissolved in aqueous sodium hydroxide solution (12 g in 15 ml water) by heating on a steam bath. The clear solution is cooled (15-20°) and aqueous solution of potassium persulphate (22.2 g, 0.082 mole in 200 ml water) is added dropwise during 4 hr with stirring. The mixture is stirred for 2 hr more and left overnight at room temperature. It is cooled and acidified to congo red with hydrochloric acid. The separated unchanged product is filtered and the filtrate extracted with ether (2 × 30 ml). The clear filtrate is heated in a boiling water bath for 30 min with sodium sulphite (1.5 g) and concentrated hydrochloric acid (100 ml). The solution is cooled, the separated product is filtered and washed with water. It is crystallized from methanol as pale brown needles. Yield 2 g (15%). M.p. 248-50° (lit.[2] m.p. 248-50°).

5.A.3 5,7-DIHYDROXYCOUMARIN

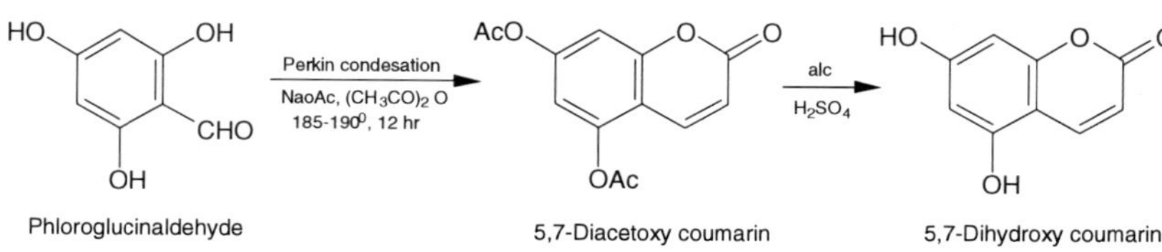

Phloroglucinaldehyde (2.14) (5 g, 0.032 mole) and fused sodium acetate (10 g) are powdered in an intimate mixture and heated at 185-190° (oil bath) with acetic anhydride (25 ml) for 12 hr. The mixture is poured over crushed ice. The separated product is filtered and washed with water. It is crystallised from methanol as colourless prisms of 5,7-diacetoxycoumarin. Yield 4 g (47%). M.p. 140-41°. (lit.[3] m.p. 140°).

The 5,7-diacetoxycoumarin (3 g, 0.011 mole) is deacetylated by heating with alcoholic sulphuric acid (7%, 30 ml) for 4 hr on a boiling water bath. The solution is diluted with water (50 ml) and alcohol removed in vacuo. The remaining solution is cooled and the separated product is filtered and washed with water. It is crystallised from methanol as almost colourless rectangular prisms. Yield 1.5 g (73.5%). M.p. 285-87° (lit.[3] m.p. 285-86°).

5.A.4 5,7-DIMETHOXYCOUMARIN

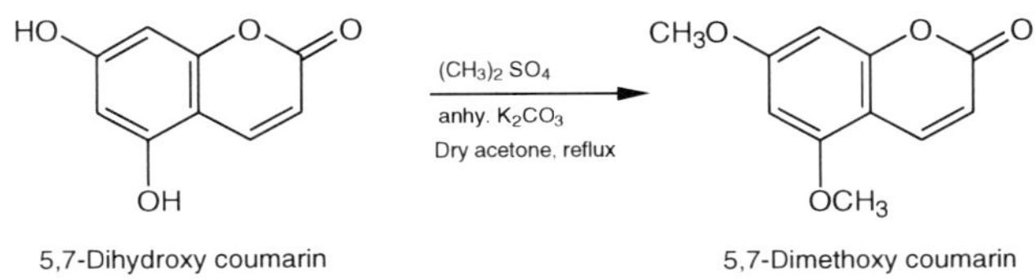

5,7-Dihydroxycoumarin (5.A.3) (1 g, 0.0056 mole) in acetone (70 ml) is refluxed for 4 hr under anhydrous conditions with dimethyl sulphate (1.14 ml, 1.52 g, 0.012 mole) and ignited potassium carbonate (5 g). The mixture is filtered and the inorganic salts washed with hot acetone (2 × 20 ml). The combined acetone solution is distilled and the residue macerated with crushed ice. The separated product is filtered and washed with water. It is crystallised from methanol as colourless needles. Yield 0.9 g (77.6%). M.p. 147° (lit.[3] m.p. 147°).

5.A.5 6,7-DIHYDROXYCOUMARIN (AESCULETIN)

A mixture of hydroxyhydroquinone triacetate (1.21.) (4 g, 0.0159 mole), malic acid (3 g, 0,02 mole) and concentrated sulphuric acid (6 ml) is

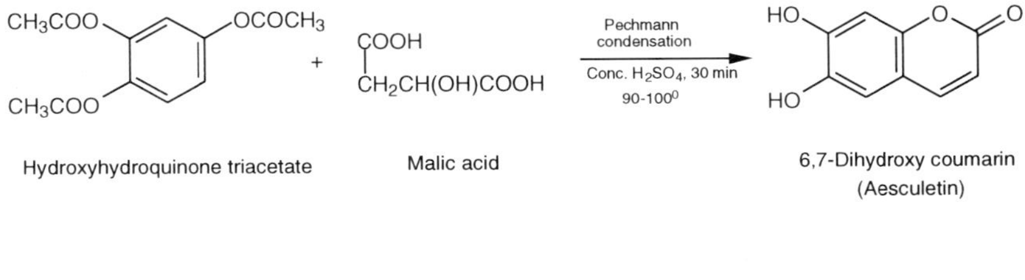

heated under anhydrous conditions in a boiling water bath for 30 min. The reaction mixture is cooled and poured over crushed ice. The separated product is filtered and washed with water. It is crystallised from alcohol as pale yellow prisms. Yield 1.5 g (53%). M.p. 270°. (lit.[4] m.p. 270°).

5.A.6 7,8-DIHYDROXYCOUMARIN (DAPHNETIN)

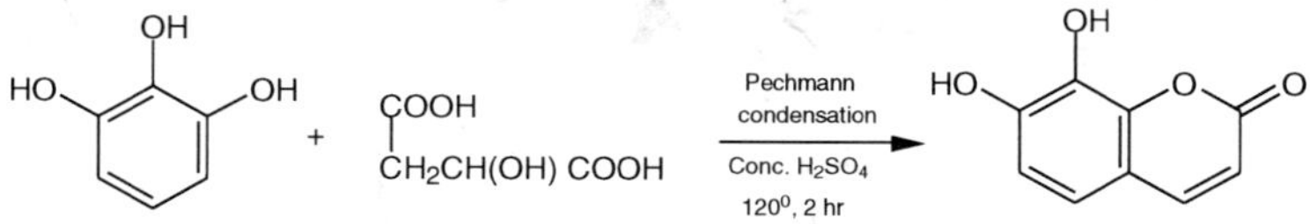

Pyrogallol — Malic acid — 7,8-Dihydroxy coumarin (Daphnetin)

An intimate mixture of pyrogallol (12.6 g, 0.1 mole), malic acid (13.4 g, 0.1 mole) and concentrated sulphuric acid (25 ml) is heated in an oil bath at 120° until effervescence ceases (about 2 hr.). The reaction mixture is cooled and poured with stirring into crushed ice (400 g) and left overnight. The solid product is filtered and washed with water. It is crystallised from dilute alcohol as pale yellow needles. Yield 6 g (33.7%). M.p. 256-57°.

5.A.7 3-HYDROXYCOUMARIN

$$\underset{\text{Glycine}}{H_2NCH_2COOH} + \underset{\text{Benzoyl Chloride}}{C_6H_5COCl} \xrightarrow{\text{aq. NaOH}} \underset{\text{Benzoyl glycine (Hippuric acid)}}{C_6H_5CONHCH_2CO\,OH}$$

Step (i) Hippuric Acid (Benzoyl Glycine)

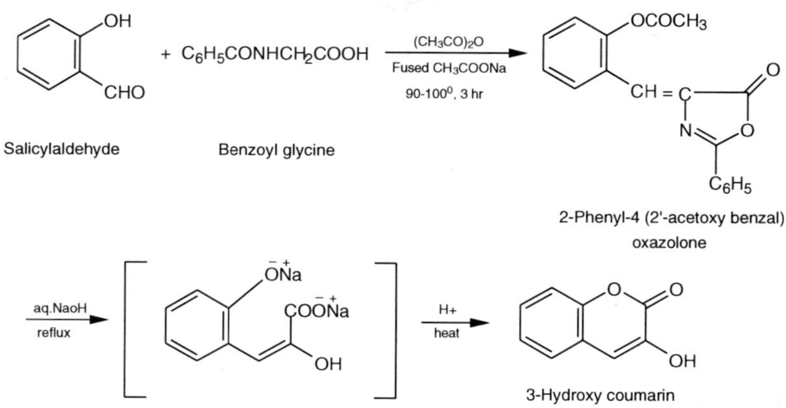

A solution of glycine (7.5 g, 0.1 mole) in sodium hydroxide solution (10%, 100 ml) is treated with benzoyl chloride (15 g, 12.4 ml, 0.11 mole) (added in two lots). The mixture is shaken vigorously until all the benzoyl chloride has reacted. The reaction mixture is treated with crushed ice and acidified with concentrated hydrochloric acid to congo red. The separated product is filtered and washed with cold water. The solid product is heated (water bath) with carbon tetrachloride (30 ml) and filtered. Carbon tetrachloride dissolves any benzoic acid present. The solid product crystallised from boiling water (with the addition of little decolorizing carbon if necessary). Hippuric acid is obtained as colourless flakes on cooling the solution. Yield 12.5 g (70%). M.p. 187° (lit.[6] m.p. 187°).

Step (ii) 2-Phenyl-4-(2′-acetoxybenzal) Oxazolone

A mixture of salicylic aldehyde (25 g, 21.8 ml, 0.2 mole), hippuric acid (42 g, 0.235 mole) and fused sodium acetate (15 g) are heated in a boiling water bath with acetic anhydride (60 ml) for 30 min under anhydrous conditions. A bright yellow crystalline solid that formed redissolves on further heating. Alcohol (20 ml) is added to the reaction mixture and cooled. The separated azlactone is filtered, washed with alcohol and then with hot water till the aqueous washings are colourless. It is crystallised from alcohol to give the required oxazolone. Yield 40 g (63.6%). It is used as such for the subsequent step.

Step (iii) 3-Hydroxycoumarin

The above oxazolone (40 g, 0.13 mole) is refluxed with aqueous sodium hydroxide (20 g in 100 ml water) till no more ammonia evolved (2 hr). The resulting solution is diluted with water (100 ml) and saturated with sulphur dioxide gas with cooling. The reaction mixture is left for 12 hr in an ice chest and the separated benzoic acid is filtered. The clear filtrate is heated for 3 hr on boiling water with concentrated hydrochloric acid (60 ml). The solution is cooled and the separated product is filtered. It is crystallised from alcohol as colourless prismatic needles. Yield 13 g (61.6%). M.p. 151-52° (lit.[7] m.p. 151-52°).

5.A.8 3-HYDROXY-7-METHOXYCOUMARIN

Step (i) 2-Phenyl-4-(2′-acetoxy-4′-methoxybenzal) Oxazolone

An intimate mixture of 2-hydroxy-4-methoxybenzaldehyde (2.6) (15.2 g, 0.1 mole), hippuric acid (5.A.7) (30 g, 0. 167 mole) and anhydrous sodium

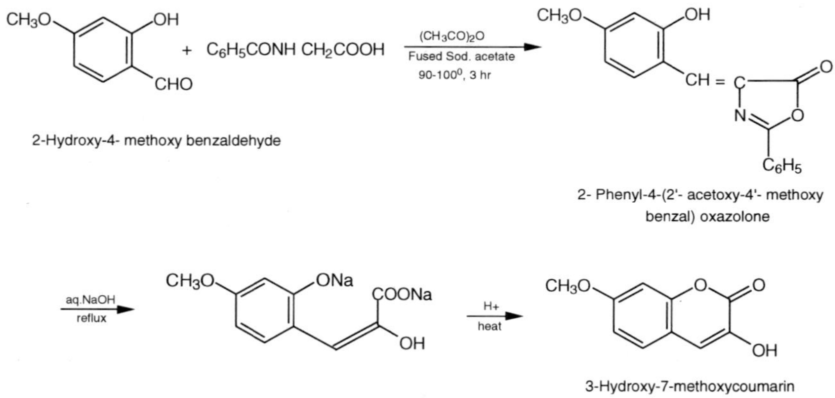

acetate (12 g) is heated on boiling water bath with acetic anhydride (40 ml) for 1 hr under anhydrous conditions with occasional stirring. A bright yellow solid is obtained in the beginning. It dissolves on further heating. Alcohol (10 ml) is added to the cooled reaction mixture. The azlactone acetate is filtered, washed with alcohol and finally with hot water till the aqueous washings are clear. It is crystallised from alcohol as bright yellow needles. Yield 20 g (59.3%). M.p. 165° (lit.[7] m.p. 165°).

Step (ii) 3-Hydroxy-7-methoxycoumarin

The above azlactone (20 g, 0.059 mole) is refluxed with aqueous sodium hydroxide (10 g in 50 ml water) for 2.5 hr. The reaction mixture is diluted with water (50 ml) and saturated with sulphur dioxide gas while cooling. It is left in ice-chest for 12 hr. The separated benzoic acid is filtered. The clear aqueous filtrate is heated on boiling water bath for 3 hr with concentrated hydrochloric acid (60 ml). The solution is cooled, the separated product filtered and washed with water. It is crystallised from alcohol as colourless prismatic needles. Yield 10 g (87.8%). m.p. 177-78°. (lit.[7] m.p. 177-78°).

5.A.9 4-HYROXYCOUMARIN

Method (i)

A solution of 2-hydroxyacetophenone (4.A.1) (2 g, 1.77 ml, 0.0147 mole) in dry diethyl carbonate (15 ml) is added to a suspension of sodium methoxide (1 g, 0.0185 mole) in diethyl carbonate (10 ml). The reaction mixture is heated on a steam bath for 5 hr under anhydrous conditions. It is cooled, diluted with water (25 ml) and extracted with ether (2 × 40 ml)

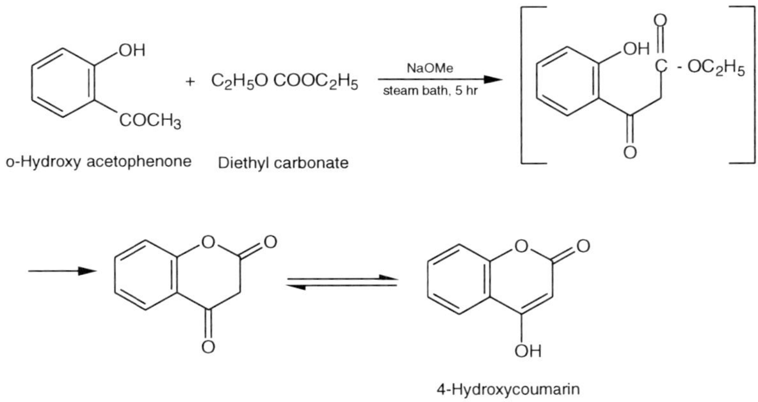

to remove excess diethyl carbonate. The clear alkaline solution is acidified with hydrochloric acid. The separated product is filtered and washed with water. It is crystallised from methanol as pale yellow needles. Yield 1.2 g (50.4%). M.p. 204-206° (lit.[8,9] m.p. 206°).

Method (ii)

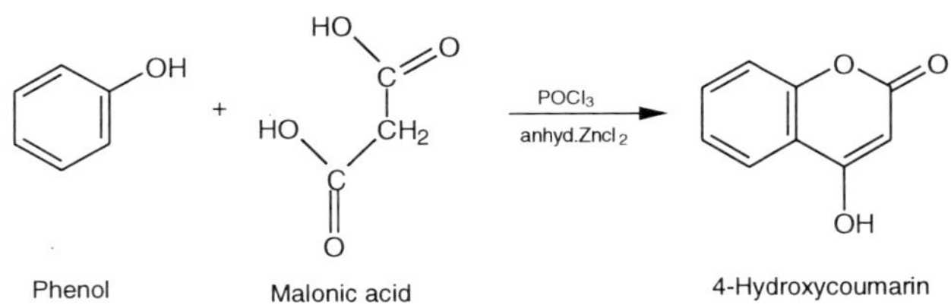

A mixture of phenol (11.2 g, 0.119 mole), malonic acid (12.4 g, 11.6 ml, 0.119 mole), fused zinc chloride (49 g, 0.36 mole) and phosphorus oxychloride (32.8 ml, 54 g, 0.35 mole) is heated with stirring for 35 hr at 64° under anhydrous conditions. The reaction mixture is cooled, crushed ice (100 g) added and the separated product filtered and washed with water. It is purified by dissolving in sodium carbonate solution (5%) and acidifying the clear alkaline solution. The separated product is filtered and washed with water. It is crystallised from dilute alcohol as pale yellow needles. Yield 6 g (31.1%). M.p. 204-206° (lit.[8,9] m.p. 206°).

5.A.10 4-HYDROXY-7-METHOXYCOUMARIN

A solution of 2-hydroxy-4-methoxyacetophenone (4.A.7) (1 g, 0.006 mole) in dry diethyl carbonate (10 ml) is added to a suspension of sodium

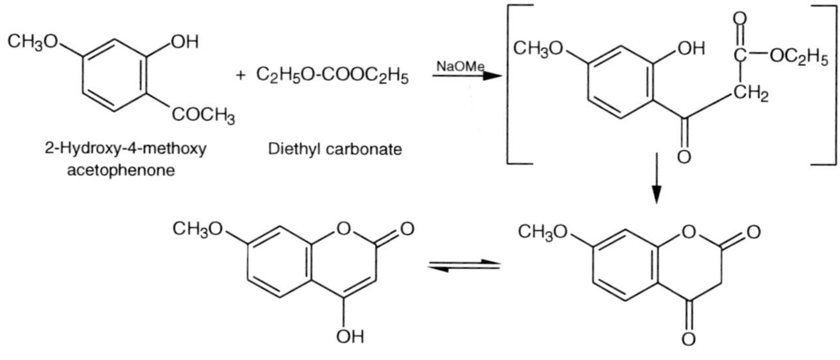

4-Hydroxy-7-methoxycoumarin

methoxide (prepared from 0.5 g sodium, 0.02 mole) in diethyl carbonate (15 ml). The reaction mixture is heated on a steam bath for 4 hr under anhydrous conditions. It is cooled, diluted with water (50 ml) and extracted with ether (2 × 50 ml). The clear alkaline solution is acidified with hydrochloric acid. The separated product is filtered and washed with water. It is crystallised from methanol as colourless small needles. Yield 0.8 g (69.6%). M.p. 256° (lit.[8] m.p. 256°).

5.A.11 4,7-DIHYDROXYCOUMARIN

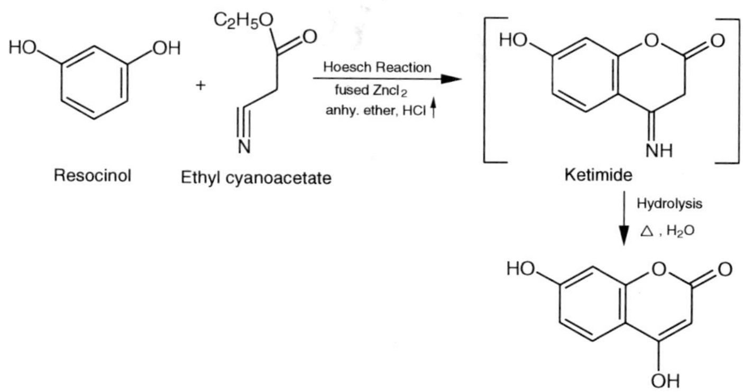

4,7-Dihydroxycoumarin

A solution of dry resorcinol (22 g, 0.2 mole) and ethyl cyanoacetate (6.A.23) (21.26 ml, 22.6 g, 0.2 mole) in dry ether (60 ml) containing freshly fused zinc chloride (10 g, 0.073 mole) is saturated with dry hydrogen chloride gas at 0-10° under anhydrous conditions. After 2 hr, the ketimine hydrochloride starts separating, the gas is passed for 4 hr more and the mixture allowed to stand overnight at room temperature. The ether is decanted and the solid product washed with dry ether (2 × 20 ml). It is hydrolysed by refluxing with a solution of concentrated sulphuric acid (20 ml) in water (100 ml) for 1 hr. The mixture is cooled, the separated product filtered and washed with cold water. It is crystallised from hot water as long slender needles. Yield 25 g (70.2%). M.p. 265° (decomp.) (lit.[10] m.p. 265° decomp.).

5.A.12 7-HYDROXY-4-METHOXYCOUMARIN[11]

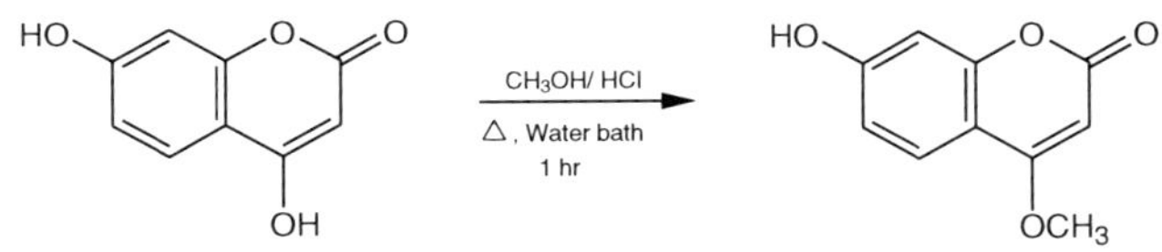

4,7- Dihydroxy coumarin → 7-Hydroxy-4-methoxycoumarin

4,7-Dihydroxycoumarin (5.A.11.) (0.8 g, 0.0045 mole) is refluxed in methanol (35 ml) containing concentrated hydrochloric acid (5 ml) for 1 hr in a water bath. Excess methanol is removed in vacuo and the solution poured over crushed ice. The separated product is filtered, washed with water and dried. TLC examination showed it to be a mixture of the required compound and some starting material. The crude product is chromatographed over silica gel coulmn (2 × 20 cm). Elution with chloroform-methanol (19:1) gives the required 7-hydroxy-4-methoxycoumarin. Yield 0.4 g (46.5%). M.p. 275-76°. It could also be obtained in 90% yield by partial methylation of 4,7-dihydroxycoumarin in acetone with equimolar amount of dimethyl sulphate in the presence of dry sodium bicarbonate (6 hr refluxing) and usual working up.

5.A.13 4,5,7-TRIHYDROXYCOUMARIN

Into a cooled solution (0-10°) of dry phloroglucinol (10 g, 0.079 mole) in dry ether (15 ml) and freshly distilled ethyl cyanoacetate (6.A.23) (10 ml, 10.63 g, 0.094 mole) containing fused zinc chloride (5 g, 0.036 mole) a rapid stream of dry hydrogen chloride gas is passed under anhydrous conditions. After 2 hr, a reddish ketimine starts separating. The flow of gas is continued for 4 hr more and the solution kept overnight in a ice-chest. The ether is decanted and the solid product washed with dry ether (2 × 25ml). It is filtered and crystallised from water to give 4,5,7-trihydroxycoumarin ketimide. Yield 9 g (58.8%). M.p. > 300°.

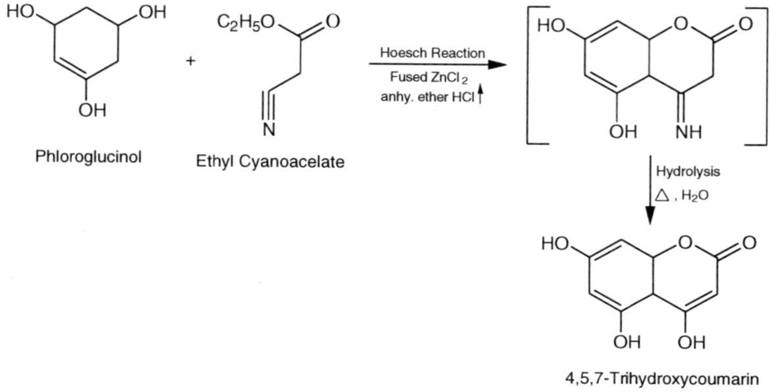

The above ketimide (8 g, 0.04 mole) is refluxed with concentrated hydrochloric acid (50 ml) and water (200 ml) for 45 min. The solution is filtered and cooled. The separated product is filtered and washed with cold water. It is crystallised from hot water as tiny yellow needles. Yield 6 g (74.6%). M.p. >300° (lit.[10] m.p. >300°).

5.A.14 4-HYDROXY-5,7-DIMETHOXYCOUMARIN

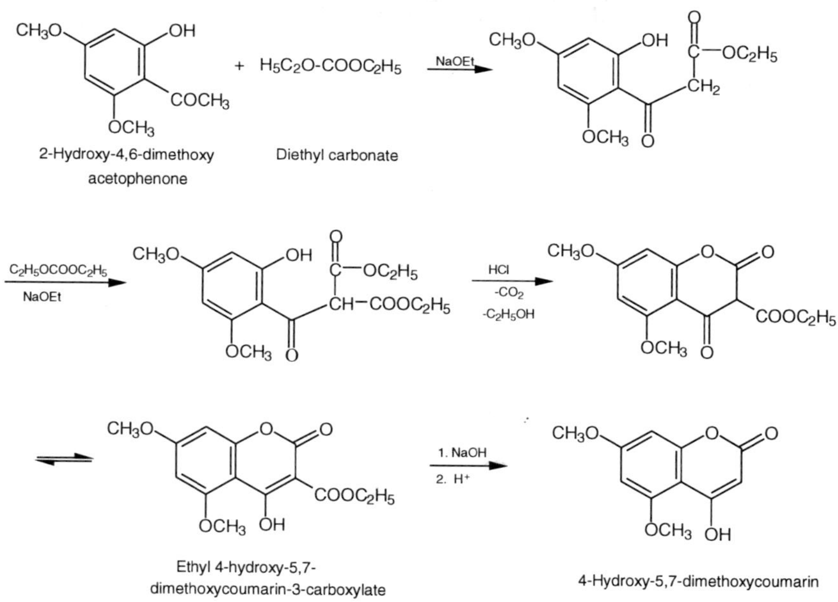

Step (i) Ethyl 4-hydroxy-5,7-dimethoxycoumarin 3-carboxylate[12]

A solution of 2-hydroxy-4,6-dimethoxyacetophenone (4.A.45) (1 g, 0.005 mole) in diethyl carbonate (10 ml) is added to a suspension of sodium methoxide (prepared from 0.5 g sodium , 0.02 mole) in diethyl carbonate (15 ml). The reaction mixture is heated on a steam bath under anhydrous conditions for 4 hr. It is cooled, diluted with cold water (25 ml) and the aqueous solution is extracted with ether (2 × 30 ml) to remove excess of diethyl carbonate. The clear alkaline solution is acidified in cold with dilute hydrochloric acid. The separated product is filtered and washed with water. It is crystallised from methanol as colourless needles. Yield 1.3 g (86.7%) m.p. 182-83°.

Step (ii) 4-Hydroxy-5,7-dimethoxycoumarin[12]

Ethyl 4-hydroxy-5,7-dimethoxycoumarin-3-carboxylate (1 g, 0.0034 mole) is heated in a water bath with aqueous sodium hydroxide (5%, 20 ml). The cooled and clear alkaline solution is acidified. The separated product is filtered and washed with water. It is crystallised from methanol as colourless needles. Yield 0.7 g (92.7%). M.p. 182-83° (lit.[8] m.p. 182-83°).

5.A.15 7-HYDROXY-3-METHYLCOUMARIN (3-METHYLUMBELLIFERONE)

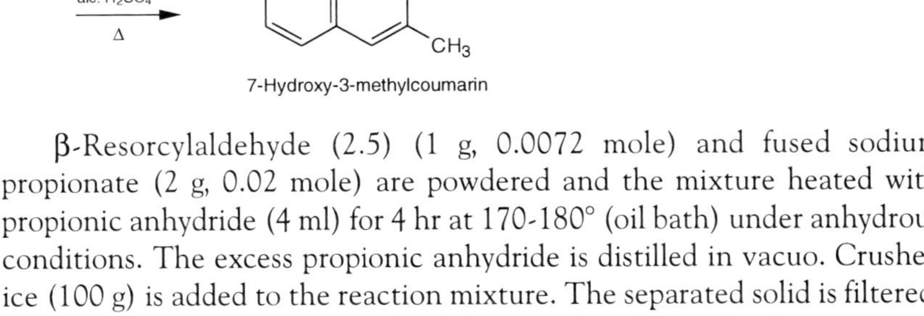

β-Resorcylaldehyde (2.5) (1 g, 0.0072 mole) and fused sodium propionate (2 g, 0.02 mole) are powdered and the mixture heated with propionic anhydride (4 ml) for 4 hr at 170-180° (oil bath) under anhydrous conditions. The excess propionic anhydride is distilled in vacuo. Crushed ice (100 g) is added to the reaction mixture. The separated solid is filtered, washed with water and crystallised from methanol as colourless prismatic needles of 3-methyl-7-propionoxycoumarin. Yield 1 g (66.7%). M.p. 132-33°.

The above propionoxycoumarin (1 g, 0.0048 mole) is hydrolysed by refluxing with alcoholic sulphuric acid (7%, 20 ml) for 4 hr on a boiling water bath. The solution is diluted with water (10 ml) and alcohol distilled under reduced pressure. The product separated on cooling is filtered and washed with water. It is crystallised from methanol as colourless prisms. Yield 0.7 g (82.4%). M.p. 218-19° (lit.[13] m.p. 218-19°).

Note: Sodium propionate is prepared by adding sodium hydroxide solution (2-3%) to a solution of propionic acid in alcohol till it gave pink colour with phenolphthalein. A very small amount of propionic acid is added till the solution became colourless. Alcohol is evaporated to dryness (water bath). The residue is macerated with ether and dried at 120°.

5.A.16 7-HYDROXY-4-METHYLCOUMARIN (4-METHYLUMBELLIFERONE)

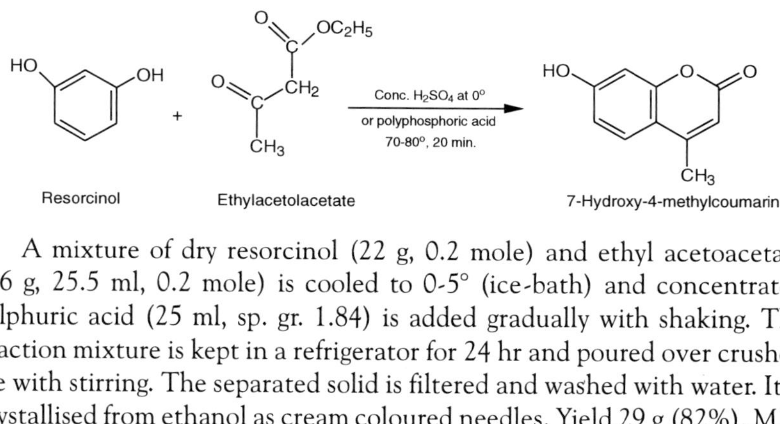

A mixture of dry resorcinol (22 g, 0.2 mole) and ethyl acetoacetate (26 g, 25.5 ml, 0.2 mole) is cooled to 0-5° (ice-bath) and concentrated sulphuric acid (25 ml, sp. gr. 1.84) is added gradually with shaking. The reaction mixture is kept in a refrigerator for 24 hr and poured over crushed ice with stirring. The separated solid is filtered and washed with water. It is crystallised from ethanol as cream coloured needles. Yield 29 g (82%). M.p. 185° (lit.[14] m.p. 185°).

It is also obtained by heating dry resorcinol (5.6 g, 0.05 mole) in ethyl acetoacetate (6.6 g, 6.5 ml, 0.05 mole) with poly-phosphoric acid (80 g) [polyphosphoric acid is obtained by dissolving 52 g phosphorus pentoxide in 28 g commercial orthophosphoric acid (d, 175)] for 20 min at 70-80°. The reaction mixture is poured into excess cold water. The separated product is filtered and washed with water. It is crystallised from ethanol as cream coloured needles. Yield 8.2 g (91.5%). M.p. 185°.

5.A.17 5-METHOXY-4-METHYLCOUMARIN[15] (WITTIG REACTION)

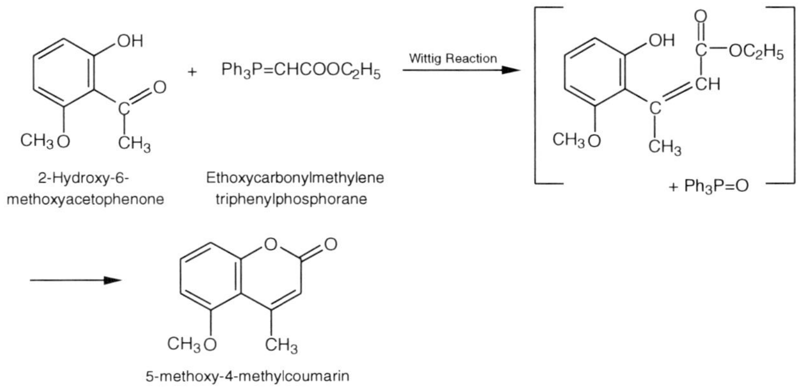

A mixture of 2-hydroxy-6-methoxyacetophenone (4.A.17.) (1 g, 0.006 mole) and ethoxycarbonylmethylenetriphenyl phosphorane (6.A.20) (2.5 g, 0.007 mole) in anhydrous benzene (25 ml) is refluxed for 36 hr. The solvent is distilled off and the residue is subjected to column chromatography over silica gel (2 × 20 cm). Eluation with benzene-petroleum ether (1:1) gave the required coumarin. It is crystallised from benzene-petroleum ether as colourless shining needles. Yield 0.82 g (71.9%). M.p. 137-38°.

5.A.18 7-HYDROXY-4-PHENYLCOUMARIN (4-PHENYLUMBELLIFERONE)

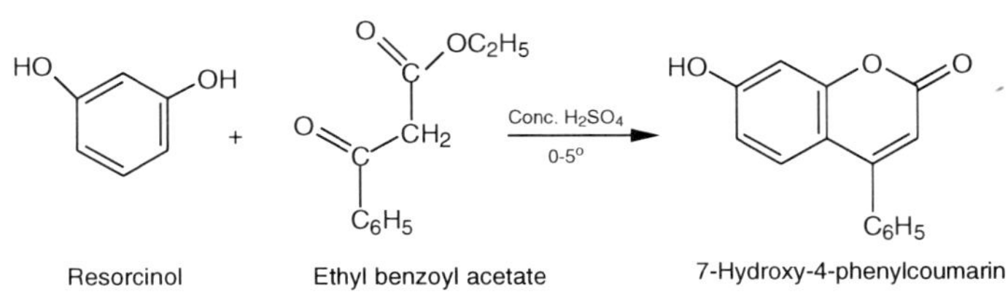

A mixture of dry resorcinol (12 g, 0.1 mole) and ethylbenzoyl acetate (6.C.41) (20 ml, 22.2 g, 0.115 mole) is cooled in ice (0-5°) and concentrated sulphuric acid (60 ml, sp. gr. 1.84) is added gradually with shaking, keeping the temperature at 0-5°. The deep red viscous solution thus obtained is kept for 24 hr in a refrigerator. It is poured over crushed ice with stirring. The separated product is filtered and washed with water. It is crystallised from alcohol as yellow needles. Yield 12 g (46%). M.p. 244-45° (lit.[16] m.p. 244°).

5.A.19 7-METHOXY-4-PHENYLCOUMARIN

Method (i)[17]

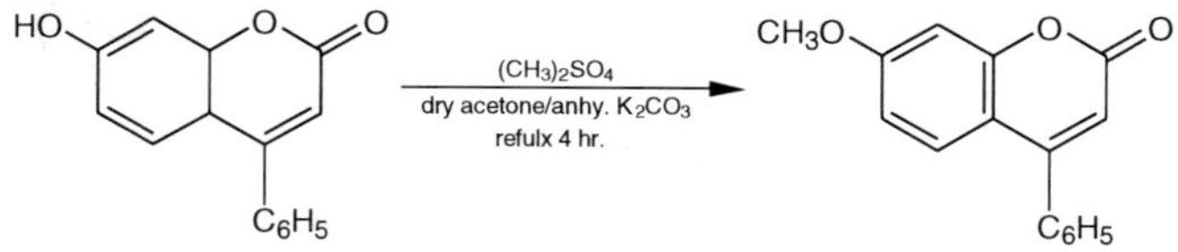

7-Hydroxy-4-Phenyl courmarin   7-Methoxy-4-phenylcourmarin

7-Hydroxy-4-phenylcoumarin (5.A.18.) (2 g, 0.0084 mole) in dry acetone (100 ml) is refluxed for 4 hr under anhydrous conditions with dimethyl sulphate (0.9 ml, 1.2 g, 0.0095 mole) in the presence of ignited potassium carbonate (8 g). The mixture is filtered and the inorganic salts washed with hot acetone (2 × 25 ml). The combined acetone solution is distilled and the residual product macerated with crushed ice. The seprated product is filtered and washed with water. It is crystallised from methanol as colourless shining needles. Yield 1.3 g (61.3%). M.p. 116-18°.

Method (ii)[18]

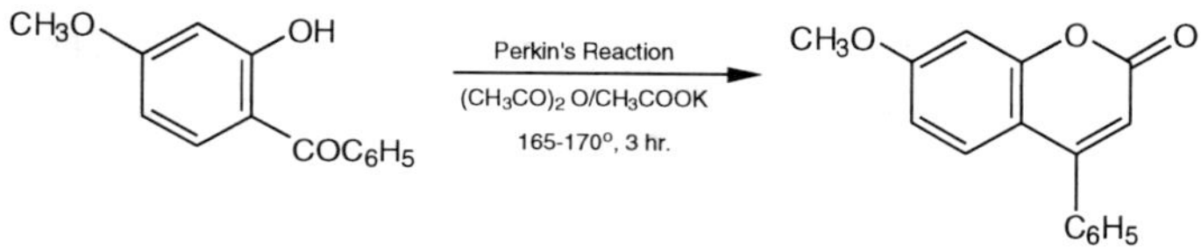

2-Hydroxy-4-methoxybenzophenone   7-Methoxy-4-phenylcoumarin

A mixture of 2-hydroxy-4-methoxybenzophenone (4.E.107) (10 g, 0.044 mole), fused potassium acetate (5 g, 0.051 mole) and acetic anhydride (30 ml) is heated for 30 hr at 165-70° under anhydrous conditions. The reaction mixture is poured over crushed ice. The separated product is filtered and washed with water. It is crystallised from methanol as colourless needles. Yield 2 g (18.1%). M.p. 116-17°.

5.A.20 5,7-DIHYDROXY-4-PHENYLCOUMARIN

Dry hydrogen chloride gas is passed through a solution of dry phloroglucinol (12.6 g. 0.1 mole) and ethyl benzoyl acetate (6.C.41) (18 ml, 20 g, 0.104 mole) in absolute ethanol (100 ml) for 3 hr under

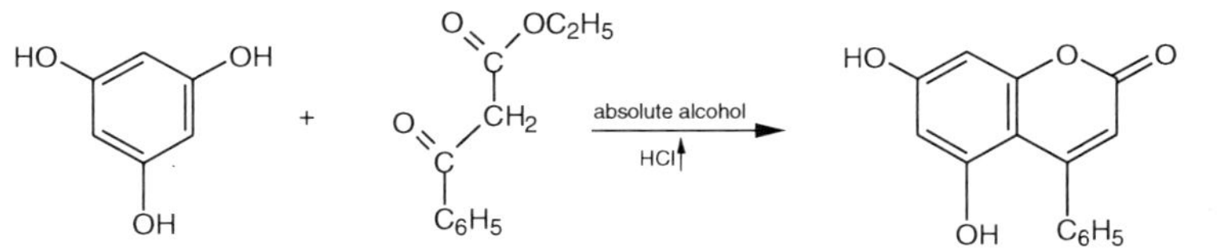

Phloroglucinol Ethyl benzoyl acetate 5,7-Dihydroxy-4-phenylcoumarin

anhydrous conditions. The mixture is kept for 2 days at room temperature and then poured over crushed ice. The separated product is filtered and washed with water. It is crystallised from methyl alcohol as pale yellow needles. Yield 13.5 g (53.1%). M.p. 213-14° (lit.[19,20] m.p. 213°).

5.A.21 5,7-DIMETHOXY-4-PHENYLCOUMARIN

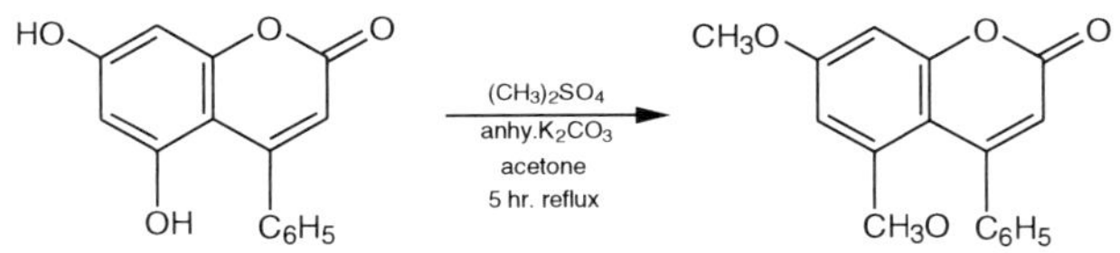

5,7-Dihydroxy-4-phenylcourmarin 5,7-Dimethoxy-4-phenylcoumarin

A solution of 5,7-dihydroxy-4-phenylcoumarin (5.A.20) (2g, 0.0079 mole) in dry acetone (100 ml) is refluxed for 5 hr under anhydrous conditions with dimethyl sulphate (1.7 ml, 2 .27 g, 0.018 mole) in presence of ignited potassium carbonate (5g). The mixture is filtered and the inorganic salts washed with hot acetone (2 × 25 ml). The combined acetone solution is distilled and the residue macerated with crushed ice. The solid product is filtered and washed with water. It is crystallised from methanol as colourless needles. Yield 1.8 g (81.1%). M.p. 166-68° (lit.[20] m.p.167°).

5.A.22 7-HYDROXY-3-PHENYLCOUMARIN (3-PHENYLUMBELLIFERONE)

An intimate mixture of β-resorcylaldehyde (2.5.) (5g, 0.036 mole), sodium phenyl acetate (7 g, 0.044 mole) and acetic anhydride (30 ml) is refluxed at 170-180° (oil bath) for 8 hr under anhydrous conditions. The reaction mixture is cooled and poured over crushed ice. The separated 7-acetoxy-3-phenyl coumarin is filtered and washed with water. It is crystallised from methanol as glistening needles. Yield 6 g (59.4%). M.p. 184-85°.

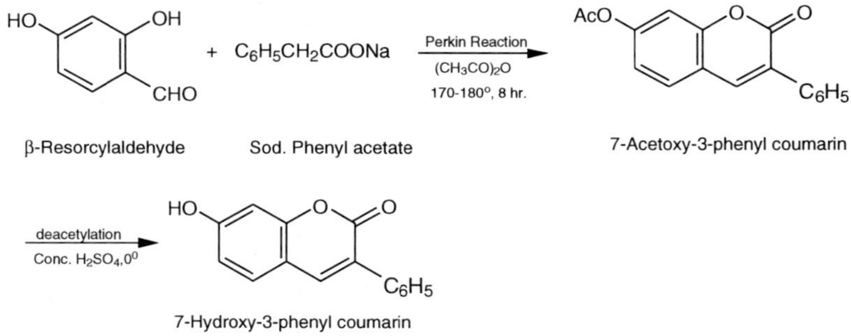

The above 7-acetoxy-3-phenylcoumarin (6 g, 0.02 mole) is treated with concentrated sulphuric acid (20 ml) at 0° for 2 hr. It is poured over crushed ice. The separated 7-hydroxy-3-phenylcoumarin is filtered and washed with water. It is crystallised from alcohol as colourless needles. Yield 4.5 g (88.2%). M.p. 207-208° (lit.[21] m.p. 207-208°).

Note: Sodium phenyl acetate is prepared by adding sodium hydroxide solution (2-3%) to a solution of phenylacetic acid in alcohol till it gave pink colour with phenolphthalein. A few crystals of phenyl acetic acid are added till the solution became colourless. Alocohol is evaporated to dryness (water-bath). The residue is macerated with ether and dried at 120° (air oven).

5.A.23 4,7-DIHYDROXY-3-PHENYLCOUMARIN

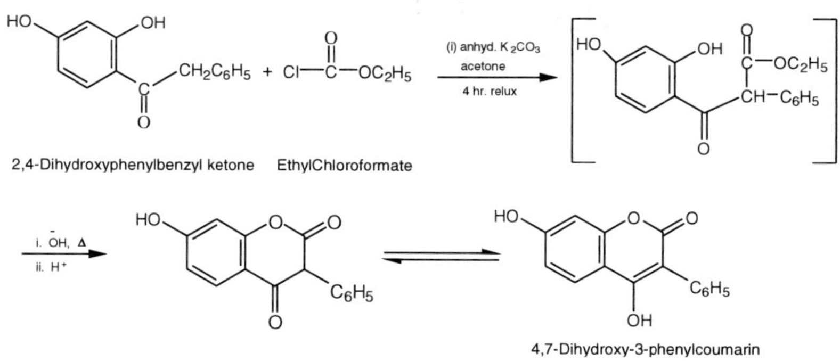

2,4-Dihydroxy phenyl benzyl ketone (4.F.117) (3 g, 0.013 mole) is refluxed in dry acetone (75 ml) with ethyl chloroformate (3.7 ml, 4.2 g, 0.039 mole) in the presence of ignited potassium carbonate (10 g) for 4 hr under anhydrous conditions. The mixture is filtered and the inorganic salts washed with hot acetone (2 × 25 ml). The combined acetone solution is distilled and the residual oily product heated for 1 hr in a boiling water bath with aqueous sodium hydroxide (1N, 100 ml). The solution is cooled and acidified with hydrochloric acid. The separated product is filtered and washed with water. It is crystallised from methanol as colourless prismatic needles. Yield 2 g (59.9%). M.p. 284-85° (descomp.) (lit.[8] m.p. 284-85°).

5.A.24 4-HYDROXY-7-METHOXY-3-PHENYLCOUMARIN

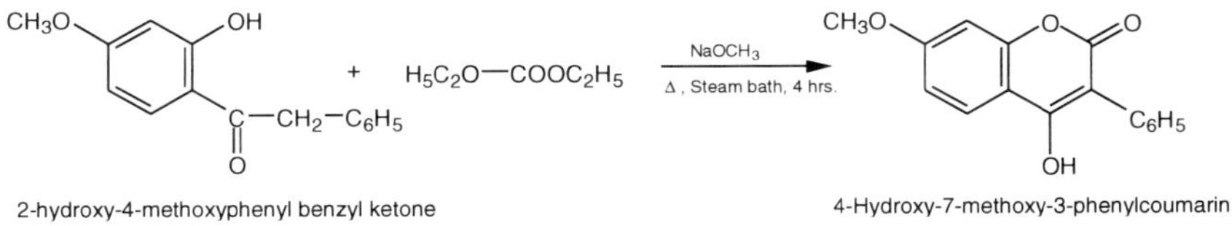

2-hydroxy-4-methoxyphenyl benzyl ketone   4-Hydroxy-7-methoxy-3-phenylcoumarin

Sodium (2 g, 0.087 mole) is dissolved in absolute methanol (25 ml) and excess methanol is removed in vacuo. A suspension of the above dried sodium methoxide in diethyl carbonate (25 ml) is treated with 2-hydroxy-4-methoxyphenyl-benzyl ketone (4.E.118) (5 g, 0.02 mole) in diethyl carbonate (25 ml). The mixture is heated on a steam bath for 4 hr under anhydrous conditions. Water (50 ml) is added and excess of diethyl carbonate removed by ether extraction (2 × 50 ml). The clear aqueous alkaline solution is acidified in cold with dilute hydrochloric acid. The separated product is filtered, washed with water and dried. It is crystallised from benzene as clusters of plates. Yield 3.5 g (63.6%). M.p. 204-205° (lit.[8] m.p. 204°).

5.A.25 3,7-DIMETHOXY-4-HYDROXYCOUMARIN[22]

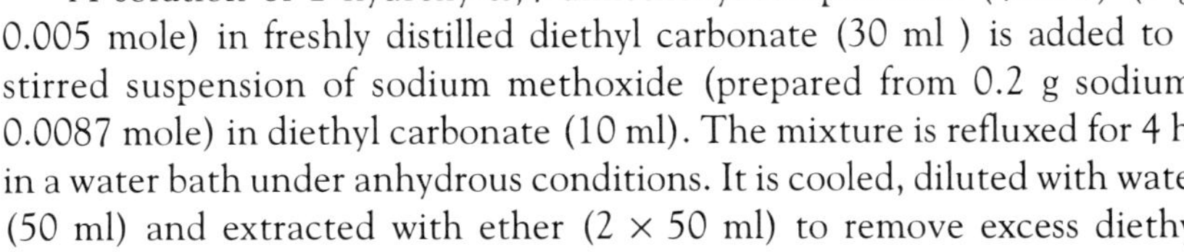

2-Hydroxy-ω,4-dimethoxyacetophenone   Diethyl cabonate   3,7-Dimethoxy-4-hydroxycoumarin

A solution of 2-hydroxy-ω,4-dimethoxyacetophenone (4.B.67) (1 g, 0.005 mole) in freshly distilled diethyl carbonate (30 ml) is added to a stirred suspension of sodium methoxide (prepared from 0.2 g sodium, 0.0087 mole) in diethyl carbonate (10 ml). The mixture is refluxed for 4 hr in a water bath under anhydrous conditions. It is cooled, diluted with water (50 ml) and extracted with ether (2 × 50 ml) to remove excess diethyl carbonate. The remaining aqueous solution is acidified with dilute hydrochloric acid. The separated product is filtered, washed with water and dried. It is crystallised from ethyl acetate-benzene as colourless rectangular prisms. Yield 0.8 g (70.8%). M.p. 167-69°.

Note: The diethyl carbonate method is applicable in case of ketones, which have only ortho hydroxyl group free; other hydroxyl groups must be protected.

5.A.26 3-METHOXY-4,7-DIHYDROXYCOUMARIN[22]

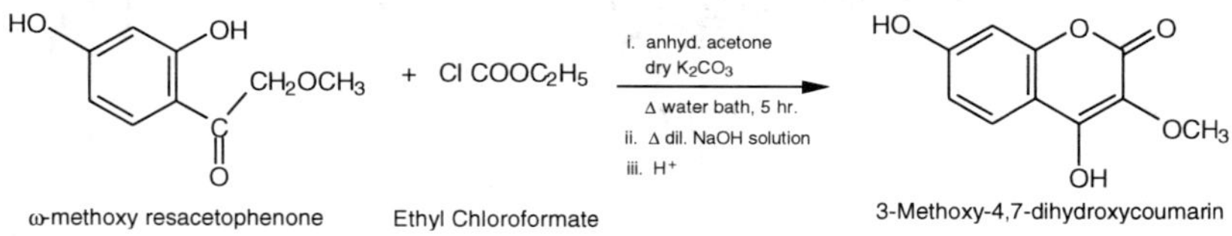

A solution of ω-methoxyresacetophenone (4.B.66.) (2 g, 0.011 mole) in acetone (75 ml) is refluxed for 5 hr under anhydrous conditions with ethyl chloroformate (2.1 ml, 2.38 g, 0.022 mole) and ignited potassium carbonate (4g). The mixture is filtered and the inorganic salts washed with hot acetone (2 × 20 ml). The combined acetone solution is distilled and the residual product heated for 2 hr (water-bath) with sodium hydroxide solution (2N, 100 ml). The solution is diluted with water (25 ml) and extracted with ether (2 × 50 ml). The clear alkaline solution acidified with hydrochloric acid. The separated product is filtered, washed with water and dried. It is crystallised from ethyl acetate-benzene as colourless neeedles. Yield 1.2 g (52.6%). M.p. 278-80°.

Note: The ethyl chloroformate method is applicable to polyhydroxy ketones.

5.A.27 7-HYDROXY-3,4-DIMETHOXYCOUMARIN[22]

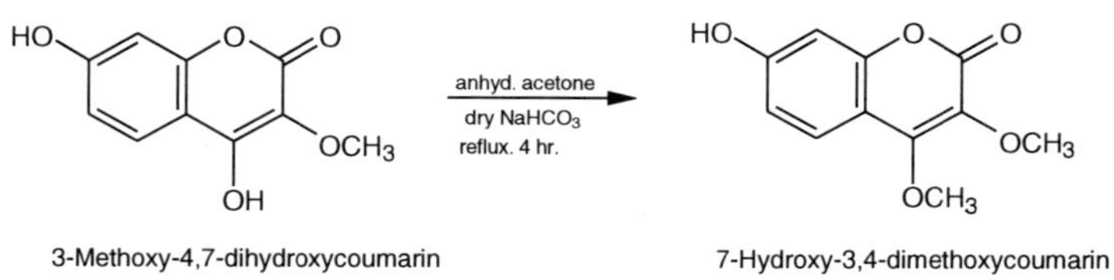

3-Methoxy-4,7-dihydroxycoumarin (5.A.26) (0.5 g, 0.0024 mole) in acetone (50 ml) is refluxed for 4 hr under anhydrous conditions with

dimethyl sulphate (0.25 ml, 0.33 g, 0.0026 mole) in the presence of anhydrous sodium bicarbonate (2 g). The reaction mixture is filtered and the inorganic salts washed with hot acetone (2 × 20 ml). The combined acetone solution is distilled and the residue macerated with crushed ice. The separated product is filtered and washed with water. It is crystallised from methanol as colourless needles. Yield 0.3 g, (56.6%). M.p. 201-202°.

5.A.28 7-HYDROXY-3,4-DIMETHYLCOUMARIN

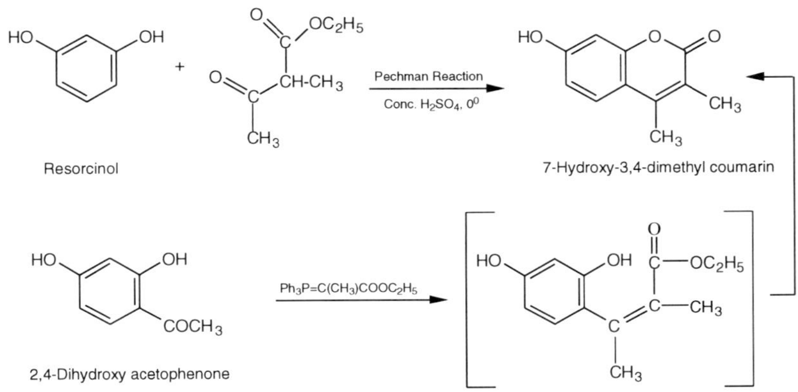

Method (i)[23]

To a cold solution (0°) of dry resorcinol (2 g, 0.018 mole) in ethyl 2-methyl acetoacetate (2.6 ml, 2.65 g, 0.0184 mole) is added gradually concentrated sulphuric acid (5 ml) with stirring. The mixture is stirred for 1 hr and kept in a refrigerator for 24 hr. It is poured over crushed ice (200 g) with stirring. The separated product is filtered and washed with water. It is crystallised from ethanol as colourless crystals. Yield 3 g (87%). M.p. 237-39°.

Method (ii)[23]

A mixture of 2,4-dihydroxyacetophenone (4.A.6.) (1 g, 0.0066 mole) and 1-carbethoxyethylidene (triphenyl) phosphorane (6.A.11) (2.5 g, 0.0069 mole) in anhydrous benzene (30 ml) is refluxed for 36 hr under anhydrous conditions. The solvent is distilled off and the residue is subjected to column chromatography over silica gel (2 × 30 cm). Elution with benzene-petroleum ether (1:1) gives the required product. Yield 0.8 g (64%). M.p. 237-39°.

5.A.29 7-METHOXY-3,4-DIPHENYLCOUMARIN[24]

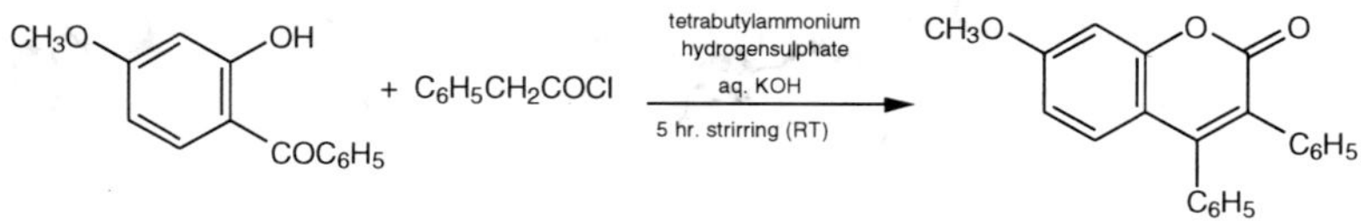

2- Hydroxy-4-Methoxybenzophenone

7-Methoxy-3,4-diphenylcourmarin

A solution of 2-hydroxy-4-methoxybenzophenone (4.E.107) (2.28 g, 0.01 mole) in dry benzene (30 ml) is stirred at room temperature with tetrabutylammonium hydrogen sulphate (100 mg) and aqueous potassium carbonate (20%, 50 ml). A solution of phenylacetyl chloride (1.5 ml, 1.75 g, 0.011 mole) in benzene (10 ml) is added dropwise to the above stirred solution. The reaction mixture is stirred for 5 hr at room temperature. The organic layer is separated, washed with water and dried (anhydrous sodium sulphate). The solvent is removed in vacuo. The residual product is crystallised from ethanol as white crystalline solid. Yield 2.2 g (67.1%). M.p. 168-69°.

Note: Phenyl acetyl chloride is prepared by heating phenylacetic acid with excess thionyl chloride for 20 min (water bath) under anhydrous conditions. Excess thionyl chloride is removed on a water bath. It is used as such.

5.A.30 BENZOFURAN (COUMARONE)

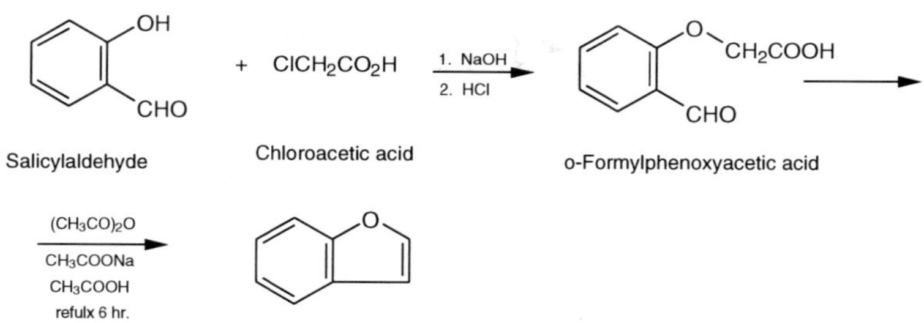

Benzofuran

Step (i) o-Formylphenoxy Acetic Acid

A solution of sodium hydroxide (13.35 g, 0.33 mole in 350 ml water) is added slowly with stirring to a mixture of salicylaldehyde (20 g, 17.5 ml, 0.164 mole), chloroacetic acid (15.75 g, 0.167 mole). The mixture is heated to boiling with stirring. The solution is refluxed for 3 hr. The red-brown coloured solution is cooled and acidified with concentrated hydrochloric acid (30 ml). The unreacted salicylaldehyde is recovered by steam distillation. The remaining mixture is cooled. The separated oily product solidifies on standing. The solid product is filtered. It is crystallised from hot water (charcoal) as almost colourless crystals. Yield 13.5 g (45.8%). M.p. 132-33° (lit.[25] m.p.132°).

Step (ii) Benzofuran

A mixture of the above o-formylphenoxyacetic acid (10 g, 0.055 mole), anhydrous sodium acetate (20 g), acetic anhydride (50 ml) and glacial acetic acid (50 ml) is refluxed for 6 hr. The mixture is poured with stirring into crushed ice (300 g) and allowed to stand for 5-6 hr with occasional shaking. It is extracted with ether (3 × 75 ml) and the combined ether extract is washed with aqueous sodium hydroxide (5%) until the aqueous layer is basic. The ether layer is washed with water (until the washing is neutral), and dried (anhydrous calcium chloride). Ether is removed in a rotary evaporator. The residue is distilled and benzofuran is collected at 170-72° (lit.[25] b.p. 170°). Yield 4.75 g (72.5%).

5.A.31 6-HYDROXYCOUMARAN-3-ONE

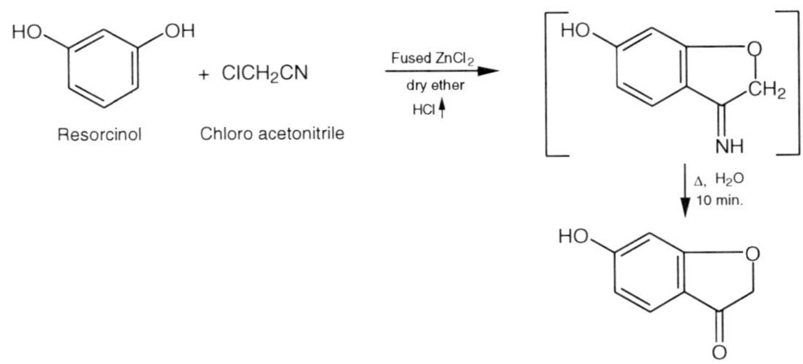

Dry hydrogen chloride gas is passed for 30 min through a well stirred mixture of resorcinol (3 g, 0.027 mole), chloroacetonitrile (6.A.12) (2 g, 1.7 ml, 0.026 mole), powdered fused zinc chloride (2 g, 0.0145 mole) and dry ether (40 ml). The precipitated ketimine hydrochloride is separated and washed by decantation with dry ether (2 × 10 ml). The product is dissolved in water (200 ml) and the solution refluxed for 10 min. The solution is cooled overnight in a refrigerator. The separated product is filtered and washed with cold water (2 × 5 ml) and dried. The crude material is added to a hot solution of potassium acetate (3 g) in absolute alcohol (20 ml). The mixture is refluxed for 15 min, cooled and poured into cold water (50-60 ml). The product obtained is filtered and washed with water. It is crystallised from hot water as glistening yellow flakes. Yield 3.4 g (83%). M.p. 243-44° (lit.[26] m.p. 243-44°).

5.A.32 4,6-DIMETHOXYCOUMARAN-3-ONE

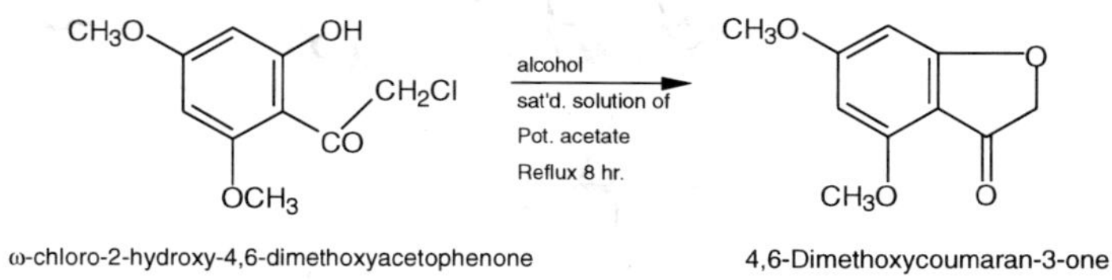

ω-Chlorophloroacetophenone dimethyl ether (4.B.92) (4 g, 0.017 mole) in alcohol (20 ml) and saturated solution of potassium acetate (20 ml) is refluxed for 8 hr. The solution is filtered and the filtrate cooled. The separated product is filtered and washed with water. It is crystallised from dilute alcohol as colourless crystalline product. Yield 2.5 g (74.2%). M.p. 137-38° (lit.[27] m.p. 137-38°).

5.A.33 2-METHYL CHROMONES 7,8-DIHYDROXY-2-METHYLCHROMONE

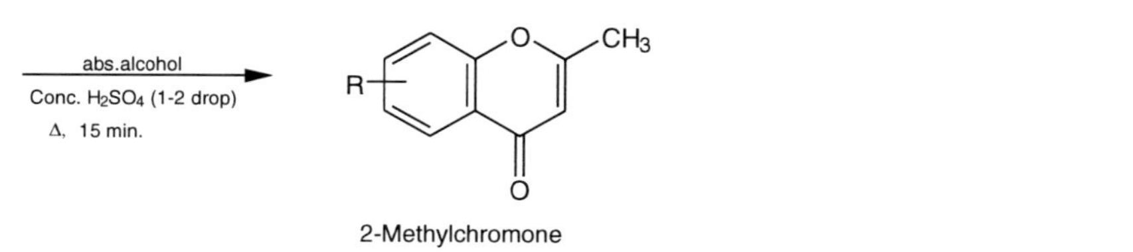

Step (i) Claisen condensation of o-hydroxyacetophenones with Ethyl acetate: Formation of o-hydroxybenzoyl acetones[28]

A solution of the appropriate o-hydroxyacetophenone (0.007 mole) in dry ethyl acetate (10 ml) is added dropwise to a suspension of pulverized sodium (0.017 mole) in dry ether (8 ml). After the initial reaction subsides, the mixture is refluxed for 1 hr. The cooled reaction mixture is diluted with cold water (caution), the upper benzene layer is separated and the aqueous alkali solution is extracted with ether to remove unreacted ethyl acetate. The clear aqueous alkaline solution is acidified in cold with dilute acetic acid. The separated diketone is filtered, washed with water and crystallised from a suitable solvent. In case an oil is obtained, it is extracted with ether, extract washed with sodium bicarbonate (5%), dried (anhydrous magnesium sulphate) and distilled; the crude oily product is used directly for the next step. Yield 60-70%.

2-Hydroxy-3,4-dimethoxybenzoyl acetone[28] is obtained by the Claisen condensation of 2-hydroxy-3,4-dimethoxyacetophenone (4.A.37). It is crystallised from benzene-petroleum ether as yellow needles. M.p. 107-108°.

2-Hydroxy-5-methoxybenzoylacetone[28] is obtained by the Claisen condensation of 2-hydroxy-5-methoxyacetophenone (4.A.11). It is crystallised from benzene-petroleum ether as yellow needles. M.p. 102-103°.

Step (ii) Cyclisation of o-hydroxybenzoylacetone: Formation of 2-methylchromones[28]

The appropriate o-hydroxybenzoylacetone (0.2 g) in absolute ethanol (5 ml) containing concentrated sulphuric acid (2 drops) is refluxed for 15 min on a water bath. The solution is cooled, diluted with water (10 ml) and rendered alkaline by adding few drops of sodium hydroxide solution. The reaction mixture is extracted with chloroform (3×10 ml), extract washed with water and dried (anhydrous magnesium sulphate). The chloroform is distilled off and the product obtained is crystallised from appropriate solvent. Yield 80-90%.

7,8-Dihydroxy-2-methylchromone[28] is obtained by the cyclisation of 2-hydroxy-3,4-dimethoxybenzoyl acetone. M.p. 101-102° (lit.[29] m.p. 102°).

6-Methoxy-2-methylchromone[28] is obtained by the cyclisation of 2-hydroxy-5-methoxybenzoylacetone. M.p. 105-106° (lit.[30] m.p. 105-107°).

5.A.34 7-HYDROXY-2-METHYLCHROMONE (K.R. ACETYLATION)

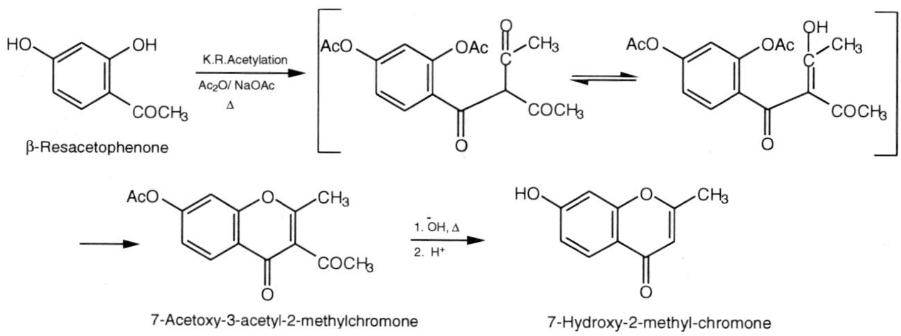

Step (i) 7-Acetoxy-3-acetyl-2-methylchromone

β-Resacetophenone (4.A.6) (10 g, 0.066 mole) is heated with fused sodium acetate (20 g) in acetic anhydride (30 ml) for 5 hr at 170-180° (oil bath) under anhydrous conditions. The reaction mixture is poured over crushed ice, stirred well and left overnight. The separated product is filtered and washed with water. It is crystallised from alcohol as light yellow needles. Yield 10 g (58.5%). M.p. 127° (lit.[31,32] m.p. 126-27°).

Step (ii)

The above 7-acetoxy-3-acetyl-2-methylchromone (1 g, 0.0038 mole) is refluxed (2 hr) with aqueous sodium carbonate solution (10%, 50 ml). The clear solution is then acidified with hydrochloric acid (1:1). The separated product is filtered and washed with water. It is crystallised from alcohol as thick prismatic crystals. Yield 0.5 g (73.5%). M.p. 249-50°. (lit.[31,32] m.p. 254-55°).

5.A.35 5,7-DIHYDROXY-2-METHYLCHROMONE

A mixture of dry phloroacetophenone (4.A.44) (12 g, 0.071 mole), fused sodium acetate (15 g) and dry distilled acetic anhydride (100 ml) is heated under anhydrous conditions for 1 hr at 140-150° (oil bath). The temperature is then raised to 180-85° and refluxing is continued for 6 hr more. The mixture is cooled, poured over crushed ice and stirred. The separated product is filtered and washed with water. It is crystallised from alcohol as colourless needles of 5,7-diacetoxy-3-acetyl-2-methylchromone. Yield 10 g. m.p. 131-32°. The diacetoxy compound is hydrolysed by adding it in small portions to boiling solution of sodium

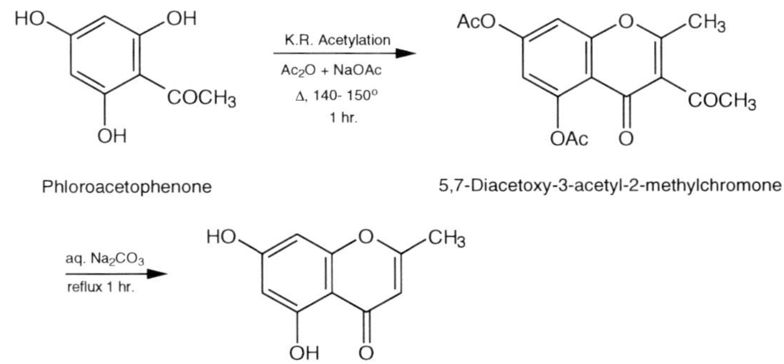

Phloroacetophenone

5,7-Diacetoxy-3-acetyl-2-methylchromone

5,7-Dihydroxy-2-methylchromone

carbonate (10%, 150 ml). The mixture is finally refluxed for 2 hr. The clear alkaline solution is cooled and acidified with hydrochloric acid. The separated product is filtered and washed with water. It is crystallised from dilute alcohol as colourless prisms. Yield 6 g (43.8%). M.p. 278-80° (lit.[33] m.p. 279°).

5.A.36 2-METHYL-3-METHOXY-5,7-DIHYDROXYCHROMONE

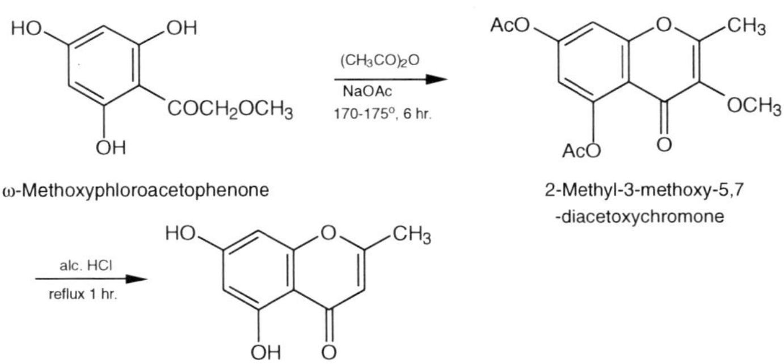

ω-Methoxyphloroacetophenone

2-Methyl-3-methoxy-5,7-diacetoxychromone

5,7-Dihydroxy-2-methylchromone

A mixture of anhydrous ω-methoxyphloroacetophenone (4.B.70) (6 g, 0.03 mole), fused sodium acetate (10 g) and acetic anhydride (25 ml) is heated for 6 hr at 170-75° (oil bath) under anhydrous conditions. The reaction mixture is cooled and poured over crushed ice (250 g) with vigorous stirring. The separated 2-methyl-3-methoxy-5,7-diacetoxy-chromone is filtered and washed with water. It is refluxed for 1 hr in a mixture of alcohol (100 ml) and concentrated hydrochloric acid (12 ml). The solution is concentrated and cooled. The separated product is filtered and washed with water. It is crystallised from alcohol as colourless long needles. Yield 5 g (74.4%). M.p. 221-22° (lit.[34] m.p. 221-22°).

5.A.37 FLAVONE

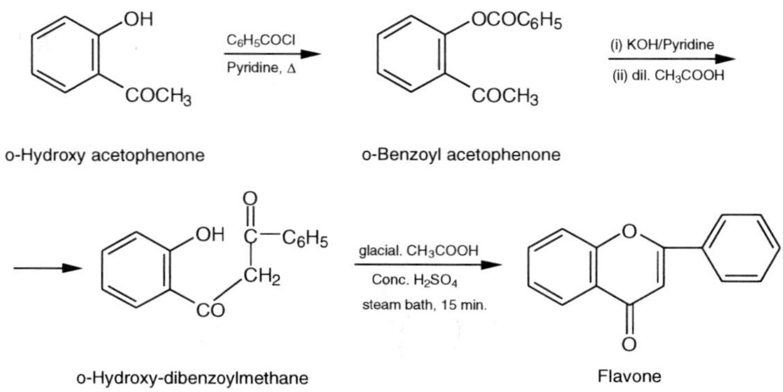

Mechanism

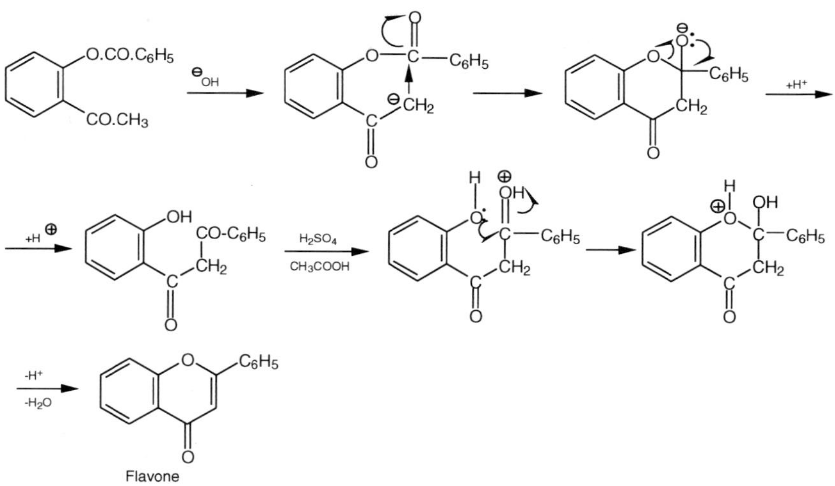

Step (i) o-Benzoyloxyacetophenone

In a 50 ml conical flask fitted with calcium chloride tube are placed o-hydroxyacetophenone (4.A.1) (6 ml, 6.79 g, 0.05 mole), benzoyl chloride (10.6 g, 8.75 ml, 0.075 mole) and dry pyridine (20 ml). The temperature of reaction mixture rises, and when no further heat is evolved (10-15 min) the mixture is heated in a boiling water bath for 15 min, cooled and poured with stirring into dilute hydrochloric acid (3%, 250 ml) containing crushed ice (100 g). The separated product is filtered washed with methanol (20 ml) and then with water. It is crystallised from methanol as white crystals. Yield 10 g (83%); m.p. 87-8° (lit.[35] m.p. 87-8°).

Step (ii) o-Hydroxydibenzoylmethane

To a warm solution (50°) of o-benzoyloxyacetophenone (10 g, 0.042 mole) in pyridine (35 ml) is added pulverized potassium hydroxide (3.5 g) and the mixture stirred for 15 min. The viscous mass so obtained is cooled to room temperature and acidified with acetic acid (50 ml, 10%). The separated diketone is filtered. It is crystallised from ethanol as light yellow crystals. Yield 9 g (90%). M.p. 119-20°. For subsequent step, crystallisation is not necessary (lit.[36] m.p. 119-20°).

Step (iii) Flavone

To a solution of the above diketone (8.8 g, 0.03 mole) in glacial acetic acid (40 ml) is added with shaking concentrated sulphuric acid (1.8 ml). The mixture is heated on a steam bath for 15 min and then poured over crushed ice (200 g) with stirring. The separated product is filtered, washed with water (until free from acid). It is crystallised from petroleum ether as white needles. Yield 7.9 g (97 %). M.p. 96-7° (lit.[35] m.p. 96-7°).

5.A.38 7-HYDROXY FLAVONE

Method (i) (Allan Robinson Method)

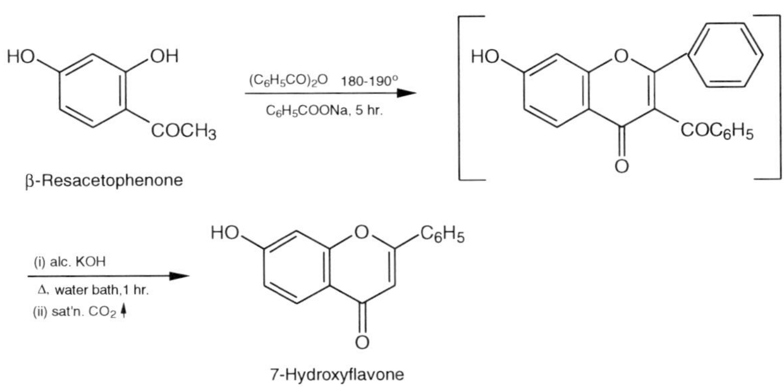

An intimate mixture of β-resacetophenone (4.A.6) (10 g, 0.066 mole), sodium benzoate (10 g, 0.069 mole) and benzoic anhydride (40 g, 33.3 ml) is heated for 5 hr at 180-190° (oil-bath) under reduced pressure. The product is cooled and dissolved in alcoholic potassium hydroxide solution (10%, 200 ml) by heating in a boiling water bath. The solution is refluxed for 1 hr and alcohol is removed under reduced pressure. Water (100 ml) is added and the clear alkaline solution is saturated with carbon dioxide gas (2 hr). The mixture is cooled and the separated product filtered and washed with water. It is crystallised from alcohol as flat needles. Yield 11.6 g (74%). M.p. 240-41°. (lit.[37] m.p. 240-41°).

Method (ii)

β-Resacetophenone (4.A.6) (1 g, 0.0066 mole) is refluxed with benzoyl chloride (1.7 ml, 2.06 g, 0.0146 mole) ignited potassium carbonate (5 g) and dry acetone (30 ml) for 30 hr. The reaction mixture is filtered and the inorganic salts washed with hot acetone (2 × 20 ml). The combined acetone solution is distilled and the residue is refluxed with aqueous sodium carbonate (5%, 50 ml) for 2 hr. The clear solution is cooled and saturated with carbon dioxide gas (1 hr). The separated product is filtered and washed with water. It is crystallised from alcohol as flat needles. Yield 0.85 g (54.1%). M.p. 239-40°. (lit.[38] m.p. 239-40°).

Method (iii)

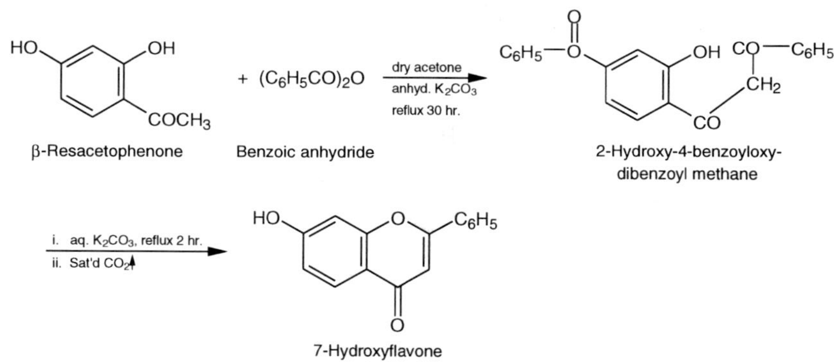

β-Resacetophenone (4.A.6) (2 g, 0.0132 mole) is dissolved in dry acetone (75 ml) and the solution refluxed with benzoic anhydride (7.5 g, 0.0332 mole) and ignited potassium carbonate (8 g) for 30 hr under anhydrous conditions. The solvent is distilled off and the residue is treated with water. A yellow potassium salt of the diketone separated out. It is filtered (filtrate A) washed with more carbonate solution and heated with acetic acid (10%) to get the free diketone. It is filtered and crystallised from benzene when 2-hydroxy-4-benzoyloxydibenzoylmethane separates as pale yellow crystals. Yield 3 g. M.p. 167°. The diketone is refluxed with aqueous potassium carbonate (5%) for 2 hr, and is mixed with the filtrate A. The solution is saturated with carbon dioxide (2-3 hr). 7-Hydroxyflavone separates out. It is crystallised from alcohol as flat needles. Yield 1.8 g (57.3%) m.p. 239-40° (lit.[38] m.p. 239-40°).

Method (iv)

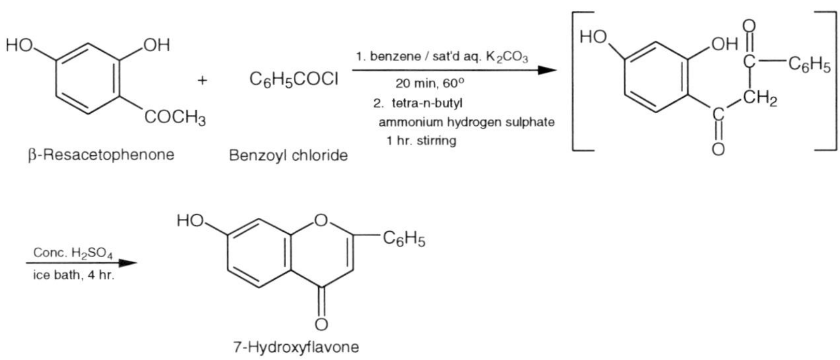

Phase Transfer Catalysed Synthesis

β-Resacetophenone (4.A.6) (1.0 g, 0.0066 mole) and benzoyl chloride (3.0 ml, 3.6 g, 0.026 mole) are dissolved in benzene (30 ml) and stirred at 60° with saturated aqueous potassium carbonate (30 ml) for 20 min until benzoylation is complete (TLC). To this mixture is added tetra n-butylammonium hydrogen sulphate (3 g, 0.0088 mole) and stirring continued for 1 hr until TLC shows the absence of the dibenzoate. The orange-yellow benzene layer is separated and the aqueous solution extracted with chloroform (2 × 25 ml). The combined benzene and chloroform solution is washed with water and distilled. The residue is dried in vacuo over phosphorous pentoxide. To the dry residual product is added concentrated sulphuric acid (7.5 ml) and the reaction mixture is kept in ice bath for 4 hr. It is poured over crushed ice. The separated product is filtered and washed with water. It is crystallised from alcohol as almost colourless needles. Yield 1.5 g (96.2%). M.p. 240° (lit.[38,39] m.p. 239-40°).

5.A.39 6-HYDROXYFLAVONE

2,5-Dihydroxyacetophenone (4.A.10.) (10 g, 0.066 mole) is refluxed for 24 hr under anhydrous conditions with benzoyl chloride (15 ml, 18.16 g,

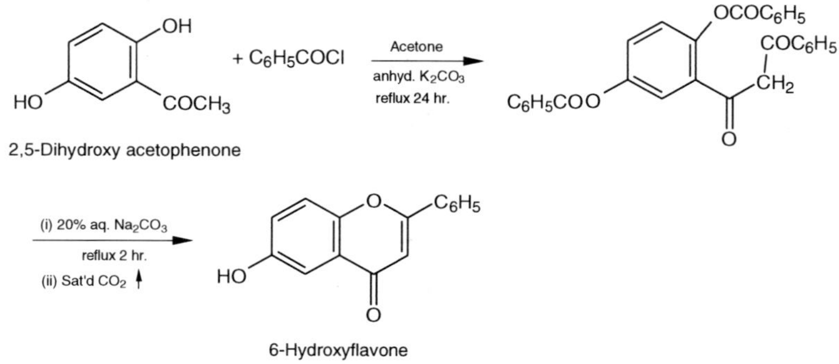

0.129 mole) in dry acetone (200 ml) in the presence of ignited potassium carbonate (40 g). The mixture is filtered and the inorganic salts washed with hot acetone (2 × 30 ml). The combined acetone solution is distilled off. The residual product is refluxed with sodium carbonate solution (10%, 100 ml) for 2 hr. The clear alkaline solution is saturated with carbon dioxide gas (1.5 hr). The separated product is filtered and washed with water. It is crystallsed from alcohol as colourless needles. Yield 6 g (38.3%). M.p. 233-34° (lit.[40] m.p. 234°).

5.A.40 5,7-DIHYDROXYFLAVONE

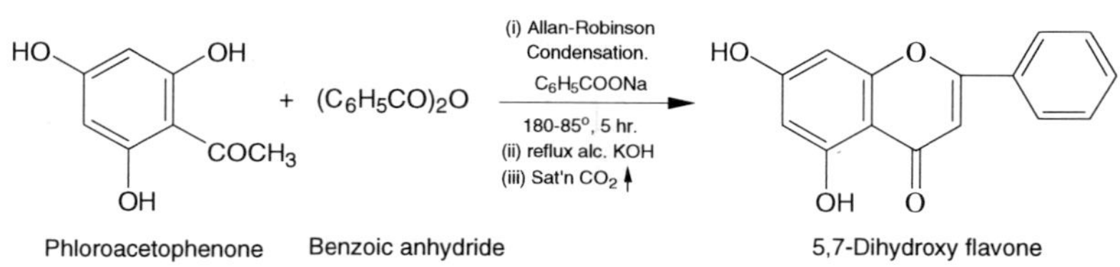

An intimate mixture of phloroacetophenone (4.A.44) (5 g, 0.03 mole), sodium benzoate (6 g, 0.04 mole) and benzoic anhydride (32 g, 26.7 ml) is heated under reduced pressure at 180-185° (oil bath) for 5 hr. The reaction mixture is cooled and dissolved in alcohol (160 ml) (approx. 20 min heating on a water bath). The alcoholic solution is refluxed for 20 min with a solution of potassium hydroxide (24 g) in water (25 ml) and alcohol (50 ml). Alcohol is distilled off under reduced pressure and water (500 ml) added to the residual solution. The clear alkaline solution is saturated with carbon dioxide gas (3-4 hr). The separated product is filtered and washed with water. It is crystallised from alcohol as bright yellow needles. Yield 7 g (92.7%). M.p. 275° (lit.[41] m.p. 275°).

5.A.41 3-METHOXY-7-HYDROXYFLAVONE

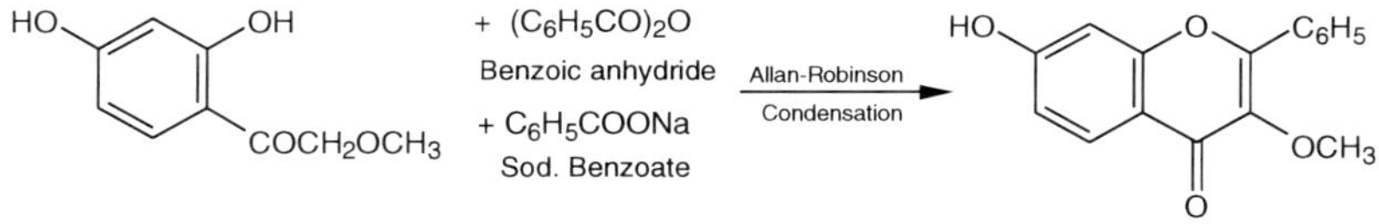

ω-Methoxyresacetophenone 3-methoxy-7-hydroxyflavone

An intimate mixture of ω-methoxyresacetophenone (4.B.66) (10 g, 0.055 mole), benzoic anhydride (40 g, 33.3 ml) and sodium benzoate (10 g, 0.069 mole) is heated under reduced pressure for 4 hr at 180-84° (oil bath). The product is cooled and refluxed with alcoholic potash (10%, 200 ml) for 30 min. Alcohol is removed in vacuo and water (500 ml) added to dissolve the residue. The clear alkaline solution is saturated with carbon dioxide gas (2-3 hr). The separated product is filtered and crystallised from acetic acid as clusters of needles. Yield 12.5 g (84.9%). M.p. 227° (lit.[42] m.p. 227°).

5.A.42 7-METHOXY-3-HYDROXYFLAVONE (7-METHOXYFLAVONOL)

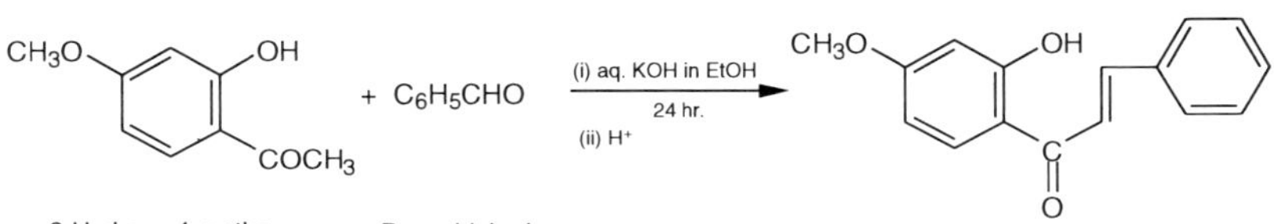

2-Hydroxy-4-methoxy acetophenene Benzaldehyde 2'-Hydroxy-4'-Methoxy chalcone

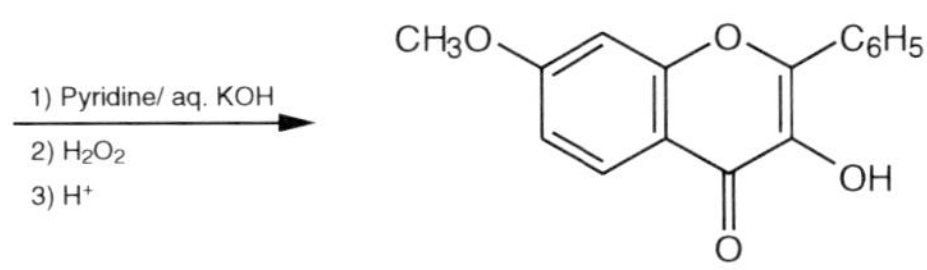

7-Methoxy-3-hydroxyflavone

Step (i) 2'-Hydroxy-4'-methoxychalcone

To a solution of 2-hydroxy-4-methoxyacetophenone (4.A.7) (4g, 0.024 mole) and benzaldehyde (2.8 g, 2.7 ml, 0.026 mole) in ethanol (30 ml) is added an aqueous solution of potassium hydroxide (6 g in 20 ml water). The clear solution is kept at room temperature for 24 hr. Crushed ice (100 g) is added and the clear alkaline solution is acidified with hydrochloric acid. The separated product is filtered and washed with water. It is crystallised from ethanol as yellow needles. Yield 4 g (65.4%). M.p. 104-105° (lit.[43] m.p. 105°).

Similar condensation of o-hydroxyacetophenone (4.A.1) with anisaldehyde gives 2'-hydroxy-4-methoxychalcone in 65% yield. M.p. 93-94° (lit.[44] m.p. 93-94°).

Step (ii) 7-Methoxy-3-hydroxy Flavonol

2'-Hydroxy-4'-methoxychalcone (1.5 g, 0.0059 mole) is dissolved in pyridine (10 ml) and aqueous potassium hydroxide (10%, 25 ml) added. The mixture is warmed to 60° to make it clear. Hydrogen peroxide (10%, 10.5 ml, 1.05 g, 0.031 mole) is added slowly with constant shaking at 10-15° and the solution left for 3 hr at room temperature. The solution is filtered and the clear alkaline solution acidified with hydrochloric acid. The separated product is filtered and washed with water. It is crystallised from methanol as light yellow needles. Yield 1 g (63.3%). M.p. 180-81° (lit.[45] m.p. 180°).

Similar reaction of 2'-hydroxy-4'-methoxychalcone with hydrogen peroxide in pyridine gives 4'-methoxyflavanol in 70% yield. M.p. 224-25° (lit.[46] m.p. 225°).

5.A.43 5,7-DIMETHOXYFLAVANONE

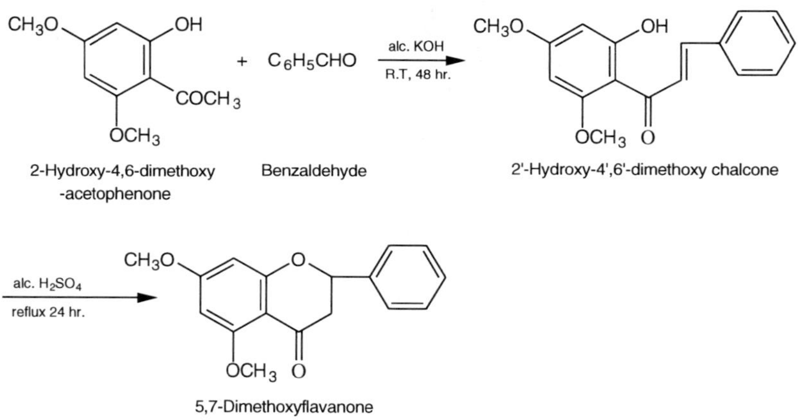

Step (i) 2'-hydroxy-4',6'-dimethoxychalcone

A cooled solution of 2-hydroxy-4,6-dimethoxyacetophenone (4.A.45) (2 g, 0.01 mole) and benzaldehyde (4 ml, 4.2 g, 0.039 mole) in ethanol (40 ml) is treated with potassium hydroxide solution (6 g in 6 ml water). The mixture is shaken and left at room temperature for 48 hr. It is diluted with water (50 ml) and extracted with ether (50 ml). The clear aqueous solution is acidified in cold. The separated product is filtered and washed with sodium bicarbonate solution and then with water. It is crystallised from alcohol as yellow prisms. Yield 2 g (69%). M.p. 92-3° (lit.[47,48] m.p. 90-1°).

Step (ii) 5,7-Dimethoxyflavanone

2'-Hydroxy-4',6'-dimethoxychalcone (0.5 g, 0.0018 mole) is refluxed in alcoholic sulphuric acid (4% , 50 ml) for 24 hr. The alcohol is distilled off in vacuo and the residue is diluted with water (50 ml). The separated product is filtered and washed with ether to remove any adhering chalcone. It is crystallised from alcohol as almost colourless needles. Yield 0.4 g (80%). M.p. 143° (lit.[48] m.p. 143-44°).

5.A.44 7-HYDROXYISOFLAVONE (FARKAS METHOD)

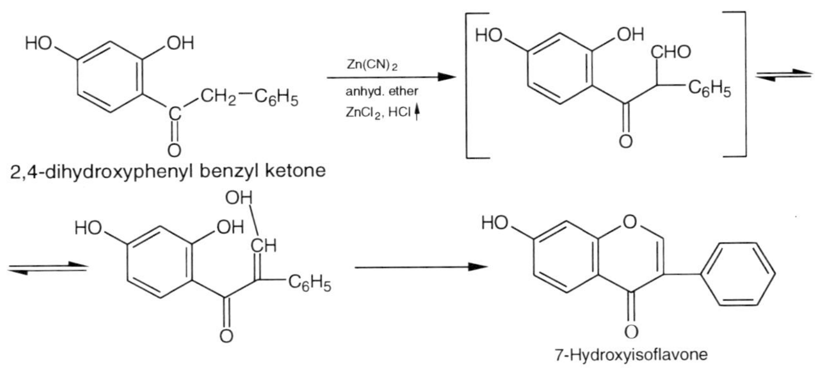

A stream of dry hydrogen chloride gas is passed for 3 hr into a mixture of 2,4-dihydroxyphenyl benzyl ketone (4.F.117) (4 g, 0.0175 mole) in dry ether (75 ml) and anhydrous zinc cyanide (4 g, 0.034 mole) at 0-5°. The reaction mixture is kept overnight at room temperature. The ether is decanted and oily product is washed with dry ether (2 × 20 ml). It is dissolved in water (150 ml) and the solution is heated for 1 hr at 80°. Water is decanted and the sticky mass macerated with ethanol (25 ml). The separated solid product is filtered. It is crystallised from ethanol as colourless prisms. Yield 0.7 g (16.6%). M.p. 213-14° (lit.[49,50] m.p. 215°).

5.A.45 7,4'-DIHYDROXYISOFLAVONE (DIADZIN)

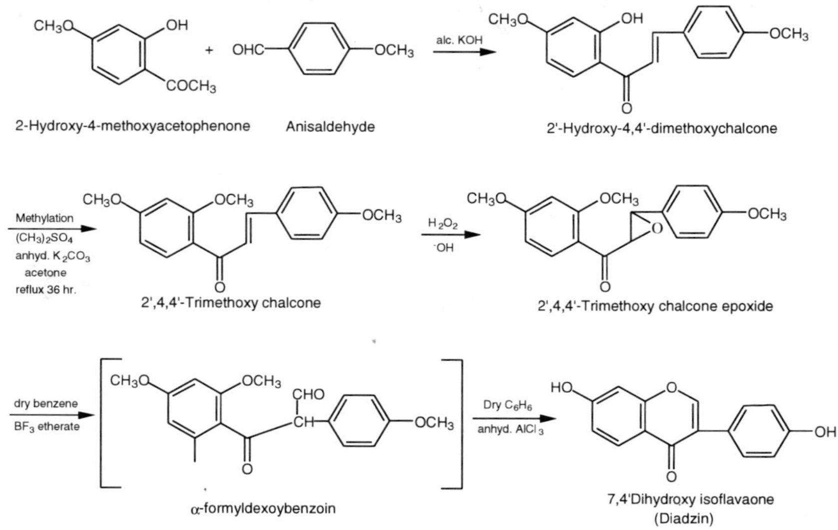

Step (i) 2'-Hydroxy-4,4'-dimethoxychalcone

To a solution of anisaldehyde (4.7 ml, 5.26 g, 0.039 mole) and 2-hydroxy-4-methoxyacetophenone (4.A.7.) (5.1 g, 0.03 mole) in ethanol (100 ml) is added aqueous potassium hydroxide (9 g in 9 ml water). The mixture is heated on a water bath till the solution becomes clear and then left for 24 hr at room temperature. It is diluted with water (200 ml) and then acidified with hydrochloric acid. The separated product is filtered, washed with sodium bicarbonate solution and then with water. It is crystallised from ethyl acetate- petroleum ether as yellow needles. Yield 6 g (68.9%). M.p. 113-14° (lit.[51] m.p. 113-14°).

Step (ii) 2',4,4'-Trimethoxychalcone

The above 2'-hydroxy-4,4'-dimethoxychalcone (6 g, 0.02 mole) in dry acetone (150 ml) is refluxed for 36 hr under anhydrous conditions with dimethyl sulphate (6 ml, 8 g, 0.06 mole) and ignited potassium carbonate (20 g). The acetone is distilled off and water (250 ml) added to the residue. The separated solid is filtered and washed with water. It is crystallised from methanol as very pale yellow small plates. Yield 5.4 g (85.7%). M.p. 88-9° (lit.[51] m.p. 88-9°).

Step (iii) 2',4,4'-Trimethoxychalcone epoxide

A solution of the above 2,4,4'-trimethoxychalcone (5 g, 0.0168 mole) in methanol (75 ml) is treated with aqueous sodium hydroxide solution (8%, 7 ml) and hydrogen peroxide (30%, 7 ml, 2.1 g, 0.062 mole) added dropwise with shaking. The mixture is refluxed with occasional shaking. For the first eight minutes a brisk reaction takes place; the refluxing is continued for 3 min more. The reaction mixture is left for 18 hr at room temperature. The separated epoxide is filtered and washed with water. It is crystallised from methanol as colourless long needles. Yield 3 g (56.6%). M.p. 105-106° (lit.[52] m.p. 105-106°).

Step (iv) 7,4'-Dihydroxyisoflavone (Diadzin)

The above epoxide (2 g, 0.006 mole) is dissolved in dry benzene (30 ml) and treated at 10° with boron trifluoride etherate (2 ml, 2.308 g, 0.0.16 mole). The solution becomes yellow, orange and finally deep red and a deep coloured complex separates. The mixture is shaken for 20 min at room temperature and then ether (100 ml) and water (50 ml) added and the mixture is shaken thoroughly. The organic layer is separated and distilled off at 35-40° in vacuo and the residue is dried (over phosphorous pentoxide in vacuum desiccator). The α-formyl deoxybenzoin is obtained as an orange solid. The above α-formyl deoxybenzoin is refluxed in dry benzene (75 ml) with anhydrous powdered aluminium chloride (4 g, 0.03 mole) for 18 hr under anhydrous conditions. Benzene is distilled and the residual product is treated with ice (100 g) and hydrochloric acid (40 ml). The mixture is left for 24 hr at room temperature and extracted with ether (3 × 50 ml). The ether extract is washed with sodium bicarbonate solution and then extracted with sodium carbonate solution (5%). The clear alkaline solution is acidified and extracted with ether (2 × 30 ml). Ether is distilled off and the residue macerated with ethanol (2.5 ml). The separated product is filtered and crystallised from alcohol as colourless small prisms. Yield 0.1 g (6.3%). M.p. 315°. (lit.[53] m.p. 315°).

5.A.46 5,7-DIHYDROXYISOFLAVONE

Method (i)

To a cooled solution of 2,4,6-trihydroxyphenylbenzyl ketone (4.F.119) (12.2 g, 0.05 mole) in dry N,N-dimethylformamide (100 ml) is added dropwise boron trifluoride etherate (21.5 ml , 24.8 g, 0.17 mole) and the solution is kept for 40 min at room temperature. The temperature is

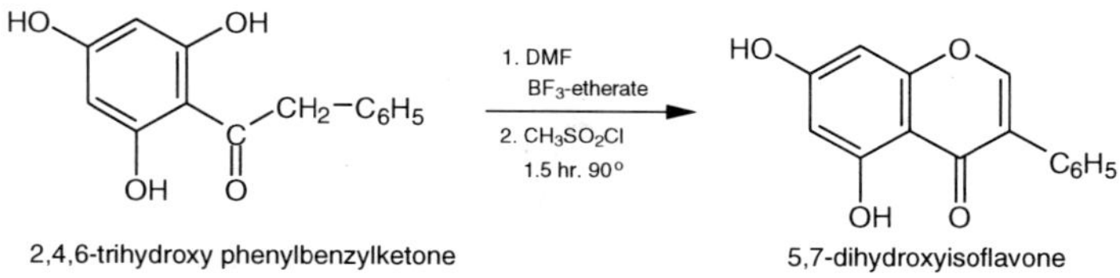

subsequently raised to 60° and methanesulphonlyl chloride (13.8 ml, 24.8 g, 0.18 mole) added in one lot. The resulting mixture is heated for 90 min on a boiling water bath, cooled and then poured into ice cold water. The separated product is filtered, washed with water and dried. It is crystallised from benzene as yellow needles. Yield 8 g (63%). M.p. 198-201° (lit.[54] m.p. 198-201°).

Method (ii)

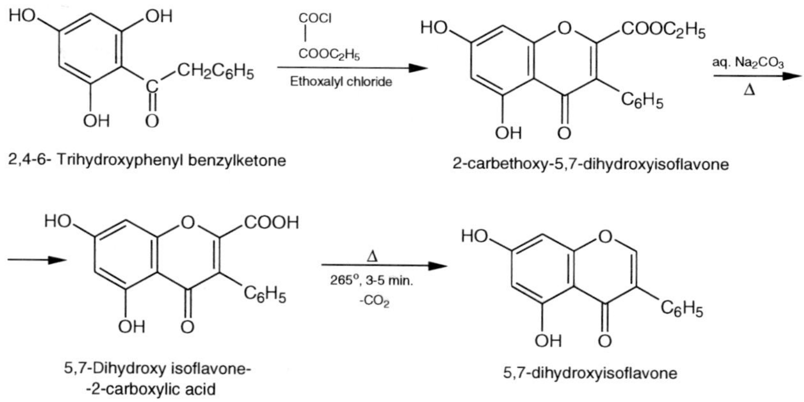

Step (i) 2-Carbethoxy-5,7-dihhdroxyisoflavone

Freshly prepared ethoxalyl chloride (22 ml, .88 g, 0.197 mole) is added to an ice-cold solution of 2,4,6-trhiydroxyphenyl benzyl ketone (4.F.119) (6 g, 0.02 mole) in dry pyridine (25 ml). The mixture is left for 24 hr at room temperature and water (100 ml) added. It is extracted with chloroform (3 × 30 ml). The chloroform extract is washed successively with dilute hydrochloric acid (to remove pyridine), water and then dried (anhydrous sodium sulphate). The solvent is evaporated and the product obtained is crystallised from alcohol as pale yellow needles. Yield 4.8 g (60%). M.p. 229-30° (lit.[55] m.p. 230°).

Step (ii) 5,7-Dihydroxyisoflavone

The above carbethoxy isoflavone (4 g, 0.012 mole) in acetone (100 ml) is refluxed for 4 hr with aqueous sodium carbonate solution (5%, 100 ml). The acetone is removed in vacuo and the remaining aqueous solution is acidified with hydrochloric acid. The separated product is filtered and washed with water. It is crystallised from dilute alcohol as yellow needles of 5,7-dihydroxyflavone–2-carboxylic acid. Yield 3.2 g (87.7%). M.p. 255° (lit.[55] m.p. 255°).

The above carboxyisoflavone (0.2 g, 0.0007 mole) is heated just above its melting point (265°) in a metal bath for 3-5 min. The solid cake left behind is extracted with chloroform (25 ml) and the solution washed with sodium bicarbonate solution to remove the unchanged product. Chloroform is distilled and the residual product is crystallised from benzene as colourless plates of 5,7-dihydroxyisoflavone. Yield 0.13 g (76.4%). M.p. 195-97° (lit.[55] m.p. 195-96°).

5.A.47 2-METHYL-7-HYDROXYISOFLAVONE

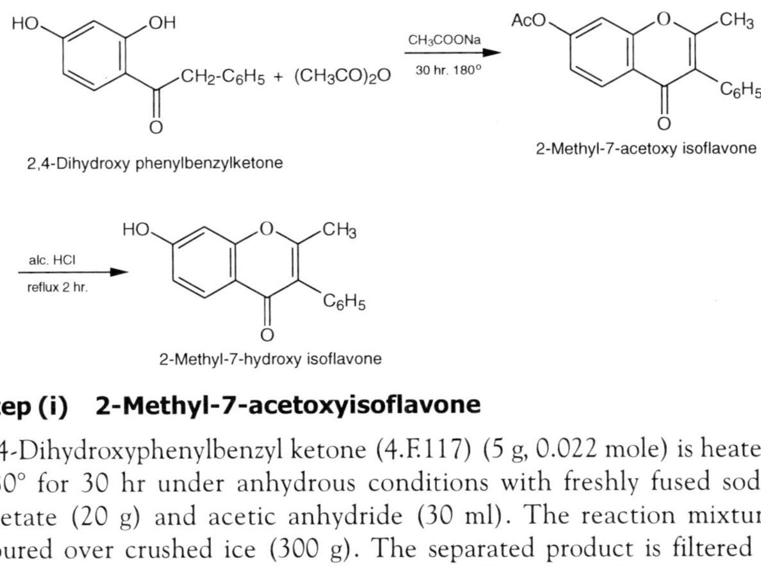

2,4-Dihydroxy phenylbenzylketone

2-Methyl-7-acetoxy isoflavone

2-Methyl-7-hydroxy isoflavone

Step (i) 2-Methyl-7-acetoxyisoflavone

2,4-Dihydroxyphenylbenzyl ketone (4.F.117) (5 g, 0.022 mole) is heated at 180° for 30 hr under anhydrous conditions with freshly fused sodium acetate (20 g) and acetic anhydride (30 ml). The reaction mixture is poured over crushed ice (300 g). The separated product is filtered and washed with water. It is crystallised from ethanol as colourless needles. Yield 5 g (77.5%). M.p. 163° (lit.[56] m.p. 163°).

Step (ii) 2-Methyl-7-hydroxyisoflavone

The above 2-methyl-7-acetoxyisoflavone (5 g, 0.017 mole) is dissolved in ethanol (100 ml) and refluxed with concentrated hydrochloric acid (40 ml) for 2 hr. Ethanol is distilled off and water (150 ml) added. The separated product is filtered and washed with water. It is crystallised from dilute alcohol as colurless small needles. Yield 3.5 g (81.4%). M.p. 240°. (lit.[56] m.p. 240°).

2-Methyl-5,7-dihydroxyisoflavone, m.p. 228° (lit.[56] m.p. 228°) obtained from 2,4,6-trihydroxyphenylbenzyl ketone (4.F.119).

5.A.48 1-HYDROXYXANTHONE

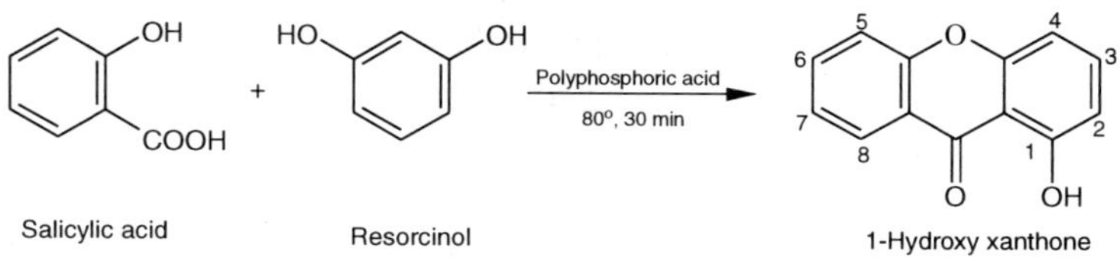

A mixture of salicylic acid (6 g, 0.043 mole) anhydrous resorcinol (6 g, 0.055 mole) and polyphosphoric acid [prepared by heating phosphorus pentoxide (40 g) with orthophosphoric acid (25 ml) at 80° for 30 min] is heated at 140° (oil-bath) for 4 hr under anhydrous conditions. The reaction mixture is cooled and poured over crushed ice. The separated product is filtered and washed with hot water to remove the unchanged starting components. The crude product is washed with sodium carbonate solution (5%) to remove the last traces of the starting materials. Finally it is treated with sodium hydroxide solution (10%). 1-Hydroxyxanthone separates out as its sodium salt. It is filtered and warmed with dilute hydrochloric acid. The product is finally filtered and washed with water. It is crystallised from alcohol as yellow orange needles. Yield 5.6 g (60.9%). M.p. 148° (lit.[57] m.p. 148°).

5.A.49 1,3-DIHYDROXYXANTHONE

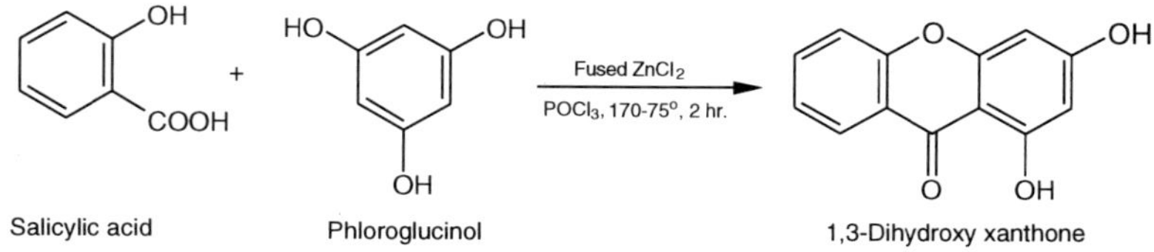

A mixture of salicylic acid (6 g, 0.043 mole), anhydrous phloroglucinol (7 g, 0.056 mole), freshly fused zinc chloride (15 g, 0.11 mole) and phosphorous oxychloride (35 ml, 57.6 g, 0.375 mole) is heated at 170-75° (oil bath) for 2 hr under anhydrous conditions. The reaction mixture is poured over crushed ice and left overnight. The separated product is filtered and washed thoroughly with water and dried. It is dissolved in anhydrous ethyl acetate and impurities precipitated with petroleum ether. The solution is filtered and the filtrate is evaporated to dryness. The solid product is crystallised from methanol as yellow needles. Yield 3 g (30.3%). M.p. 256-58° (lit.[58] m.p. 256-58°).

5.A.50 1-HYDROXY-3-METHOXYXANTHONE

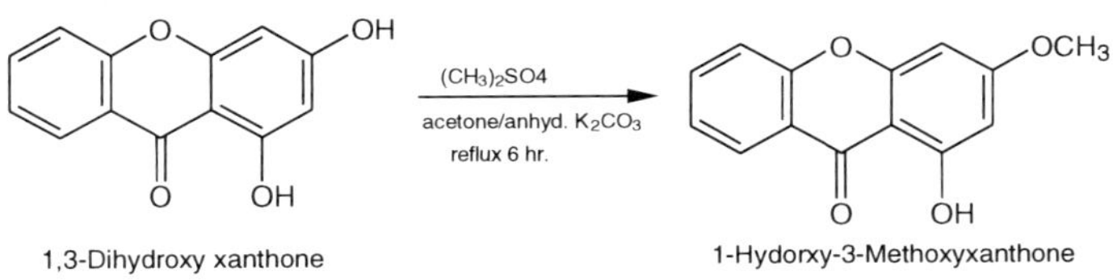

1,3-Dihydroxyxanthone (3.0 g, 0.0132 mole) in dry acetone (100 ml) is refluxed for 6 hr under anhydrous conditions with dimethyl sulphate (1.4 ml, 1.86 g, 0.0148 mole) in the presence of ignited potassium carbonate (10 g). The mixture is filtered and the inorganic salts washed with hot acetone (2 × 20 ml). The combined acetone solution is distilled and the residue macerated with crushed ice. The solid product is filtered and washed with water. It is crystallised from ethyl alcohol as almost colourless needles. Yield 2 g (62.9%). M.p. 146-47° (lit.[59] m.p. 146-47°).

5.A.51 1,3-DIHYDROXY-5-METHOXYXANTHONE

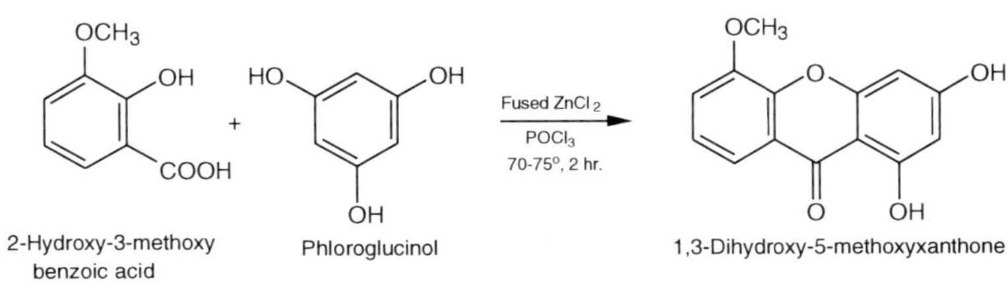

A mixture of 2-hydroxy-3-methoxybenzoic acid (o-vanillic acid) (3.3) (10 g, 0.06 mole), dry phloroglucinol (12 g, 0.095 mole) freshly fused zinc chloride (30 g, 0.22 mole) and phosphorus oxychloride (75 ml, 123.4 g,

0.8 mole) is heated at 70-75° for 2 hr under anhydrous conditions. The reaction mixture is poured over crushed ice and left overnight. The separated product is filtered, washed with sodium bicarbonate solution (5%) and then with hot water and dried. It is dissolved in ethyl acetate and the impurities precipitated with petroleum ether. The solution is filtered and the filtrate evaporated to dryness. The product is crystallised from methanol as yellow needles. Yield 3 g (19.5%).m.p. 303-304° (lit.[59] m.p. 303-304°).

5.A.52 1,3-DIHYDROXY-6-METHOXYXANTHONE

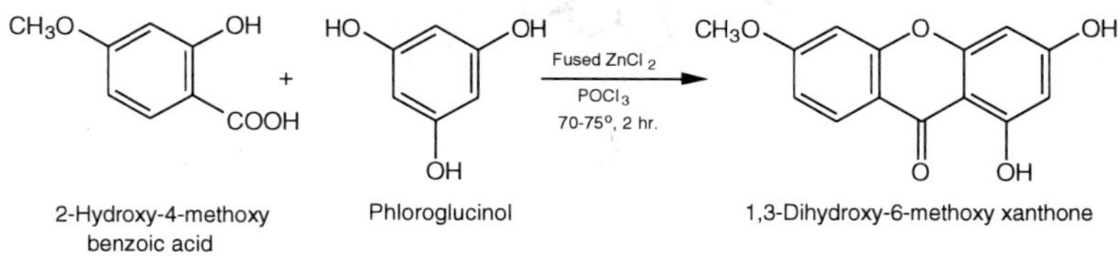

It is obtained from 2-hydroxy-4-methoxybenzoic acid (3.8.) (10 g, 0.06 mole) phloroglucinol (10 g, 0.079 mole), freshly fused zinc chloride (30 g, 0.22 mole) and phosphorous oxychloride (105 ml, 172.7 g, 1.13 mole) (heating for 2 hr at 70-75°). The reaction mixture is worked up as in the case of 1,3-dihydroxy-5-methoxyxanthone (5.A.51). It is crystallised from methanol as orange prisms. Yield 6 g (39%). M.p. 257-58° (lit.[58a] m.p. 257-58°).

5.A.53 1,3-DIHYDROXY-7-METHOXYXANTHONE

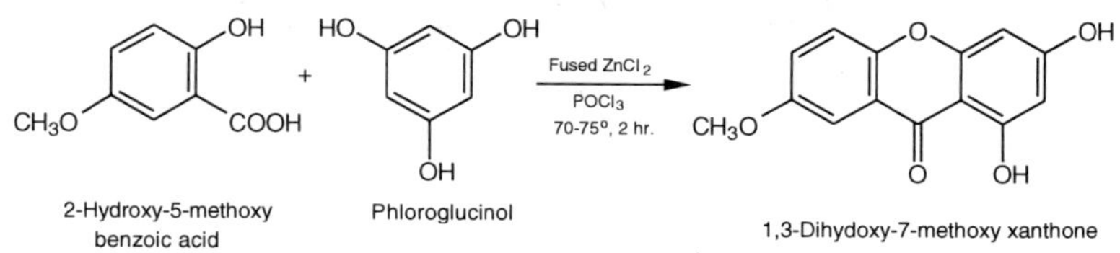

It is obtained from 2-hydroxy-5-methoxybenzoic acid (3.11) (1 g, 0.006 mole), dry phloroglucinol (1.1 g, 0.0085 mole), freshly fused zinc chloride (3 g, 0.022 mole) and phosphorous oxychloride (12 ml, 19.7 g, 0.129 mole) (heating for 2 hr at 70-80°) under anhydrous conditions. The reaction mixture is worked up as in the case of 1,3-dihydroxy-5-methoxyxanthone (5.A.51). It is crystallised from dilute alcohol as yellow needles. Yield 0.25 g (16.3%). M.p. 239-41° (lit.[58] m.p. 239-40°).

5.B.54 2,5-DIMETHYLTHIOPHEN

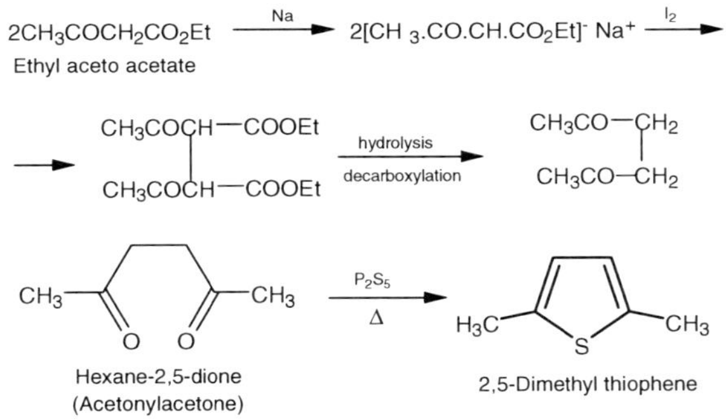

Step (i) Hexane-2,5-dione (Acetonylacetone), $CH_3COCH_2\,CH_2COCH_3$

Ethyl acetoacetate (65 g, 63.5 ml, 0.5 mole) in dry ether (250 ml) is added under anhydrous conditions to an stirred suspension of pulverised sodium (11.5 g, 0.5 mole) and dry ether (250 ml) placed in a three necked flask (1 L capacity) fitted with a reflux condenser and a dropping funnel. If the reaction becomes too vigorous, the reaction mixture is cooled in ice-water. The stirring is continued till all the sodium has dissolved. A solution of iodine (63.5 g, 0.25 mole) in dry ether (350 ml) is added till the colour persists. The separated sodium iodide is filtered, washed with ether and the combined filtrate and washings evaporated (rotary evaporator). The crude 2,3-diacetylsuccinate[61] is refluxed for 1 hr with potassium carbonate solution (20%, 250 ml). The cooled clear yellow solution is treated with potassium carbonate (50 g). The organic phase is separated and the aqueous layer extracted with ether (2 × 50 ml). The combined organic phase and ether extracts are dried (anhydrous sodium sulphate) and distilled (rotary evaporator). Acetonyl acetone is collected by distilling the oily product (185-192°). (lit.[60] b.p. 187-88°). Yield 7.5 g (26%).

Step (ii) 2,5-Dimethylthiophen

Acetonylacetone (23 g, 24 ml) is added to phosphorus pentasulphide (55 g, 0.25 mole) placed in a three necked flask (500 ml capacity) fitted with reflux condenser, a dropping funnel and a stopper. The reaction mixture is heated with a small flame until the exothermic reaction starts. The flame is removed and acetonylacetone (91 g, 94 ml total 1 mole) is added dropwise during 60 min so that the reaction mixture refluxes gently. If the reaction becomes too vigorous, the flask is cooled in ice- water. After the addition is complete and the reaction has subsided, the mixture is refluxed for 1 hr. It is cooled and poured into ice-water (200 ml) (Fume cupboard). The aqueous mixture is extracted with ether (5 × 40 ml). The combined ethereal solution is washed with aqueous sodium carbonate (5%), dried (anhydrous sodium sulphate) and distilled. 2,5-Dimethylthiophen is obtained by distillation. Yield 52 g (46%). B.p. 135-36° (lit.[62] b.p.136-36.5°).

5.C.55 2-AMINOPYRIDINE

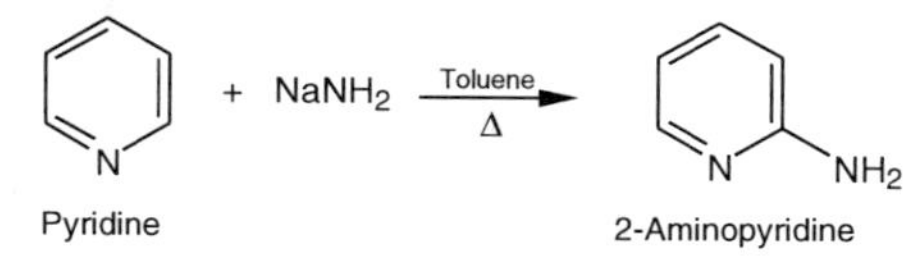

Pyridine (25 g, 25.6 ml, 0.316 mole) is added dropwise during 1 hr to a vigorously stirred mixture of anhydrous sulphur-free toluene (75 ml), granular sodamide (22.5 g, 0.577 mole) placed in a three necked R.B.flask (500 ml capacity) fitted with a sealed stirrer, reflux condenser, a thermometer and a inlet tube for nitrogen. The reaction is carried out in nitrogen atmosphere at 130-140° (oil bath). After some time, the reaction mixture turns brown and vigorous reaction occurs. The reaction mixture is heated for 5 hr at 110°. The reaction mixture is cooled and water (50 ml) is added dropwise through the condenser during 60 min, the nitrogen being bubbled through the addition of water and the stirring continued. The temperature of the reaction mixture rises to 50-55°. The lower aqueous layer is separated (separatory funnel) and extracted with toluene (2 × 35 ml). The combined upper layer and the toluene extracts are dried (anhydrous potassium carbonate). The toluene is removed by distillation. The remaining viscous liquid is distilled in vacuo at 120-130° (oil bath temperature). 2-Amino-pyridine distils at 95°/10 mm. It solidifies to a colourless solid on cooling. Yield 20 g (67%). M.p. 55° (lit.[63] m.p. 55°).

5.C.56 PYRIDINE-2-CARBOXYLIC ACID (PICOLINIC ACID)

Potassium permanganate (225 g, 1.42 mole) is added in ten equal portions to mixture of 2-picoline (50 g, 53 ml, 0.54 mole) and water (500 ml) at 70°. Total time required for addition is 3-3.5 hr. The temperature is maintained at 20° for the first five additions and then at 85-90° for the last five. Each

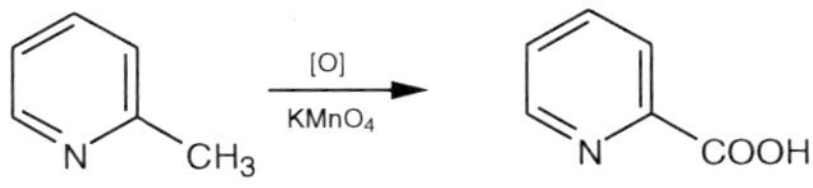

2-Picoline

Pyridine-2-Carboxylic acid (Picolinic acid)

successive addition of potassium permanganate is made only after the preceding amount is decolourised, the temperature is raised to 95° and then filtered while hot with suction. The manganese dioxide cake is washed with hot water (4 × 100 ml). The combined filtrate is evaporated (rotary evaporator) to about 150 ml. The solution is cooled and acidified with concentrated hydrochloric acid (about 65 ml) to pH 3.2 (the isoelectric point). The solution is continuously extracted with chloroform (500 ml) in a liquid-liquid extractor for 6 hr. The chloroform layer is dried (anhydrous sodium sulphate) and solvent evaporated in vacuo (rotary evaporator). The residual solid is dried (vacuum desiccator). It is extracted in a thimble placed in a Soxhlet apparatus with benzene (600 ml). The extraction is stopped till no picolinic acid is obtained on cooling the benzene concentrate. The benzene solution is cooled and the separated picolinic acid is filtered. The filtrate is used again in the Soxhlet extraction. Total yield 40g (61%). M.p. 137-38° (lit.[64] m.p. 136-37°).

5.C.57 PYRIDINE-3-CARBOXYLIC ACID (NICOTINIC ACID)

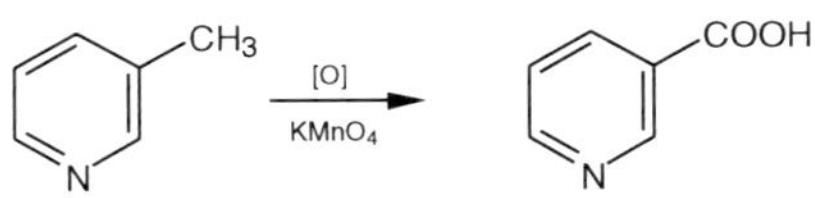

3-Picoline

Pyridine-3-Carboxylic acid (Nicotinic acid)

3-Picoline (50 g, 52.2 ml, 0.54 mole) is dissolved in water (500 ml). It is oxidised with potassium permanganate (225 g, 1.42 mole) as in the case of picolinic acid (5.C.56). The manganese dioxide cake is washed with water (4 × 250 ml) and the combined filtrate and washings evaporarted to about 600 ml. The solution is acidified to pH 3.4 (the isoelectric point) with the help of narrow range indicator paper (60-65 ml concentrated hydrochloric acid is required). The solution is cooled and the separated nicotinic acid is filtered and washed with water (3 × 25 ml). The filtrate and washings are concentrated to about 300 ml and cooled (5°). A second crop of nicotinic acid is obtained. It is filtered and washed with cold water. The combined nicotinic acid is crystallised from hot water as colourless needles. Yield 33 g (50%). M.p. 235° (lit.[64] m.p. 236°).

5.C.58 PYRIDINE-4-CARBOXYLIC ACID (ISONICOTINIC ACID)

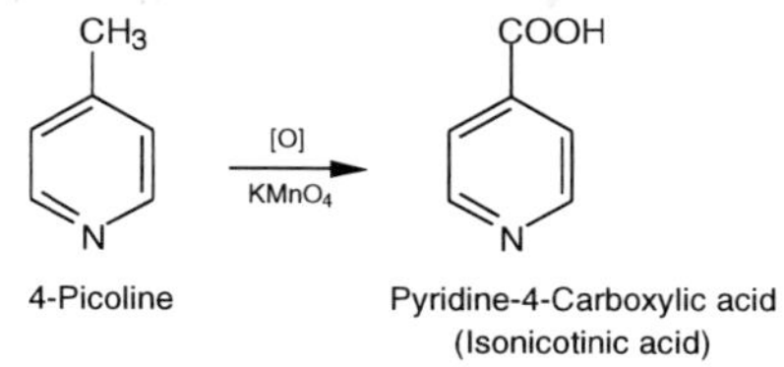

It is obtained by oxidiation of 4-picoline (50 g, 50.25 ml, 0.54 mole) with potassium permanganate (225 g, 1.42 mole) as in the case of picolinic acid (5.C.56.). The combined filtrate is evaporated to about 200 ml and acidified to pH 3.6 with concentrated hydrochloric acid. Isonicotinic acid precipitates out. The combined solution is heated to 90-95° to dissolve most of the isonicotinic acid. The mixture is allowed to crystallise slowly. The product is filtered and washed well with water. The combined mother liquor is concentrated to about 300 ml and cooled. A second crop of the product is obtained. The combined product is crystallised from hot water as colourless crystals. Yield 42 g (63.7%). M.p. 311° (sealed tube) (lit.[65] m.p. 315°).

5.C.59 6-HYDROXYNICOTINIC ACID

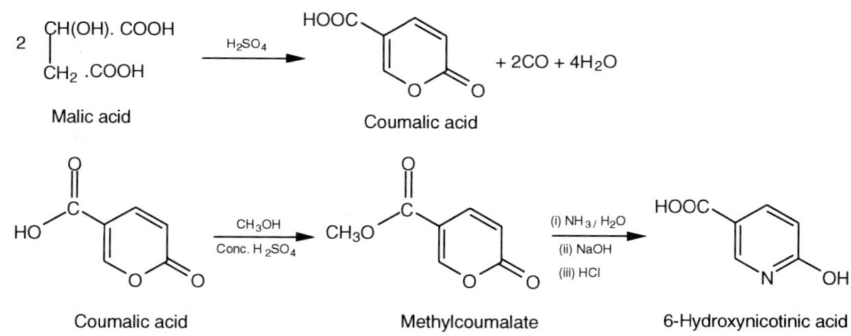

Step (i) Coumalic Acid

Concentrated sulphuric acid (90 ml) is added to finely powdered dry malic acid (101 g, 0.75 mole). At intervals of 30 min, three 50 ml portions of fuming sulphuric acid (25% SO_3) are added. The reaction is exothermic and a steady evolution of gas sets in. Excess foaming is contained by swirling the mixture. After the evolution of gas has slackened (1 hr), the reaction mixture is heated for 2 hr (boiling water bath). The mixture is cooled and poured over crushed ice (400 g) with stirring. The mixture is cooled in a refrigerator for 24 hr. The separated coumalic acid is filtered and washed with cold water. It is crystallised from methanol as light yellow needles. Yield 32.5 g (62%). M.p. 206-208° (lit.[66] m.p. 206-208°).

Step (ii) Methylcoumalate

Finely powdered coumalic acid (17.5 g, 0.125 mole) is added in portions to concentrated sulphuric acid (50 ml); the mixture is shaken occasionally. The temperature rises and is maintained below 30° (ice-water). Methyl alcohol (25 ml) is added slowly with shaking to the above mixture; the temperature is kept below 35°. The reaction mixture is heated on a boiling water bath for 1.5 hr, cooled and poured over ice-water (250 ml) with stirring. Sodium carbonate (about 75 g)is added as a slurry in water until the mixture is just neutral. In case excess sodium bicarbonate is added, the supernatant liquid acquires red colour; in such event a few drops of sulphuric acid are added. The final solution should be an orange colour. The mixture is filtered and the solid product macerated 4-5 times with cold water (40 ml each time), containing a drop of concentrated sulphuric acid. After the final washing the crude product is filtered. Yield 9 g (47%). This is used as such for the next stage. However, it can be purified by vacuum sublimation. M.p. 69-70° C (lit.[67] m.p. 69-70°).

Step (iii) 6-Hydroxynicotinic acid

Methyl coumalate (9 g, 0.058 mole) is added in small portions during 5-10 min to a stirred and cooled solution (ice-water) of ammonia solution (10 ml, d 0.880) and water (15 ml). The temperature during the addition is kept below 20°. The resulting dark red solution is stirred for 30 min more at 20°. The mixture is added to a boiling solution of aqueous sodium hydroxide (20 g sodium hydroxide in 125 ml water). The reaction mixture is boiled for 5 min and then cooled to 10°. The required product is precipitated by adding concentrated hydrochloric acid (temp.$<$ 30°) and the solution cooled. The separated bright yellow product is filtered, washed with water and dried in air. Yield 4.3 g (58%). M.p. 299-300° (decomp.) (lit.[67a] m.p. 299-300°).

5.C.60 QUINOLINE

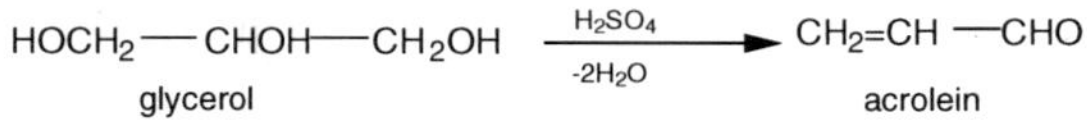

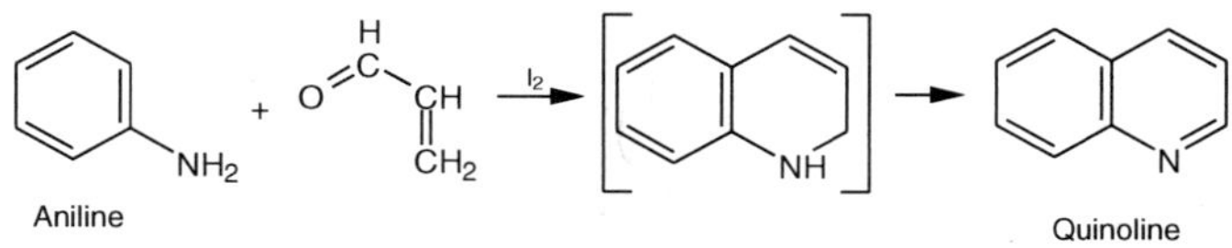

Aniline (10 g, 9.8 ml, 0.107 mole), glycerol (15 g, 11.9 ml, 0.163 mole) and iodine (0.5 g) are placed in a three necked flask (250 ml capacity) fitted with a sealed stirrer, reflux condenser and a thermometer. The mixture is stirred. Concentrated sulphuric acid (30 g, 16.4 ml, 0.306 mole) is added dropwise through a separatory funnel fitted on the top of the condenser. The reaction commences and the temperature rises to 100-105°. The reaction mixture is heated gradually to 140° (oil-bath). It is heated for 1 hr at 170°, cooled and rendered alkaline (5M sodium hydroxide, about 8-8.5 ml). The mixture is steam distilled until no more oily drop passes over. The distillate contains quinoline along with some unreacted aniline. It is purified as follows. Concentrated sulphuric acid (8 ml) is added carefully to the steam distillate obtained above. The solution is cooled (0-5°) and concentrated solution of sodium nitrite is added until a drop of the reaction mixture colours potassium iodide- starch paper deep blue instantly (approximately 3 g of sodium nitrite is required). The diazotised solution is heated on a boiling water bath for 1 hr it is treated with sodium hydroxide solution (15 g in 50 ml water) and the mixture is steam distilled. Quinoline is obtained by ether extraction of the distillate. Yield 6.9 g (50%). B.p. 236-38°; 118-120°/20 mm (lit.[68] b.p. 236-38°).

5.C.61 2-METHYLQUINOLINE (QUINALDINE)

Concentrated hydrochloric acid (60 ml) is added slowly to cold (ice bath) aniline (31 g, 30.3 ml 0.333 mole) placed in a R.B.flask (500 ml capacity) fitted with a reflux condenser, subsequently paraldehyde (45 g, 1.02 mole) is added. The flask is shaken so that the contents are thoroughly mixed. The flask is removed from the ice bath and shaking at room temperature continued for 1-2 hr. It is heated carefully to boiling point. In case the reaction becomes vigorous, it is cooled in ice-bath. The mixture is refluxed for 3 hr and cooled. The solution rendered alkaline with sodium hydroxide

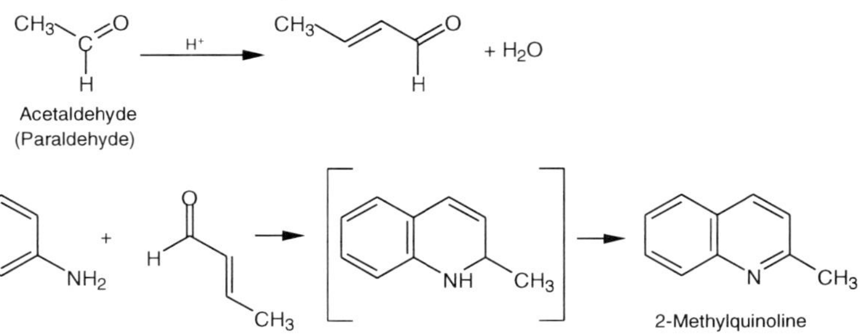

solution (12 M, 100 ml). It is then steam distilled till the oily substance stops distilling over. The steam distillate is extracted with ether (2 × 100 ml). The ether extract is dried (magnesium sulphate) and solvent removed. The residual product is refluxed for 20 min with acetic anhydride (10 ml). The solution is cooled and rendered alkaline with sodium carbonate solution. It is again steam distilled till the only product stops coming over. The distillate is extracted with toluene (2 × 50 ml). The toluene solution is distilled. Quinaldine is collected at 245-48°. Yield 20 g (42.5%) (lit.[69] b.p. 246-47°).

5.C.62 8-HYDROXYQUINOLINE (OXINE)

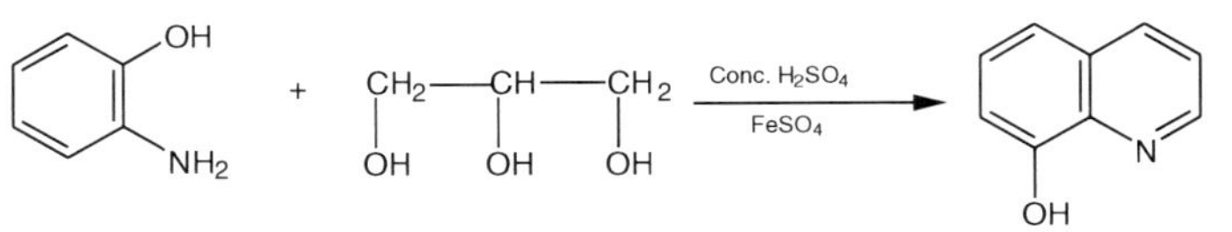

Concentrated sulphuric acid (85 ml) is placed in a three necked R.B. flask (500 ml capacity) fitted with a stirrer. o-Aminophenol (56.25 g, 0.515 mole) is added to the sulphuric acid. This is followed by the addition of anhydrous glycerol (143.5 g, 1.56 mole). The temperature of the mixture is kept below 80° (by cooling if necessary). The mixture is maintained in fluid state by placing it on a steam bath. In a separate three necked flask (1000 ml capacity) fitted wih a stirrer, reflux condenser and a thermometer is placed o-nitrophenol (36.25 g, 0.26 mole), crystallised iron (II) sulphate (5 g). The mixture is stirred and heated to 100-120°. This above mixture of o-aminophenol-sulphuric acid and glycerol is added in 8-10 portions during 2-2.5 hr to the stirred mixture. The reaction is allowed to proceed at 130-135° before adding the subsequent portions. The mixture is heated for 4 hr more. The solution is cooled and neutralised with sodium hydroxide solution (125 g in 125 ml water) with rapid stirring and addition of ice so that the temperature does not rise above 40°. The solution attains a pH of about 7. The separated product is filtered along with tarry by - products and dried. It is distilled in vacuo. 8-Hydroxyquinoline is collected at 100-110°/ 5 mm. It solidifies on cooling to a white solid. Yield 70 g (94%). M.p. 74-5° (lit.[70] m.p. 76°).

5.C.63 8-NITROQUINOLINE

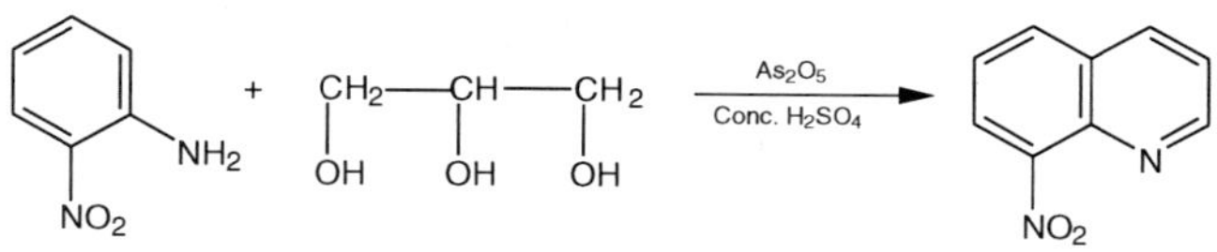

o-nitro aniline glycerol 8-Nitroquinoline

A mixture of o-nitroaniline (34.50 g, 0.25 mole), arsenic pentoxide (43 g, 0.187 mole) and anhydrous glycerol (92 g, 73 ml, 1 mole) is stirred in a three necked flask (as in the case of quinoline) (5.C.60.). The mixture is heated to 100° (oil bath) and concentrated sulphuric acid (110 g, 60 ml, 1.12 mole) is gradually added through the condenser at such a rate that the temperature does not rise above 120° (20-25 min). A calcium chloride drying tube is fitted on to the condenser. The temperature of the reaction mixture is raised gradually to 130-35° and maintained at this temperature for 8 hr. In case the reaction becomes vigorous during the first hour, the oil bath is removed temporarily. The reaction mixture is cooled and poured into water (700 ml). It is treated with decolourising carbon (10 g), stirred and heated to 90° for 1 hr and filtered. The cooled filtrate is neutralised slowly with aqueous ammonia solution (1 part of concentrated ammonia, d 0.88 and 1 part water). The separated nitro product is filtered and washed with cold water. It is crystallised from methanol. Yield 22.6% (52%). M.p. 92° (lit.[71] m.p. 89-90°).

5.C.64 5,8-DIMETHOXYQUINOLINE

Step (i) 2,5-Dimethoxynitrobenzene

A mixture of concentrated nitric acid (5 ml) and water (5 ml) is added dropwise with stirring to a solution of 1,4-dimethoxybenzene (1.6) (5 g, 0.036 mole) in glacial acetic acid (15 ml). The acid is added at such a rate that the reaction mixture is maintained below 40°. The separated nitro compound is filtered and washed with water. It is crystallised from

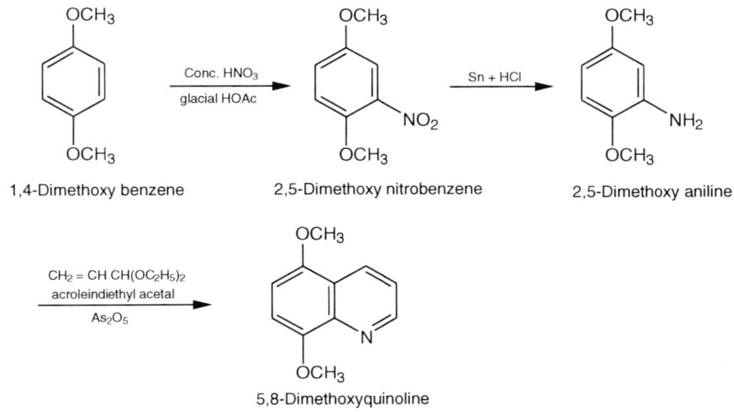

methanol as yellow coloured needles. Yield 5.8 g (87%). M.p. 72° (lit.[72] m.p. 72°).

Step (ii) 2,5-Dimethoxyaniline

To a mixture of 2,5-dimethoxynitrobenzene (4 g, 0.022 mole) and tin (9 g, 0.076 mole), concentrated hydrochloric acid (20 ml) is added dropwise with occasional shaking. The mixture is rendered slightly alkaline by the addition of dilute solution of sodium hydroxide. The reaction mixture is extracted with ether (3 × 50 ml). The ether extract is washed with water, dried (anhydrous sodium sulphate) and distilled. The product obtained is crystallised from ether-petroleum ether. Yield 2 g (60%). M.p. 78° (lit.[72] m.p. 80°).

Step (iii) 5,8-Dimethoxyquinoline

A mixture of the above 2,5-dimethoxyaniline (3.06 g, 0.02 mole), arsenic pentoxide (5.75 g, 0.025 mole) and phosphoric acid (25 ml) is heated at 75-80° (oil bath). Acrolein diethyl acetal (3 g, 3.5 ml, 0.023 mole) is added dropwise to the above mixture with stirring. The mixture is stirred for 1 hr more at 75-80°. It is cooled to room temperature and poured into ice cold water (100 ml). The aqueous solution is rendered alkaline (pH 8) with ammonium hydroxide. The precipitated solid is filtered, dried and extracted with cyclohexane in a Soxhlet apparatus. The resulting product is purified by column charomatography on silica gel column (2 × 20 cm) using benzene as the eluant. Finally, it is crystallised from petroleum ether. Yield 0.61 g (16%). M.p. 74-5° (lit.[73] m.p. 75-6°).

5.C.65 5,8-DIMETHOXY-2-METHYLQUINOLINE

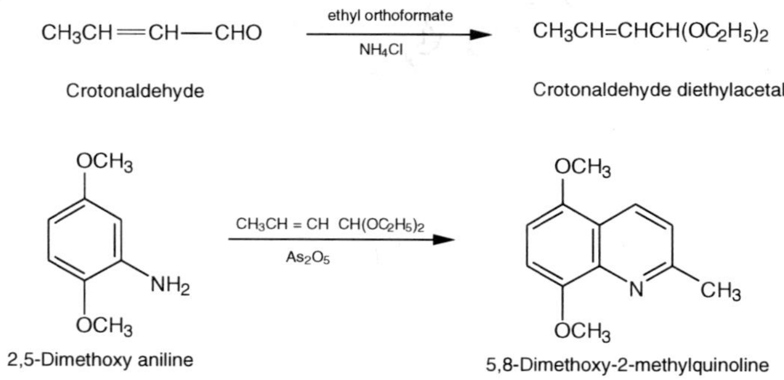

Step (i) Crotonaldehyde Diethylacetal

To a solution of crotonaldehyde (14 g, 16.55 ml, 0.2 mole) in ethyl orthoformate (36 g) is added ammonium nitrate (0.7 g) in ethanol (25 ml). The mixture is refluxed for 10 min. The reaction mixture is cooled, filtered, diluted with ether and washed with ammonia solution and then with water. The ether solution is dried (anhydrous sodium sulphate) and the solvent removed in vacuo. The residual product is distilled to get crotonaldehyde acetal. Yield 21.6 g (53%). B.p. 75°/45 mm (lit.[74,75] b.p.71°/43 mm).

Step (ii) 5,8-Dimethoxy-2-methylquinoline

It is prepared by condensing 2,5-dimethoxyaniline (5.C.64) with crotonaldehyde diethyl acetal following the same procedure as in the case of 5,8-dimethoxyquinoline (5.C.64). Yield 22%. M.p. 84° (lit.[76] m.p. 84-5°).

5.C.66 2,4-DIMETHYL-5,8-DIMETHOXYQUINOLINE

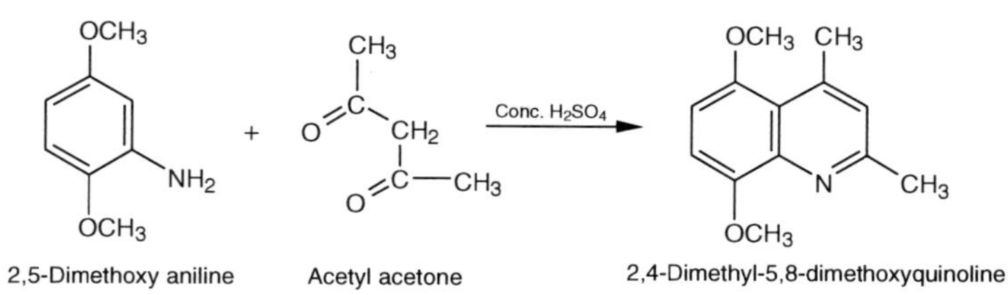

Acetyl acetone (5.C.76) (4 g, 4.1 ml, 0.04 mole) is added to a mixture of 2,5-dimethoxyaniline (5.C.64) (5.1 g, 0.033 mole) and anhydrous calcium sulphate (5 g). The reaction mixture is refluxed for 6 hr. It is cooled and calcium sulphate filtered off. The product is washed with water and extracted with ether (2 × 30 ml). The ether is distilled and to the brown residue obtained is added carefully concentrated sulphuric acid (35 ml) while cooling in an ice-bath. The solution is heated on a steam bath for 10 min and then poured into cold water. The bright yellow sulphate of the quinoline separates out. The mixture is rendered alkaline with aqueous sodium hydroxide. The quinoline separates out as grey solid. It is filtered, washed with water and dried. It is crystallised from ethyl acetate. Yield 2.8 g (38.8%). M.p. 105-106° (lit.[77] m.p. 107°).

5.C.67 2,4,6-TRIMETHYLQUINOLINE

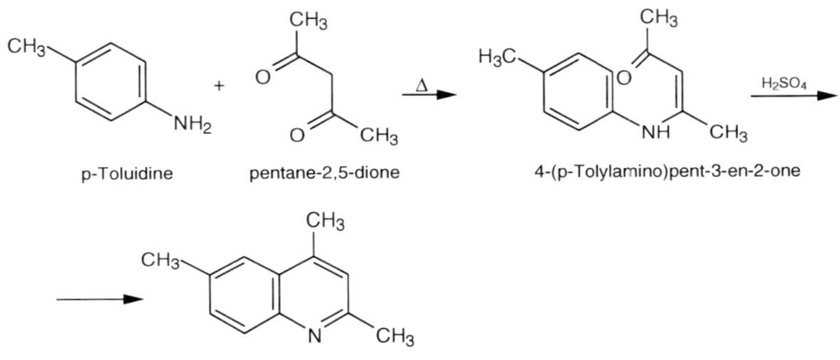

2,4,6-Trimethyl quinoline

Step (i) 4-(p-Tolylamino) pent-3-ene-2-one

A mixture of p-toluidine (5.32 g, 0.05 mole), granular anhydrous calcium sulphate (10 g) and pentane-2,4-dione (5.C.76) (5.1 g, 0.51 mole) is heated for 1 hr on a steam bath under anhydrous conditions with occasional shaking. The reaction mixture is cooled, ether (40 ml) added and filtered. The residual calcium sulphate is washed with more ether (2 × 20 ml). The combined ether solution is evaporated. The residual product is crystallised from hexane (about 20 ml) as deep straw coloured plates. Yield 7.2 g (76%). M.p. 67-8° (lit.[78] m.p. 69°).

Step (ii) 2,4,6-Trimethylquinoline

The above 4-(p-tolylamino) pent-3-ene-2-one (6 g, 0.032 mole) is added in portions to concentrated sulphuric acid (25 ml, d 1.84) in a conical flask (250 ml capacity). The flask is shaken. The temperature of the reaction mixture rises to 60-70° and a complete solution is obtained. The mixture is heated for 30 min (steam bath), cooled and ice-water (250 ml) added slowly. The solution is rendered alkaline with sodium carbonate solution.

Sometimes quinoline salt separates. In case this happens, the cake is broken with the help of a spatula. Eventually the quinoline separates as an oil. The mixture is cooled, the product soldifies. It is filtered and washed with cold water. Finally it is crystallised from dilute alcohol (charcoal) as long glistening white needles. Yield 5.2 g (96%). M.p. 66-7° (lit.[78] m.p. 68-9°).

5.C.68 2-HYDROXY-4-METHYLQUINOLINE (4-METHYLCARBOSTYRIL)

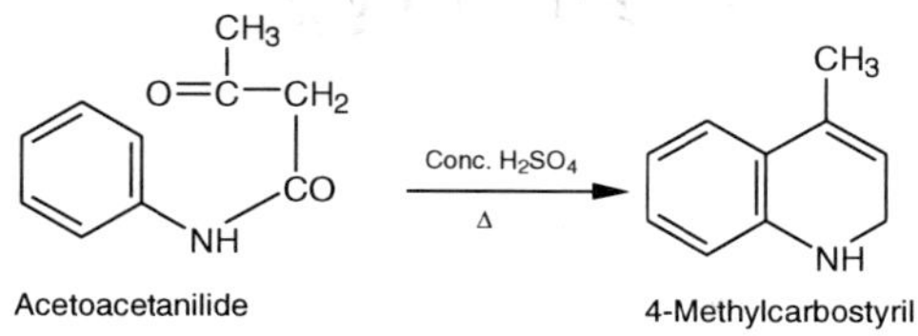

Concentrated sulphuric acid (45 ml) is taken in a three necked R.B. flask (250 ml capacity) fitted with a mechanical stirrer and a thermometer. The acid is heated to 75°, the source of heat is removed and acetoacetanilide (44.5 g, 0.25 mole) is added in portions (25-30 min). The temperature is maintained at 70-75° by occasional cooling (if necessary) until nearly all the acetoacetanilide has been added. The last portion (7-10 g) is added without cooling. The temperature rises to about 95°. The reaction mixture is now heated to maintain this temperature for about 30 min. The reaction mixture is cooled to 60-65° and poured into water (1.2 litre) with vigorous stirring. The resultant mixture is cooled, the product is filtered, washed with water (2 × 100 ml) and then with methanol (2 × 50 ml) and dried in air. The 4-methylcarbostyril is crystallised from alcohol. Yield 35 g (87%). M.p. 223-24° (lit.[79] m.p. 224°).

5.C.69 2-PHENYLQUINOLINE-4-CARBOXYLIC ACID (ATOPHAN)

Pure benzaldehyde (12.5 g, 12 ml, 0.118 mole) is placed in a R.B.flask (500 ml capacity) fitted with a reflux condenser. Freshly distilled pyruvic acid

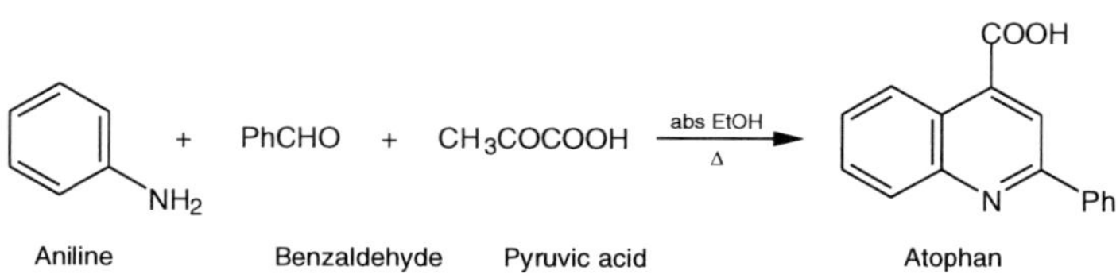

(11 g, 0.125 mole) and absolute ethanol (100 ml) are added. The mixture is heated on a boiling water bath. Aniline (11.25 g, 11.0 ml, 0.121 mole) in absolute ethanol (50 ml) is added slowly (during 1 hr) with frequent shaking. The mixture is refluxed for 3 hr (water-bath) and left overnight. The separated product is filtered and washed with little ether. It is crystallised from alcohol. Yield 15 g (45%). M.p. 216-17° (lit.[80] m.p. 210-12°).

Note: Pyruvic acid is obtained by grinding together tartaric acid (100 g, 0.66 mole) and freshly fused potassium hydrogen sulphate (150 g, 1.1 mole). The mixture is distilled (at 210-220°, oil bath) until liquid no longer distils. The distillate is redistilled. Pyruvic acid distils at 75-80°/25 mm. Yield 30 g (51%).

5.C.70 QUINOLINE-2-CARBOXYLIC ACID (QUINALDINIC ACID)

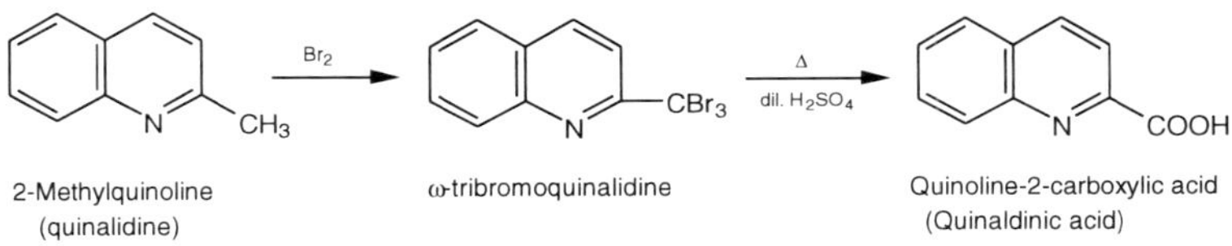

A solution of bromine (24 g, 7.75 ml, 0.15 mole) in glacial acetic acid (30 ml) is added dropwise at 70° to a mixture of quinaldine (2-methylquinoline) (5.C.61) (7 g, 0.05 mole), anhydrous powdered sodium acetate (25 g, 0.305 mole) and glacial acetic acid (100 ml). The bromine solution is added in 10 min. The reaction mixture is heated to boiling for a few minutes until the separation of sodium bromide causes bumping. The reaction mixture is then heated for 30 min on a boiling water bath and then cooled. It is poured into ice water (300 ml). The separated ω-tribromoquinaldine is filtered and washed with water. It is crystallised from alcohol. Yield 16.2 g (78.6%). M.p. 128-29°.

The ω-tribromoquinaldine is hydrolysed by refluxing with excess of dilute sulphuric acid (1:1) until a test portion, on neutralisation gives no unchanged halogen compound. The solution is cooled and neutralised with dilute sodium hydroxide. It is treated with excess copper sulphate solution. The separated green copper quinaldinate is filtered and washed with water. It is suspended in hot water and treated for considerable time with hydrogen sulphide gas. The separated copper sulphide is filtered and the filtrate is evaporated to dryness on a water bath. It is crystallised from glacial acetic acid. Yield 6.9 g (81.6%). M.p. 156-57° (lit.[81] m.p. 156°).

5.C.71 QUINOLINE-5-CARBOXYLIC ACID

m-Aminobenzoic acid — Quinoline-5-carboxylic acid

A mixture of m-aminobenzoic acid (20 g, 0.146 mole), nitrobenzene (12 g, 10 ml, 0.097 mole), glycerol (44 g, 34.9 ml, 0.478 mole), boric acid (5 g) and concentrated sulphuric acid (28 ml) is heated under reflux until the reaction commenced. When the reaction has subsided, the whole mass is boiled for 14 hr, diluted with water (about 50 ml) and strongly basified with sodium hydroxide solution. The black tar is decanted and the clear alkaline solution acidified with acetic acid to give quinoline-5-carboxylic acid. Yield 24 g (95%). M.p. 350° (decomp.) (lit.[82] m.p. 342°).

5.C.72 5,8-DIMETHOXYISOQUINOLINE

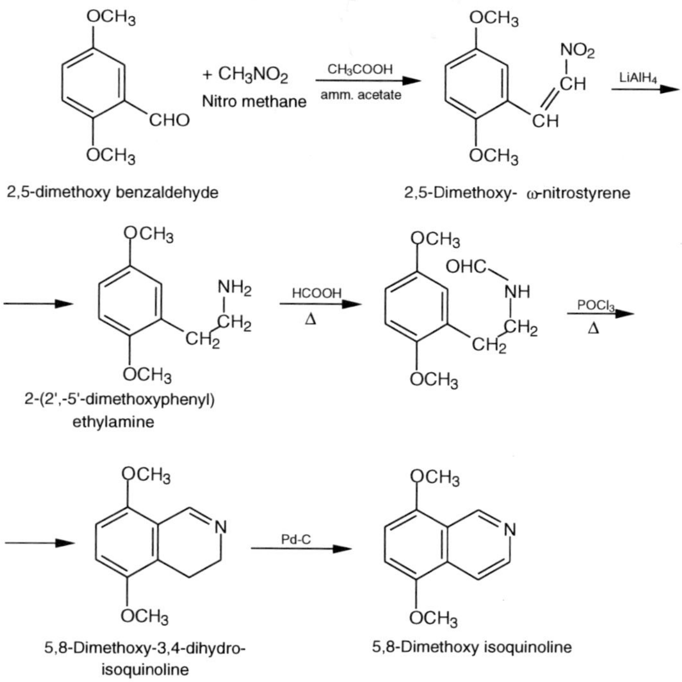

Step (i) 2,5-Dimethoxy-ω-nitrostyrene

A mixture of 2,5-dimethoxybenzaldehyde (2.9) (10 g, 0.06 mole), nitromethane (10 ml, 11.27 g, 0.188 mole), glacial acetic acid (50 ml) and ammonium acetate (4 g, 0.052 mole) is refluxed for 3 hr. The reaction mixture is cooled to room temperature and poured into crushed ice. The yellow product is filtered and washed with water. It is crystallised from dilute acetic acid. Yield 12 g (95%). M.p. 118-119° (lit.[83] m.p. 120°).

Step (ii) 2-(2′,5′-dimethoxyphenyl) ethylamine

A solution of 2,5-dimethoxy-ω-nitrostyrene (4.5 g, 0.021 mole) in tetrahydrofuran (75 ml) is added dropwise to a stirred suspension of lithium aluminium hydride (2 g, 0.053 mole) in tetrahydrofuran (75 ml) during 2 hr. The reaction mixture is stirred for 2 hr more. Water (100 ml) is added carefully dropwise to the reaction mixture. It is extracted with hydrochloric acid (2N, 3 × 50 ml). The combined acid extract is basified with aqueous sodium hydroxide solution and finally extracted with ether (2 × 100 ml). The ether extract is dried over potassium hydroxide and the solvent evaporated to give the required compound as an oil. Yield 2.26 g (58%). B.p. 163°/10 mm. (lit.[83] b.p. 160°/10 mm).

Step (iii) 5,8-Dimethoxyisoquinoline

Excess formic acid is added dropwise to the above amine (2.1 g, 0.012 mole) and the reaction mixture refluxed for 3 hr. The formic acid is removed in vacuo and the residue neutralised with aqueous sodium bicarbonate (10%). The mixture is extracted with ethyl acetate (2 × 75 ml). The extract is dried over anhydrous sodium sulphate and the solvent removed in vacuo to give the formyl derivative. Yield 1.53 g (63%). M.p. 71° (lit.[83] m.p. 70°).

The above compound (1.5 g, 0.007 mole) is dissolved in a mixture of dry toluene (15 ml) and phosphorous oxychloride (6 ml). The solution is refluxed for 2 hr and poured over crushed ice. The aqueous solution is extracted with ether (2 × 75 ml). The ether extract is dried (sodium sulphate) and evaporated to give 5,8-dimethoxy-3,4-dihydroisoquinoline. Yield 0.083 g (60%). M.p. 158° (lit.[83] m.p. 160°).

Pd-C (5%, 1g) is added to the solution of the above dihydroisoquinoline (0.5 g, 0.0026 mole) in decalin (50 ml). The reaction mixture is heated at 180-90° (oil bath) for 7 hr under nitrogen with stirring. The stirring is continued till no more hydrogen evolves. The reaction mixture is then filtered and the catalyst washed with benzene (2 × 20 ml). The combined filtrate is extracted with hydrochloric acid (4N, 3 × 25 ml). The clear acid extract is cooled and basified with dilute sodium hydroxide. It is finally extracted with ether (2 × 50 ml), ether extract dried (sodium sulphate) and the solvent evaporated in vacuo. The required isoquinoline is crystallised from benzene-petroleum ether. Yield 0.28 g (57%). M.p. 58° (lit.[83] m.p. 58°).

5.C.72A 5,8-DIMETHOXY-1-METHYLISOQUINOLINE

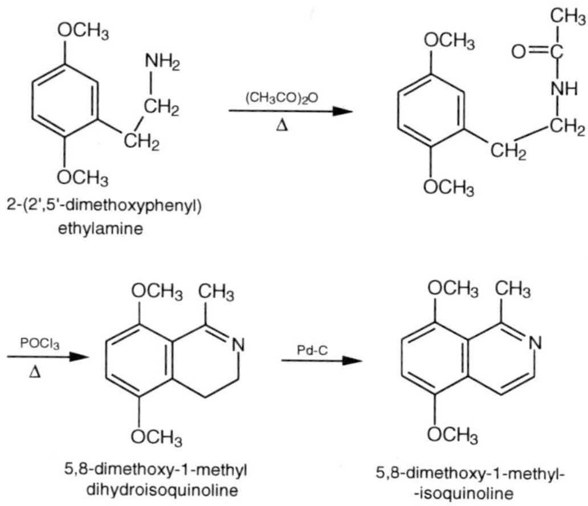

It is similarly prepared from 2-(2',5'-dimethoxyphenyl) ethylamine (5.C.72) (5.85 g, 0.03 mole), acetic anhydride (2.5 ml) and acetic acid (5 ml) (1 hr refluxing). The acetamide obtained (57% yield, m.p. 102°) is refluxed in dry toluene and phosphorus oxychloride to give the dihydroisoquinoline (60% yield, m.p. 189-90°) which on dehydrogenation with Pd/C (as in the above case) gave the required 5,8-dimethoxy-1-methylisoquinoline in 56% (3.67 g). Yield M.p. 55° (lit.[83] m.p. 54°).

5.C.73. 2,3-DIPHENYLQUINOXALINE

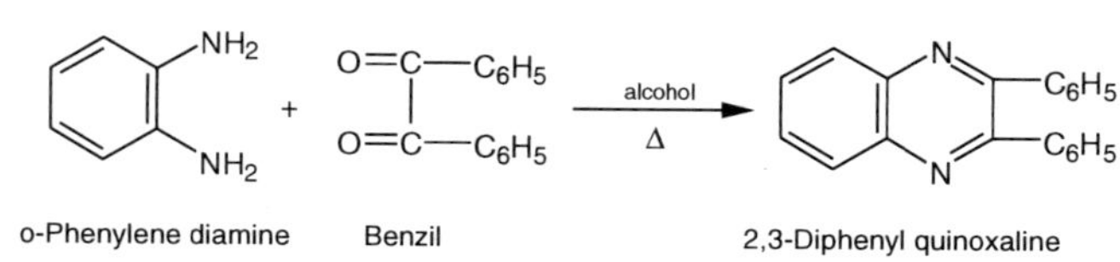

o-Phenylenediamine (2.2 g, 0.02 mole) in alcohol (15 ml) is added to a warm solution of benzil (4.2 g, 0.02 mole) in alcohol (15 ml). The resulting solution is warmed on a water bath for 30 min. Water is added until a slight cloudiness persists. The solution is made clear by heating and it is cooled. The separated product is filtered. It is crystallised from dilute alcohol. Yield 2.86 g (51%). M.p. 125-26°. (lit.[84] m.p. 125-26°).

5.C.74 PIPERAZINE-2,5-DIONE (DIKETOPIPERAZINE)

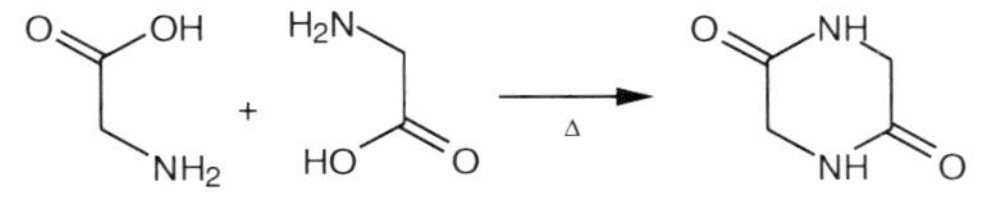

Glycine

Piperazine -2,5-dione (Diketopiperazine)

A mixture of glycine (50 g, 0.66 mole) and ethyleneglycol (250 ml) is taken in a 500 ml three neck flask fitted with an air condenser, sealed mechanical stirrer and a thermometer. The mixture is stirred and heated to 170-75° (oil bath) for 1 hr. It is cooled and left in a refrigerator for 12 hr. The separated product is filtered (or centrifuged) and washed with cold water (150 ml). It is crystallised from boiling water (using decolourising carbon); the hot aqueous solution is filtered through a preheated Buchner funnel. The product separates on cooling in ice chest. It is filtered, washed with little ice-cold water and dried (air -oven at 50°). Yield 17 g (45%). Its m.p. is indefinite and depends on the rate of heating. If it is introduced into a rapidly heating melting point apparatus at 270°, piperazine-2,5-dione melts at 310-12° (decomp.) (lit.[85] m.p. 311-12°, decomp.).

5.C.75 2,5-DIMETHYLPYRROLE

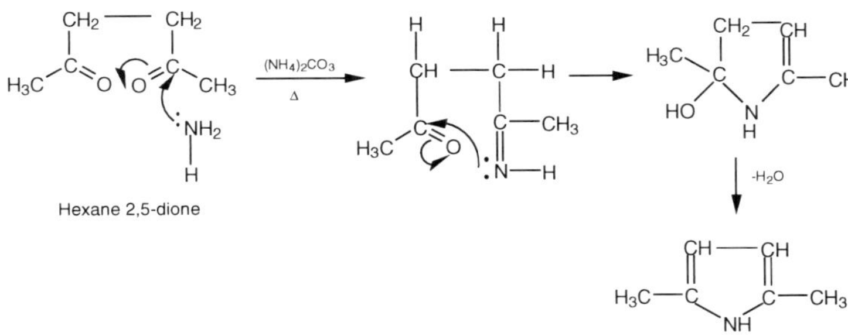

Hexane 2,5-dione

2,5-Dimethyl pyrrole

A mixture of hexane-2,5-dione (5.B.54) (25 g, 25.70 ml, 0.22 mole) and ammonia carbonate (100 g) is heated at 100° (oil-bath) in a R.B. flask (250 ml capacity) fitted with an air condenser until effervescence ceases (60-90 min). The ammonium carbonate which sublimes in the condenser is pushed back into the reaction mixture. The air condenser is replaced by a reflux condenser and the mixture refluxed for 30 min (bath temperature 115°). The reaction mixture is cooled and hot water (10 ml) added through the condenser to dissolve any ammonium carbonate. The upper yellow layer of dimethyl pyrrole is separated (separatory funnel) and the lower layer extracted with chloroform (2 × 10 ml). The combined organic phase and chloroform extract are dried (anhydrous sodium sulphate). The dried liquid is distilled in nitrogen atmosphere. The dimethylpyrrole is obtained at 80-84°/25 mm. Yield 18 g (86%). (lit.[86] b.p. 80-85°/25 mm).

5.C.76 3,5-DIMETHYLPYRROLE

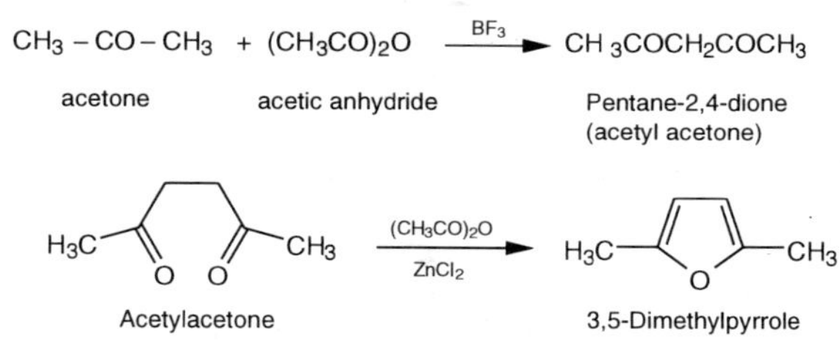

Step (i) Pentane 2,4-dione (Acetyl acetone)

Pure anhydrous acetone (29 g, 36.7 ml, 0.5 mole) and acetic anhydride (127.5 g, 118 ml, 1.25 mole) are placed in a three necked R.B.flask (500 ml capacity) fitted with a gas inlet tube and a gas outlet leading to gas absorption device charged with aqueous alkali to trap excess of boron trifluoride (the third neck is stoppered). The flask is cooled in freezing mixture (ice-salt) and a current of boron trifluoride (from a gas cylinder) is bubbled slowly (2 bubbles per second). The gas is passed for 5 hr. The increase in weight of the reaction mixture is approximately 125 g. The reaction mixture is poured into a solution of sodium acetate (400 g) in water (1.2 litre). The reaction mixture is steam-distilled and the distillate collected (about 1 litre). The distillate is shaken with clear solution of copper (II) acetate (60 g) in water (750 ml) at about 80° (if the solution is not clear, it is filtered). The solution is shaken well and allowed to stand overnight in ice-chest. The separated copper complex is filtered and shaken with sulphuric acid (20%, 200 ml) and ether (200 ml). The ether layer is separated and the aqueous layer extracted with more ether (2 × 100 ml). The combined ether extract is dried (anhydrous sodium sulphate) and distilled. The residue is distilled to give pentane-2,4-dione at 134-36°. Yield 40 g (80%) (lit.[87] m.p. 134-36°).

Step(ii) 3,5-Dimethylpyrrole

Pentane-2,4-dione (25 g, 25.6 ml, 0.25 mole) is added dropwise to a stirred solution (at 15°) of hydrazine sulphate (37.5 g, 0.29 mole) in sodium hydroxide solution (2.5M, 200 ml) placed in a three necked R.B. flask fitted with a mechanical stirrer, the thermometer and a dropping funnel. During addition, the temperature of the reaction mixture is maintained at 15°. After the addition is complete (20 min), the mixture is stirred for 45 min more at 15°. Water (100 ml) is added and the solution extracted with ether (3 × 75 ml). The ether extract is dried (anhydrous potassium carbonate) and ether removed (rotatory evaporator). The separated product is crystallised from petroleum ether as pale yellow solid. Yield 18 g (76%). M.p. 107-108° (lit.[87] m.p. 107-108°).

5.C.77 PYRAZOLONES (PYRAZOLIN-5-ONES)

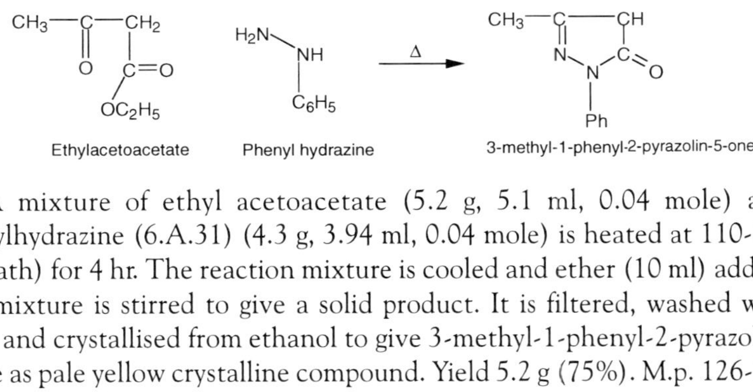

A mixture of ethyl acetoacetate (5.2 g, 5.1 ml, 0.04 mole) and phenylhydrazine (6.A.31) (4.3 g, 3.94 ml, 0.04 mole) is heated at 110-20° (oil bath) for 4 hr. The reaction mixture is cooled and ether (10 ml) added. The mixture is stirred to give a solid product. It is filtered, washed with ether and crystallised from ethanol to give 3-methyl-1-phenyl-2-pyrazolin-5-one as pale yellow crystalline compound. Yield 5.2 g (75%). M.p. 126-27° (lit.[88] m.p. 127°).

Following pyrazolin-5-ones are similarly obtained.

1,3-Diphenyl-2-pyrazolin-5-one, m.p. 136-37° (lit.[88] m.p. 136-37°) is prepared by the condensation of ethyl benzoyl acetate with phenylhydrazine.

1-Phenyl-3-propyl-2-pyrazolin-5-one, m.p. 110-11° (lit.[89] m.p. 110-11°) is prepared by the condensation of ethyl butyryl acetate with phenylhydrazine.

5.C.78 4-ETHOXYMETHYLENE-3-METHYL-1-PHENYL-2-PYRAZOLIN-5-ONE

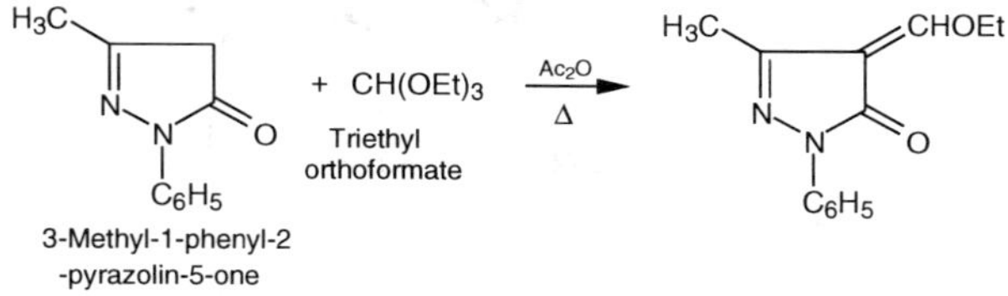

A mixture of 3-methyl-1-phenylpyrazolin-5-one (5.C.77) (0.522 g, 0.003 mole) and triethyl orthoformate (6.A.24) (0.442 g, 0.5 ml, 0.003 mole) in acetic anhydride (8 ml) is heated at 70-80° for 3 min. The reaction mixture is cooled. The separated orange product is filtered. Yield 0.50 g (72.5%). M.p. 185°. (lit.[90] m.p. 185°).

5.C.79 5-CHLORO-3-METHYL-1-PHENYL-1H-PYRAZOLE-4-CARBOXALDEHYDE

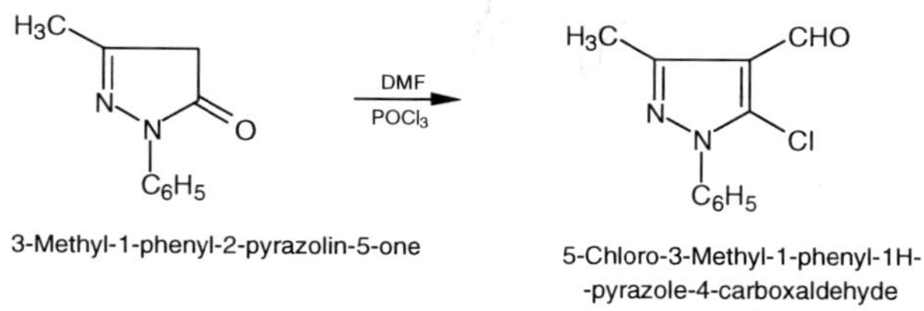

3-Methyl-1-phenyl-2-pyrazolin-5-one (5.C.77) (2.205 g, 0.018 mole) in dry dimethyl formamide (10 ml, 9.44 g, 0.13 mole) is cooled to 0° and treated dropwise with phosphorous oxychloride (4.6 g, 2.8 ml, 0.03 mole), maintaining the temperature between 10-15°. The reaction mixture is heated on a steam bath for 1.5 hr, cooled and poured into ice-water with stirring. The separated product is filtered and washed with water. It is crystallised from ethanol as yellow needles. Yield 1.87 g (67%). M.p. 144-45° (lit.[91] m.p. 148°).

5.C.80 2,3-DIMETHYL-1-PHENYL-3-PYRAZOLIN-5-ONE (ANTIPYRIN)

Dimethyl sulphate (18 g, 13.5 ml, 0.143 mole) is added dropwise to a stirred mixture of 3-methyl-1-phenyl-2-pyrazolin-5-one (5.C.77) (21.75 g, 0.125 mole), sodium hydroxide solution (10 g sodium hydroxide in 10 ml water) and methanol (20 ml). The mixture is stirred for 30 min and then

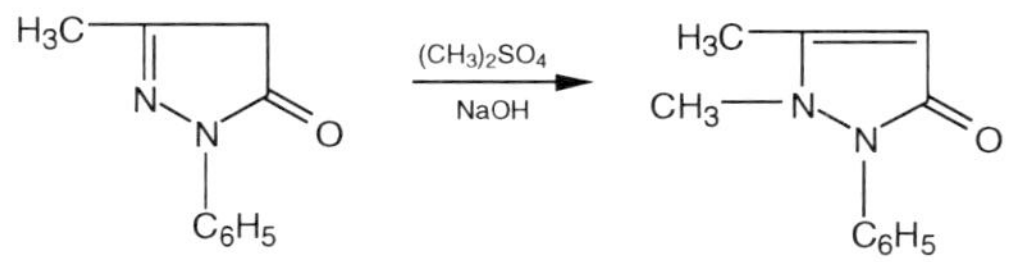

3-methyl-1-phenyl--2-pyrazolin-5-one   2,3-dimehyl-1-phenyl-3-pyrazolin-5-one (Antipyrin)

heated on a boiling water bath for 1 hr. Methanol is removed under reduced pressure and hot water added to the residual solution. The impurities are removed by filtration and the filtrate is extracted with benzene (3 × 50 ml). The solvent is evaporated and the product crystallised from hot water (charcoal) as white crystalline solid. Yield 17.4 g (74%). M.p. 113° (lit.[92] m.p. 113°).

5.C.81 BENZIMIDAZOLE

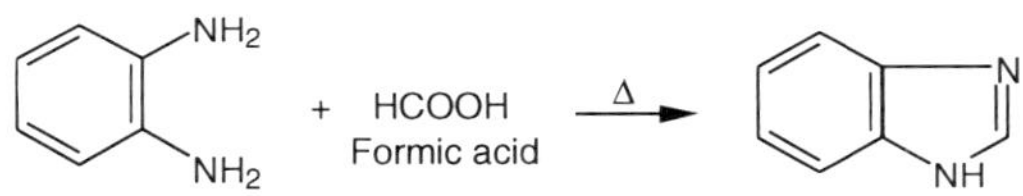

O-Phenylene diamine   Benzimidazole

A mixture of o-phenylene diamine (13.5 g, 0.125 mole) and formic acid (90%, 8.7 ml, 7.83 g, 0.17 mole) is heated for 2 hr on a water bath at 100°. The mixture is cooled and basified with stirring with sodium hydroxide solution. The separated product is filtered and washed with water. It is crystallised from hot water (charcoal). Yield 12.50 g (85%). M.p. 170-71° (lit.[93] m.p. 170-71°).

5.C.82 2-METHYL BENZIMIDAZOLE

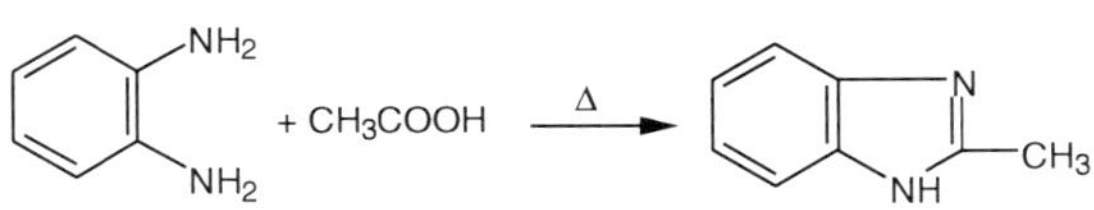

o-Phenylene diamine   2-methylbenzimidazole

A mixture of o-phenylenediamine dihydrochloride (10.86 g, 0.06 mole), water (40 ml) and acetic acid (5.4 g, 5.15 ml, 0.09 mole) is refluxed for 1 hr. The reaction mixture is cooled and basified (with sodium hydroxide solution). The separated product is filtered and washed with water. It is crystallised from dilute alcohol. Yield 4.4 g (55.7%). M.p. 176-77° (lit.[94] m.p. 176°).

5.C.83 INDOLE

$$O{-}CH_3C_6H_4NH_2 + HCO_2H \longrightarrow O{-}MeC_6H_4NHCHO + H_2O$$

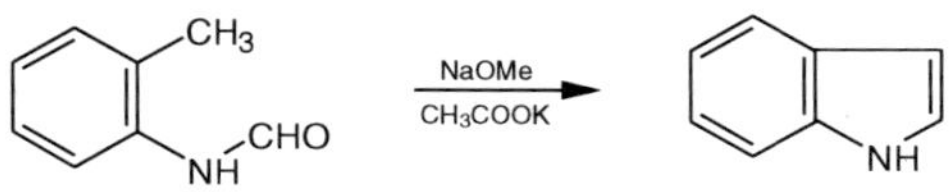

N-formyl o-toluidine Indole

Step (i) N-Formyl-o-toluidine

A mixture of o-toluidine (21.5 g, 21.4 ml, 0.2 mole) and formic acid (90%, 17.5 ml, 15.75 g, 0.34 mole) is heated on a boiling water bath for 3 hr. The reaction mixture is distilled under reduced pressure. The N-formyl-o-toluidine is collected at 173-175^0/25 mm. It soldifies on cooling. Yield 21.5 g (79.3%). It is crystallised from benzene-petroleum ether, m.p. 60-1° (lit.[95] m.p. 60-1°).

Step (ii) Indole

Sodium (5.75 g, 0.25 mole) is dissolved in anhydrous methanol (120 ml) by heating under anhydrous conditions. N-formyl-o-toluidine (34 g, 0.25 mole) is added followed by addition of freshly fused powdered potassium acetate (50 g, 0.51 mole). The mixture is refluxed till a clear solution is obtained. Methanol is removed (rotary evaporator). The product is distilled (fume cupboard) by heating the flask at 300-350° using a metal bath (woods metal). When no further distillation takes place, the residual o-toluidine is removed by applying vacuum. The remaining mixture is cooled, water (100 ml) added and steam distilled. Colourless crystals of indole separate from the steam distillate; the steam distillate is rendered slightly acidic (hydrochloric acid) and filtered. It is crystallised from petroleum ether. Yield 5 g (17%). M.p. 51-2°.

5.C.84 2-PHENYLINDOLE

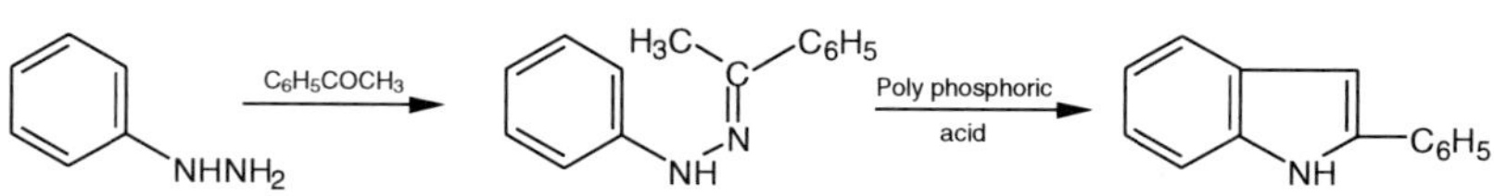

Phenyl hydrazine Acetophenone phenylhydrazone 2-Phenyl indole

Step (i) Acetophenone Phenylhydrazone

A mixture of acetophenone (20 g, 19.4 ml, 0.166 mole) and phenylhydrazine (18 g, 16.4 ml, 0.166 mole) in ethanol (50 ml) and a few drops of glacial acetic acid is warmed (water bath, 20 min). The mixture is cooled, filtered and the product washed with dilute hydrochloric acid and cold alcohol (10 ml). It is crystallised from ethyl alcohol as white crystalline product. M.p. 105-106° (lit.[96] m.p. 106°).

Step (ii) 2-Phenylindole

A mixture of acetophenone phenylhydrazone (28 g, 0.133 mole and polyphosporic acid (180 g) is heated on a boiling water bath. The mixture is stirred and kept at 100-120° for 10 min. It is treated with water (400 ml) and stirred. The separated product is filtered and washed with water. It is crystallised from alcohol (charcoal). The 2-phenylindole thus obtained is filtered and dried in a vacuum desiccator over calcium chloride. Yield 20 g (79%). M.p. 188-89° (lit.[97] m.p. 188-89°).

5.C.85 BENZOTRIAZOLE

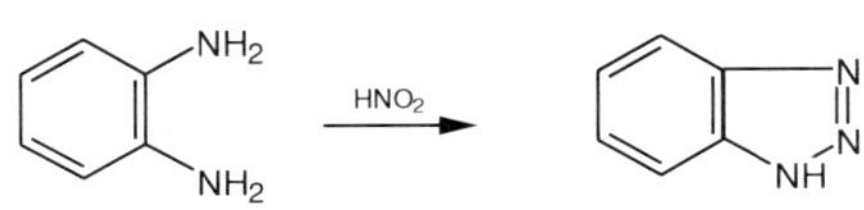

o-Phenylene diamine   Benzotriazole

Pure o-phenylenediamine (10.8, 0.1 mole) is dissolved in a mixture of glacial acetic acid (12 g, 11.5 ml, 0.2 mole) and water (30 ml) (slight warming is necessary). The solution is cooled to 15° and then a solution of sodium nitrite (7.5 g, 0.11 mole) in water (15 ml) is added in one lot to the stirred solution. The reaction mixture warms and within 3-4 min attains a temperature of 80-85° and then begins to cool; the colour of the mixture changes from deep red to pale brown. The mixture is stirred for 15 min. It is chilled in ice-water for 30 min and the separated product is filtered and washed with ice-cold water (3 × 30 ml). Some more product is obtained from the mother liquor by concentration and cooling. Total yield is 8 g (67%). It is crystallised from benzene. M.p. 99-100° (lit.[98] m.p. 99-100°).

5.D.86 2,4-DIMETHYLTHIAZOLE

A mixture of finely powdered dried acetamide (75 g, 1.27 mole), powdered phosphorous pentasulphide (50 g) and dry benzene (50 ml) is taken in two

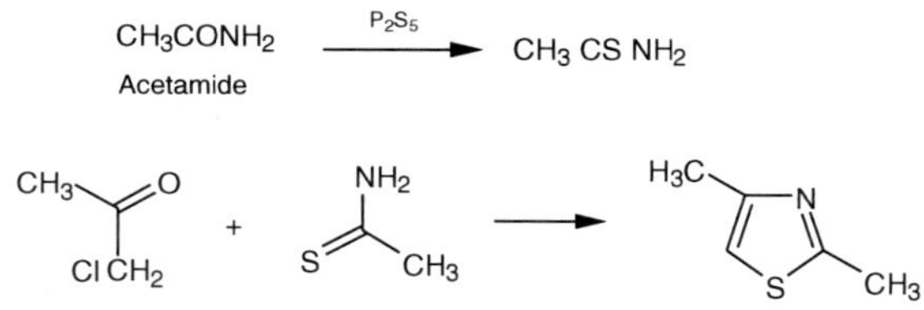

Acetamide

2,4-Dimethylthiazole

necked R.B.flask (500 ml capacity) fitted with reflux condenser and a dropping funnel (the reaction is performed in a fume cupboard). A solution of chloroacetone (116 g, 100 ml, 1.25 mole) in dry benzene (37 ml) is separately made. Chloroacetone-benzene solution (5 ml) is added to the reaction mixture and warm gently (water-bath). The water bath is removed as soon as the exothermic reaction starts. The remaining chloroacetone-benzene solution is added in 5 ml portions at such intervals that the reaction remains under control. After the addition is complete, the mixture is shaken for 15-20 min. water (200 ml) is added to the mixture and the lower aqueous layer separated (separatory funnel). The lower layer is rendered alkaline by adding sodium hydroxide solution (20%). The mixture is extracted with ether (5 × 75 ml). The ether extract is dried (anhydrous sodium sulphate) and ether removed (rotary evaporator). The residual product is distilled and the fraction between 140-150° is collected. It is redistilled. Yield 57 g (40%). B.p. 143-45° (lit.[99] b.p. 143-45°).

5.D.87 2-AMINOTHIAZOLE

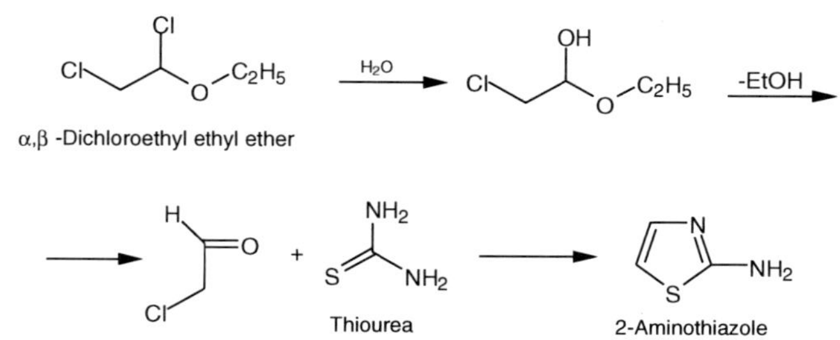

α,β -Dichloroethyl ethyl ether

Thiourea

2-Aminothiazole

α,β-Dichloroethyl ethyl ether ($ClCH_2.CHCl.O.C_2H_5$) (71.5 g, 61.3 ml, 0.5 mole) is added to a stirred solution of thiourea (38 g, 0.5 mole) in warm water (100 ml) in a three necked flask (250 ml capacity) fitted with a sealed mechanical stirrer, a dropping funnel and a reflux condenser. The mixture is stirred and heated gently for 2 hr. The cooled solution is rendered alkaline by the addition of solid sodium hydroxide. 2-Amino-thiazole is liberated from its salt. It is extracted with ether (3 × 100 ml), the ether extract dried (anhydrous sodium sulphate) and distilled. The residual crude product is crystallised from alcohol as yellow crystalline solid. Yield 40 g (80%). M.p. 91-2° (lit.[100] m.p. 93°).

5.D.88 2-AMINO-4-METHYLTHIAZOLE

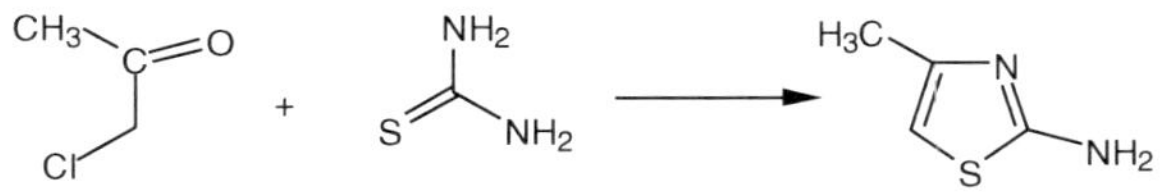

Chloroacetone Thiourea 2-Amino-4-methylthiazole

Chloroacetone (46.25 g, 40 ml, 0.5 mole) is added during 30 min to a stirred suspension of thiourea (38 g, 0.5 mole) in water (100 ml) in a three necked flask (250 ml capacity) fitted with sealed stirrer, a reflux condenser and a dropping funnel. Thiourea dissolves as the reaction proceeds and the temperature rises. The solution is refluxed for 2 hr. The solution is cooled and rendered alkaline by adding sodium hydroxide pellets (100 g). The mixture is extracted with ether (3 × 75 ml). The ether extract is dried (anhydrous sodium sulphate) and evaporated (rotary evaporator). 2-Amino-4-methylthiazole is obtained by distilling in vacuo. Yield 42 g (73.7%). B.p. 130-33°/18 mm. It solidifies on cooling in ice. M.P. 44-5° (lit.[101] M.P. 45-6°).

5.D.89 4-PHENYL SUBSITUTED-2-AMINOTHIAZOLES

Method (i)

General Procedure[18] 2-Amino-4-(p-Chlorophenyl) Thiazole

A mixture of p-chloroacetophenone (30.5 g, 25.6 ml, 0.2 mole) in dry benzene (100 ml), thiourea (29.6 g, 0.39 mole) and iodine (0.2 mole) is refluxed for 40 hr on a steam bath. The solvent is removed under reduced pressure. The residual product is washed with ether (20 ml) and extracted

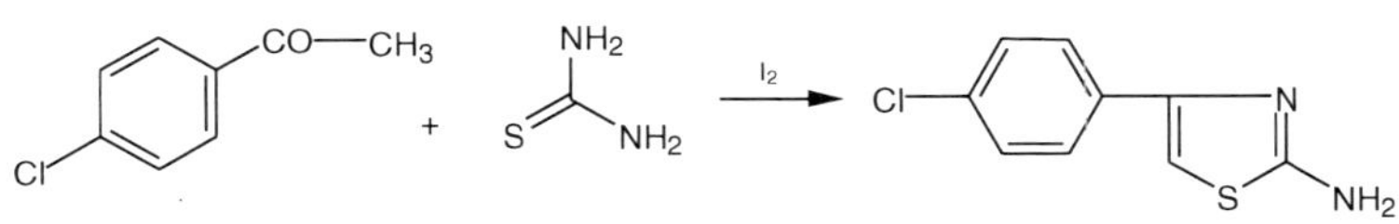

p-chloroacetophenone Thiourea 2-Amino-4-(p-chlorophenyl)thiazole

with hot water. The aqueous solution containing the hydroiodide is treated with liquor ammonia. The separated solid is filtered and washed with water. It is crystallised from alcohol as almost colourless needles. Yield 27.2 g (65.4 %). M.p. 173°.

Following thiazoles are similarly obtained.
2-Amino-4-(p-bromophenyl) thiazole, m.p. 177° (lit.[102] m.p. 177°);
2-Amino-4-(p-methoxyphenyl)thiazole, m.p. 204° (lit.[102] m.p. 204°);
2-Amino-4-phenylthiazole[18], m.p. 150-51°,
2-Amino-4-(o-methoxyphenyl) thiazole[18], m.p. 189-90°
2-Amino-4-(p-methylphenyl) thiazole, m.p. 123-24° (lit.[102] m.p. 123-24°).

Method (ii)

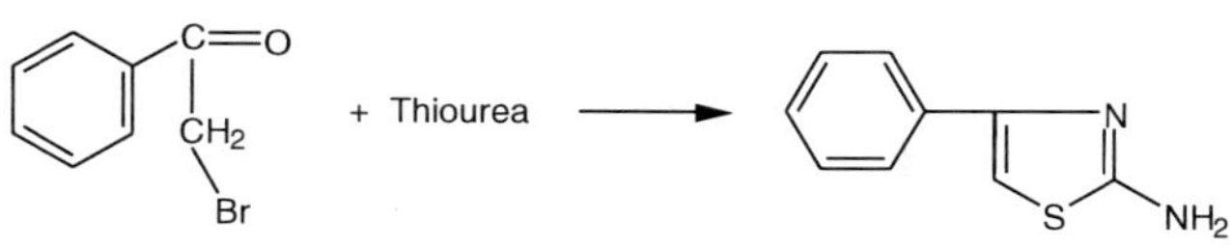

General procedure This method consists in the reaction of phenacyl bromides (α-haloketones) with thiourea and is widely used and gives better yields.

Step (i) α-Halo ketones (phenacyl bromides)

A solution of p-chloroacetophenone (2 g, 1.7 ml, 0.0129 mole) in carbon tetrachloride (100 ml) is treated dropwise (1 hr) with a solution of bromine (2.3 g, 0.014 mole) in carbon tetrachloride (30 ml) with stirring at 30-35°. The solvent is distilled off and the residual product is crystallised from benzene-petroleum ether as shining crystals. Yield 3 g (99%). M.p. 93° (lit.[103] m.p. 93°).

Following phenacyl bromides, p-$RC_6H_4COCH_2Br$ are obtained similarly.

R	m.p.	lit.m.p. (Ref)	% Yield
H	51-2°	49-51° (104)	64
CH_3	50°	51° (105)	61
Br	109°	109° (106)	70
OCH_3	72°	75-6° (107)	60
NO_2	96°	98° (108)	55

Note: This is a general procedure for the preparation of α-halo ketones. It is normally observed that the product obtained is invariably a mixture of mono and di- bromo compounds. For the synthesis of thiazoles the mixture can be profitably used, since both mono- and di-bormo compounds give the same product. However pure monobromo compound can be obtained by column chromatography over silica gel column and eluting with petroleum ether-benzene (10:1).

Step (ii) 4-Phenylsubstituted-2-aminothiazoles

General procedure[18] 2-Amino-4-(p-chlorophenyl) thiazole

p-Chlorophenacyl bromide (1 g, 0.0043 mole) in absolute alcohol (25 ml) is refluxed with thiourea (0.35 g, 0.0046 mole) for 4 hr. The solvent is distilled off. Crushed ice is added to the residual product followed by the addition of liquor ammonia. The separated product is filtered and washed with water. It is crystallised from alcohol as pure yellow needles. Yield 0.62 g (70%). M.p. 173°.

5.D.90 2-AMINOBENZOTHIAZOLE

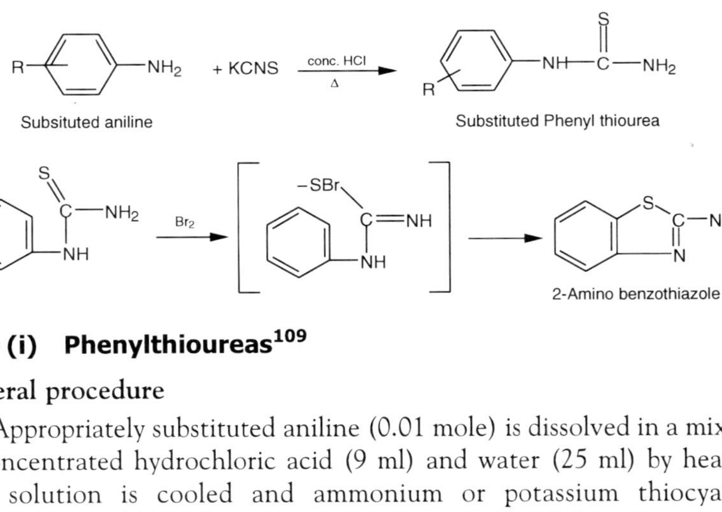

Step (i) Phenylthioureas[109]

General procedure

Appropriately substituted aniline (0.01 mole) is dissolved in a mixture of concentrated hydrochloric acid (9 ml) and water (25 ml) by heating. The solution is cooled and ammonium or potassium thiocyanate (0.1 mole) is added. The reaction mixture is refluxed for 4 hr and then cooled. The separated product is filtered and washed with water. It is crystallised from alcohol.

Following thioureas, 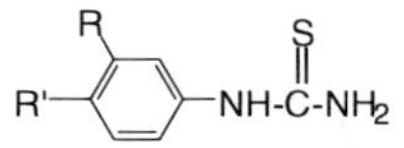 are prepared.

R	*R_1*	*M.p. (°C)*	*Lit. m.p. (°C)*	*Yield (%)*
CH_3	H	157	157	85
OCH_3	H	144	144	80
OC_2H_5	H	115	115	78
Cl	H	130	130	88
H	CH_3	195	195	90
H	OCH_3	224	224	75
H	Cl	178	178	82
H Br	154	154	65	

Step (ii) 2-Aminobenzothiazole[109]

General procedure

Appropriately substituted thiourea (0.1 mole) in chloroform (100 ml) is treated dropwise with a solution of bromine (0.1 mole) in chloroform while stirring the reaction mixture over a period of 2 hr. The temperature of the reaction mixture is maintained below 5°. Stirring is continued for 1 hr more and the solution is refluxed on a water bath till the evolution of hydrogen bromide vapours cease (3-4 hr). The reaction mixture is concentrated and the resulting solid product is treated with a solution of sulphur dioxide in water. The clear solution is neutralized with aqueous ammonia. The separated product is filtered and washed with water. It is crystallised from alcohol.

Following 2-aminobenzothiazoles are prepared.

R	**R_1**	**m.p. (°C)**	**Lit. m.p. (°C)**	**Yield (%)**
CH_3	H	190	190	83
OCH_3	H	198	198	70
OC_2H_5	H	154	154	60
Cl	H	183	183	80
H	CH_3	132	132	75
H	OCH_3	162	162	60
H	Cl	200	200	82
H	Br	202	202	70

5.D.91 4-PHENYLSUBSTITUTED 2-MERCAPTOTHIAZOLES

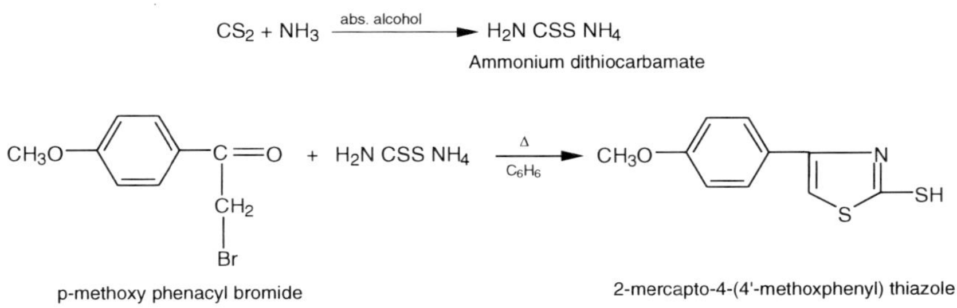

Step (i) Ammonium Dithiocarbamate NH_2CSSNH_4

General procedure

Ammonia gas is passed into absolute alcohol (250 ml) at 0-5° until the gain in weight is 39 g (2.3 moles). To this solution (still at 0-5°)' is added a well cooled solution of carbon disulphide (60 ml) and ether (200 ml). The reaction mixture is kept at 0-5° for 3 hr. The mixture is kept overnight and again cooled (0-5°). The separated lemon yellow shining crystals are filtered, washed with ether (2 × 10 ml) and dried (vacuum desiccator). Yield 80 g, m.p. 99° (decomp.) (lit.[110] m.p. 99°).

Step (ii) 2-Mercapto-4-(4'-methoxyphenyl) thiazole

General procedure

Ammonium dithiocarbamate (1.5 g, 0.0136 mole) is added to a solution of p-methoxyphenacyl bromide (5.D.89) (2 g, 0.0087 mole) in absolute ethanol (50 ml) with stirring. The mixture is refluxed for 3 hr and the solvent distilled off. The residue is refluxed with benzene for 15 min and filtered. The filtrate is cooled. The separated product is filtered. The required compound is obtained as pale yellow shining prisms. Yield 0.82 g (42%). M.p. 195-96° (lit.[111] m.p. 196°).

Following thiazoles are similarly obtained.

2-Mercapto-4-(4'-methylphenyl)thiazole is obtained by the condensation of ammonium dithiocarbamate (1.6 g, 0.145 mole) with p-methyl phenacyl bromide (2 g, 0.0094 mole). Yield 50%. M.p. 190-91° (lit.[111] m.p. 190°).

2-Mercapto-4-(4'-bromophenyl)thiazole is obtained by the condensation of p-bromophenacyl bromide (2g, 0.0072 mole) with ammonium dithiocarbamate (1.2 g, 0.0109 mole) as pale yellow shining prisms. Yield 49%. Mp. 219-20° (lit.[111] m.p. 220-22°).

2-Mercapto-4-phenylthiazole is obtained by the condensation of phenacyl bromide (2 g, 0.01 mole) with ammonium dithiocarbamate (1.7 g, 0.015 mole) as lemon yellow shining crystals. Yield 48%. M.p. 174-75° (lit.[111] m.p. 173°).

5.D.92 2-HYDROXY PYRIMIDINE

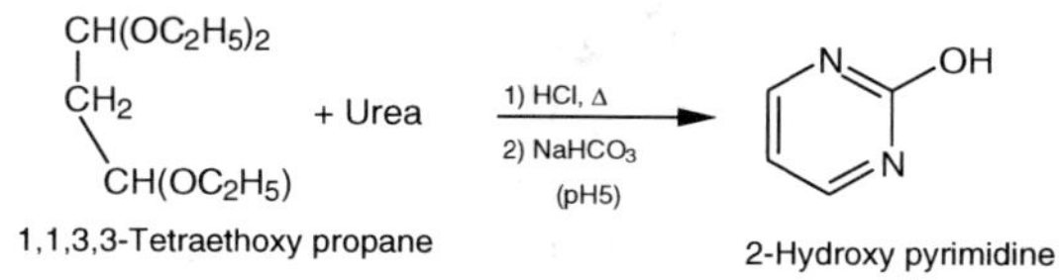

To a warm solution of urea (1.5 g, 0.025 mole, in 10 ml ethanol), 1,1,3,3,-tetraethoxypropane (5.5 g, 6 ml, 0.025 mole) and concentrated hydrochloric acid (5 ml) are added. The mixture is stirred for 1 hr at 30-40° and the solution cooled to 0°. The separated 2-hydroxypyrimidine hydrochloride is filtered. The mother liquor is treated with ether (100 ml) and shaken. By this process some more of the hydrochloride salt is obtained. The combined salt is dissolved in sodium bicarbonate solution and the clear solution acidified to pH 5 with dilute hydrochloric acid. The solvent is evaporated to dryness and the residue is extracted with ethyl acetate in a soxhlet apparatus. The crude product obtained by distillation of ethyl acetate solution is crystallised from chloroform to give crystalline needles. Yield 0.8 g (33%). M.p. 179° (lit.[112] m.p. 179°).

5.D.93 BARBITURICACID

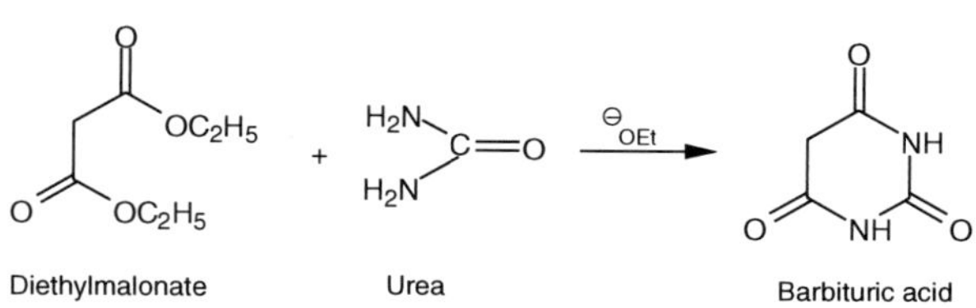

Sodium (11.5 g, 0.5 mole) is added to absolute ethanol (250 ml) in a R.B.flask fitted with a reflux condenser and a calcium chloride drying tube. If the reaction becomes too vigorous, it is cooled in ice-water. When all the sodium has reacted, diethyl malonate (80 g, 76 ml, 0.5 mole) is added. Subsequently a solution of dry urea (30 g, 0.5 mole) in hot absolute alcohol

(250 ml, 70°) is added to the reaction mixture. The mixture is shaken well and refluxed for 8 hr at 110° (oil bath). A white solid separates. To the reaction mixture is added hot water (400 ml) and then concentrated hydrochloric acid is added with stirring (about 45 ml) until the solution is acidic. The solution is filtered and the filtrate is left in a refrigerator for 12 hr. The separated barbituric acid is filtered, washed with cold water and dried (100°, 3 hr). Yield 50 g (78%). M.p. 244-45° (decomp.) (lit.[113] m.p. 244-45°).

5.D.94 AMINOBARBITURIC ACID (URAMIL)

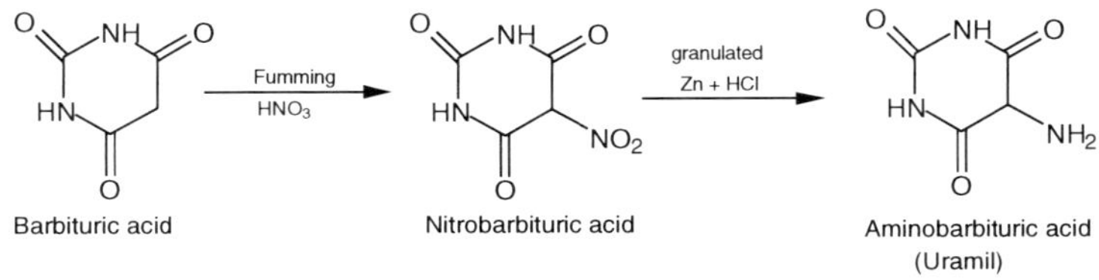

Step (i) Nitrobarbituric Acid

Barbituric acid (5.D.93) (25 g, 0.195 mole) is added during 2 hr to stirred and cooled (ice-bath) fuming nitric acid (36 ml, d 1.52). The temperature is kept below 40° during the addition. The mixture is stirred for 1 hr more. Water (100 ml) is added to the stirred mixture and the solution cooled (10°). The separated product is filtered (sintered funnel) and washed with cold water. It is dissolved in boiling water (about 220-240 ml), filtered and cooled slowly. The separated nitrobarbituric acid is filtered, washed with water and dried (2 hr, 90°). It separates as trihydrate, m.p. 181-3° (decomp.). Yield 35 g. It is dried at 110-15° for 3 hr. m.p. 176° (decompn.). (yield of dried sample is 23.5 g (69.5%) (lit.[114] m.p. 176°).

Step (ii) Uramil

5-Nitrobarbituric acid (19 g, 0.11 mole) is heated (water bath) with concentrated hydrochloric acid (100 ml), granulated tin (62 g) and concentrated hydrochloric acid (100 ml) are added during 30 min. The mixture is heated (water bath) with occasional shaking until the yellow coloured nitrocompound is not visible. More concentrated hydrochloric acid (750 ml) is added and the mixture is heated until all the white solid dissolves. A little decolourising charcoal is added and the hot mixture is filtered (sintered funnel). The solution is cooled for 12 hr at 0°. The separated product is filtered and washed with dilute hydrochloric acid and finally with water. A second crop is obtained by concentrating the mother liquor in a rotary evaporator to about 200 ml and cooling. The uramil is dried in vacuum desiccator over concentrated sulphuric acid. Total yield 12.5 g (79.6%). It does not melt below 400° (lit.[115] m.p. > 400°).

5.D.95 1,3-DIARYLTHIOBARBITURIC ACIDS (1,3-DIARYL-1,2,3,4,5,6-HEXAHYDRO-4,6-DIOXO-2-THIOXOPYRIMIDINE

These are obtained by the condensation of appropriately substituted thiourea with malonic acid in presence of acetyl chloride.

Step (i)

(a) Diarylthioureas (Symmetrical)

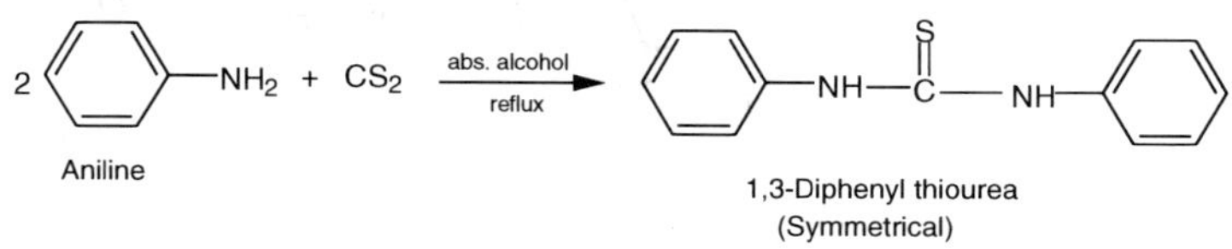

1,3-Diphenylthiourea, General Procedure

A mixture of aniline (20 g, 19.6 ml, 0.215 mole) and carbon disulphide (25 g, 19.7 ml, 0.329 mole) in absolute alcohol (50 ml) is refluxed on a steam bath for 8 hr or until the reaction mixture soldifies. The separated product is filtered, washed with excess of dilute hydrochloric acid (1:1) to remove unreacted aniline and finally with cold alcohol. The 1,3-diphenylthiourea thus obtained is crystallised from ethanol, as white shining crystals. Yield 20 g (40.8%). M.p. 153-54° (lit.[116] m.p. 154-55°).

Following thioureas are prepared

1,3-Di (2-methylphenyl)thiourea, m.p. 161-62° (lit.[117] m.p. 161-62°)
1,3-Di (3-methylphenyl)thiourea, m.p 111-12° (lit.[117] m.p. 111-12°);
1,3-Di (4-methylphenyl)thiourea, m.p 174-75° (lit.[117] m.p. 174-75°);
1,3-Di (2-methoxyphenyl) thiourea, m.p 176-77° (lit.[118] m.p. 176-77°)
1,3-Di (3-methoxyphenyl) thiourea, m.p.[18] 135-36°
1,3-Di (4-methoxyphenyl) thiourea, m.p 185-86° (lit.[117] m.p. 185-86°)
1,3-Di (3-chlorophenyl) thiourea, m.p. 131-32° (lit. [117] m.p. 131-32°)
1,3-Di (4-chlorophenyl) thiourea, m.p. 176-77° (lit [117] m.p. 176-77°)
1,3-Di (4-bromophenyl) thiourea, m.p. 188° (lit.[117] m.p. 188°)

(b) 1,3-Diarylthioureas (Unsymmetrical)

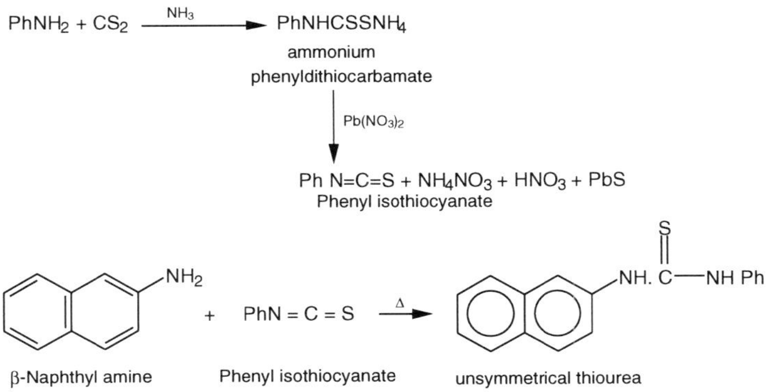

Step (i) Phenylisothiocyanate

General procedure

Carbon disulphide (54 g, 42.7 ml, 0.71 mole) and concentrated ammonia (90 ml, sp. gr. 0.9) is mixed and cooled in ice bath. Aniline (56 g, 55 ml, 0.6 mole) is then added to the mixture with stirring in 20 min and the stirring is continued for 30 min more. The reaction mixture is allowed to stand for 30 min. A heavy precipitate of ammonium phenyldithiocarbamate separates out. It is filtered, suspended in water (800 ml) and treated with a solution of lead nitrate (200 g, 0.6 mole) in water (400 ml) with constant stirring. A heavy brown precipitate of lead sulphide separate out; it soon turns black. The mixture is steam distilled and the distillate is collected in a receiver containing sulphuric acid (1N, 10 ml) till no more oily liquid distils. The oil is extracted with ether (100 ml). The ether solution is dried (anhydrous calcium chloride) and distilled. Phenyl isothiocyanate distills at 121-22°/35 mm (lit.[119] b.p. 121-22°/35 mm; b.p. 221°).

Following isothiocyanates are obtained

p-Tolyl isothiocyanate, m.p. 26°

3-Chlorophenylisothiocyanate, oil,

4-Chlorophenylisothiocyanate, m.p. 43°,

4-Methoxyphenylisothiocyanate, oil,

4-Bromophenylisothiocyanate, m.p. 60-61°

Step (ii) 1,3-Diarylthiourea (Unsymmetrical)

General procedure 1-(2'-Naphthyl)-3-phenylthiorea

β-Naphthylamine (1.43 g, 0.01 mole) and phenylisothicyanate (1.35 g, 1.2 ml, 0.01 mole) are separately dissolved in minimum quantity of ethanol, both the solutions are mixed and heated on a steam bath for 20 min. The solution is cooled and the separated product is filtered, washed with dilute hydrochloric acid and water. It is crystallised from ethanol- acetic acid as almost colourless needles. Yield 2.7 g (97.1%). M.p. 158-59° (lit.[120], m.p. 159°).

1-(2'-Naphthyl)-3(4"-methoxyphenyl) thiourea is similarly prepared from β-naphthylamine and p-methoxyphenylisthiocyanate in 80% yield. M.p. 170° (lit.[121] m.p. 170°).

Step II 1,3-Diarylthiobarbituric Acids (1,3-Diaryl- 1,2,3,4,5,6,-hexahydro –4,6-Dioxo-2-Thioxopyrimidine

General procedure

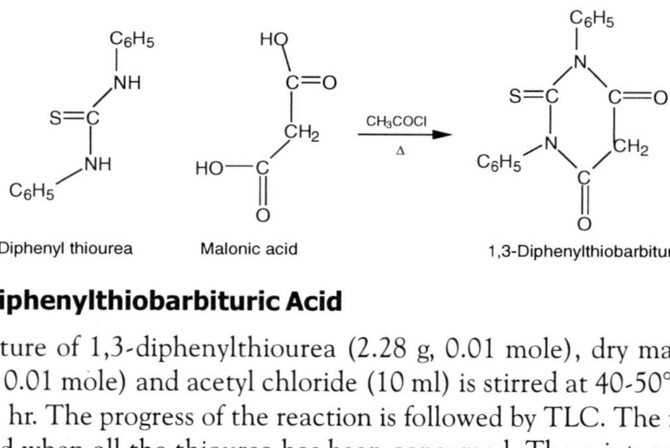

Diphenyl thiourea Malonic acid 1,3-Diphenylthiobarbituric acid

1,3-Diphenylthiobarbituric Acid

A mixture of 1,3-diphenylthiourea (2.28 g, 0.01 mole), dry malonic acid (1.1 g, 0.01 mole) and acetyl chloride (10 ml) is stirred at 40-50° (oil bath) for 1-2 hr. The progress of the reaction is followed by TLC. The reaction is stopped when all the thiourea has been consumed. The mixture is treated with crushed ice and the separated product filtered. It is washed with water and crystallised from acetic acid. Yield 2 g (66.7%). M.p. 245-46° (lit.[118] m.p. 245-46°).

Following 1,3-diarylthiobarbituric acids are prepared.

1,3-Di-o-methylphenyl thiobarbituric acid, m.p. 191-92° (lit.[118] m.p. 191-92°)

1,3-Di-m-methylphenyl thiobarbituric acid, m.p. 262-63° (lit.[118] m.p. 262-63°)

1,3-Di-p-methylphenyl thiobarbituric acid, m.p. 233-34° (lit.[118], m.p. 233-34°)

1,3-Di-o-methoxyphenyl thiobarbituric acid m.p. 202-04° (lit.[118] m.p. 202-04°)

1,3-Di-m-methoxyphenyl thiobarbituric acid m.p. 240-41° (lit.[118] m.p. 240-41°)

1,3-Di-p-methoxyphenyl thiobarbituric acid m.p. 202-03° (lit.[122] m.p. 202-03°)

1,3-Di-m-chlorophenyl thiobarbituric acid m.p.[18] 262-63°

1,3-Di-p-chlorophenyl thiobarbituric acid m.p. 203-04° (lit.[118] m.p. 203-04°)

5.D.96 1,2,3,4,7-PENTAHYDRO-5,7,7-TRIMETHYL-4-OXO-1,3-DIPHENYL-2-THIOXOPYRANO [2,3-D] PYRIMIDINE[123]

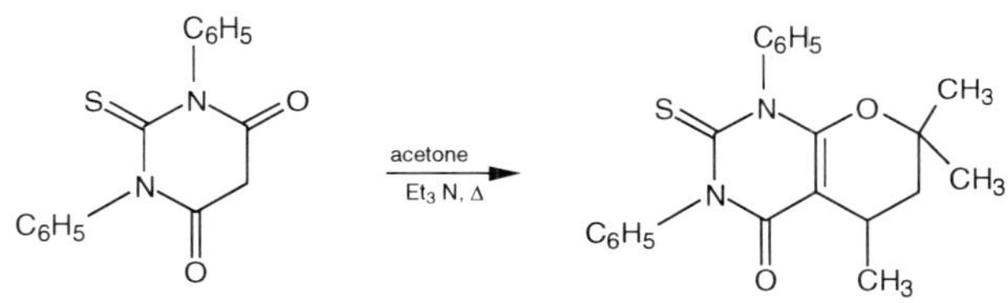

1,3-Diphenyl thiobarbituric acid

1,2,3,4,7-pentahydro-5,7,7-trimethyl-4-oxo-1,3-diphenyl-2-thioxopyrano [2,3-d] pyrimidine

1,3-Diphenylthiobarbituric acid (5.D.95) (1.48 g, 0.005 mole), dry tri ethylamine (0.2 ml) in dry acetone (50 ml) is refluxed for 10 hr. The solvent is removed in vacuo and the residue is treated with crushed ice. The separated product is filtered, washed with water and dried. It is crystallised from benzene-petroleum ether as light yellow needles. Yield 1.7 g (89.9%). M.p. 206-207°.

Using appropriate symmetrical 1,3-diarylthiourea, following [2,3-d] pyrimidines are prepared[123].

Ar	*m.p.*
o-$CH_3C_6H_4$ -	158-59°
p-$CH_3C_6H_4$ -	190-91^0
m-ClC_6H_4 -	200-201°
p-ClC_6H_4 -	214-15°
p-$OCH_3C_6H_4$ -	245-46°

5.D.97 1,3-BIS(4-CHLOROPHENYL)-6-CHLORO-5-FORMYL-2-THIOBARBITURIC ACID[18] (1,3-BIS (4-CHLOROPHENYL)-5-CARBOXALDEHYDE-6-CHLORO-1,3-DIHYDRO-4-OXO-2-THIOXOPYRIMIDINE

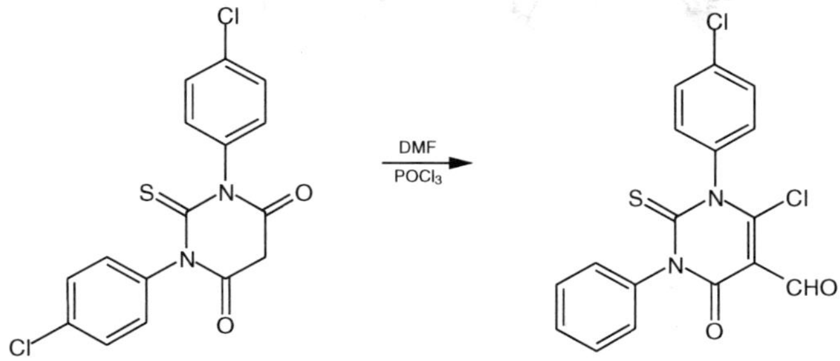

1,3-Di(4-chlorophenyl)thiobarbituric acid

1,3-Bis(4-chlorophenyl)-5-carboxaldehyde-6-chloro -1,3-dihydro-4-oxo-2-thioxopyrimidine

A solution of 1,3-di(4-chlorophenyl)thiobarbituric acid (5.D.95) (3.65 g, 0.01 mole) in dimethylformamide (11 ml, 10.38 g, 0.142 mole) is cooled to 0° (ice-salt bath). Phosphorous oxychloride (4.62 ml, 7.6 g, 0.05 mole) is then added dropwise, maintaining the temperature between 10-20°. After the addition is over, the reaction mixture is refluxed for 45 min. The solution is cooled and poured over crushed ice. It is left overnight at room temperature. The separated product is filtered and dried. It is crystallised from methanol as pale yellow flakes. Yield 2.78 g (68%). M.p. 190-91°.

5.D.98 5-ETHOXYMETHYLENE-1,3-BIS (2-METHOXYPHENYL)-2-THIOBABITURIC ACID

General procedure[124]

A mixture of 1,3-bis(2-methoxyphenyl)thiobarbituric acid (5.D.95) (0.356 g, 0.001 mole) and triethylorthoformate (6.A.24) (5 ml) is heated for 1 hr at 120° with continouous stirring. The reaction mixture is cooled and the separated yellow product is filtered and washed with hexane. It is crystallised from benzene-petroleum ether as bright yellow powder. M.p. 203°. Yield 95%.

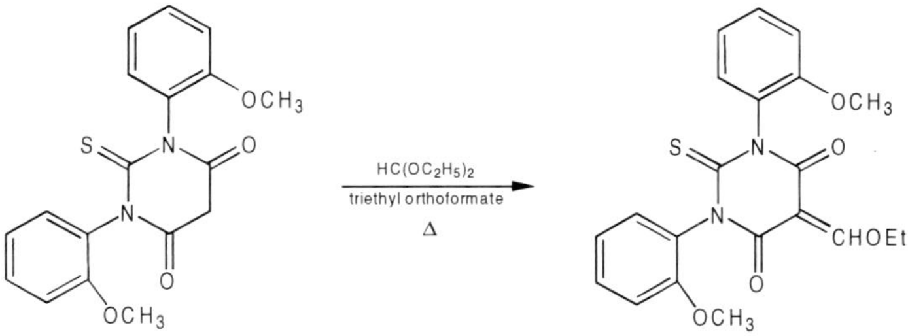

1,3-bis(2-methoxyphenyl)-thiobarbituric acid

5-Ethoxymethylene-1,3-bis-(2-methoxyphenyl)-2-thiobarbituric acid

Following compounds are similarly obtained[124].

Ar	m.p. (°C)
2-methyl phenyl	221
3-methylphenyl	205
4-bromophenyl	261
4-chlorophenyl	259
phenyl	248

References

1. W. Bridge, A.J. Crocker, T. Cubin and A. Robertson, J. Chem. Soc., 1937, 1533.
2. V.D. Nageswara Sastri, N. Narasimhachari, P. Rajagopalan, T.R. Seshadri and T.R. Thiruvengadam, Proc. Indian Acad. Sci., 1953, 37A, 681.
3. R.G. Heyes and A. Robertson, J. Chem. Soc., 1936, 1832.
4. G. Amiard and A. Allais , Bull. Soc. Chim. France, 1947, 512.
5. H.V. Pechmann, Ber., 1884, 17, 929.
6. J. Baum, Ber., 1886, 19, 502; Z. Physiol. Chem., 1885, 9, 465.
7. Hans A. Offe and H. Jakakewitz, Ber., 1947, 80, 469.
8. J. Boyd and A. Robertson, J. Chem. Soc., 1948, 174.
9. V.R. Shah, J.L. Bose and R.C. Shah, J. Org. Chem., 1960, 25, 677.
10. A. Sonn, Ber., 1917, 50, 1292.
11. V.K. Ahluwalia and Devendra Kumar, Indian J. Chem., 1977, 15B, 945.
12. V.K. Ahluwalia and Devendra Kumar, Indian J. Chem., 1979, 17B, 167.
13. A. Michael, Ber., 1905, 38, 2099.
14. H.Von Pechmann and C. Duisberg, Ber., 1883, 16, 2122.
15. V.K. Ahluwalia, Vimal D. Mehta and C.H. Khanduri, Indian J. Chem., 1988, 27B, 68.

16. H. Von Pechmann and C. Duisberg, Ber., 1883, 16, 126.
17. V.K. Ahluwalia, A.C. Mehta and T.R. Seshadri, Proc. Indian Acad. Sci., 1957, 45A, 15.
18. V.K. Ahluwalia et al., unpublished work.
19. V. Kostenecki and Weber, Ber, 1893, 26, 2906
20. A. Ulubelen, R.R. Kerr and T.J. Mabry, Phytochemistry, 1982, 21, 1145.
21. D.B. Dey and K.K. Row, J. Indian Chem. Soc., 1924, 1, 122.
22. V.K Ahluwalia and Chandra Prakash, Indian J. Chem., 1975,13, 791.
23. V.K. Ahluwalia, Vimal D. Mehta and C. H. Khanduri, Indian J. Chem., 1988, 27B, 68.
24. V.K. Ahluwalia and C.H. Khanduri, Indian J. Chem., 1989, 28B, 599.
25. A. Rossing, Ber., 1884, 17, 2988.
26. E.C. Horning and D.B. Reisner, J. Am. Chem. Soc., 1948, 70, 3619.
27. P. Friedlaender and L.C. Schnell, Ber., 1897, 30, 2153.
28. V.K. Ahluwalia and Devendra Kumar, Indian J. Chem., 1977, 15B, 514.
29. S. Alexander and A. Sinha , J. Chem. Soc., 1950, 3345.
30. P.F. Wiley, J. Am. Chem. Soc., 1951, 4205.
31. V. Kostanecki and A. Rozycki, Ber., 1901, 35, 105.
32. Ch. Bheemashankara Rao, G. Subramanyam and V. Venkateswarlu, J. Org. Chem, 1959, 24, 683.
33. K.C. Gulati, S.R. Seth and K. Venkataraman, J. Chem. Soc., 1934, 1765.
34. Kalff and Robinson, J. Chem. Soc., 1925, 1968.
35. W. Baker, J. Chem. Soc., 1933, 1386.
36. Wheeler, J. Chem. Soc., 1939, 1681.
37. S. Rangaswami and T.R. Seshadri, Proc. Indian Acad. Sci., 1939, 10A, 6.
38. S.K. Grover, A.C. Jain, S.K. Mathur and T.R. Seshadri, Indian J. Chem., 1963, 1, 382.
39. S. Saxena, J.M. Makrandi and S.K. Grover, Synthesis, 1985, 697.
40. T.C. Chadha and K. Venkataraman, J. Chem. Soc., 1933, 1073.
41. R. Robinson and K. Venkataraman, J. Chem. Soc., 1926, 2347.
42. J. Allen and R. Robinson, J. Chem. Soc., 1924, 2192.
43. T. Emillewicz and V. Kostanecki, Ber., 1899, 32, 311.
44. F. Herstein and V. Kostanecki, Ber., 1899, 32, 318.
45. Kostanecki and Stoppane, Ber., 1904, 37, 1181.
46. A. Edelstein and K. Kostanecki, Ber., 1905, 38, 1509.
47. Kostanecki and Tamber, Ber., 1899, 32, 2260.
48. K.V. Rao and T.R. Seshadri, Proc. Indian Acad. Sci., 1946, 23A, 213.

49. L. Farkas, Chem & Ind., 1957, 1212; Ber., 1957, 90, 2940.
50. H.S. Mahal, H.S. Rai and K. Venkataraman, J. Chem. Soc., 1934, 1121.
51. G. Bergellini and M. Finkelstein, Gazzetta, 1912, 42(ii), 417.
52. W. Baker and R. Robinson, J. Chem. Soc., 1932, 1798.
53. S.K. Grover, A.C. Jain & T.R. Seshadri, Indian J. Chem., 1963, 1, 517.
54. R.J. Bass, J. Chem. Soc. Chem. Comm., 1976, 78.
55. W. Baker, J. Chadderton, J.B. Harborne and W.D. Ollis, J. Chem. Soc., 1953, 1852.
56. W. Baker and R. Robinson, J. Chem. Soc., 1925, 1985.
57. B.M. Desai, P.R. Desai and R.D. Desai, J. Indian Chem. Soc., 1960, 37, 53.
58. P.K. Grover, G.D. Shah and R.C. Shah, J. Chem. Soc. (C), 1955, 3982.

58a. A.K. Tehim, Ph.D. Thesis, Delhi University 1984, P. 119.

59. V.V. Kane, A.B. Kulkarni and R.C. Shah, J. Sci. Industrial Res., 1959, 18B, 28 & 75.
60. Paal, Ber., 1885, 18, 58.
61. 2,3-Diacetosuccinate can be crystallised from dilute acetic acid. M.p. 88°.
62. G.N. Jean and F.F. Nord, J. Org. Chem., 1955, 20, 1363.
63. Tschitschibabin, Chem. Zentr., 1915, 1, 1065,
64. S. Parakewas, Synthesis, 1974, 819.
65. Dictionary of organic compounds by Chapman & Hall, 5th Edn., P-02822.
66. V. Pachmann, Ann., 1891, 264, 262, 272.
67. A.K. Bahl & W. Kemp, J. Chem. Soc.(c), 1971, 2268.
68. H. Horino and N. Inoue, Tetrahedron Lett., 1979, 2403.
69. Gattermann-Wicland, Praxis des Organischen Chemikers (De Gruyter, Berlin, 40th ed., 1961) P. 316.
70. Z.H. Skraup, Montash, 1880, 1, 316; 1882, 3, 536.
71. F. Misani and M.T. Bogert, J. Org. Chem., 1945, 10, 347.
72. A. Blackhall and R.H. Thomson, J. Chem. Soc., 1954, 3917.
73. C.E. Kaslow and V.V. Young, J. Am. Chem. Soc., 1950, 72, 5325.
74. D. Todd and S. Teich, J. Am. Chem. Soc., 1953, 75, 1895.
75. J. Reisch, R. Pagnucco and N. Jantos, Phytochemistry, 1968, 7, 997.
76. P. Nickel and E. Find, Ann., 1976, 367.
77. F. Lions, W.H. Perkin, Jr. and R. Robinson, J. Chem. Soc., 1925, 1158.
78. G. Saint-Ruf, Ashish De and J.C. Perche, Bull. Soc. Chim. Fr., 1973, 2514.
79. L. Knorr, Just. Liebigs Ann. Chem., 1885, 236, 69.
80. R.N. Du Puis and H.G. Hindwall, J. Am. Chem. Soc., 1934, 56, 471.
81. A. Reissert, Ber., 1905, 38, 1610.

82. Dictionary of organic compounds by Chapman & Hall, 5th Edition, Q-00061.
83. T.R. Govindachari and M.V. Lakshmikantham, Proc. Indian Acad. Sci., 1957, 46A, 406.
84. O. Hinsberg et al., Ber., 1894, 27, 2181.
85. Fischer, Ber., 1906, 39, 2930.
86. Ajello and Cusmano, Gazz. Chim. Ital., 1939, 69, 210.
87. Meerwein and Vossen, J. Prakt. Chem., 1934, 141, 149.
88. L. Knorr, Ber., 1883, 2597.
89. Zoss and Hennion, J. Am. Chem. Soc., 1941, 63, 1151.
90. P.B. Tripathy and M.K. Rout, J. Sci. Industrial Res., 1961, 20B, 177.
91. Yu. Kvitko and B.A. Porai-Koshits, Zh. Organ. Khim., 1966, 2(1), 169; K.A. Pawar and A.A. Patel, Indian J. Chem., 1994, 33B, 156.
92. Muller et al., Montash, 1958, 89, 23.
93. Wundt, Ber., 1878, 11, 826.
94. M.A. Phillips, J. Chem. Soc., 1928, 2392.
95. Galat and Friedman, J. Am. Chem. Soc., 1948, 70, 1280.
96. Dictionary of organic compounds by Chapman and Hall, 5th Edition, A-00121.
97. H.M. Kissman, D.W. Farnsworth and B. Witkop, J. Am., Chem., Soc., 1952, 74, 3948.
98. Ladenburg, Ber., 1876, 9, 219.
99. Hantzsch, Ann., 1889, 250, 265.
100. H. Erlenmeyer, L. Herzfeld and B. Prijs et al., Helv. Chim. Acta, 1955, 1293.
101. J.M. Sprague, A.H. Land and C. Ziegler, J. Am. Chem. Soc., 1946, 68, 2155; Rossi, Gazz-Chim. Ital., 1955, 85, 898.
102. G.N. Mahapatra, J. Indian Chem. Soc., 1956, 33, 527.
103. M.A. Collect. Bull. Soc. Chem. Fr., 1899, 21, 68.
104. L. Yanovskaya et al, Zh. Obsheh. Khim., 1952, 22, 1598.
105. G.A. Olah, Synthesis, 1974, 12, 895.
106. M.I. Shevchuk et al., Zh. Obsheh Khim., 1963, 33, 1135.
107. M.L. Malik and S.K. Grover, Indian J. Chem., 1976, 14, 513.
108. J.W. Baker, J. Chem. Soc., 1931, 2416.
109. R.R. Gupta, K.G. Ojha and M. Kumar, J. Heterocyclic Chem., 1980, 17, 1325.
110. Julian and Sturgis, J. Am. Chem. Soc., 1935, 57, 1126.
111. G. Vernin and J. Metzer, Bull. Soc. Chim. France, 1963, 2498.

112. D.H. Rammler, Ann. N.Y. Acad. Sci., 1975, 243, 46.
113. Biltz and Wittek, Ber., 1921, 54, 1035; Michael, J. Prakt. Chem., 1887, 35, 456.
114. Ceresole, Ber., 1883, 16, 1134, Baeyer, Ann., 1864, 130, 140.
115. H.C. Koppel, et al., J. Am. Chem. Soc., 1958, 80, 2751.
116. S.C. Nigam, G.S. Saharia and H.R. Sharma, J. Inst. Chem., 1980, 52, 75.
117. G.M. Dyson and H.J. George, J. Chem., Soc., 1924, 125, 1702.
118. I.N.S. Das and S. Dutt, Proc. Indian Acad. Sci., 1938, 8A, 145.
119. F.B. Dains, R.Q. Breuster and C.P. Olander, Chem. Abstr., 1923, 17, 543, A.W.F. Hotmann, Ber., 1882, 15, 985.
120. D.J. Beaver, D.P. Roman and P.J. Stottel, J. Am. Chem. Soc, 1957, 79, 1236.
121. L. Ram, Agric. Biol. Chem., 1982, 46(2), 557.
122. S.C. Nigam, G.S. Saharia and H.R. Sharma, Def. Sci. J., 1981, 31(1), 15.
123. V.K. Ahluwalia, H.R. Sharma and Renu Tyagi, Tetrahedron, 1986, 42, 4045.
124. V.K. Ahluwlia, Rashmi Sharma and Mukesh K. Sharma, Indian J. Chem., 1992, 31B, 688.

6

Miscellaneous Compounds

6.A.1 ALUMINIUM ISOPROPOXIDE

$$3(CH_3)_2CHOH + Al \xrightarrow{HgCl_2} Al\left[OCH(CH_3)_2\right]_3 + 3H$$

A mixture of aluminium wire or aluminium turning (13.5g, 0.5 mole), anhydrous isopropyl alcohol (250 ml) and mercuric chloride (1g) is heated to gentle reflux on a water bath for 10-20 min (under anhydrous conditions) until a vigorous reaction sets in. The reaction mixture is cooled if it becomes too vigorous. The mixture is refluxed for 5-6 hr., distilled in vacuo using an oil bath. The aluminium isopropoxide distills as colourless viscous liquid at 140-150° (1.2 mm) (lit.[1] b.p. 145.50°/5 mm). It solidifies to a colourless solid, yield 95 g (93%). M.P. 118°.

6.A.2 ALUMINIUM TERT-BUTOXIDE

$$(CH_3)_3COH + Al \xrightarrow[HgCl_2]{\text{aluminium isopropoxide}} Al\left[OCH(CH_3)_3\right]_3$$

Dry tert-butyl alcohol (50 g; 63.6 ml, 0.675 mole) and aluminium isopropoxide(2-3 g) are added to aluminium shavings (16 g, 0.59 gram atoms) under anhydrous conditions in a R.B. flask (500 ml capacity) protected by a calcium chloride tube. The mixture is heated to boiling (steam bath) and mercuric chloride (0.1 g) added followed by vigorous shaking. As the reaction proceeds, the colour of the reaction mixture changes from milky to black. The reaction is allowed to proceed for 30 min without heating. Additional dry tert-butyl alcohol (78 ml, 3.3 mole) and dry benzene (50 ml) are added. The reaction again sets in on gentle heating and continues without further heating. The reaction is allowed to proceed for 2 hr and then refluxed for 9-10 hr. The benzene and unreacted tert-butyl alcohol are removed by distillation (steam-bath); the final traces are removed in vacuo (30 mm). Dry ether (250 ml) is added and the solid aluminium tert-butoxide is dissolved by refluxing (10-15 min). After cooling undried ether (10 ml) is added followed by vigorous shaking. The mixture is allowed to stand for 2 hr and the unused aluminium, aluminium hydroxide and mercury are removed by centrifuging for 20-25 min. The solvent is then removed (steam bath); the final traces are removed in vacuo (15-30 mm). The flask is cooled and the aluminium tert-butoxide crushed with a spatula and transferred to bottles sealed against moisture. Yield 98 g (67.2%)

Note: The method described is a modification of the literature[2] method.

6.A.3 2-AMINOBENZENETHIOLS

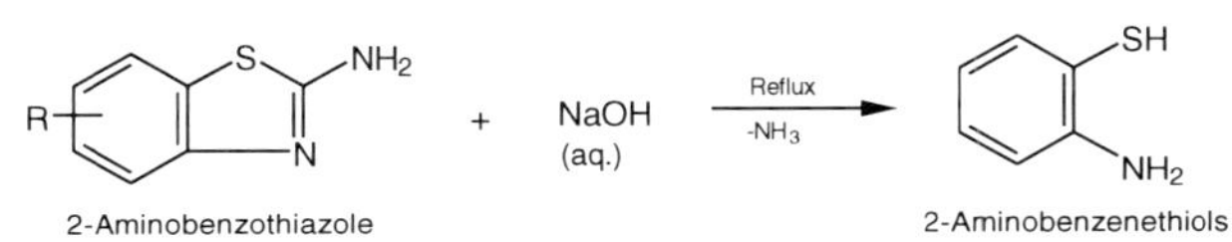

Appropriately substituted 2-aminobenzothiazole (5.D. 90) and sodium hydroxide (5 times by weight of 2-aminobenzothiazole) and water (10 times by weight of 2-aminobenzothiazole) is refluxed till the evolution of ammonia gas ceases (3-4 hr). The mixture is filtered, diluted with cold water (150 ml) and neutralised with acetic acid (5N) with vigorous stirring. During acidification the temperature of the reaction mixture is kept below 5°. The yellowish precipitate of 2- aminobenzenethiol is extracted with ether (3 × 25 ml). The solvent is dried (anhydrous sodium sulphate) and distilled to give the desired product. Following 2-aminobenzenthiols are prepared.

Substituent at Position

3	5	M.P. (°C)	Lit m.p. (°C) (Ref)	% yield
CH_3	H	48	48(3)	34
OCH_3	H	110-12 (Hydrochloride)	110-12(3) (Hydrochloride)	38
OC_2H_5	H	134 (Hydrochloride)	134(3) (Hydrochloride)	15
Cl	H	78	78(3)	30
H	CH_3	90	90(3)	40
H	OCH_3	105	105(3)	36
H	Cl	110	110(3)	28
H	Br	113	113(3)	31

6.A.4 P-ANISYL ALCOHOL

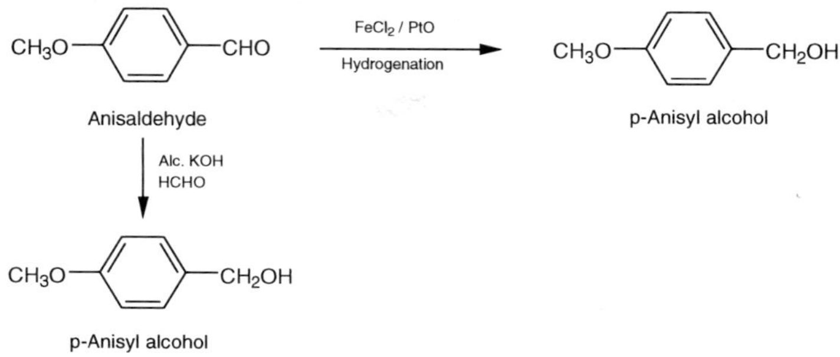

Method (i)

Anisaldehyde (13.6 g, 12.15 ml, 0.1 mole) is dissolved in ethyl alcohol (95%, 50 ml). To this solution are added ferrous chloride solution (0.2M, 0.3 ml), and platinum oxide catalyst (0.17 g). The mixture is taken in a bottle, which is fixed to the shaker connected with hydrogen tank. Hydrogen is admitted at three atmospheres (total pressure) and the reduction is completed in 22 min. Sodium hydroxide solution (N, 0.4 ml) is added to the reaction mixture and filtered. The filtrate is distilled in vacuo to yield anisyl alcohol. B.p. 127-30°/8 mm (lit.[4] bp 127-30°/8 mm), Yield 25.2 g (91.3%).

Method (ii)

To a stirred solution of potassium hydroxide (4 g) in methanol (10 ml) kept at 60-65° is added a mixture of anisaldehyde (16.6 g, 14.8 ml, 0.12 mole) and formaldehyde (40%, 1.5 ml) and methanol (30 ml) at such a rate that the reaction temperature remains at 60-70°. The reaction mixture is kept for 3 hr at 60-70°, solvent distilled off and the water (25 ml) added. The mixture is extracted with ether (2 × 50 ml). The ether extract is washed with water, dried (anhydrous sodium sulphate) and distilled. Yield 10 g (59.4%). B.p. 127-30°/8 mm.

6.A.5 BENZAL CHLORIDE[5] ($C_6H_5CHCL_2$) (BENZYLIDENE CHLORIDE)

A mixture of thionyl chloride (35 ml) and hexamethyl phosphoramide (HMPA) (1 ml, catalytic amount) is cooled to –5° to –10° and stirred slowly. Benzaldehyde (6.89 g, 6.6 ml, 0.065 mole) is added dropwise to the above stirred solution during 30 min keeping the temperature below –5° during addition. The temperature of the reaction mixture is allowed to rise to room temperature and the mixture stirred for 4 hr. It is poured over crushed ice and extracted with ether (3 × 25 ml). The ethereal extract is dried (anhydrous magnesium sulphate) and the solvent removed (rotary evaporator). The residual product is distilled under reduced pressure. Yield 9.4 g (90%). B.p. 104-105°/30 mm; 206°/760 mm.

Note: This is an excellent method[5] for the preparation of geminal dichlorides. Following carbonyl compounds are converted into geminal dihalides (given name of carbonyl compound, reaction time and % yield).

Cinnamaldehyde, 4hr, 90%
1-Napthaldehyde, 6hr, 87%
9-Anthraldehyde, 6hr, 88%
4-Nitrobenzaldehyde, 8 hr, 26%
4-Anisaldehyde, 8hr, 22%
Benzophenone, 8hr, 18%
9-Fluorenone, 12hr, 32%.

6.A.6 BENZOYL PEROXIDE

To stirred hydrogen peroxide (40 volumes, 50 ml, 1.63 mole) are added gradually in alteration sodium hydroxide solution (3N, 30 ml) and redistilled benzoyl chloride (30 g, 0.21 mole). The temperature during the addition is maintained at 5-8°. Care is taken to see that the solution remained slightly alkaline throughout the addition, which takes about 30 min. The reaction mixture is stirred for 30 min more and kept aside. When the smell of benzoylchloride disappears, the formed product is filtered and washed with cold water. It is dried between the folds of filter paper and crystallised from chloroform—methanol mixture (1:2) as colourless needles. Yield 12 g (24.9%). M.p. 106° (lit.[6] m.p. 106-8°).

6.A.7 1-BROMO-3-METHYLBUT-2-ENE (PRENYL BROMIDE)

Hydrogen bromide required for the preparation of prenyl bromide is obtained as follows: Phenol (50 g, 46.7 ml, 0.53 mole) in carbon

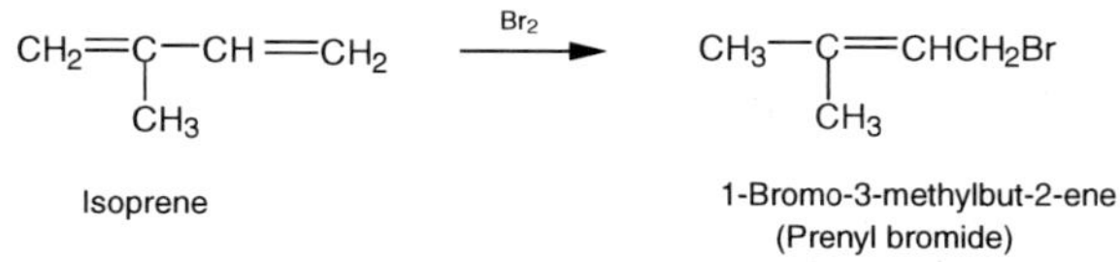

tetrachloride (50 ml) is stirred in flask fitted with a dropping funnel and a delivery tube connected to a gas trap containing a small amount of phenol in carbon tetrachloride to remove excess of bromine from hydrogen bromide vapours. Bromine (approx. 50 ml) is added dropwise at a regular rate and hydrogen bromide evolved is collected into acetic acid (100 ml) till the increase in weight of acetic acid is about 30%.

Isoprene (50 g, 73.4 ml, 0.735 mole) is cooled in an ice-chest. Hydrogen bromide solution (61 g HBr, 48.3 ml, 30% in acetic acid solution, 0.754 mole) is added to the solution with continuous shaking. The mixture is left in a refrigerator for 48 hr and then poured into ice cold water (approx. 1l). A dark coloured oil separates out. It is separated (separatory funnel), dried (calcium chloride) and distilled under reduced pressure. The fraction distilling at 26-33°/12 mm is collected (lit.[7] b.p. 26-33°/12 mm), Yield 81 g (74%).

6.A.7A 3-BROMO-3-METHYL-1-BUTYNE

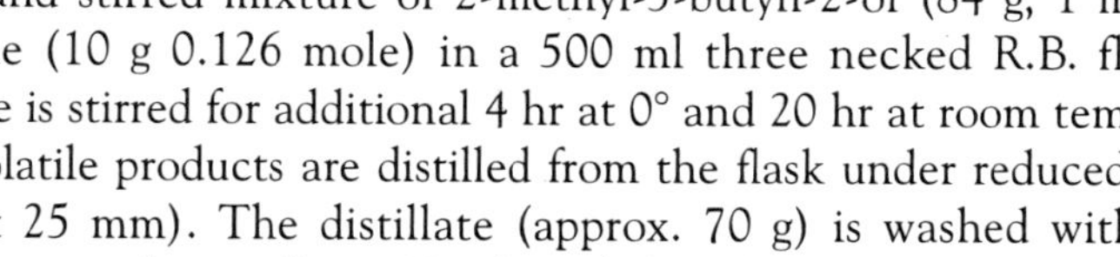

Phosphorus tribromide (102.4 g, 0.378 mole) is added to a cooled (ice-bath) and stirred mixture of 2-methyl-3-butyn-2-ol (84 g, 1 mole) and pyridine (10 g 0.126 mole) in a 500 ml three necked R.B. flask. The mixture is stirred for additional 4 hr at 0° and 20 hr at room temperature. The volatile products are distilled from the flask under reduced pressure (75° at 25 mm). The distillate (approx. 70 g) is washed with sodium bicarbonate solution (2 × 100 ml) and then dried (anhydrous magnesium sulphate). The product obtained is distilled under reduced pressure through an 18" column packed with glass helices to yield 5 g (8%) of 2-methyl-1-buten-3-yne, B.p. 33° and 37 g (25%) of 3-bromo-3-methyl-1-butyne. B.p. 42.4° (106 mm) (lit.[8] b.p. 42.2°/106 mm).

6.A.8 2-BROMOCYCLOHEXANONE

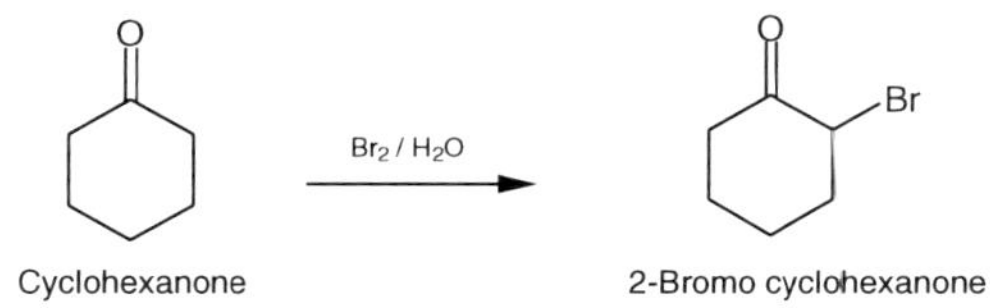

Cyclohexanone (5 g, 5.3 ml, 0.05 mole) and water (30 ml) are placed in a three necked flask equipped with a stirrer and a dropping funnel. Bromine (8 g, 2.58 ml, 0.05 mole) is added dropwise during 1 hr at 5° to the mixture while stirring. After the addition is complete, the stirring is continued at room temperature until the reaction mixture becomes colourless (about 1 hr); the reaction mixture is allowed to warm during this period. The upper layer of bromocyclohexanone is separated by ether extraction (3 × 20 ml). The ether extract is washed with water, saturated sodium chloride solution and dried (anhydrous sodium sulphate). It is distilled. 2-Bromocyclohexanone is collected by distillation in vacuo (83°/ 6 mm) (lit.[9], b.p. 83°/6 mm). Yield 6 g (60.7%).

6.A.9 N-BROMOSUCCINIMIDE

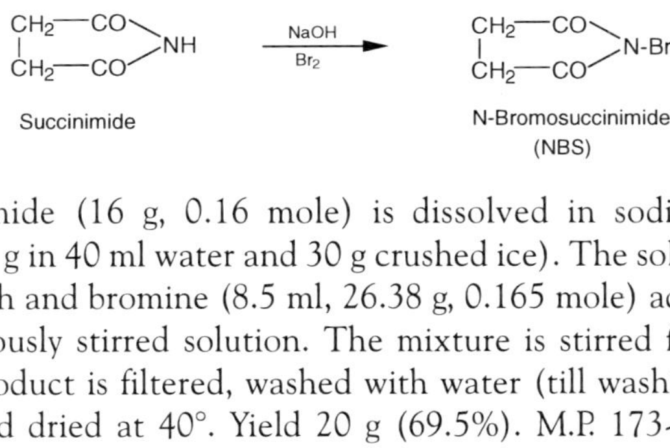

Succinimide (16 g, 0.16 mole) is dissolved in sodium hydroxide solution (6.4 g in 40 ml water and 30 g crushed ice). The solution is cooled in an ice-bath and bromine (8.5 ml, 26.38 g, 0.165 mole) added in one lot to the vigorously stirred solution. The mixture is stirred for 5 min. The separated product is filtered, washed with water (till washings are free of bromine) and dried at 40°. Yield 20 g (69.5%). M.P. 173-75° (decomp.) (lit.[10] M.p. 173-75°).

Caution: It is highly irritating to eyes, skin, mucous membrane.

6.A.9A n-BUTYL LITHIUM

Under an atomosphere of argon, a solution of n-butyl bromide (68.5 g, 0.5 mol) in anhydrous ether (250 ml) is added with stirring to small chips of lithium containing 1-2% of sodium (7.75 g, 1.1 g-atom) in ether (75 ml). The reaction starts after the addition of about 40 ml of butyl bromide solution at room temperature. The temperature rises and the lithium metal becomes bright. If the reaction does not start, the addition of a small amount of 1,2-dibromoethane (0.5 ml) is often effective. Then the reaction mixture is cooled (–5°C to –10°C) and addition of the butyl bromide solution is continued slowly (about 4 hr). At the end of the addition, the solution is stirred for 2 hr at –5°C to –10°C ; then the reaction mixture is allowed to warm to room temperature. After 2 hr, excess lithium metal is removed. For many purposes, the use of a clear solution, obtained after the reaction mixture has stood overnight at 0° to –5°C is preferable. n-Butly lithium in ether can be stored under an argon atmosphere without decomposition for 15 days at 0°C or 2 months at –15°C.

6.A.10 2-CARBETHOXYCYCLOPENTANONE

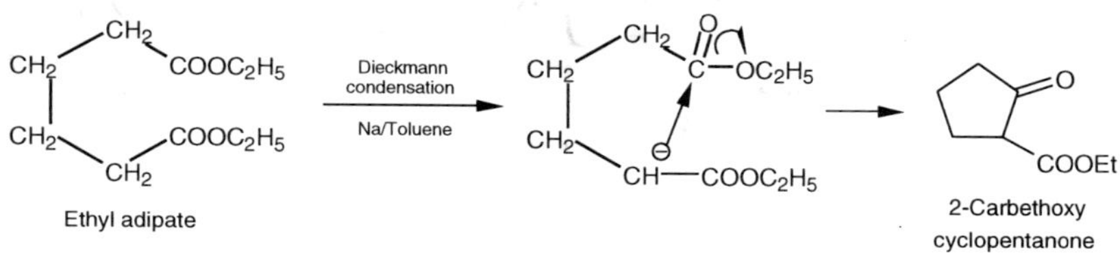

In a 1 l, three necked R.B. flask fitted with a mercury seal stirrer, a dropping funnel and a reflux condenser having a calcium chloride tube is placed pulverised sodium (12.5 g, 0.54 gram atom) and dry toluene (100 ml). Ethyl adipate (101 g, 103 ml, 0.58 mole) is added slowly to the stirred solution (during 2 hr). The reaction flask is maintained at 100-110° (oil bath) during addition and for about 4 hr more. The reaction mixture becomes more viscous. Dry toluene (500 ml) is added to keep the reaction mixture fluid enough for stirring. The resulting mixture is cooled to 0° (ice salt bath) and poured into dilute acetic acid (10%, 500 ml). The toluene layer is separated, washed with water, sodium carbonate solution (5%) and again with water. Toluene is removed in vacuo and the residue is distilled to give 2-carbethoxycyclopentanone, B.p. 108-111°/15 mm. (lit.[11] b.p. 102°/ 11 mm). Yield 60 g (66.3%).

6.A.11 1-CARBETHOXYETHYLIDENETRIPHENYL PHOSPHORANE

Triphenyl phosphine (15 g, 0.057 mole) in dry benzene (30 ml) is added to a solution of ethyl 2-bromopropionate (7.5 ml, 10.45 g, 0.0577 mole) in dry benzene (20 ml) at room temperature. A salt is precipitated with

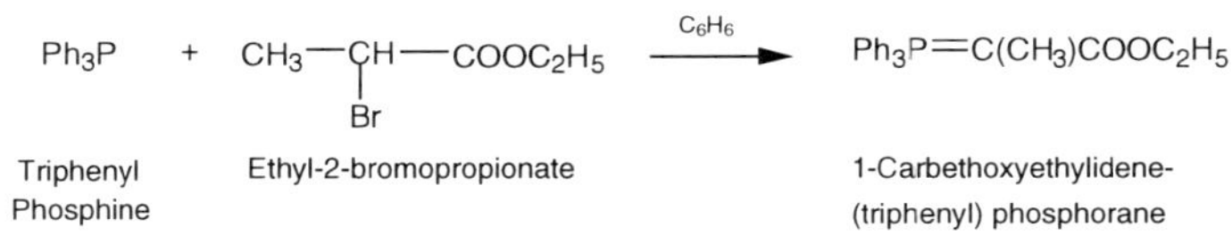

simultaneous rise in temperature. The mixture is vigorously shaken and allowed to cool to room temperature. It is left for 12 hr at room temperature. The separated product is filtered, washed with dry benzene and dried. A stirred solution of the above salt in water (150 ml) and benzene (100 ml) is neutralised with aqueous sodium hydroxide (10%) to phenolphthalein end point. The benzene layer is separated, dried (anhydrous sodium sulphate) and concentrated to one-third to its volume. Petroleum ether is added to the above concentrate. The separated crystalline phosphorane is filtered and dried. Yield 14 g (67.5%). M.p. 152-53° (lit.[12] m.p. 152-53°).

6.A.12 CHLOROACETONITRILE

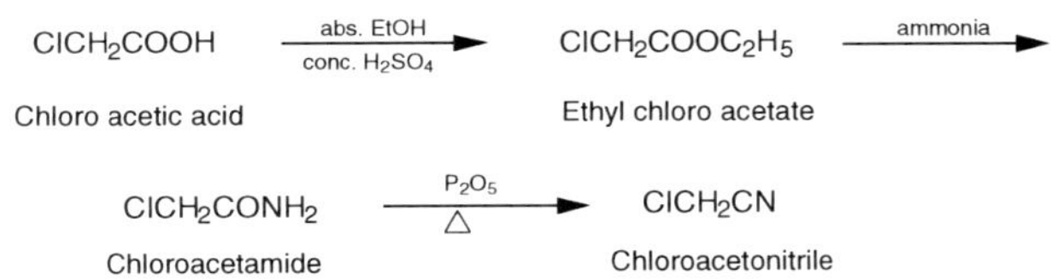

Step (i) Ethyl Chloroacetate

A mixture of pure dry monochloroacetic acid (47.5 g, 0.5 mole) absolute ethyl alcohol (50 ml) and concentrated sulphuric acid (3 ml) is refluxed for 36 hr under anhydrous conditions. The mixture is poured into excess water (300 ml) and the ester layer separated. It is washed with dilute sodium hydroxide solution (5%), water (3 × 30 ml) and dried (anhydrous magnesium sulphate). Distillation of the product gives ethyl chloroacetate. Yield 31 g (50%). B.p. 143°. (lit.[13] b.p. 142°/751 mm).

Step (ii) Chloroacetamide

Chilled ammonia (100 ml, sp. gr. 0.9) is added slowly to a vigorously stirred ethyl chloroacetate (108 g, 0.88 mole) (cooled in ice-salt bath). The solution is stirred for 20 min in cold. More ammonia (100 ml) is again added and mixture stirred for 15 min more. The mixture is allowed to stand for 30 min, the separated chloroacetamide is filtered and washed with cold water (25 ml) (to remove ammonium chloride) and air dried. Yield 64 g

(78%). M.p. 118-19°. The product contains traces of ammonium chloride. It is further purified by crystallisation from water. M.p. of the pure compound 119-20°. (lit.[13a] m.p. 121°).

Step (iii) Chloroacetonitrile

An intimate mixture of chloroacetamide (56.1 g, 0.6 mole) and phosphorous pentoxide (93 g, 0.655 mole) is heated for 1.5 hr at 160° (oil-bath) in a RB flask fitted with a reflux condenser. Chloroacetonitrile is obtained by distillation of the reaction mixture. Yield 28 g (62%). B.p. 124-28°. It is purified by distilling over phosphorous pentoxide (5 g). Pure chloroacetonitrile distils at 123-24°. (lit.[14,15] b.p. 123-24°).

Alternatively, the mixture of chloroacetamide and phosphorous pentoxide is heated in a suitable solvent (trimethyl benzene, boiling range 166-174°) in an oil bath at reflux temperature (180°). The mixture is cooled and then distilled. Chloroacetonitrile is obtained as above in 82% yield.

6.A.13 3-CHLORO-3-METHYLBUT-1-YNE

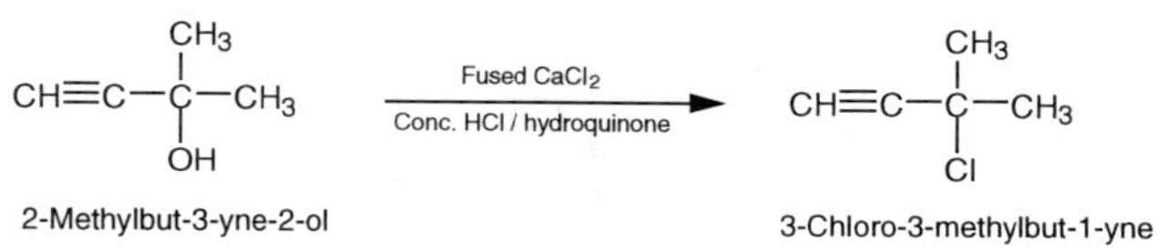

To 2-methylbut-3-yne-2-ol (24.5 ml, 21.27 g, 0.253 mole) cooled at 15° is added anhydrous calcium chloride (30 g), chemically pure hydrochloric acid (105 ml) and hydroquinone (0.5 g). The whole mixture is shaken vigorously for 1 hr. A green coloured layer separates out. It is separated (separatory funnel), washed with saturated calcium chloride solution (3 × 20 ml) and dried over anhydrous potassium carbonate. It is distilled and the fraction boiling at 75-6° is collected (lit.[16] b.p. 72.6-75.8°). Yield 13 g (50.2%).

6.A.14 M-CHLOROPERBENZOIC ACID

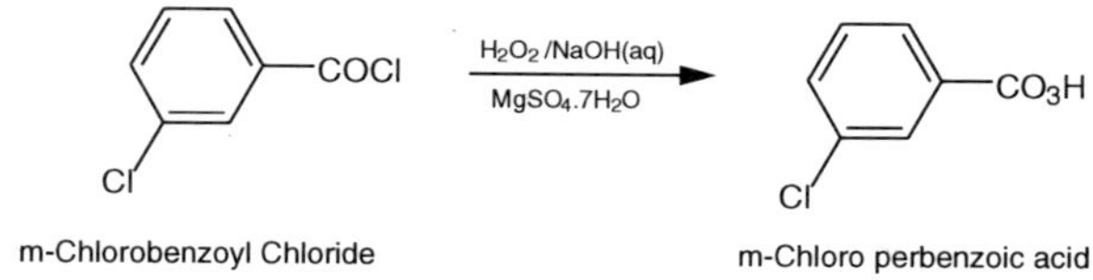

m-Chlorobenzoyl chloride (26.5 g, 0.150 mole) is added in one portion with vigorous stirring to a cooled solution (10-15°) of magnesium sulphate heptahydrate (18 g), a solution of sodium hydroxide (18 g) in water (180 ml), hydrogen peroxide (30%, 45 ml, 13.5 g, 0.4 mole) and dioxane (225 ml). The temperature of the mixture is not allowed to rise above 25°, if necessary some ice pieces are added to maintain the temperature. The reaction mixture is stirred for 25 min at this temperature. Cold dilute sulphuric acid (20%, 450 ml) is added to the reaction mixture and the organic phase separated in a separatory funnel. The aqueous layer is extracted with dichloromethane (4 × 75 ml) and the combined extracts are dried (anhydrous magnesium sulphate) and distilled in vacuo at 25-35°. Finally a white pasty solid is obtained. The residual solvent is removed under full vacuum for 2 hr until a white powder remains. It is stored in dark coloured tight bottles. Yield 22.5 g (87.2%). Sodium thiosulphate analysis indicates 80-85% active oxygen present.

Note: 1. m-Chlorobenzoyl chloride is prepared by refluxing m-chlorobenzoic acid with excess thionyl chloride. The mixture on distillation gives required product, b.p. 112°/11 mm.

2. The procedure followed is a modification of the reported[17] method.

6.A.15 P-CHLOROPHENYLMERCURYL CHLORIDE[18]

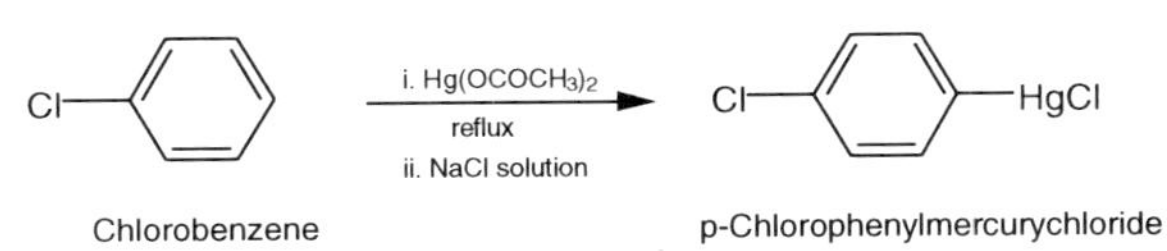

To boiling chlorobenzene (28 g, 25.3 ml, 0.25 mole) is added mercuric acetate (8 g, 0.025 mole) and the solution is refluxed for 4-5 hr. The hot clear solution is poured into boiling solution of sodium chloride (20%, 100 ml). The formed product is filtered, washed with water and dried. Yield 6 g (69%). M.p. 224° (lit.[18] m.p. 225°).

6.A.16 CINNAMYL BROMIDE

Method (i)[19,20]

Triphenylphosphine (53.2 g, 0.203 mole) and dry acetonitrile (180 ml) are taken in a 500 ml three necked R.B. flask. The flask is cooled (ice-water bath) and bromine (32g, 10.3 ml, 0.20 mole) is added dropwise under

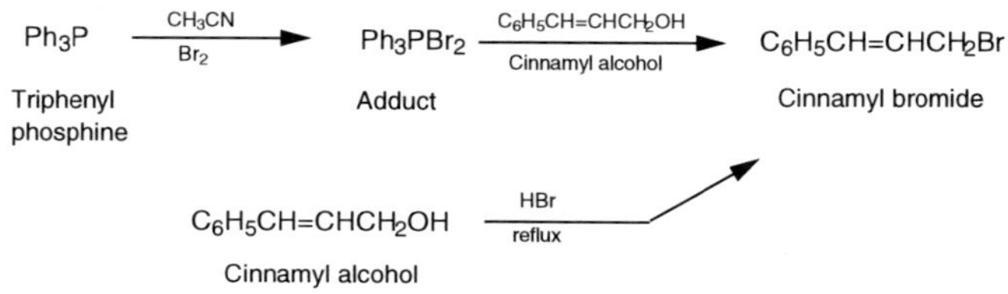

anhydrous conditions during 20-25 min. The solid triphenylphosphine disappears and a white adduct, Ph_3PBr_2 separates out. To the above stirred reaction mixture a solution of cinnamyl alcohol (27g, 25.9 ml, 0.20 mole) in dry acetonitrile (25 ml) is added during 5-10 min. All the precipitate must dissolve (if necessary the reaction mixture is warmed). The solvent is removed by distillation in vacuo. Cinnamyl bromide is finally obtained by distilling at 91-80° (2-4 mm). Yield 30 g. The product is dissolved in ether (100 ml), solution washed with aqueous sodium carbonate, dried (anhydrous magnesium sulphate) and distilled. The pure product is finally obtained by distilling at 66-8°/0.07 mm. Yield 23.8 g (60%). (lit.[21,22] b.p. 106-08°/4 mm).

Method (ii)

A solution of cinnamyl alcohol (27 g, 0.20 mole) is refluxed with hydrobromic acid (48%, 33 ml) for 1 hr. the reaction mixture is cooled and organic layer separated. The aqueous layer is extracted with ether and the combined organic layer and ether distilled. The residual oily product is distilled to give cinnamyl bromide. Yield 23.5 g (60%). B.p. 108-10° (4 mm) (lit.[21,22] b.p. 106-08°/4 mm).

6.A.17 DIAZOMETHANE

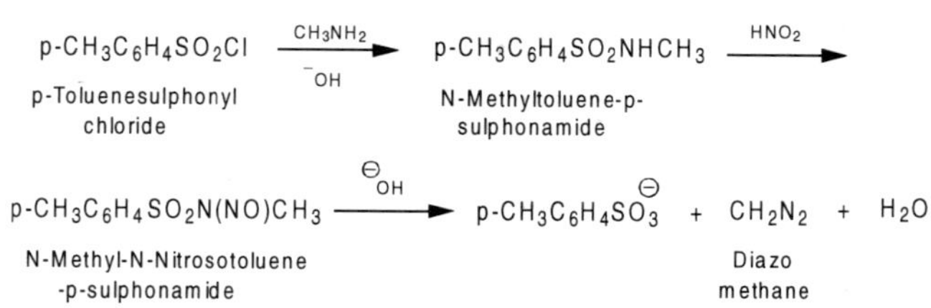

Step (i) p-Tolylsulfonylmethylnitrosamide

p-Toluenesulphonyl chloride (9.5 g) is added in small amount at a time to stirred methylamine solution (8.7 ml of 40% aqueous solution) in a R.B. flask. The mixture becomes warm; the temperature is not allowed to rise above 90°. To this stirred mixture is added sodium hydroxide solution (50%, 2.5 ml) followed by addition of p-toluenesulphonyl chloride (4.5 g) in portions. To the above acidic mixture is added sodium hydroxide solution (50%, 1.5 ml) followed by addition of p-toluenesulphonyl chloride (2 g). The mixture is rendered alkaline by the addition of sodium hydroxide solution. It is heated in a water bath with vigorously stirring for 15 min. The hot reaction mixture is poured into acetic acid (75 ml). The solution is cooled in ice bath to 5° and to this cooled solution is added dropwise with stirring sodium nitrite solution (6.3 g in 12.5 ml water). The temperature during this addition is maintained below 10°. The nitroso derivative separates out as yellow crystalline solid. Cold water (100 ml) is added to mixture, the product filtered and washed with water. Yield 16 g. M.p. 58-60° (lit.[23] m.p. 58-60°). It is used as such for the preparation of diazomethane.

Step (ii) Diazomethane

A solution of the above nitroso derivative (2.14 g) in ether (30 ml) is added dropwise (separatory funnel) to a solution of potassium hydroxide (0.5 g in 10 ml ethanol) in a distillation flask. The formed diazomethane is collected in ether in a flask which is cooled externally by ice-salt mixture (Precaution: Only rubber corks should be used for the assembly of the apparatus. Diazomethane explodes by use of ground glass apparatus or by coming in contact with rough surface).

6.A.18 2,3-DICHLORO-5,6-DICYANOBENZOQUINONE (DDQ)

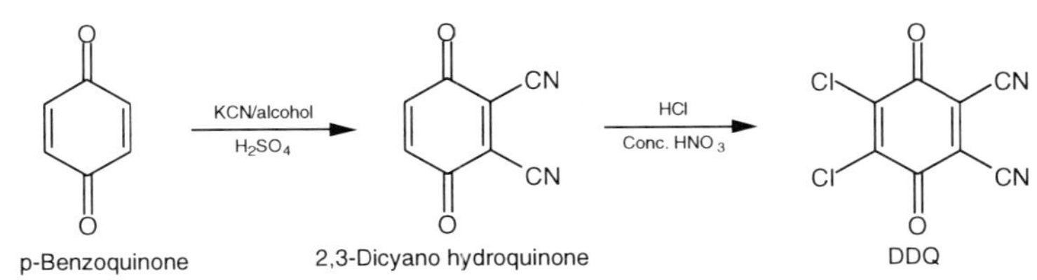

Step (i) 2,3-Dicyanohydroquinone

To a stirred solution of p-benzoquinone (1.20.) (10 g, 0.093 mole) in ethanol (300 ml) is added concentrated sulphuric acid (13 ml) in ethanol (50 ml). The temperature is adjusted to 24-28° and maintained at that range while potassium cyanide solution (20 g, 0.305 mole in 50 ml water) is added during 30-40 min (use a hood or fume cupboard). The reaction mixture becomes dark brown with green fluorescence and alkaline to litmus. It is just acidified. The precipitated potassium sulphate is filtered and the cake washed with hot alcohol (40 ml). The combined filtrate and washings are evaporated under reduced pressure to a small bulk (75 ml) without prolonged heating (use a rotary evaporator). The mixture is cooled and the separated product filtered and washed with ice-cold water. It is crystallised from hot water to give 2,3-dicyanohydroquinone. Yield 4.7 g (63%). M.p. 226-28° (lit.[24], m.p. 230°).

Step (ii) 2,3-Dichloro-5,6-dicyanobenzoquinone (DDQ)

A slurry of 2,3-dicyanobenzoquinone (2.5 g, 0.015 mole) in water (20 ml) and concentrated hydrochloric acid (17 ml) is treated with concentrated nitric acid (4.7 g of 70%) at 35 ± 3°. The mixture is stirred for 45 min at this temperature. The reaction mixture foams initially and turns yellow after nitric acid has been added. The yellow suspension is stirred for 1 hr. The separated product is filtered and washed with carbon tetrachloride and dried. Yield 3.2 g, (89.1%). M.p. 212-14° (lit.[24a] m.p. 212-13°).

6.A.18A O,O-DIETHYLCHLOROTHIOPHOSPHONATE

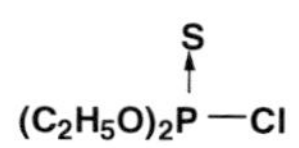

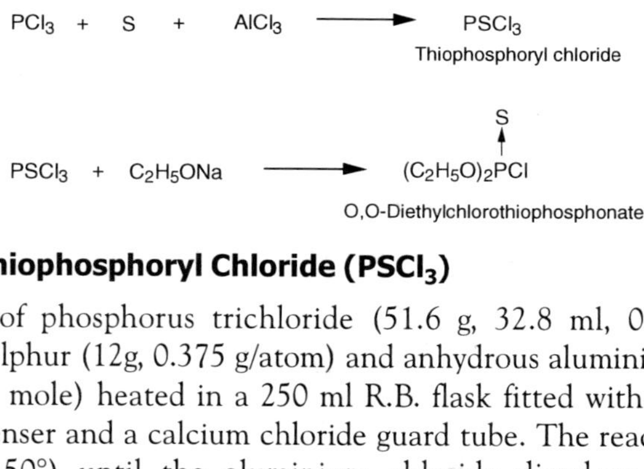

Step(i) Thiophosphoryl Chloride ($PSCl_3$)

A mixture of phosphorus trichloride (51.6 g, 32.8 ml, 0.375 mole), powdered sulphur (12g, 0.375 g/atom) and anhydrous aluminium chloride (1 g, 0.0075 mole) heated in a 250 ml R.B. flask fitted with an efficient reflux condenser and a calcium chloride guard tube. The reaction flask is heated (40-50°) until the aluminium chloride dissolves and strong exothermic reaction commences (caution). The flask is heated until the reaction is complete. The resulting clear brownish solution is cooled and then distilled under anhydrous conditions. Yield 56.5g (89%). B.p. 122-23° (lit.[25a], b.p. 122-23°).

Note: The ratio of the reactants used is critical.

Step (ii) O,O-Diethylchlorothiophosphate $(C_2H_5O)_2P(\rightarrow S)-Cl$

Sodium ethoxide (from 23 g sodium) in absolute ethanol (600 ml) is added during 3-4 hr to a stirred solution of thiophosphoryl chloride (84.75 g, 0.5 mole) in dry benzene (250 ml); the reaction mixture is kept at 5-10°. It is allowed to stand for 5-6 hr at room temperature and then concentrated in vacuo to a thick slurry. Benzene (100 ml) and water (250 ml) is added to the mixture. The organic layer is separated (separatory funnel) and the aqueous layer is extracted with more benzene (2 × 75 ml). The combined benzene extract is washed with water, dried (anhydrous magnesium sulphate) and solvent removed by distillation. Final distillation gave 47 g (50%) of colourless O,O-diethylchlorothiophosphate, B.p. 71.5-72°/7 mm. (lit.[25b], b.p. 71.7-72°/7mm).

Following O,O-dialkyl chlorothiophosphates are similarly obtained. O,O-Dimethylchlorothiophosphate, B.p. 68°/12 mm, (lit.[25c] b.p. 68°/12 mm), O,O-Di-n-butylchlorothiophosphate, B.p. 95-8°/7 mm (lit.[25d] b.p. 95-8°/7 mm).

6.A.18B 2,4-DIMETHOXYBENZONITRILE

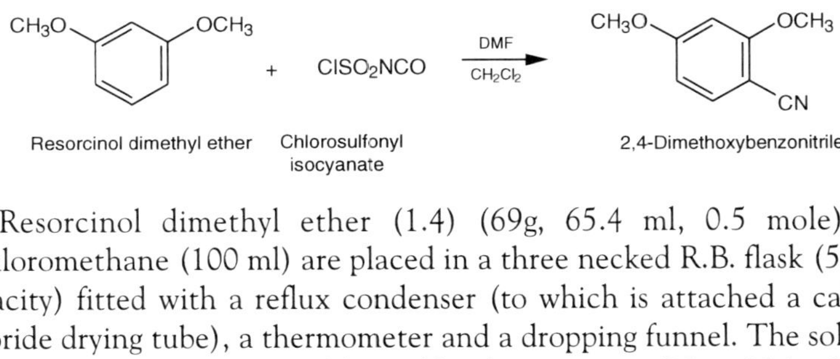

Resorcinol dimethyl ether   Chlorosulfonyl isocyanate   2,4-Dimethoxybenzonitrile

Resorcinol dimethyl ether (1.4) (69g, 65.4 ml, 0.5 mole) and dichloromethane (100 ml) are placed in a three necked R.B. flask (500 ml capacity) fitted with a reflux condenser (to which is attached a calcium chloride drying tube), a thermometer and a dropping funnel. The solution is stirred and a solution of chlorosulfonyl isocyanate (75 g, 46.1 ml, 0.53 mole) in dichloromethane (50 ml) is added at 15-20° during 20 min. The mixture is stirred for 1 hr at room temperature, cooled (10-15°) and N,N-dimethylformamide (77 g, 81.6 ml, 1.05 mole) is added during 5 min. The flask is removed from cooling bath, the temperature gradually rises to about 30° and then falls. After 1 hr the separated amide N-sulfonyl chloride dissolves and the clear solution is poured over crushed ice (about 100 g). The mixture is extracted with dichloromethane (2 × 100 ml). The extract is washed with water (50 ml) and distilled. The residual white solid is triturated with cold water (100 ml) and filtered. The formed 2,4-dimethoxybenzonitrile (78.5 g, 96%) melts at 91°.

Note: 1. Chlorosulfonyl isocyanate can either be procured from American Cyanamide Co or prepared as described in organic synthesis, Collective Volume VI, p 226-228. The preparation in the laboratory is extremely risky.

2. This laboratory preparation is a simple method for introducing the cyano substituents into compounds that can readily undergo electrophilic substitution.

3. The procedure described is a modification of the method of G. Lohaws, Chem. Ber., 1967, 100, 2719.

6.A.19 DIPHENYL CARBINOL

$$C_6H_5COC_6H_5 + Zn / alcohol \longrightarrow (C_6H_5)_2CHOH$$

Benzophenone → Diphenyl Carbinol

A mixture of benzophenone (5 g, 0.027 mole), alcohol (50 ml), sodium hydroxide (5 g) and zinc powder (5 g) are stirred in a R.B. flask. The temperature rises to 70°. After 3 hr the temperature starts falling. The reaction mixture is filtered and the residue washed with hot alcohol (2 × 2.5 ml). The combined filtrate is poured over crushed ice (200 g). Acidification of the clear alkaline solution with concentrated hydrochloric acid gives diphenyl carbinol. It is filtered, washed with water and crystallised from alcohol by cooling in freezing mixture of ice and salt. Pure diphenyl carbinol separates in the form of colourless crystals. Yield 3.6 g (71.3%). M.p. 65° (lit.[26] m.p. 64-6°).

6.A.20 ETHOXYCARBONYLMETHYLENETRIPHENYL-PHOSPHORANE

$$Ph_3P + BrCH_2COOEt \xrightarrow{\text{dry benzene}} Ph_3P{=}CHCOOC_2H_5$$

Triphenyl Phosphine + Ethyl bromo acetate → Ethoxycarbonyl methylene triphenyl phosphorane

A solution of triphenylphosphine (15.7 g, 0.06 mole) in dry benzene (30 ml) is added to a solution of ethylbromoacetate (7 ml, 10.54 g, 0.063 mole) in dry benzene (20 ml) at room temperature. A salt is precipitated and the temperature of the reaction mixture rises to 70°. The mixture is thoroughly shaken and is allowed to cool to room temperature.

It is left for 12 hr at room temperature. The separated solid is filtered and washed with dry benzene and dried.

A solution of the above salt in water (150 ml) and benzene (100 ml) is neturalised with aqueous sodium hydroxide (10%) to a phenolphthalein end point. The benzene layer is separated, dried (anhydrous sodium sulphate) and concentrated to about one third of its volume. Addition of petroleum ether to the above concentrate gives the crystalline phosphorane. It is filtered and dried. Yield 14.6 g (70%). M.p. 125-127.5°. (lit.[27] m.p. 125-127.5°).

6.A.21 ETHOXYACETYL CHLORIDE

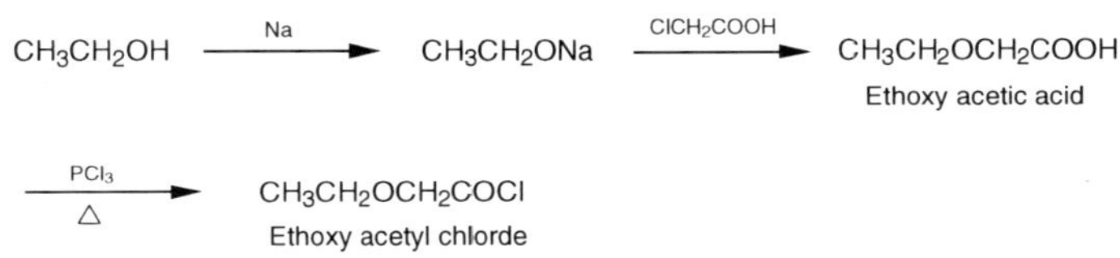

Step (i) Ethoxy Acetic Acid

Metallic sodium (35 g, 1.52 mole) is added in lots to absolute alcohol (625 ml) in a R.B. flask fitted with a reflux condenser and a calcium chloride tube. When all the sodium has dissolved, a solution of dried chloro acetic acid (71g, 0.75 mole) in absolute alcohol (90 ml) is added slowly in 10 ml portions. The mixture is gently heated for 10 min in a water bath. Excess ethyl alcohol is removed as completely as possible by distillation (steam bath) and finally by passing steam in the residue. The aqueous solution is cooled and concentrated hydrochloric acid (70 ml) added. The mixture is cooled and separated sodium chloride filtered and the inorganic salts washed with ether (2 × 50 ml). The total filtrate is extracted with ether (5 × 50 ml). The ether layer is dried (anhydrous sodium sulphate) and distilled. The residual oily product is distilled to give ethoxyacetic acid distilling at 109-11°/17-18 mm. (lit.[28] b.p. 109-11°/17-18 mm). Yield 58 g (74.3%).

Step (ii) Ethoxyacetyl Chloride

Phosphorus trichloride (10 g, 6.35 ml, 0.073 mole) is added slowly in cold to pure ethoxy acetic acid (20 g, 18.2 ml, 0.192 mole). The mixture is heated on a steam bath until the evolution of hydrogen chloride ceases. The mixture is cooled. The upper layer of ethoxyacetyl chloride is poured from phosphorus acid by decantation. It is distilled and the fraction boiling at 125-27° is collected. (lit.[29], b.p. 127°). Yield 21 g (89.2%).

6.A.22 ETHOXYMETHYL CHLORIDE

An ice cold solution of freshly distilled ethyl alcohol (60 ml, 47.1 g, 1.02 mole) and formalin (40%, 90 ml, 36 g, 1.2 mole) is saturated with hydrogen chloride gas. The contents are transferred to a separatory funnel. The upper layer of ethoxymethyl chloride is separated, dried (anhydrous sodium sulphate) and distilled. The fraction boiling at 80-3° is collected. (lit.[30] b.p. 81-2°). Yield 56 g (57.8%).

6.A.23 ETHYL CYANOACETATE

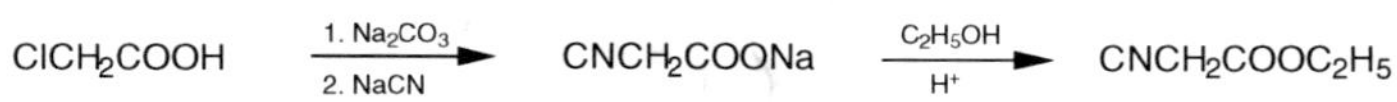

Ethyl cyano acetate

A solution of chloroacetic acid (100 g, 1.06 mole) in water (140 ml) is warmed to 50° and neutralised with anhydrous sodium carbonate (58 g, 0.54 mole). To this solution at room temperature is added a solution of sodium cyanide (59 g, 1.16 moles) in water (150 ml), warmed to 55°. The temperature of the reaction mixture is kept below 95° (by the use of ice bath or by adding 40 ml cold water). When the temperature does not rise any more, the reaction mixture is boiled for 5 min and then cooled to room temperature. The solution is filtered (if not clear) and then acidified with hydrochloric acid (139 g, 120 ml, 1.16 moles, sp. gr. 1.156) (in a fume cupboard). The resulting solution is evaporated on a water bath at 60-70° in vacuo (20-30 mm) till no more distillate comes over. To the residue is added alcohol (120 ml, 95%), the solution filtered from the separated sodium chloride and the residue washed with alcohol (100 ml). The combined alcoholic solution is evaporated under reduced pressure at 50-60°. The residue thus obtained (10.8 g) is treated with absolute alcohol (120 ml) and concentrated sulphuric acid (2 ml) and refluxed for 3 hr. The excess of alcohol and water formed are removed by distillation under reduced pressure and the residue again heated for 2 hr with absolute alcohol (60 ml) and concentrated sulphuric acid (1 ml). The alcohol is removed by distillation under reduced pressure, the residual solution is cooled and neutralised with a concentrated solution of sodium carbonate.

The upper layer of ethyl cyanoacetate is separated and the aqueous solution extracted with benzene. The upper layer and the benzene extract are combined and distilled in vacuo. After the solvent, alcohol and water have been removed, the ester is collected at 97-98°/16 mm. (lit.[31] b.p. 97-8°/16mm). Yield 98 g (80%).

6.A.24 TRIETHYL ORTHOFORMATE, $CH(OC_2H_5)_3$

$$C_2H_5OH + Na + CHCl_3 \rightarrow CH(OC_2H_5)_3$$

Clean sodium cut into small pieces (103.5 g, 4.5 g atoms) is added to a solution of absolute alcohol (1.5 litre) and dry chloroform (164 ml, 244.7 g, 2.05 mole) (time required 2 hr; the flask can be cooled by external cooling). When all the sodium has reacted, the separated sodium chloride is filtered under anhydrous conditions; the solid product is washed with absolute alcohol (10 ml). The combined filtrate and washing are distilled (steam bath) to remove most of the alcohol. The liquid in the flask is decanted from the separated sodium chloride and the mixture is distilled fractionally at atmospheric pressure. The fraction boiling below 85° is mainly alcohol and is discarded. The required orthoformic ester is collected at 140-46°, (Lit[32], b.p. 145-47°). Yield 60-65 g (20-21.4%).

6.A.25 o- AND p-HYDROXYPHENYLMERCURYL CHLORIDE

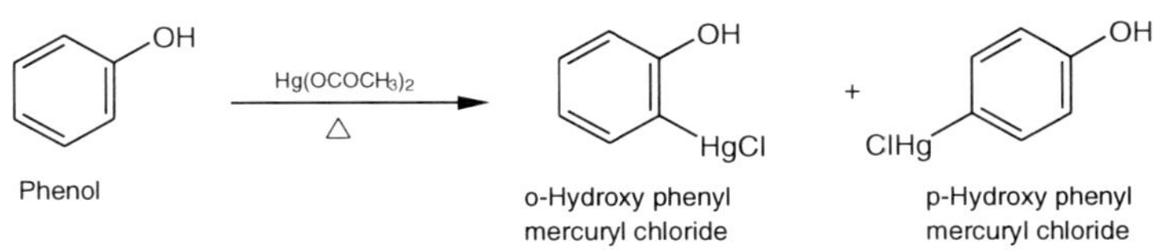

Phenol (12 g, 11.2 ml, 0.127 mole) is heated on a water bath and mercuric acetate (25 g, 0.078 mole) is added with stirring during 10 min. When whole of mercuric acetate has dissolved, boiling water (30 ml) is added and the mixture is heated with constant stirring for 5 min more. The mixture is then poured into an aqueous solution of sodium chloride (5%, 50 ml). A white precipitate is formed. The whole mixture is heated on a boiling water bath for 2 min more. The solid product is filtered. It is p-hydroxyphenyl mercuryl chloride. Yield 9 g (35%) m.p. 224° (lit[33] m.p. 224°). The filtrate is kept at 10° for 6 hr when o-hydroxyphenyl mercuryl chloride is obtained. Yield 3 g (12%). M.p. 150° (lit.[33] m.p. 148°).

6.A.25A JONES REAGENT[33A]

Chromimum trioxide (66.75 g, 0.67 mole)is dissolved in water (100 ml) and concentrated sulphuric acid (57.5 ml) is added with cooling. The cold solution is diluted with water up to 250 ml to form an 8N reagent. The reagent is added dropwise to a solution of the compound to be oxidized in acetone (distilled from potassium permanganate) at 20°C.

6.A.25B LEAD TETRAACETATE

It is prepared by treatment of lead oxide with acetic acid in the presence of acetic anhydride.

$$Pb_3O_4 + 8CH_3COOH \xrightarrow[\Delta]{(CH_3CO)_2O} \underset{\text{Lead tetraacetate}}{(CH_3COO)_4Pb} + 2(CH_3COO)_2Pb + 4H_2O$$

The filtrate which contains lead acetate is treated with chlorine.

$$2(CH_3COO)_2Pb + Cl_2 \longrightarrow (CH_3COO)_4Pb + PbCl_2$$

Procedure

In a three necked flask (500 ml) equipped with a thermometer and a sealed stirrer, glacial acetic acid (110 g, 1.83 mole) and acetic anhydride (37 g, 0.36 mole) are placed and heated to 55-60°C with vigorous stirring. Dry red lead powder (60 g, 0.088 mole) is added in portions of 4.5 g. A fresh portion is added only after the colour due to the previous potions has largely disappeared and maintaining the temperature at 65°C during addition. After the additions are completed the temperature rises to about 80°C in order to complete the reaction. The thick and dark coloured solution is then cooled and filtered and washed with glacial acetic acid. The crude lead tetraacetate is dissolved in hot glacial acetic acid containing a little acetic anhydride and treated with a decolorizing carbon. The hot solution is filtered, cooled and the crystalline product thus obtained is filtered, dried in a vacuum desiccator over potassium hydroxide pellets. Yield 30 g (78%).

The mother liquor contains lead acetate which is taken in original flask and heated to about 75°C with stirring and passed a stream of dry chlorine gas through the solution. When the reaction is complete, decolorizing carbon is added and maintained the temperature at 75°C for a few minutes. The suspension is filtered through a preheated Buchner funnel. The colourless needles of lead tetraacetate are obtained on cooling the filtrate.

6.A.26 ACTIVE MANGANESE DIOXIDE

$$KMnO_4 + MnSO_4.4H_2O \xrightarrow{NaOH} MnO_2$$

To a hot stirred solution of potassium permanganate (25 g, 0.158 mole in 150 ml water) is added a solution of manganese sulphate tetrahydrate (27 g, 0.121 mole in 50 ml water) and a solution of sodium hydroxide (40%, 30 ml). Both solutions are added simultaneously during 1 hr. The mixture is stirred for 1 hr more. The precipitated manganese dioxide is centrifuged and washed with water until the washings are colourless and free of sodium hydroxide. The product is then dried at 110° for 24 hr. It is ground to a fine powder. Yield 28 g.

Note: The procedure described is a modification of literature method[34].

6.A.27 METHYL IODIDE

$$(CH_3)_2SO_4 + 2CaCl_2 + 2KI \rightarrow 2CH_3I + CaSO_4 + K_2SO_4 + 2Cl_2\uparrow$$

A solution of potassium iodide (400 g, 2.41 mole) in water (220 ml) is placed in a 2 litre three necked flask fitted with a thermometer, a separatory funnel and mercury sealed mechanical stirrer and small fractionating column (6") with arrangement for downward distillation. Cold water is passed through the condenser and the receiver is also cooled in cold water to avoid loss of methyl iodide. The thermometer is suspended through the top of the fractionating column down to flask. Calcium chloride (30 g) is then added and the mixture heated to 60-65° (water bath) with stirring. Dimethyl sulphate (315 g, 236.3 ml, 2.5 mole) is then gradually added through the separatory funnel; the rate of addition is such that methyl iodide distils briskly. The addition requires 1.5 hr. After all the dimethyl sulphate has been added, the temperature is raised to 65-70° for about 45 min to complete the distillation of methyl iodide. The product is separated from the distillate (separatory funnel). It is dried (anhydrous calcium chloride), decanted and a few crystals of potassium iodide are added. The material is distilled to give pure methyl iodide. Yield 308 g (90%). B.p. 41-3° (lit.[35] b.p.41-3°).

6.A.28 METHYL THIOGLYCOLLATE

To a solution of thioglycollic acid (10 g, 7.55 ml, 0.108 mole) in dry chloroform (25 ml) and anhydrous methanol (30 ml) is added concentrated sulphuric acid (1.5 ml). The solution is stirred at room

$$HSCH_2COOH + CH_3OH \xrightarrow[\Delta]{\text{Conc. } H_2SO_4} HSCH_2COOCH_3$$

temperature for 2 hr. The solution becomes turbid and after 2 hr separates into two layers. Water (30 ml) is then added to the mixture, it is shaken well and the chloroform layer separated. It is dried (sodium sulphate) and distilled. The residue is distilled azeotropically with dry benzene (100 ml) to remove last traces of moisture. It is finally distilled to give methyl thioglycollate. B.p. 42-3°/10 mm; yield 10 g (87%) (lit.[36] b.p. 130-31°/ 27 mm).

6.A.29 PALLADIUM CATALYSTS

i. Palladium on Barium Sulphate (5% Pd)

To a vigorously stirred, hot (80°) solution of reagent grade barium hydroxide octahydrate (63.1 g, 0.2 mole) in distilled water (600 ml) contained in a 2 litre beaker is added in one lot sulphuric acid (6N, 120 ml). Some more sulphuric acid (6N) is added to make suspension just acidic to litmus. To this hot stirred suspension of barium sulphate is added a solution of palladium chloride (8.2 g, 0.021 mole) in concentrated hydrochloric acid (20 ml, 0.12 mole) and water (25 ml) (this palladium chloride solution is prepared by heating the mixture for 1.5–2 hr on a steam bath till dissolution is complete) and formaldehyde solution (37%, 4 ml, 1.48 g, 0.05 mole). The suspension is made slightly alkaline to litmus with sodium hydroxide solution (30%) (constant stirring is maintained). The suspension is stirred for 5 min more and then allowed to settle. The clear supernatant liquid is decanted. Distilled water (200 ml) is added, catalyst resuspended and allowed to settle. The catalyst is washed with water 10-12 times by decantation. It is then filtered (sintered glass funnel) and the cake washed with water (5 × 250 ml). After the last washing the water is removed as completely as possible by filtration. The funnel along with catalyst is dried in an air oven at 80°. The catalyst is powdered and stored in a tightly closed bottle. Yield 45 g.

Note: The procedure described is a modification of literature[37] method.

ii. Palladium on Barium Carbonate (5% Pd)

In the above preparation of palladium on barium sulphate an equal weight of precipitated barium carbonate (46.5 g) is substituted for the barium hydroxide and sulphuric acid to give palladium on barium carbonate catalyst. The amount of hydrochloric acid is reduced to 4.1 ml. Other details of experimental procedure are same (as in i).

Note: The procedure described is a modification of literature[37] method.

iii. Palladium on Carbon (5% Pd)

The carbon (Norite, Darco or other carbons) to be used is first heated with nitric acid (10%) on a steam bath for 2-3 hr. It is washed free of acid with water and dried at 100-110° (air oven) before use.

A suspension of carbon (46.5 g) in water (1000 ml) contained in a 2 litre beaker is heated to 80° (steam bath). To this suspension is added a solution of palladium chloride (4.2 g, 0.011 mole) in concentrated hydrochloric acid (20 ml, 0.12 mole) and distilled water (25 ml). Formaldehyde solution (37%, 4 ml, 14. 8 g, 0.05 mole) is then added and the suspension is made slightly alkaline to litmus with sodium hydroxide solution (30%), constant stirring being maintained. The stirring is continued for 5 min more. The catalyst is filtered and washed with water (12 × 120 ml). The excess water is removed by filtration and the cake is dried at room temperature; it is finally dried over potassium hydroxide or calcium chloride in a desiccator. The dry catalyst is stored in tightly closed bottle. Yield 46 g.

Note: The procedure described is a modification of literature[38] method.

iv. Palladium Chloride on Carbon (10% Pd)

A solution of palladium chloride (4.16 g, 0.011 mole) in concentrated hydrochloric acid (2.75 ml) and water (20 ml) is prepared by heating the mixture on a steam bath. The solution is poured into a solution of sodium acetate trihydrate (62.5 g , 0.46 mole) in distilled water (250 ml) contained in a 500 ml reduction bottle. Carbon (Norite, purified and dried as given in iii above) (22.5 g) is added and the mixture is hydrogenated until absorption ceases (1-2 hr). The catalyst is collected on a Buchner funnel and washed with water (5 × 120 ml). The catalyst, after removal of most of water at the pump is dried first in air and then in a desiccator over calcium chloride. The catalyst is stored after being powdered in a tightly closed bottle. Yield 25 g.

Note: The procedure described is a modification of literature[39] method.

v. Palladium Chloride on Carbon (5% Pd)

A solution of palladium chloride (4.1 g, 0.011 mole) in concentrated hydrochloric acid (20 ml, 0.12 mole) and distilled water (25 ml) is prepared by heating on a steam bath for 1.5-2 hr till the dissolution is complete. The solution is diluted with distilled water (70 ml) and poured over purified carbon (purified as in iii above) in an 6" evaporating dish. The suspension is stirred and heated on a steam bath till the whole mixture is dry. It is finally dried at 100° (air oven). The catalyst is powdered and stored in a closed bottle. Yield 48 g.

The required quantity of palladium chloride on carbon is transferred to a hydrogenation bottle and reduced with hydrogen in a solvent, which is to be used for hydrogenation. When no more hydrogen is absorbed, the catalyst is filtered on a sintered funnel and washed with more of the solvent to remove the hydrogen chloride. It is transferred back to the hydrogenation bottle. The material to be hydrogenated is then added and hydrogenation is completed in the usual way.

Note: The procedure described is a modification of literature[40] method.

6.A.30 PERACIDS

a. Performic Acid

Perfomic acid is prepared by treating hydrogen peroxide solution (30%, 100 volume) with excess formic acid (88-90%).

b. Peracetic Acid

Peracetic acid can be prepared in situ by adding hydrogen peroxide (30%, 1 part) to glacial acetic acid (3 parts) in the presence of a catalytic amount of sulphuric acid.

c. Perfluoroacetic Acid

Trifluoroacetic anhydride (50 ml, 0.36 mol) is added dropwise to a stirred suspension of hydrogen peroxide (86%, 8.2 ml, 0.3 mol) in ice cold dichloromethane (140 ml.). On completion of the addition, the stirring is continued at 0°C for further 10 minutes. The reaction mixture is dried over anhydrous sodium sulphate—and use it without delay.

d. Perbenzoic Acid

$$C_6H_5COOCOC_6H_5 \xrightarrow{CH_3ONa} C_6H_5CO_3H$$

A solution of pure benzoyl peroxide (25 g, 0.103 mole) in chloroform (100 ml) is cooled to 0° and added with shaking to a cooled solution (–5°)

of sodium (2.6 g, 0.11 gram atom) in absolute methyl alcohol (50 ml). Temperature of the mixture should not rise above 0°. The mixture is kept for 10 min in ice-salt bath, transferred to a separatory funnel and extracted with ice cold water (200 ml). The chloroform layer is separated and the aqueous layer extracted with more chloroform (2 × 50 ml). The aqueous solution is acidified with dilute sulphuric acid (1N, 125 ml) and extracted with chloroform (3 × 50 ml). The combined chloroform extract is washed with water and the clear chloroform layer separated (this is a solution of perbenzoic acid in chloroform and can be used as such).

However, pure crystalline perbenzoic acid is obtained by drying the chloroform solution over anhydrous sodium sulphate and distilling off the chloroform in vacuo. The remaining pale yellow residue is dried at 30-35°/ 10 mm pressure for 5-6 hr. Yield 10.6 g (75%) m.p. 42-3°. It is fairly stable when kept in dark in a cold place. Lit.[41], m.p. 42-3°.

6.A.31 PHENYLHYDRAZINE

$$C_6H_5NH_2 + NaNO_2 + HCl \longrightarrow [C_6H_5N_2^+Cl^-] \xrightarrow[\text{metabisulphite}]{\text{Sod.}} C_6H_5NHNH_2.HCl$$

An ice cold solution (0-3°) of sodium nitrite (35 g, 0.51 mole) in water (75 ml) is added dropwise (separatory funnel, the stem of the funnel reaching nearly to the bottom of the flask) during 30 min to vigorously stirred solution of pure aniline (47.5 g, 46.5 ml, 0.51 mole) in concentrated hydrochloric acid (130 ml) and water (60 ml) at 0°. The solution is kept at 0° throughout the addition by cooling in ice-salt bath. It is ascertained that there should be slight excess of nitrous acid (potassium iodide-starch paper) at the end of 15 min after the last portion of sodium nitrite solution is added.

Sodium metabisuphite (112.5 g, 0.59 mole) is added to a solution of sodium hydroxide (50 g, 1.25 mole) in water (500 ml). The mixture is stirred (mechanical stirrer) until all the sodium metabisulphite dissolves. The mixture is cooled to 25° and phenolphthalein indicator (few drops) is added. More sodium metabisulphite is added with stirring until the pink colour just disappears. Add little more sodium metabisulphite (12 g) (total amount of sodium metabisulphite used approximately 135-140 g, 0.71- 0.74 mole). This solution is cooled at 5° (ice bath), crushed ice (60 g) added and the ice-cold diazonium chloride solution (prepared above) is added as rapidly as possible; the reaction mixture is vigorously stirred during the addition. The mixture acquires bright orange-red colour. It is slowly heated (water bath) to 60-70° and kept at this temperature for 60 min. Until the colour becomes quite dark. The solution is acidified to litmus with concentrated hydrochloric acid (approx. 40-50 ml). The heating (boiling water bath) is continued until the colour becomes much lighter (4-5 hr). The solution is filtered to remove any solid present. To the hot clear filtrate is added with stirring concentrated hydrochloric acid (500 ml) and the total solution cooled to 0° (freezing mixture). The separated phenylhydrazine hydrochloride is filtered and washed with dilute hydrochloric acid (1:3). This can be further purified by crystallisation from water (charcoal) (the clear solution is cooled to 0°) as pure white crystals. Yield 85-90%.

Phenyl hydrazine is obtained by adding sodium hydroxide solution (25%, 75 ml) to the phenylhydrazine hydrochloride (50 g). The free base is extracted with toluene (2 × 70 ml). The toluene extract is dried (sodium hydroxide pellets) and toluene distilled in vacuo. Phenyldrazine is collected at 137-138°/8 mm. Yield 33 g-36 g (60-65%).

Note: The sodium metabisulphite solution can also be prepared by passing sulphur dioxide gas into sodium hydroxide solution (100 g sodium hydroxide in 125 ml water and diluting to 750 ml) containing phenolphthalein indicator (few drops) till the pink colour just disappears and then for 2 min more.

6.A.32 SUBSITUTED PHENYLHYDRAZINE

The substituted phenylhydrazines are obtained in the form of hydrochlorides and are prepared by diazotisation of the amine hydrochlorides followed by reduction of the diazonium salts to the hydrazines. For phenyldrazines having no ortho substituent, sodium sulphite is a satisfactory reducing agent (see the preparation of phenyl hydrazine) and the hydrochlorides are generally obtained in nearly pure state. Stannous chloride reduction is satisfactory for all compounds including those having an ortho substituent, but it is necessary to recrystallise the product to remove tin which is carried down with the precipitate.

Three procedures are carried out for effecting the reduction of the diazonium salts as illustrated below.

Method A (Reduction with stannous chloride)

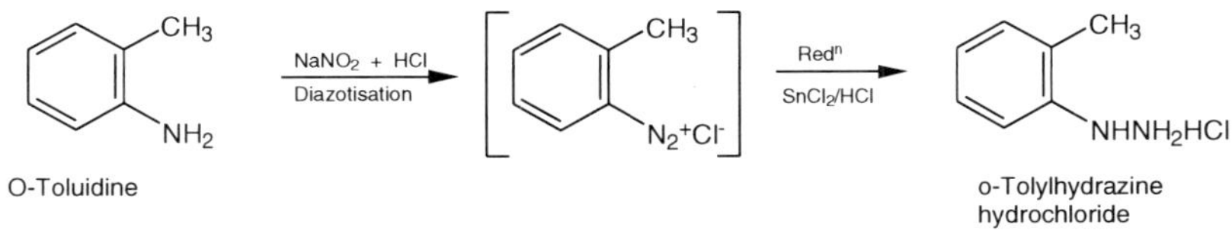

0-Tolylhydrazine hydrochloride

o-Toluidine (107.2 g, 106.7ml, 1 mole) and water (400 ml) are placed in a 3 litre, 3 necked flask equipped with a stirrer, thermometer and a dropping funnel. The mixture is stirred while concentrated hydrochloric acid (400 ml) added. The reaction flask is cooled rapidly to –20 ° by using dry ice acetone bath. The temperature is maintained to –15 to –20° while a solution of sodium nitrite (71 g, 1.03 mole) in water (250 ml) is added through a dropping funnel, the tip of which extending nearly to the bottom of the flask. The diazotisation required about 10 min. The reaction mixture is allowed to warm to 0° and poured with stirring into a solution of stannous chloride dihydrate (457.5 g, 2.03 mole) in hydrochloric acid (500 ml) at 0°. The resulting mixture is stored at 0 ° overnight. The separated product is filtered, washed with hydrochloric acid and then with ether. The crude product (181 g) contained appreciable amount of stannic ion. It is recrystallised from water (600 ml). The pure o-tolylhydrazine hydrochloride (90 g, 56.8% yield) melts at 195° (decomp) (lit.[43] m.p. 195°).

Note: It is obtained only in 13.3 % yield by reduction of the diazonium salt with sodium sulphite (Method B).

Method B (Reduction with sodium sulphite)

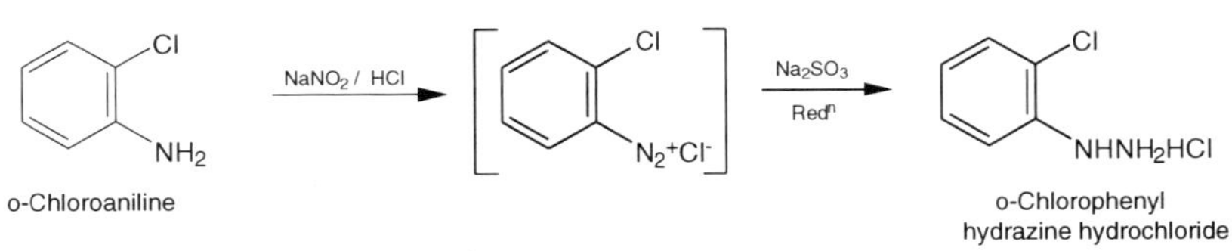

o-Chlorophenylhydrazine hydrochloride

o-Chloroaniline (63.8 g, 52.6 ml, 0.5 mole) and water (150 ml) are placed in a 1 litre, 3 necked flask fitted with a thermometer, stirrer and a dropping funnel. The mixture is stirred and cooled in an ice-bath while concentrated hydrochloric acid (125 ml) is added as rapidly as the temperature could be kept below 60°. The reaction mixture is cooled rapidly to –5° in a dry ice-alcohol bath maintained at –5 to –10° while a solution of sodium nitrite (34.5 g, 0.5 mole) in water (150 ml) is added through a dropping funnel, the tip of which extending nearly to the bottom of the reaction flask. The diazotization required about 10 min. The reaction mixture is allowed to warm to –3° and filtered rapidly through a glass wool plug into a solution of sodium sulphite at 5° which is freshly prepared by neutralizing to phenolphthalein, a solution of sodium hydroxide (97%, 100 g, 2.5 moles) in water (750 ml) with sulphur dioxide gas. The resulting mixture is warmed slowly to 60° and maintained at 60-70° for 1 hr. The solution is acidified to litmus with hydrochloric acid and heated on a steam bath for 6 hr. The reaction mixture is treated with activated charcoal and filtered through a layer of celite filtering aid, hydrochloric acid (500 ml, sp. gr., 1.19) added and the resulting solution allowed to cool. The separated o-chlorophenylhydrazine hydrochloride is filtered, washed with 3N hydrochloric acid and ether. Yield 44.5 g (50%). M.p. 198-200° (lit.[43] m.p. 198-200°).

Method C (Reduction with sodium sulphite-sodium hydroxide)

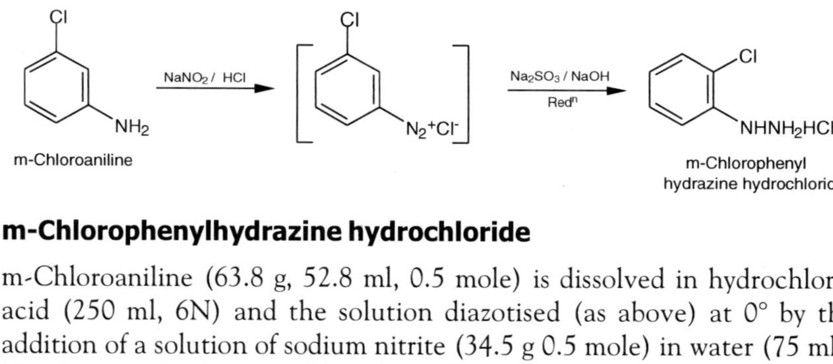

m-Chloroaniline

m-Chlorophenyl hydrazine hydrochloride

m-Chlorophenylhydrazine hydrochloride

m-Chloroaniline (63.8 g, 52.8 ml, 0.5 mole) is dissolved in hydrochloric acid (250 ml, 6N) and the solution diazotised (as above) at 0° by the addition of a solution of sodium nitrite (34.5 g 0.5 mole) in water (75 ml). The diazotised solution is filtered rapidly and stirred with cold solution of sodium sulphite which is prepared by adding sodium bisulphite (130 g) to a solution of sodium hydroxide (50 g in 500 ml water), cooling to about 25° and adding more of sodium bisulphite until the pH of the solution is about 8 and finally adding more sodium bisulphite (12 g). The reaction mixture is warmed on a steam bath for 1 hr, acidified to litmus and heated for 4 hr more. The hot solution is treated with decolorizing charcoal and filtered. Concentrated hydrochloric acid (500 ml) is added and the solution allowed to cool. The separated m-chlorophenyl hydrazine hydrochloride is filtered washed with hydrochloric acid (150 ml, 3 N) and crystallised from a mixture of water and hydrochloric acid. Yield 65.7 g (73.5%). M.p. 245-46° (decomp.) (lit.[43] m.p. 245-46°).

Following phenylhydrazine hydrochlorides are similarly prepared.

Hydrochloride	*Method*	*Yield (%)*	*m.p.*
o-Tolylhydrazine	A	56.5	195°(d)
p-Tolylhydrazine	A	63	155-60°
2,4-Dimethylphenyl hydrazine	A	30-40	182-84°
o-Chlorophenyl hydrazine	B	50	198-200°
m-Chlorophenyl hydrazine	C	73.5	224-26°(d)
p-Chlorophenylhydrazine	B	67	221-23°(d)
2,4-Dichlorophenylhydrazine	A	50.3	193-94°(d)
3-Chloro-2-methylpheylhydrazine	A	70	239°(d)
4- Chloro-2-methylphenylhydrazine	A	54.2	201-202°(d)
5-chloro-2-methylphenylhydrazine	A	75.5	215-16°(d)
5-Chloro-2-methoxyphenylhydrazine	A	58.6	192-93°(d)

Note: The m.p's are same as reported[43]

Following phenylhydrazines have also been prepared Phenylhydrazine, B.p. 137-38°/18 mm (Lit[44] b.p. 137-38^0/18 mm). 2,4-Dinitrophenylhydrazine, m.p. 200°(d) (Lit[45] m.p. 200-202°). N,N-Diphenylhydrazine, m.p. 35° (lit.[46] m.p. 35°).

Substituted benzoylhydrazines

These are obtained by refluxing the corresponding esters with hydrazine hydrate.

p-Methylbenzoylhydrazine, (p) CH_3-C_6H_4-CONHNH$_2$

$$C_6H_5COOC_2H_5 + NH_2NH_2 \rightarrow C_6H_5CONHNH_2$$

A mixture of ethyl benzoate (61.4 g, 0.1 mole) and hydrazine hydrate (20 ml) is refluxed for 6 hr. The separated solid is filtered, washed with water and crystallised from alcohol as colourless needles. Yield 41 g (67%). m.p. 115° (lit.[47] m.p. 117°).

Following benzoylhydrazines are similarly obtained.

p-Chlorobenzoylhydrazine, m.p. 163° (lit.[48], m.p. 163°).

p-Bromobenzoylhydrazine m.p.[49] 164°

p-Nitrobenzoylhydrazine, m.p.[49], 209°

6.A.33 PHENYL MERCURYL CHLORIDE (PhHgCl)

$$C_6H_6 + CH_3COOH + HgO + BF_3 \text{ etherate} \rightarrow C_6H_5HgCl$$

To a mixture of acetic acid (30 ml) and red mercuric oxide (8 g, 0.037 mole) is added boron trifluoride ethereate (1 ml) and the mixture is warmed to 50-60° (water bath) with frequent stirring. A clear solution is obtained. To this, dry benzene (40 ml, 34.96 g, 0.45 mole) is added and the mixture is refluxed for 3 hr. Unreacted benzene is removed in vacuo and to the residual product is added a boiling solution of sodium chloride in water (20%, 100 ml). The separated product is filtered and washed several times with water to remove sodium chloride. The phenyl mercuryl chloride thus obtained is dried. Yield 8 g (69.2%) m.p. 256° (lit.[50] m.p. 258°).

6.A.34 PYRIDINIUM BROMIDE[49]

$$C_6H_5N + HBr \rightarrow C_6H_5N^+HBr^-$$

Dry hydrogen bromide gas is bubbled through dry benzene until it is saturated. Dry pyridine is then added to the saturated solution. The precipitated pyridinium bromide is filtered (taking care to protect it from moisture), washed with dry benzene, stored in a vacuum desiccator and used directly.

Pyridinium bromide is used for complete demethylation of methyl ethers. The reaction is clean and yields are good. For demethylation, the methyl ether (1 g) is intimately mixed with freshly prepared pyridinium bromide (5 g). The mixture is carefully heated in a dry test tube over a small flame till it melts. It is kept in melting state for 1 min, cooled to room temperature and treated with ice-cold water. The separated product is filtered and crystallised.

6.A.34A PYRIDINIUM BROMIDE PERBROMIDE

Pyridine (7.5 ml, 7.33 g, 0.093 mole) is mixed with hydrobromic acid (48%, 15 ml). The solution is cooled (ice-water) and bromine (12.5 g, 4.03 ml, 0.078 mole) is added with swirling and cooling. The separated product is filtered and washed with ice-cold acetic acid (15 ml). Without drying it is crystallised from acetic acid as orange needles. Yield 16.5 g (55.6%).

Note: The procedure described is a modification of literature[51] method.

6.A.34B PYRIDINIUM CHLORO CHROMATE[51a]

It is an useful reagent for oxidation of primary alchohols to aldehydes.

The reagent is prepared by adding chromium (VI) oxide (50 g, 0.5 mole) rapidly to hydrochloric acid (6 M, 92 ml, 0.55 mole) with stirring. After 5 minutes, the homogeneous solution is cooled to 0°C and pyridine (39.55 g, 0.5 mole) is added carefully over 10 minutes. Re- cooling to 0 °C gives a yellow-orange solid which is collected on a sintered glass funnel and dried for 1 hour in vacuo (yield 90.4 g, 84%). The solid is not appreciably hygroscopic and can be stored for extended period at room temperature.

6.A.34C PYRIDINIUM CHLOROCHROMATE ADSORBED ON ALUMINA[51b]

Pyridine (9.5 g, 0.12 mole) is added to a solution of chromium trioxide (12 g, 0.12 mole) in hydrochloric acid (6N, 22 ml) within ten minutes at 40°C. The mixture is kept at 10°C until a yellow-orange solid forms and then reheated to 40°C to dissolve the solid. Alumina (100 g) is added with stirring at 40°C. After evaporation in a rotary evaporator, the orange solid is dried in vacuo for 2 hr at room temperature. The reagent can be kept under vacuum for several weeks in the dark without losing its activity.

6.A.35 RANEY NICKEL

$$\underset{\text{Alloy}}{\text{Ni - Al}} \xrightarrow{\text{NaOH}} \underset{\text{Raney}}{\text{Ni}}$$

Nickel - aluminum alloy (150 g containing about 50% nickel) is added in small portions to a stirred and cooled solution (10°, ice bath) of sodium hydroxide (180 g) in distilled water (750 ml). The temperature of the reaction mixture is not allowed to rise above 25°. When all the alloy has been added (1.5–2 hr), the container is removed from ice bath and contents allowed to attain room temperature. When the evolution of hydrogen has slowed down, the reaction mixture is heated on a steam bath until the evolution of hydrogen becomes slow (8-10 hr). During this period the volume of the solution is maintained constant by adding distilled water. The nickel is allowed to settle down and most of the liquid is decanted. It is again stirred with distilled water, allowed to settle and water decanted. A solution of sodium hydroxide (25 g) in distilled water (250 ml) is added, the catalyst is stirred, allowed to settle and alkali is decanted. The nickel is washed a number of times with distilled water by suspension in distilled water and decantation until the washings are neutral to litmus. It is washed 20-30 times more till the alkali is completely removed. The washing process is repeated with ethanol (3 × 100 ml, 95%) and then with absolute alcohol (3 × 100 ml). The catalyst is then stored under absolute alcohol and tightly closed. The product is highly pyrophoric and must be kept under a liquid at all times. Yield about 75 g.

Note: A number of procedures have been described[52,53] for the preparation of Raney Nickel from nickel-aluminum alloy. These differ from one another in the concentration of sodium hydroxide, in the temperature, duration of digestion and in the method of washing the catalyst from sodium aluminate and alkali. Various types of Raney nickel obtained have been designated as W-1, W-2, W-3, W-4, W-5, W-6 and W-7.

6.A.35A SODAMIDE

In a 500 ml three necked flask, equipped with a removable glass stopper, sealed stirrer unit of Hershberg type and a reflux condenser with soda lime guard tube, is added anhydrous liquid ammonia (150 ml) and sodium (0.25 g) to the stirred liquid, just sufficient to produce a permanent blue colour. Then powdered iron (III) nitrate (0.25 g) is added to catalyse the conversion of sodium into sodamide followed by small pieces of clean sodium metal (6.65 g) over a period of 30 minutes. When the sodium has been converted into sodamide (as indicated by the change from a blue colour to a grey suspension), the ammonia is allowed to evaporate and sufficient anhydrous ether is added through a dropping funnel to keep the volume of liquid at about 150 ml. After nearly all the ammonia has evaporated, the suspension of sodamide is stirred and heated under reflux for 5 minutes and then cooled to room temperature. A suspension of sodamide (11. 7 g) in dry ether is thus obtained.

6.A.35B SODIUM AMALGAM (CONTAINS 3% OF SODIUM)

In a 500 ml round bottom flask equipped with a dropping funnel (containing 375 g of mercury) in a central socket and inlet and outlet tubes for dry nitrogen in two side socket, clean sodium (11.4 g) is placed and air is displaced by nitrogen. Then mercury (about 10 ml) is added and the flask is warmed gently with a free flame until reaction commences. The flame is then removed and the reaction is maintained by the slow addition of the mercury. When about half the mercury has been introduced, the amalgam will commence to solidify; it should be kept molten by heating with occasional shaking. After the addition of all the mercury, the hot molten amalgam is poured on to a uralite board, powdered in a mortar (the mortar should be provided with a tightly fitting rubber cover to protect the powdered material from oxidation in air) and stored in a tightly stoppered bottle. Amalgam of 1 and 2 percent strength may be prepared similarly.

6.A.35C TETRABUTYLAMMONIUM CHROMATE [53a]

An aqueous solution of tetrabutylammonium chloride (1.46 g, 5.25 mmole) in water (25 ml) is rapidly added to a stirred solution of chromium trioxide (0.5 g, 5 mmole) in water (12 ml) at room temperature. A yellow-orange solid precipitates immediately. The mixture is cooled to 0°C and the solid is filtered with suction through a sintered glass funnel. It is carefully washed with cold water, dried under vacuum over phosphorous pentoxide and stored over calcium chloride. Yield 1.38 g (77%).

6.A.36 P-TOLUENESULFONYLHYDRAZIDE ($P\text{-}CH_3C_6H_4SO_2NHNH_2$)

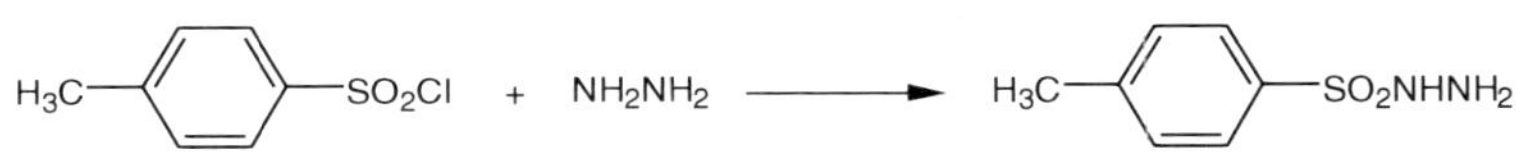

A solution of hydrazine hydrate (62.75 ml of 85% hydrazine hydrate) is added dropwise through a separating funnel to a vigorously stirred mixture of p-toluenesulfonyl chloride (100 g, 0.52 mole) and tetrahydrofuran (180 ml) at 10-15°. The temperature of the reaction mixture is maintained between 10-20°, stirring is continued for 20 min after the addition is complete. The upper layer of tetrahydrofuran is separated in a separatory funnel; it is filtered to remove any suspended impurities. The clear, colurless filtrate is stirred vigorously during the slow addition of twice the volume of distilled water. The separated p-toluenesulfonylhydrazide is filtered, washed with water and air-dried. Yield 87.5 g (89.6%). M.p. 109-110°. If necessary, it can be further purified by dissolving in hot methanol (approx. 4 ml per gm of hydrazide), filtering to remove suspended impurities followed by addition of distilled water (8-10 ml) (lit[54] m.p. 112°).

6.A.36A TRI-N-BUTYLTIN HYDRIDE[54a]

It is an useful reducing agent for the conversion of alkyl halides to hydrocarbons

Procedure

A three necked round bottom flask (500 ml) equipped with a mechanical stirrer, a condenser and a dropping funnel is taken and placed a solution of lithium aluminum hydride (2.38 g, 0.062 mole) in anhydrous ether (200 ml). The system is flushed with nitrogen and tributyl tin chloride (50 g, 0.154 mole) is added dropwise through the dropping funnel with gentle reflux of ether solution. After the addition has completed, the mixture is stirred for 7 hours. The water is then added carefully to destroy the excess lithium aluminium hydride. The ether layer is separated, washed thrice with water (100 ml) and dried over calcium sulphate. The ether is evaporated and then distillation under vacuum produced tributyltin hydride (34 g, 78 %) b.p. 69°C/0.3 mm Hg, which is stored under argon.

6.B.37 P-METHOXYBENZYL CYANIDE (P-METHOXYPHENYLACETONITRILES)

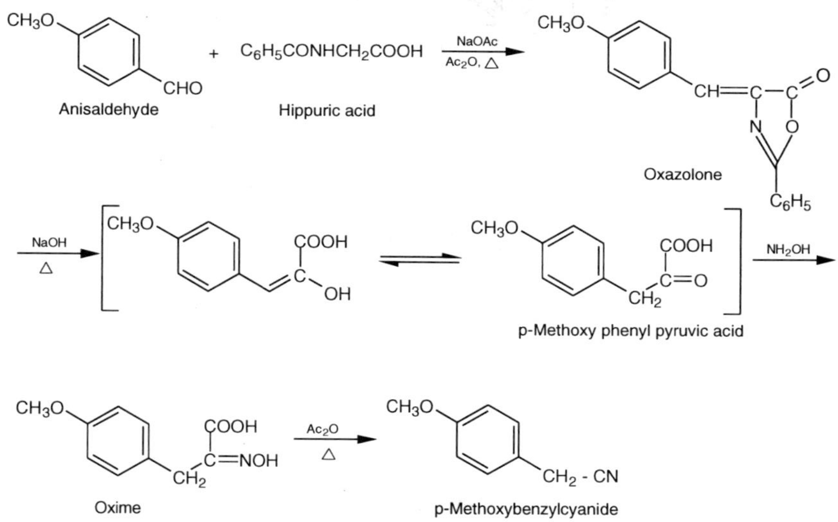

Step (i) 2-Phenyl-4-(4′- methoxybenzal)-5-oxazolone

An intimate mixture of anisaldehyde (2.2) (34 g, 30.4 ml, 0.25 mole), hippuric acid (5.A.7) (44 g, 0.245 mole), freshly fused sodium acetate

(14 g) and acetic anhydride (80 ml) is heated under anhydrous conditions in a boiling water bath for 1 hr. Alcohol (40 ml) is added to the reaction mixture while still warm and the mixture left overnight. The separated product is filtered, washed successively with hot water, alcohol and benzene. The azlactone is finally crystallised from alcohol as bright yellow needles. Yield 40 g (58.3%). M.p. 155-56° (lit.[55] m.p. 155-56°).

Step (ii) p-Methoxybenzyl Cyanide

The above azlactone (20 g, 0.07 mole) is refluxed with sodium hydroxide solution (12 g in 60 ml water) till no more ammonia is evolved (5-6 hr). The solution is cooled, diluted with water (50 ml) and saturated with sulphur dioxide gas (2 hr) and kept in a refrigerator for 12 hr. The separated benzoic acid is filtered and the filtrate is heated with concentrated hydrochloric acid (60 ml) for 1 hr in a boiling water bath. The solution is cooled and the separated p-methoxyphenyl pyruvic acid is filtered.

Hydroxylamine hydrochloride solution (12 g, in 150 ml of 10% sodium hydroxide) is added to the p-methoxyphenylpyruvic acid during 15 min. The temperature rises to 50°. The mixture is kept for 24 hr at room temperature and acidified with dilute hydrochloric acid. The separated oxime is filtered and crystallised from methanol. Yield 15 g. m.p. 149-50° (lit.[55] m.p. 149-50°).

The above dry oxime (15 g) is heated with acetic anhydride (10 ml) at 40-45°. The reaction becomes vigorous after sometime. It is heated for 10 min on a boiling water bath to ensure completion of the reaction. The reaction mixture is cooled, poured over crushed ice and extracted with ether. The ether solution is washed with aqueous sodium bicarbonate (5%), water, dried (anhydrous sodium sulphate) and distilled. The residue of p-methoxybenzylcyanide is distilled. Yield 10 g (95%). B.p. 295-98° (lit.[56] b.p. 285-98°).

6.B.38 2,4-DIMETHOXY BENZYL CYANIDE

Method I

Step (i) 2,4-Dimethoxyphenyl acetic acid[57]

2,4-Diemthoxyacetophenone (4.A.8) (13.5 g, 0.075 mole) morpholine (13 ml, 12.9 g, 0.15 mole) and sulphur (4.75 g, 0.15 mole) is refluxed for 8 hr. The mixture is diluted with chloroform (50 ml) and the chloroform

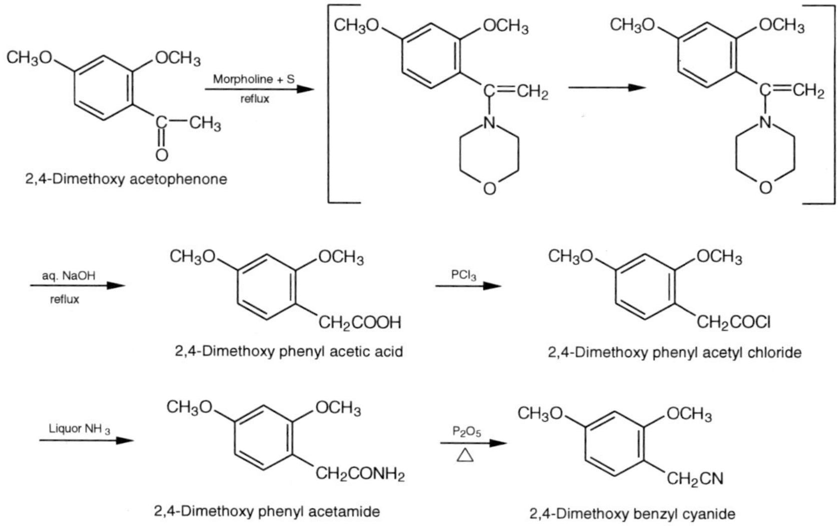

solution extracted with dilute hydrochloric acid to remove excess of morpholine. The remaining chloroform solution is distilled off. The crude morpholide obtained is refluxed with aqueous sodium hydroxide solution (10 g in 100 ml water) for 8 hr. The cooled solution is acidified with hydrochloric acid and again cooled. The separated product is filtered and washed with cold water. It is crystallised from hot water (or from benzene-petroleum ether) as colourless needles. Yield 3.9 g (26.5%). M.p. 111-12° (lit.[57] m.p. 111-12°).

Step (ii) 2,4-Dimethoxyphenyl acetyl chloride

The above 2,4-dimethoxyphenyl acetic acid (1 g, 0.005 mole) in dry chloroform (8 ml) is treated with phosphorous trichloride (0.4 ml, 0.63 g, 0.0046 mole). The mixture is kept for 2 hr at room temperature and then warmed at 60° for 1 min. The solution is decanted from phosphorus acid and evaporated in vacuum. The acid chloride is obtained as pale yellow oil. It is used as such for the next stage.

Step(iii) 2,4-Dimethoxyphenyl acetamide

To an ice-cold solution of liquor ammonia (50 ml, d, 0.9) is gradually added during 15-20 min with constant stirring, the above acid chloride (11 g, 0.05 mole). The separated product is filtered and washed with cold water. It is crystallised from hot water as aggregates of colourless broad prisms. Yield 10 g (100%). M.p. 133-34° (lit.[58] m.p. 133-34°).

Step (iv) 2,4- Dimethoxybenzyl cyanide

2,4-Dimethoxyphenylacetamide (5 g, 0.0256 mole) is mixed with phosphorus pentoxide (10 g) in a small Claisen flask. The mixture is distilled under reduced pressure. The nitrile distils over at 165-170°/5 mm. Yield 2.2 g (48.9%). It is crystallised from ether as long needles. M.p. 75-6° (lit.[59] m.p. 76°).

Method II

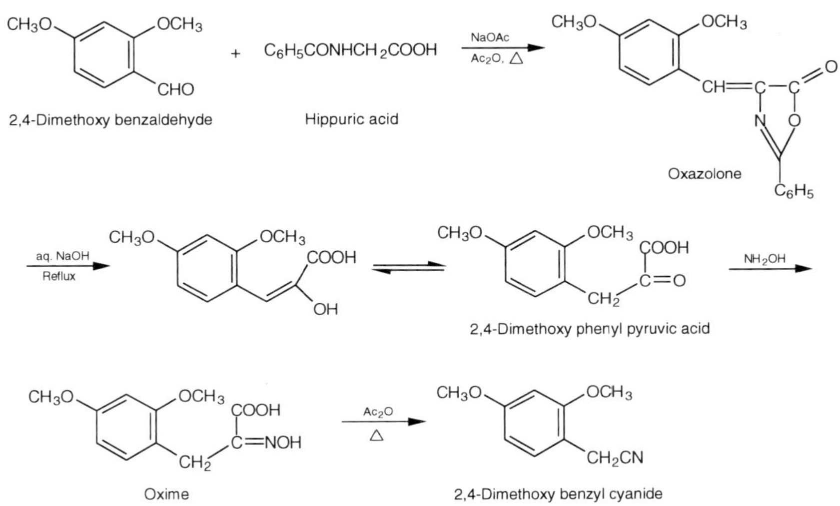

Step (i) 2-Phenyl-4-(2',4'-dimethoxybenzal)-5-oxazolone

2,4-Dimethoxybenzaldehyde (2.7) (16 g, 0.096 mole), hippuric acid (5.A.7) (17.5 g, 0.098 mole) and fused sodium acetate (8 g) are powdered to an intimate mixture and heated with acetic anhydride (30 min) for 1 hr in a boiling water bath. Alcohol (40 ml) is then added to the reaction mixture while still warm. The separated product is filtered, washed successively with water, cold alcohol and benzene. It is crystallised from ethanol as yellow needles. Yield 17.5 g (58.8%). M.p. 168° (lit.[60] m.p. 168°).

Step (ii) 2,4-Dimethoxybenzyl cyanide

A solution of the above azlactone (25 g, 0.08 mole) in sodium hydroxide solution (15 g in 70 ml water) is refluxed for 8 hr till no more ammonia is evolved. The solution is cooled, acidified with hydrochloric acid and cooled. The separated product is filtered. It consisted of 2,4-dimethoxyphenyl pyruvic acid and benzoic acid. The crude product is added to a solution of hdydroxylamine hydrochloride (27 g) in sodium hydroxide solution (10%, 100 ml). The solution is allowed to stand overnight at room temperature. It is acidified to congo red with hydrochloric acid. A mixture of 2,4-dimethoxyphenyl pyruvic acid oxime and benzoic acid separates out. It is filtered, washed with cold water and dried (vacuo desiccator over phosphorous pentoxide). The dried crude solid mixture is treated with acetic anhydride (25 ml). It is heated to 40-45° when a vigorous reaction occurs. The reaction mixture is heated at 40-45° for 10 min more. The solution is cooled and poured over crushed ice. It is extracted with ether (2 × 75 ml). The ether extract is washed with sodium carbonate solution (5%) and then with water. It is dried (sodium sulphate) and distilled. The residual product is crystallised from petroleum ether as pale yellow needles. Yield 8 g (55.9%). M.p. 75-6° (lit.[59] m.p. 76°).

6.B.39 3,4- METHYLENEDIOXY BENZYL CYANIDE

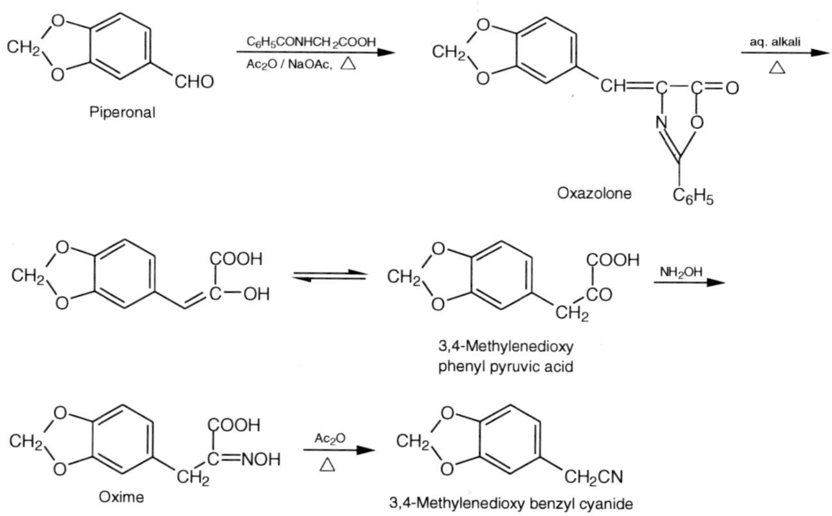

Step (i) 2-Phenyl-4-(3′,4′-methylenedioxybenzal)-5-oxazolone

An intimate mixture of piperonal (35 g, 0.23 mole), hippuric acid (5.A.7.) (40 g, 0.22 mole), freshly fused sodium acetate (16 g) and acetic anhydride (70 ml) is heated in a boiling water bath for 4 hr under anhydrous conditions. Alcohol (40 ml) is added and the reaction mixture kept in a refrigerator for 12 hr. The solid product is filtered, washed with hot water, cold alcohol and benzene. It is crystalliesd from ethanol as orange needles. Yield 40 g (61.1%). M.p. 198° (lit[61] m.p. 198°).

Step (ii) 3,4-Methylenedioxy Benzyl Cyanide

The above azlactone (20 g, 0.068 mole) is refluxed with sodium hydroxide solution (12 g in 60 ml water) till no more ammonia evolved (5-6 hr). The solution is cooled and hydroxylamine hydrochloride (20 g in 100 ml of 10% sodium hydroxide solution) is added and the reaction mixture left for 24 hr at room temperature. It is acidified to congo red with dilute hydrochloric acid. A mixture of 3,4-methylenedioxyphenyl pyruvic acid oxime and benzoic acid separates out. It is filtered, washed with cold water and dried in vacuum desiccator. The crude mixture is gently warmed with acetic anhydride (20 ml) at 40-50°. The reaction becomes vigorous after some time. It is heated for 10 min in a water bath to ensure completion of the reaction. The resulting solution is cooled and poured over crushed ice. It is extracted with ether (2 × 50 ml). The ether extract is washed with sodium hydroxide solution (5%, 3 × 25 ml), ether dried (anhydrous magnesium sulphate) and solvent distilled. The residual product is crystallised from methanol as colourless cubes. Yield 8 g (72.9%). M.p. 49° (lit.[62] m.p. 49°).

6.B.40 2,4,5-TRIMETHOXY BENZYL CYANIDE

Method I

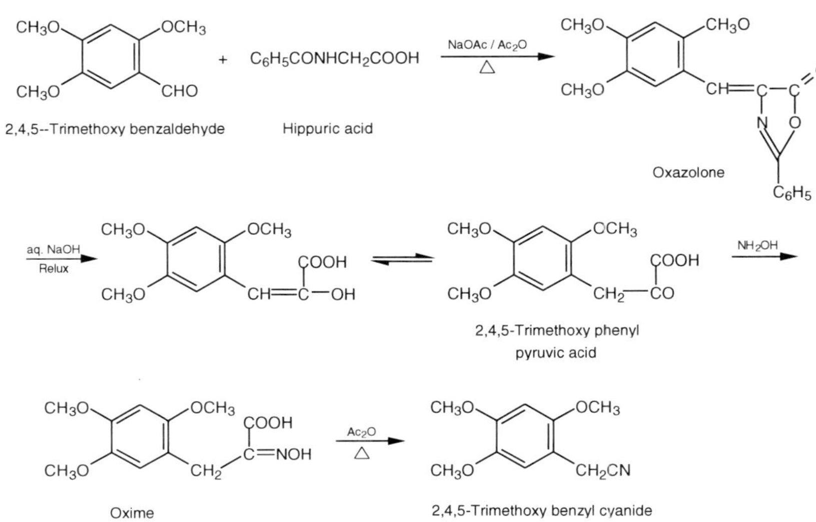

Step (i) 2-Phenyl-4-(2′,4′,5′- trimethoxybenzal)-5-oxazolone

An intimate mixture of 2,4,5-trimethoxybenzaldehyde (2.25) (45 g, 0.229 mole), hippuric acid (5.A.7) (40 g, 0.22 mole), fused sodium acetate (16 g) and acetic anhydride (70 ml) is heated in a boiling water bath for 4 hr under anhydrous conditions. Alcohol (35 ml) is added and the reaction mixture is kept in a refrigerator for 12 hr. The solid product is filtered, washed with hot water. It is crystallised from ethanol as yellow needles. Yield 40 g (52.7%). M.p. 204° (lit.[63] m.p. 204°).

Step (ii) 2,4,5-Trimethoxyphenyl pyruvic acid

The above azlactone (35 g, 0.1 mole) is refluxed with sodium hydroxide (56 g) in water (350 ml) till the evolution of ammonia ceased (2 hr). The solution is cooled, saturated with sulphur dioxide and filtered. The filtrate is treated with concentrated hydrochloric acid (200 ml) and the solution heated on a boiling water bath for 1 hr. It is cooled and the separated pyruvic acid filtered. It is crystallised from alcohol as colourless cubes and needles. Yield 15 g (57.2%). M.p. 192-94° (decomp.) (lit.[64] m.p. 192-4°).

Step (iii) 2,4,5-Trimethoxybenzyl cyanide

The above pyruvic acid (3 g, 0.013 mole) and hydroxylamine hydrochloride (2 g) are dissolved in sodium hydroxide solution (10%, 25 ml). The solution is left for 24 hr at room temperature and then acidified with concentrated hydrochloric acid and cooled. The separated 2,4,5-trimethoxyphenyl pyruvic acid oxime is filtered. It is crystallised from hot water as colourless plates. Yield 2.5 g (78.7%). M.p. 128-30° (decomp.)

The foregoing dry oxime (3.5 g, 0.013 mole) is warmed cautiously with acetic anhydride (3 ml) on a water bath. After the vigorous reaction has subsided, water (30 ml) is added and the mixture cooled. After 12 hr the separated product is filtered, washed with sodium bicarbonate solution and then with water. It is crystallised from dilute alcohol as needles. Yield 2.1 g (78%). M.p. 88.5° (lit.[64] m.p. 88.5°).

Method (ii)

Step (i) 2,4,5-Trimethoxy benzyl alcohol

Sodium borohydride (1 g) in aqueous alcohol (25%, 20 ml) is added dropwise to a stirred solution of 2,4,5-trimethoxybenzaldehyde (2.25) (2 g, 0.01 mole) in alcohol (50 ml). When the starting compound disappears

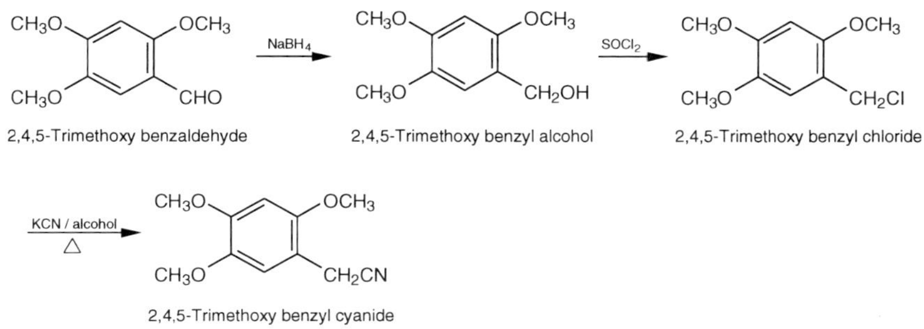

(TLC, DNP test), the solution is acidified with acetic acid and extracted with ether (3 × 30 ml). The ether extract is washed with saturated sodium chloride solution, dried (anhydrous sodium sulphate) and distilled. The residual product is crystallised from alcohol as colourless shinning needles. Yield 1.6 g (79.2%) m.p. 67-8° (lit.[65] m.p. 67-8°).

Step (ii) 2,4,5-Trimethoxybenzyl chloride

2,4,5-Trimethoxybenzyl alcohol (1 g, 0.005 mole) in dry ether (30 ml) is treated dropwise with stirring with a solution of thionyl chloride (2 g, 1.2 ml, 0.0168 mole) in dry ether (20 ml). After the addition is over, the solution is stirred for 1 hr more. The ether is distilled and the residual product is crystallised from alcohol as colourlesss needles. Yield 1 g (91.7%). M.p. 76-7° (lit.[65], m.p. 76-7°).

Step (iii) 2,4,5-Trimethoxybenzyl cyanide

2,4,5-Trimethoxybenzyl chloride (2 g, 0.009 mole) in alcohol (5 ml) is added dropwise during 1 hr to a stirred solution of potassium cyanide (1 g, 0.015 mole) in water (1 ml). The mixture is refluxed for 4 hr, cooled and filtered. The clear filtrate is concentrated in vacuo. The residual product is crystallised from alcohol as colourless needles. Yield 1.2 g (63%). M.p. 88-9° (lit. [64] m.p. 88.5°).

6.C.41 ETHYL BENZOYL ACETATE

Method (i)

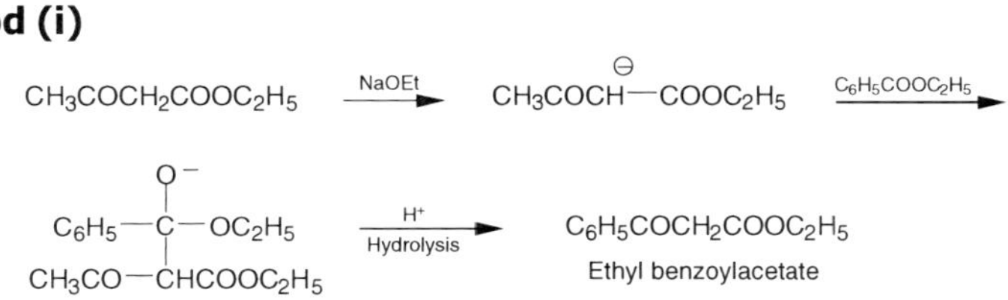

Clean sodium (46 g, 2 mole) cut into small pieces is added gradually to absolute alcohol (600 ml) placed in a 2 litre three necked flask mounted on a steam bath and fitted with a reflux condenser and a calcium chloride drying tube. The sodium ethoxide solution is stirred and cooled to room temperature. Ethyl acetoacetate (260 g, 254.6 ml, 2 mole) is then added through a separatory funnel. The reflux condenser is replaced by a short still head and the alcohol is removed by distillation at approximately room temperature under the pressure of a water pump. When approximately half the alcohol has been removed, stirring has to be discontinued due to separation of sufficient sodium enolate. The residue appeared dry after about 2 hr. The last traces of alcohol are removed by heating for 1 hr on a steam bath under 2 mm pressure. The flask is allowed to cool to room temperature under vacuum and to the cooled residue is added ethyl benzoate (570 ml, 599 g, 4 mole). The mixture is heated in an oil bath (temperature raised to 140-150°) for 6 hr. The temperature of the oil bath is gradually raised to 180° in the next 1 hr. During this period approximately 220 ml of the distillate consisting of ethyl alcohol and ethyl acetate are collected. The reaction mixture is cooled, water (250 ml) added and acidified by adding a cooled solution of sulphuric acid (100 g) in water (200 ml). Crushed ice is added to keep the mixture cold. The upper layer of ester is separated and the aqueous layer extracted with ether (2 × 200 ml). The combined ether and ester layer are washed with sodium bicarbonate (5%), water and dried (anhydrous sodium sulphate). The ether is distilled. Excess of ethyl benzoate and ethyl acetoacetate are removed by distillation under reduced pressure through a 15 cm fractionating column. Finally ethyl benzoyl acetate is distilled at 147-49°/ 12 mm. Yield 200 g (52%).

Note: Ethyl benzoate is prepared as follows. Dry hydrogen chloride gas is passed into a cooled (ice-bath) absolute alcohol (145 ml excess) under anhydrous conditions until the increase in weight is 6 g. Benzoic acid (30 g, 0.0246 mole) is added to the solution and the mixture is refluxed for 4 hr (water-bath). Excess alcohol is removed (rotary evaparator) and the remaining solution is poured over crushed ice (200 g). The mixture is extracted with ether (2 × 100 ml). The ether extract is washed with sodium bicarbonate solution (5%), water and dried (anhydrous sodium sulphate). The ether is distilled and ethyl benzoate is obtained by distillation. Yield 32 g (86.7%). B.p. 212-14° (lit.[65a] b.p. 165-69°/20 mm).

Method (ii)

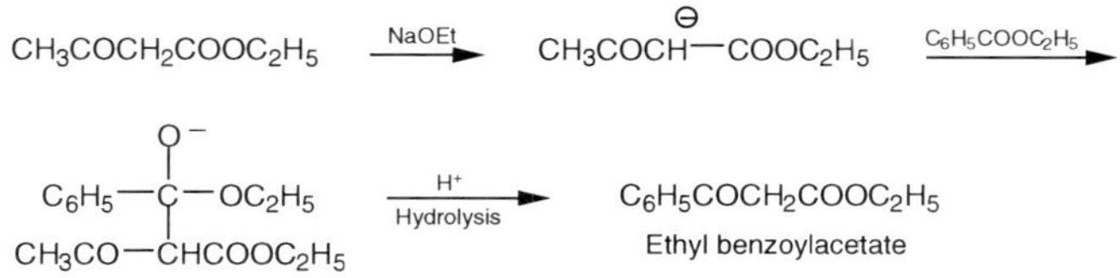

A mixture of ethyl acetoacetate (97.5 g, 95.5 ml, 0.75 mole), dry benzene and clean sodium (17.25 g, 0.75 mole) is refluxed under anhydrous conditions in a steam bath. The refluxing is continued for 18-20 hr with stirring until all the sodium has reacted (use of pulverised sodium reduces the reaction time considerably to 3-4 hr). The mixture is cooled and benzoyl chloride (109 ml, 132 g, 0.94 mole) is added during 3 hr while keeping the mixture stirred. The mixture is refluxed for 6 hr more, cooled and crushed ice (200 g) added. The benzene layer is separated, washed with sodium bicarbonate solution (5%) dried and distilled. The formed ethyl benzoylacetoacetate is collected at 177-81°/20 mm. Yield 131.6 g (75%).

A solution of ammonium chloride (16 g, 0.3 mole) in water (75 ml) containing ammonia (5 ml, sp. gr. 0.9) is warmed to 40° (water bath) and ethyl benzoyl acetoacetate (29.75 g, 0.127 mole) is added in one lot. The mixture is shaken. The reaction mixture is kept at 40° for 10 min with occasional shaking. It is extracted with ether (2 × 75 ml), the ether solution dried (anhydrous sodium sulphate) and distilled. Ethyl benzoyl acetate is collected at 147-49°/12 mm (lit.[65a] b.p. 165-169°/20 mm). Yield 18.7 g (76.6%).

6.C.42 ETHYL o-METHOXYBENZOYL ACETATE (o-METHOXYBENZOYL ACETIC ESTER) (GENERAL METHOD OF PREPARATION)

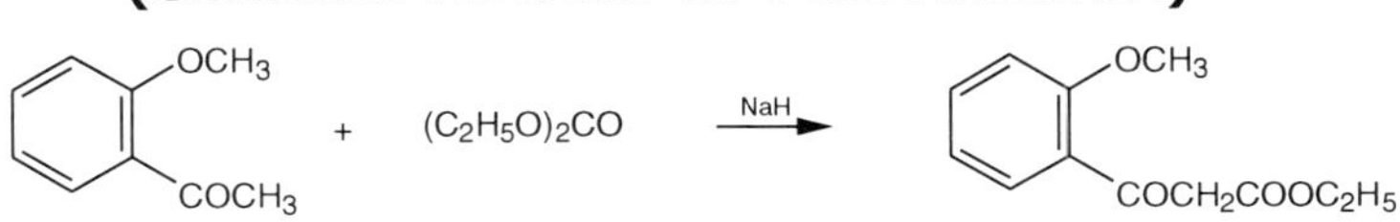

o-Methoxy acetophenone   Diethyl Carbonate   Ethyl-o-methoxy benzoyl acetate

A vigorously stirred suspension of sodium hydride (2.9 g, 0.12 mole) (the sodium hydride in oil dispersion is washed with dry ether and used) in dry benzene (50 ml) containing dry diethyl carbonate (7.5 ml, 7.3 g, 0.112 mole) is treated dropwise with o-methoxyacetophenone (4.A.2) (4.5 g,

4.13 ml, 0.03 mole) in dry benzene (50 ml) during the course of 8 hr. The mixture is refluxed for 24 hr, cooled and treated with ice (caution). The organic layer is separated, washed with aqueous sodium bicarbonate (5%), water and dried (anhydrous sodium sulphate). The benzene is distilled and the oily residue is finally distilled in vacuo on a water bath at 10 mm to remove excess of diethyl carbonate. The residual brown coloured oil is distilled under reduced pressure to give ethyl o-methoxybenzoyl acetate. Yield 3g (45%). B.p. 115°/1 mm (lit.[66] b.p. 115°/1mm).

Following esters are similarly prepared.

Ethyl p-methoxybenzoyl acetate is prepared from sodium hydride (3 g), diethyl carbonate (7.5 ml) and p-methoxyacetophenone (4.A.4) in 50% yield. B.p. 155°/1 mm (lit.[66], b.p. 115°/1mm).

Ethyl 2,4-dimethoxybenzoyl acetate is prepared from sodium hydride (3.5 g) diethyl carbonate (8.5 ml) and 2,4-dimethoxyacetophenone (4.A.8) (6.3 g) in 50% yield. B.p. 155°/1 mm. (lit.[66] b.p. 155°/1 mm).

Ethyl 3,4-dimethoxy benzoyl acetate is prepared from sodium hydride (3.5 g), diethyl carbonate (8.5 ml) and 3,4-dimethoxyacetophenone (4.A.19) (6.3 g) in 55% yield. B.p. 125°/1 mm. (lit.[66] b.p. 125°/1 mm).

Ethyl m-methoxybenzoyl acetate is prepared[67] from sodium hydride (6 g), diethyl carbonate (15 ml) and m-ethoxyacetophenone[67] (4.A.4) in 57% yield. b.p. 176-78°.

Ethyl m-ethoxybenzoyl acetate is prepared[67] from sodium hydride, diethyl carbonate and m-methoxyacetophenone[67] in 50% yield. B.p. 135-40°/74 mm.

Note: This is a general method of preparing ethyl benzoyl acetates from the corresponding acetophenones.

6.C.43 ETHYL PROPIONYL ACETATE[68]

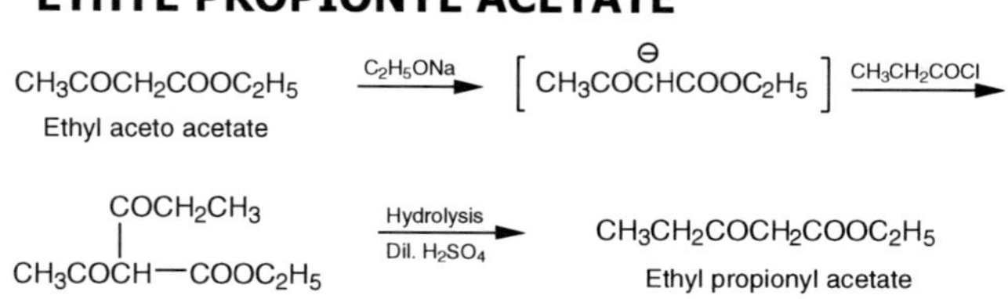

Ethyl acetoacetate (27 g, 26.4 ml, 0.207 mole) is added to a solution of sodium (4.9 g, 0.21 mole) in absolute ethyl alcohol (100 ml) under anhydrous conditions. Alcohol is removed under reduced pressure and the dry sodium salt of ethylacetoacetate is suspended in dry benzene (150 ml). A solution of propionyl chloride (20 g, 18.8 ml, 0.216 mole) in dry benzene (50 ml) is added to the above suspension during 2 hr. with constant stirring. The reaction mixture is left overnight, poured over crushed ice and acidified with cold dilute sulphuric acid. The organic layer is separated, washed with sodium bicarbonate solution (5%) and benzene is distilled off. The residue showed (TLC) the presence of ethyl acetoacetate and the required ester. It is repeatedly washed with a saturated solution of sodium bisulphite to remove ethyl acetoacetate. The remaining oily product is extracted with ether, extract washed with water and dried (anhydrous sodium sulphate). The ether is distilled to give ethyl propionyl acetate as an oil. It is used as such for further reaction. Yield 9 g (30%).

6.D.44 CHALCONE (BENZYLIDENEACETOPHENONE), $C_6H_5CH{=}CHCOC_6H_5$

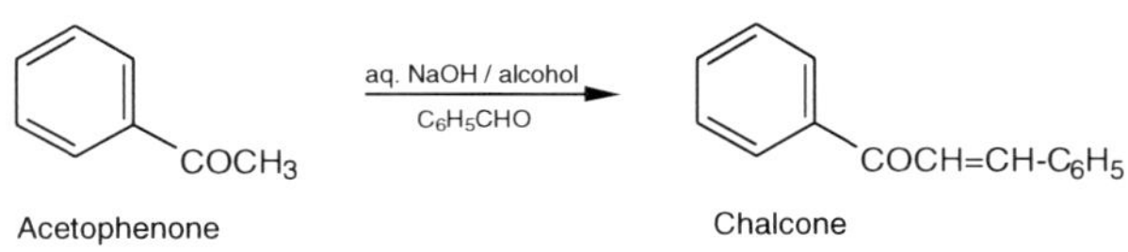

A solution of sodium hydroxide (22 g, 0.55 mole in 200 ml water) and alcohol (120 ml) is cooled (crushed ice) and the solution stirred. Freshly distilled acetophenone (52 g, 50.5 ml, 0.43 mole) is added followed by addition of pure benzaldehyde (44 ml, 45.9 g, 0.43 mole). The reaction mixture is kept at 25° and stirring continued until stirring is no longer possible (2-3 hr). The reaction mixture is left overnight in a refrigerator. The separated product is filtered on a Buchner funnel or a sintered glass funnel and washed with cold water (until the washings are neutral to litmus) and then with ice cold alcohol (20 ml).

It is crystallised from alcohol as pale yellow solid. Yield 77 g (85.5%), M.P. 56-57° (lit.[69] m.p. 56-57°).

Note: Chalcone acts as a skin irritant and so should be handled with care.

4'Methoxychalcone [3-(4'methoxyphenyl)-1-phenylpropen-3-one], m.p. 107-8° (lit.[70] m.p. 107°) is similarly prepared from p-methoxyacetophenone (4.A.4) and benzaldehyde.

6.D.45 BENZYLIDENE ACETONE (4-PHENYLBUT-3-EN-2-ONE), $C_6H_5CH=CHCOCH_3$ (BENZAL ACETONE)

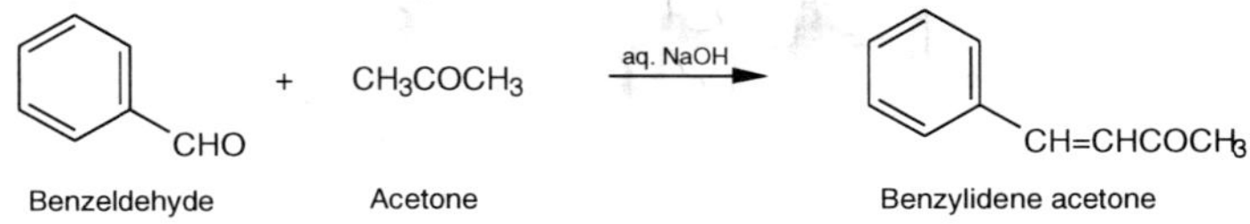

Sodium hydroxide solution (10%, 10 ml) is added dropwise to a vigorously stirred (mechanical stirring) solution of benzaldehyde (80 ml, 83.5 g, 0.787 mole) and pure acetone (80 ml, 63.28 g, 1.1 mole), keeping the temperature of the reaction mixture between 25-30°. The mixture is stirred for 2 hr at room temperature. Alternatively, the flask is stoppered and shaken for 2 hr. The mixture is acidified by the addition of dilute hydrochloric acid. It is transferred to a separatory funnel and the upper organic layer removed. The lower aqueous layer is extracted with toluene (2 × 20 ml). The combined upper layer and the toluene extract are washed with water (20 ml), dried (anhydrous magnesium sulphate) and distilled using a short fractionating column (4"). Toluene is removed by distillation at atmospheric pressure. The residual product is distilled under reduced pressure. Benzylidene acetone distills at 133-43°/16 mm. Yield 88.5 g (77.05%). It solidifies to a crystalline mass on standing. m.p. 38-9°. It is further purified by crystallization from petroleum ether. The pure product melts at 42° (lit.[71] m.p. 41°).

6.D.46 4'-METHOXYBENZYLIDENE ACETONE [4-(4'-METHOXYPHENYL)BUT-3-EN-2-ONE)] p-$OCH_3C_6H_4CH=CHCOCH_3$

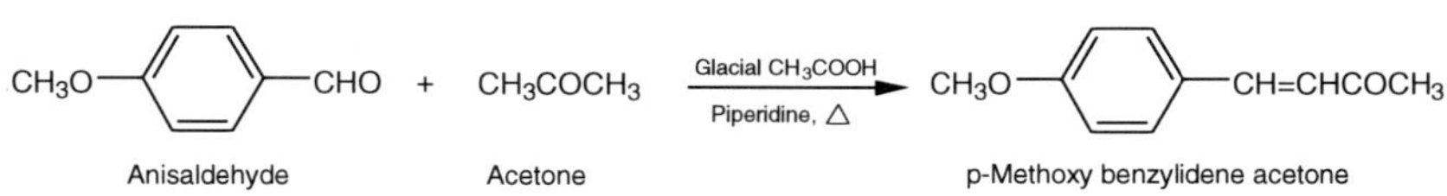

A mixture of dry acetone (75 ml), piperidine (5 ml, 4.3 g, 0.05 mole) and glacial acetic acid (5 ml, 5.25 g, 0.087 mole) is heated to reflux on a water bath. Anisaldehyde (25g, 22.3 ml, 0.18 mole) is then added to the refluxing solution over a period of 4 hr. The mixture is refluxed for 8 hr more. Excess acetone is distilled off and the sticky mass is macerated with crushed ice. The separated product is filtered, washed with water and dried. It is crystallised from benzene-petroleum ether as light yellow needles. Yield 24.1 g (90%), m.p. 72-4° (lit.[72] m.p. 72-4°).

3',4'-Dimethoxybenzylidene acetone [4-(3',4'-dimethoxyphenyl)but-3-en-2-one] m.p. 85-6° (lit.[72] m.p. 85-6°) is similarly obtained in 85% yield from acetone, piperdine and vetraldehyde.

4'-Methylbenzylideneacetone [4-(4'-methylphenyl)but-3-en-2-one] is similarly obtained as light yellow oil from acetone, piperidine and p-tolualdehyde in 80% yield.

6.E.47 1,2-NAPHTHOQUINONE

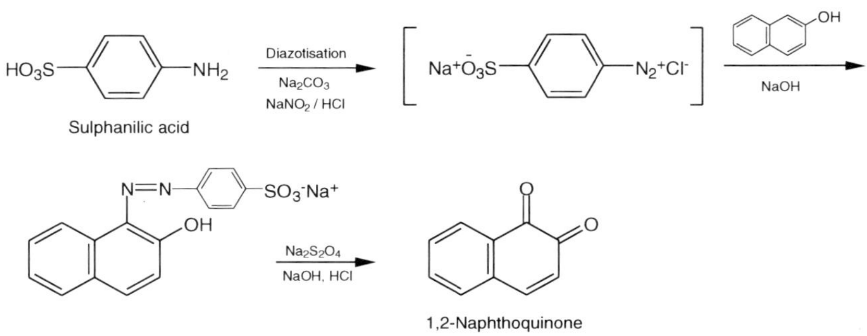

Step (i) β-Naphthol Orange (Orange II)

Sulphanilic acid dihydrate (10.5 g, 0.05 mole) and sodium carbonate (2.65 g, 0.025 mole) are dissolved in water (100 ml) by warming. The solution is cooled to 15° and a solution of sodium nitrite (3.7 g, 0.054 mole) in water (15 ml) is added slowly. The solution is poured into concentrated hydrochloric acid (10.5 ml) and crushed ice (60 g). The well mixed suspension of diazotised sulphanilic acid thus obtained is added with stirring at 5° to solution of β-naphthol (7.2 g, 0.05 mole) in sodium hydroxide solution (10%, 40 ml). The mixture is stirred for 15 min and then heated till all the separated product dissolves. Sodium chloride (20 g) is added and the solution warmed so as to dissolve the sodium chloride. The mixture is cooled in ice. The separated product is filtered and washed with saturated sodium chloride solution. It is crystallised by dissolving in minimum volume of boiling water, cooled to 80° and rectified spirit (approximately twice the volume of water) added. The solution is cooled. The separated pure dyestuff is filtered and dried in air. Yield 14 g(80%).

Note: The procedure described is a modification of literature method[73].

Step (ii) 1,2-Naphthoquinone

Sodium dithionite (25 g, 0.144 mole) is added with stirring to a solution of orange II (20 g, 0.057 mole) in water (250 ml) at 40-50°. The colour is discharged and precipitate of 1-amino-2-naphthol is obtained. The mixture is heated approximately to boiling and then cooled in ice. The separated product is filtered and washed with cold water. The precipitate is transferred to a beaker containing a solution of tin (II) chloride (0.25 g) in concentrated hydrochloric acid (5 ml) diluted with water (100 ml). The mixture is stirred, most of the amino phenol dissolves. The solution is stirred for 5 min with decolourising carbon (5 g) and filtered. To the clear solution is added concentrated hydrochloric acid (25 ml) and the solution warmed until the solid dissolves. The solution is cooled to 0°. The separated 1-amino-2-naphthol hydrochloride is filtered and washed with dilute hydrochloric acid (1:4, 25 ml).

A solution of iron (III) chloride (30 g) in a mixture of concentrated hydrochloric acid (10 ml) and water (25 ml) is prepared by heating. The solution is cooled quickly by addition of ice (30-40 g) and filtered. It is added in one lot to a clear solution of the above 1-amino-2-naphthol hydrochloride in water (150 ml) and a few drops of concentrated hydrochloric acid (this solution is obtained by dissolving and warming to 35-40° and filtration to remove any trace of insoluble material). The mixture is shaken vigorously. The separated crystalline yellow precipitate of 1,2-naphthoquinone is filtered and washed with water at 30° to remove all traces of acid. The product is dried in folds of filter papers. Yield 7 g (70.7%), M. P. 115-16° (d) (lit.[74] m.p. 115-20°(d)).

6.E.48 1,4-NAPHTHOQUINONE

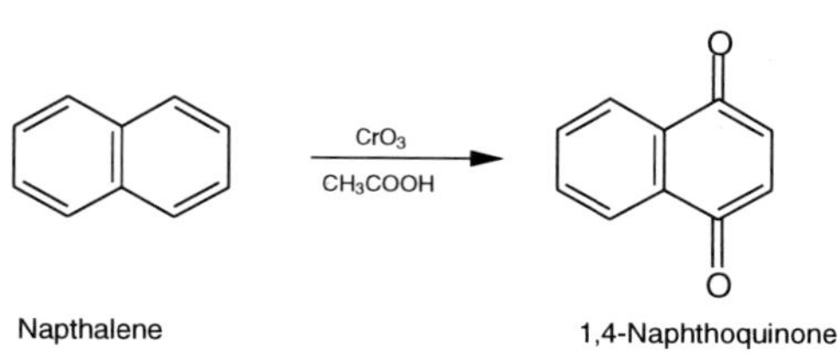

Chromium trioxide (40 g, 0.4 mole) is dissolved in a 1 litre three necked R.B. flask (fitted with a mechanical stirrer, a thermometer and a dropping funnel) in acetic acid (80% aqueous solution, 60 ml). The reaction mixture is cooled in freezing mixture. A solution of naphthalene (16 g, 0.125 mole) in glacial acetic acid (150 ml) is gradually added at 0°

with constant stirring during 2 hr. The temperature of the reaction mixture is maintained at 10-15°. The mixture is stirred overnight; during this time the cooling bath and the reaction mixture attain room temperature. It is allowed to stand for 70 hr with occasional stirring. The mixture is poured in water (1.5 litre) and the precipitated product is filtered and washed with water (75 ml) and dried in a desiccator. It is crystallized from petroleum ether (about 150-200 ml) as long yellow needles. Yield 4.25 g (21.5%), M.P. 124-25° (lit.[75] m.p. 124-25°).

6.E.49 2-HYDROXY-1,4-NAPHTHOQUINONE

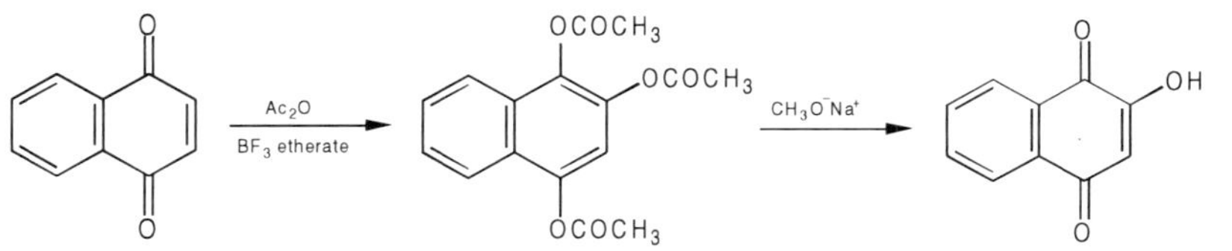

1,2,4-Triacetoxy naphthalene   2-Hydroxy-1,4-Naphthoquinone

A solution of pure 1,4-naphthoquinone (6.E.48) (32 g, 0.2 mole) in dry acetic anhydride (80 ml) is treated with freshly distilled boron trifluoride etherate (4 ml). The quinone goes into solution. After 2 hr crystals of triacetate separate out. At this point, the mixture is warmed gently to bring the rest of the quinone into solution. The solution is allowed to stand overnight at room temperature. The separated 1,2,4-triacetoxynapthalene is filtered and washed with little acetic acid and finally with alcohol. It is crystallised from alcohol as colourless needles. Yield 39 g (63.7%) M.P. 134-35° (lit.[76] m.p. 134-35°).

A suspension of the above 1,2,4-triacetoxynapthalene (45 g, 0.149 mole) and sodium methoxide (38 g) in methanol (375 ml) is stirred into a flask, cooled in ice bath. A thick paste is obtained. The salt is filtered and washed with methanol. It is dissolved in hot water (1 litre), solution filtered and acidified with dilute hydrochloric acid. The solution on cooling gives 2-hydroxy-1,4-naphthoquinone. It is filtered and crystallised from hot water as yellow needles. Yield 24 g (92.7%), M.P. 195-96° (lit.[76] m.p. 195-96°).

[On methylation in acetone with dimethyl sulphate (1.1 equivalent) in presence of ignited potassium carbonate (40 hr refluxing), the methyl ether, viz. 2-methoxy-1,4-naphthoquinone, m.p. 183-84° is obtained in 92% yield.]

6.E.50 5-HYDROXY-1,4-NAPHTHOQUINONE (JUNGLONE)

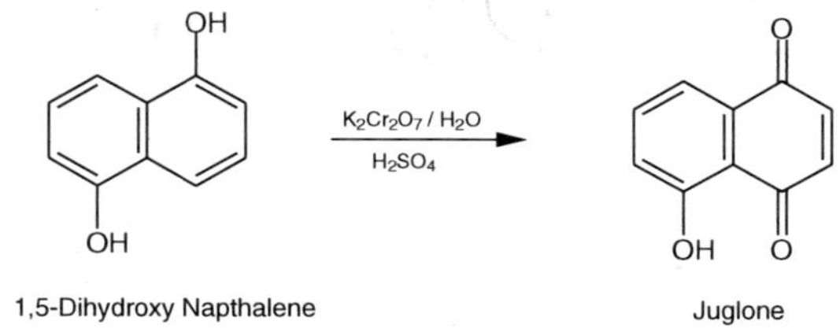

1,5-Dihydroxynaphthalene (10 g, 0.625 mole) is dissolved in water (200 ml). A solution of sodium dichromate (48 g, 0.16 mole), water (480 ml) and concentrated sulphuric acid (48 ml) is added to the above solution. The colour of the reaction mixture turns violet. The mixture is kept for 12 hr at room temperature with occasional shaking. It is filtered and the product washed with water and dried in air. The solid product is extracted with petroleum ether. The clear petroleum ether extract is evaporated. The solid product obtained is crystallised from alcohol as yellow needles. Yield 1 g (9.2%), M.P. 154° (lit.[77] m.p. 154°).

6.E.51 3-HYDROXYMETHYL-5-HYDROXY-1,4-NAPHTHOQUINONE

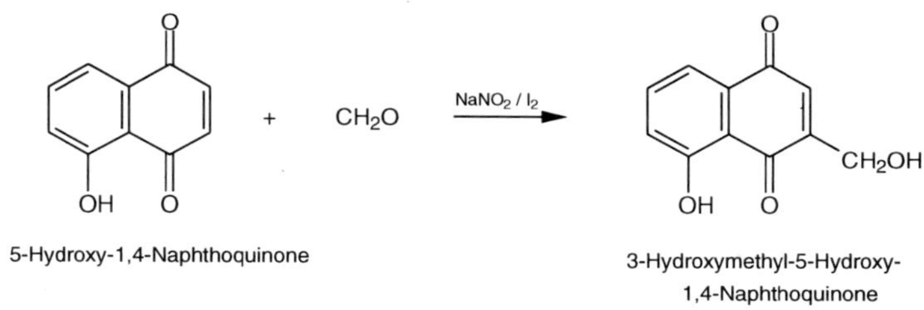

5-Hydroxy-1,4-naphthoquinone (6.E.50) (174 mg, 0.001 mole) is dissolved in aqueous formalin (30%, 20 ml, 6 g, 0.2 mole). To this solution is added sodium nitrite (69 mg, 0.001 mole) and iodine (10 mg). The reaction mixture is heated for 2 hr at 50-60°. The progress of the reaction is monitored on TLC. On disappearance of the starting compound, the solution is filtered and the filtrate is diluted with water. It is extracted with dichloromethane (3 × 20 ml). The organic layer is dried over anhydrous magnesium sulphate and distilled. The residue is chromatographed over silica gel (2 × 20 cm). Elution with benzene-ethyl acetate gives the required product. Yield 72 mg (35.3%). M.p. 115° (lit.[78] m.p. 115°).

6.E.52 2-METHYL-1,4-NAPHTHOQUINONE

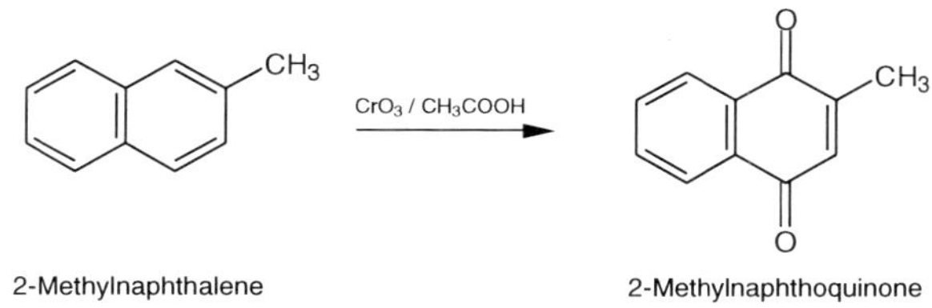

A solution of chromium trioxide (25 g, 0.25 mole) is prepared by dissolving it in water (20 ml); it is diluted with glacial acetic acid (20 ml) and then added in portions from a separatory funnel to a solution of 2-methylnapthalene (7.1 g, 0.05 mole) in glacial acetic acid (75 ml). During the addition, the solution is stirred and the temperature allowed to rise to 60°. The temperature of the reaction mixture is maintained at 60° throughout the addition which is completed in 10 min. The reaction mixture is then heated for 1 hr on a steam bath at 85-99° to complete the oxidation. It is cooled, diluted with water (300 ml) and stirred for few minutes. The separated product is filtered and washed with water. It is crystallised from methanol as yellow needles. Yield 3.5 g (40.7%), M.P. 104-106° (lit [79] m.p. 105-106°).

6.E.53 2-BROMO-1,4-NAPHTHOQUINONE

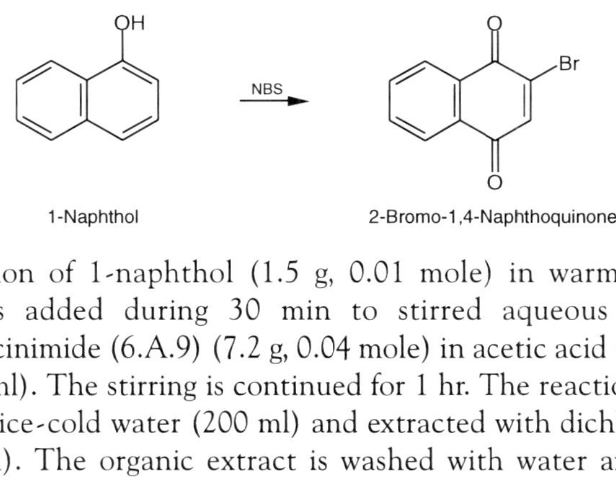

A solution of 1-naphthol (1.5 g, 0.01 mole) in warm acetic acid (100 ml) is added during 30 min to stirred aqueous solution of n-bromosuccinimide (6.A.9) (7.2 g, 0.04 mole) in acetic acid (200 ml) and water (100 ml). The stirring is continued for 1 hr. The reaction mixture is poured into ice-cold water (200 ml) and extracted with dichloromethane (4 × 100 ml). The organic extract is washed with water and saturated sodium bicarbonate solution (100 ml) and dried over anhydrous magnesium sulphate. Distillation of the solvent gives a solid product. It is crystallised from ether-petroleum ether as yellow needles, yield 0.7 g (28.2%), M.P. 131° (lit.[80] m.p. 130°).

6.E.54 2-CHLORO-1,4-NAPHTHOQUINONE

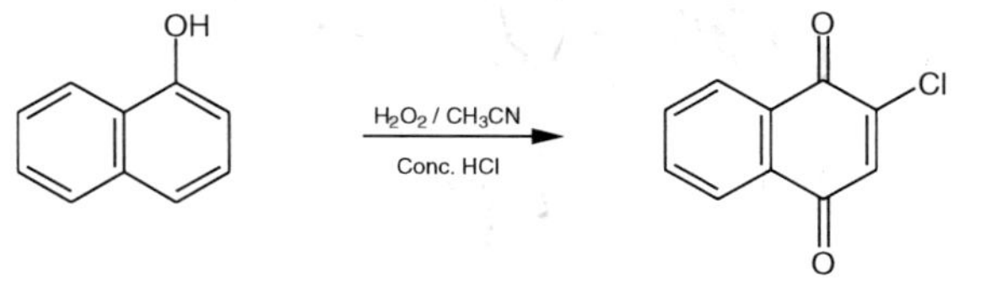

1-Naphthol — 2-Chloro-1,4-Naphthoquinone

Hydrogen peroxide (30%, 150 ml, 45 g, 1.3 mole) is added dropwise to a stirred solution (80°) of 1-naphthol (5 g, 0.0347 mole) in a mixture of acetonitrile (150 ml) and concentrated hydrochloric acid (10 N, 50 ml). The mixture is stirred for 2 hr more at 80°. It is filtered hot, the filtrate cooled to room temperature and water (200 ml) added. The separated product is filtered. Yield 3 g (44.8%), M.P. 112° (lit.[81] m.p. 112°).

6.E.55 2-BROMO-5-HYDROXY-1, 4-NAPTHOQUINONE [2-BROMOJUGLONE]

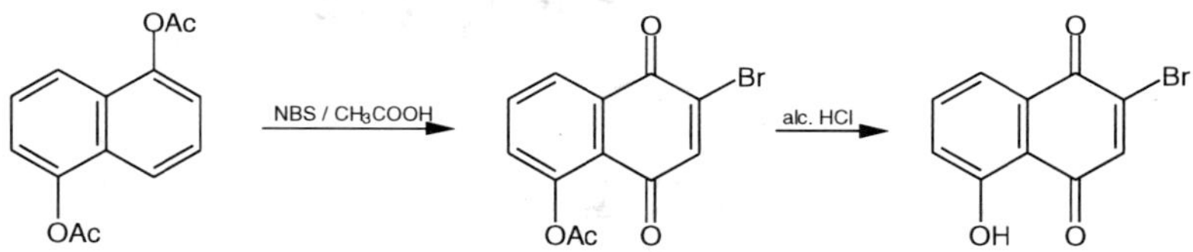

1,5-Diacetoxy naphthalene — 2-Bromo-5-acetoxy-1,4-naphthoquinone — 2-Bromo-5-Hydroxy-1,4-naphthoquinone

A solution of 1,5-diacetoxynaphthalene (1.22 g, 0.005 mole) in acetic acid (50 ml) is added during 5 min to a stirred solution of N-bromosuccinimide (6.A.9) (3.56 g, 0.02 mole) in acetic acid (50 ml) and water (100 ml). The solution is stirred at 55-60° for 45 min, diluted with water (100 ml) and extracted with dichloromethane (5 × 50 ml). The extract is washed with water, saturated sodium bicarbonate solution, dried (anhydrous magnesium sulphate) and distilled. The product obtained is 2-bromo-5-acetoxy-1,4-napthoquinone. Yield 1.13 g (76.8%) M.P. 154-55° (lit[82] m.p. 154.5-156°). It is hydrolysed with alcoholic hydrochloric acid (10%, 30 min heating on water bath) to give 2-bromojuglone. M.P. 136° (lit.[83] m.p. 136°).

6.E.56 3-CHLORO-5-HYDROXY-1, 4-NAPHTHOQUINONE (3-CHLOROJUGLONE)

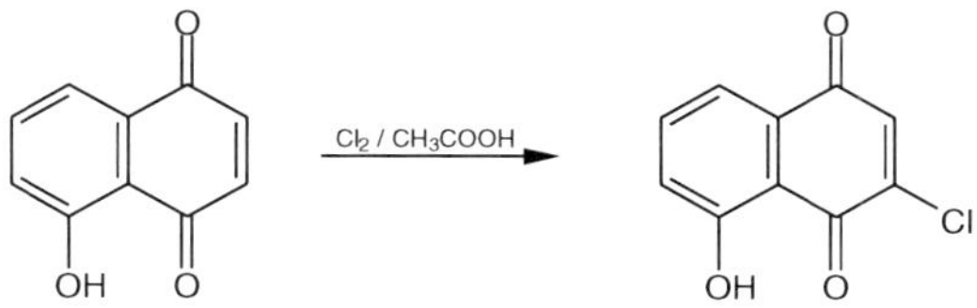

5-Hydroxy-1,4-naphthoquinone   3-Chloro-5-Hydroxy-1,4-naphthoquinone

Dry chlorine gas is passed through a solution of 5-hydroxy-1,4-naphthoquinone (6.E.50) (0.348 g, 0.002 mole) in cold acetic acid for 2-3 hr. The formed addition product is filtered and washed with cold water. It is refluxed for 20 min in ethanol to give 3-chlorojuglone. Yield 0.312 g (75%), M.P. 166° (lit.[83] m.p. 166°).

6.E.57 2-METHYLBENZOQUINONE

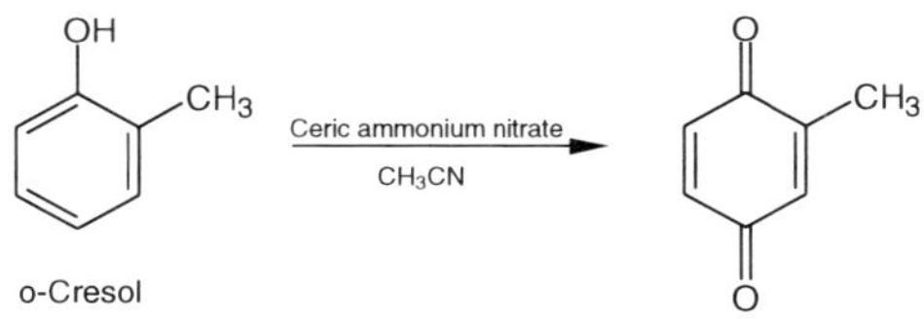

o-Cresol   2-Methylbenzoquinone

A solution of ceric ammonium nitrate (4.84 g, 0.061 mole) in water (240 ml) is added dropwise to a solution of o-cresol (4 g, 0.037 mole) in acetonitrile (200 ml). The reaction mixture is stirred for 12 hr at room temperature. It is poured into ice-water and extracted with dichloromethane (2 × 100 ml). The organic phase is dried over anhydrous magnesium sulphate and distilled in vacuo. A solid product is obtained. It is crystallised from benzene as yellow plates. Yield 2.7 g (60%), M.P. 67°, (lit.[84,85] m.p. 67-8°).

6.F.58 TRANS STILBENE (E)

A mixture of powdered benzoin (2 g, 0.0094 mole) and thionyl chloride (1 ml, 1.63 g, 0.0137 mole) is warmed gently on a steam bath till all the solid dissolves. It is heated in boiling water bath for 5 min. The excess thionyl chloride is removed in vacuo. The residual product is dissolved in ethanol (20 ml) and sodium borohydride (180 mg) added with stirring. The

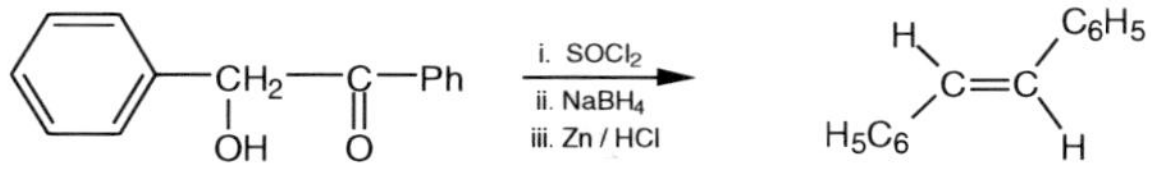

Bebzoin Trans Stilbene

mixture is stirred at 20° for 10 min. Zinc dust (1 g) and acetic acid (2 ml) are added and the reaction mixture refluxed for 1 hr. It is cooled, ether (25 ml) added and filtered. The residual zinc is washed with more ether (10 ml). The combined ethereal solution is washed with water containing hydrochloric acid, sodium carbonate solution (5%) and saturated sodium chloride solution. It is dried (anhydrous magnesium sulphate) and distilled. Yield 1.1 g (64.7%), M.P. 124-25° (lit.[86] m.p. 124-25°).

6.F.59 CIS-STILBENE (Z)

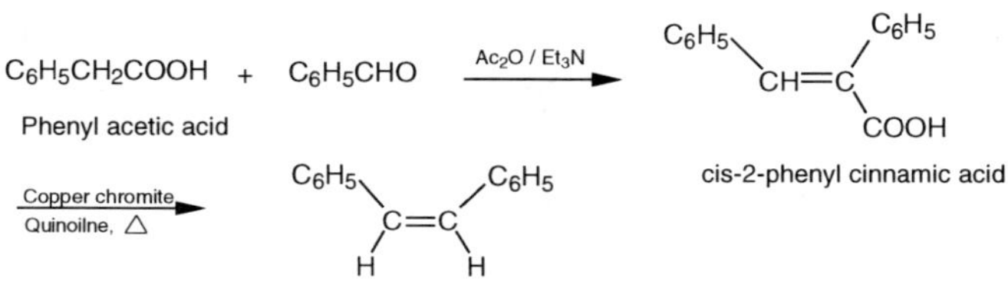

Cis Stilbene

Step (i) cis-α-phenylcinnamic Acid

A mixture of benzaldehyde (6 ml, 6.25 g, 0.059 mole) phenylacetic acid (5.5 g, 0.04 mole), acetic anhydride (13 ml) and tripropylamine (4 ml) is heated for 10 hr (steam bath). The reaction mixture is poured into dilute hydrochloric acid solution (10%). The separated solid is filtered. It is dissolved in benzene (70 ml) and the benzene solution extracted with sodium hydroxide solution (5%, 3 × 40 ml). The clear alkaline solution is acidified with hydrochloric acid. The separated product is filtered, washed with water and dried. It is crystallized from benzene-petroleum ether as colourless needles. Yield 4.5 g (50%), M.P. 172°, (lit.[87] m.p. 172°).

Note: In place of tripropylamine, triethylamine can be used[88] and the time of refluxing is reduced to 4 hr.

Step (ii) cis-stilbene (Z)

A mixture of cis- α-phenyl cinnamic acid (2.5 g, 0.011 mole), copper chromite catalyst (0.2 g) and quinoline (3 ml) is refluxed gently at 235-45° for 10 min. The mixture is cooled and ether (30 ml added). Decolourising charcoal (1 g) is added, mixture shaken and filtered; the residue is washed with ether (2 × 10 ml). The combined filtrate is extracted with dilute hydrochloric acid (5%, 4 × 20 ml) and then washed with saturated sodium chloride solution. The ether extract is dried over anhydrous magnesium sulphate and ether evaporated. The residual brownish oil (1.5 g, 75%) is crude cis-stilbene (lit.[86] m.p. 6°) containing a little trans-isomer formed by rearrangement during heating.

6.F.60 4-HYDROXY STILBENE[89]

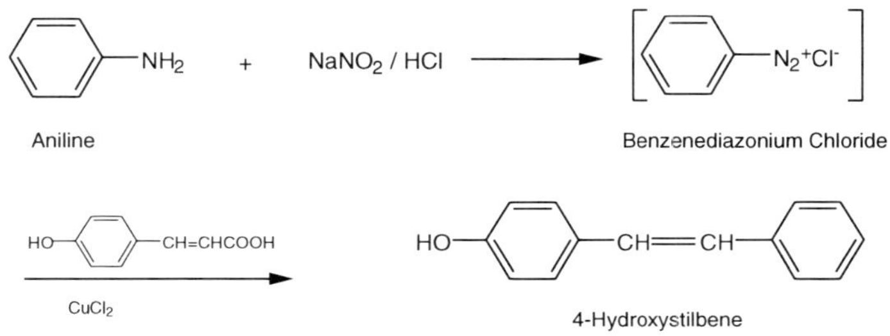

Aniline (4.7 g, 4.6 ml, 0.05 mole) is dissolved in concentrated hydrochloric acid (12.5 ml) by warming. The solution is cooled by addition of crushed ice (5-10 g) and diazotised by the dropwise addition of a cooled solution of sodium nitrite (3.5 g in 10 ml water) at 5-10°. The diazotised solution is filtered through glass wool. The clear solution is added dropwise to a mixture of p-coumaric acid (3.52) (8.2 g, 0.05mole), acetone (150 ml), sodium acetate (11.5 g) and cupric chloride (2 g dissolved in a small quantity of water). The reaction mixture is constantly stirred and gradually heated to 30° when the gas evolution started. It is kept at this temperature for 8 hr with stirring. It is allowed to stand for 2 days at room temperature. The green aqueous layer is decanted and the dark resinous mass is triturated with a small quantity of cold water. The separated product is filtered and washed with water. It is crystallised from methanol as colourless plates. Yield 4.6 g (46.9%), M.P. 184-85° (lit.[90] m.p. 184-85°).

Following stilbenes are similarly obtained using the above procedure.

4-Methoxystilbene, m.p. 140-41°

3-Methoxy-4-hydroxystilbene, m.p. 134-35°

3,4-Dimethoxystilbene, m.p. 111-12°

4,4'-Dimethoxystilbene, m.p. 208-209°

3,5-Dimethoxystilbene, m.p. 156-57°

6.F.61 3,4-DIMETHOXY STILBENE

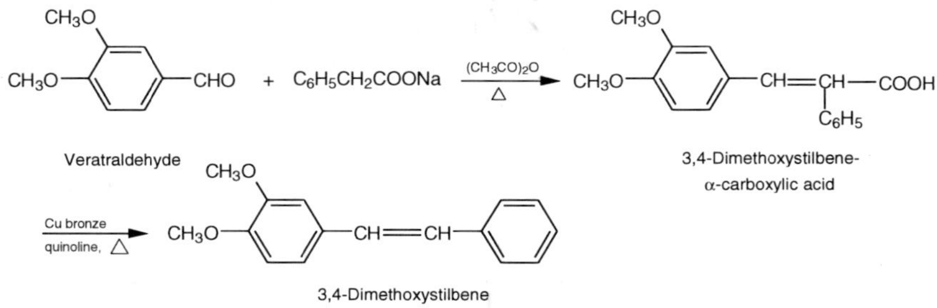

Step (i) 3,4-Dimethoxystilbene-α-carboxylic acid

$(3,4\text{-}(OCH_3)_2C_6H_3CH{=}C(COOH)C_6H_5)$

A mixture of freshly distilled veratraldehyde (2.11) (7 g, 0.042 mole), sodium phenyl acetate (5.A.22) (9 g, 0.057 mole) and acetic anhydride (35 ml) is heated in an oil bath at 150-160°. Excess acetic anhydride is removed under reduced pressure. The dark-brown mixture is poured on to crushed ice and mixture stirred. The separated product is filtered. It is purified by dissolving in sodium carbonate solution (5%) and acidifying the clear solution in cold. It is crystallised from acetic acid as long yellow needles. Yield 9 g (75.6%), M.P. 224 °.

Step (ii) 3,4-Dimethoxystilbene

The above acid (2 g, 0.007 mole) in freshly distilled quinoline (15 ml) containing copper bronze powder (1 g) is refluxed for 45 min till there is no more evolution of carbon dioxide gas. The mixture is cooled, diluted with ether (50 ml) and filtered. The ether layer is washed with dilute hydrochloric acid to remove all quinoline. The ether solution is distilled and the oily product is dissolved in dioxane (20 ml), a crystal of iodine added and the mixture is refluxed for 30 min. It is cooled, treated with dilute sodium thiosulphate solution (2 × 20 ml) to remove iodine. The dioxane is then removed under reduced pressure. The residue is dissolved in acetone and passed through a short column of alumina (2 × 10 cm). Evaporation of the acetone eluate gives a colourless product. It is crystallised from petroleum ether as wooly needles. Yield 0.45 g (26.6%), M.P. 111-12°.

6.F.62 o-CARBOXYSTILBENE (2-COOH $C_6H_4CH=CHC_6H_5$)

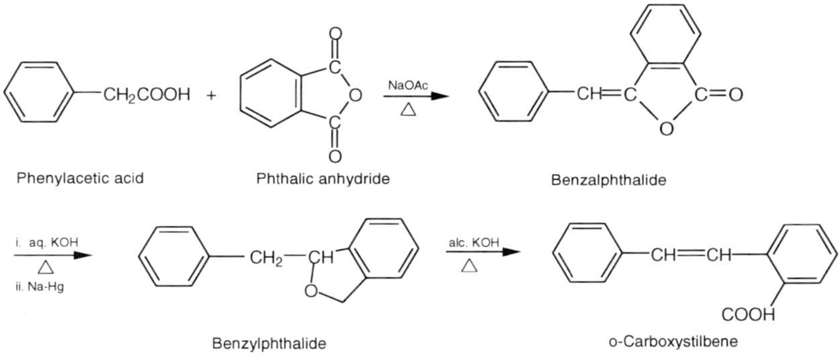

Step (i) Benzalphthalide

A mixture of freshly sublimed phthalic anhydride (25 g, 0.169 mole), phenylacetic acid (27.5 g, 0.2 mole) and freshly fused sodium acetate (0.65 g) is heated on a sand bath at 230° and then the temperature slowly raised to 240° during 2-3 hr; the water produced in the reaction distils over. The brown mass left behind is dissolved in boiling alcohol (100 ml), solution filtered and cooled. The separated yellow product is filtered, washed with cold water and dried. Yield 25g (66.7%), M.P. 98-9° (lit.[91] m.p. 98-9°).

Step (ii) Benzyl Phthalide

The above phthalide (11.5 g, 0.05 mole) is dissolved in a solution of potassium hydroxide (3 g) in water (5 ml) and heated. The dark red solution is diluted with water (100 ml) and poured on to sodium amalgam (prepared by adding 10.3 g mercury to 3.5 g molten sodium under mineral oil with stirring). The solution is vigorously stirred until the colour completely disappears (2 hr). Mercury is removed by decanting the solution. It is extracted with ether and the remaining solution is acidified with hydrochloric acid. The separated product is filtered and washed with water. It is crystallised from alcohol as colourless needles. Yield 10 g (80%), M.P. 61° (lit.[91] m.p. 61°).

Step (iii) o-Carboxystilbene

The benzyl phthalide (20 g, 0.083 mole) is dissolved in alcohol (40 ml) by warming. Potassium hydroxide (3 g) is added and mixture is gradually heated on a sand bath until all the alcohol is removed and the remaining brown residue begins to foam. The temperature is then raised to 180° and kept at this temperature until the foaming ceases (2-3 hr). The solid product is dissolved in water (75 ml), cooled and extracted with ether. The clear alkaline solution is acidified with hydrochloric acid and cooled. The separated product is filtered. It is crystallised from dilute alcohol as colourless needles. Yield 16.5 g (90%), M.P. 159-60° (lit.[91] m.p. 160°).

References

1. Young et al., J, Am. Chem. Soc., 1936, 58, 100.
2. Adkins and Cox, J. Am. Chem. Soc., 1938, 60, 1151.
3. R.R. Gupta, K.G. Ojha and M.Kumar, J. Heterocyclic Chem., 1980, 17, 1325.
4. H.C. Wallace and A. Roger, J. Am. Chem. Soc., 1924, 46, 1675.
5. J.M. Khurana and Sanjay Mehta, Indian J. Chem., 1988, 27B, 1128.
6. S. Gambarjan, Ber., 1909, 42, 4003; Dictionary of organic compounds by Chapman & Hall, 5th Edition , D-01366.
7. H. Staudinger, W. Kreis and W. Schilt, Helv. Chim. Acta, 1922, 5, 750.
8. T.L. Jacob and W.L. Petty, J. Org. Chem., 1963, 1363.
9. J. Allinger and N.L. Allinger, Tetrahedron, 1958, 2, 64.
10. K. Ziegler et al., Ann., 1942, 109, 551.

10 a. M.L. Semmehack (ed.) Organic synthesis, 1984, 62, 4, New York: Wiley.

11. R.P. Linstead and E.M. Meade, J. Chem. Soc., 1934, 940.
12. V.O. Isler, H. Gutmann, M. Montavon, R. Ruego, G.Ryse and P.Zeller, Helv. Chim. Acta, 1957, 40, 1242.
13. A.I. Vogel, J. Chem. Soc., 1948, 644.

13a. Willm, Ann., 1857, 102, 110; Dictionary of organic compounds by Chapman & Hall, 5th Edition, C-00665.

14. School Ber., 1896, 29, 2417; A. Sonn Ber., 1917, 50, 1267.
15. E.C. Horning and D. B. Reinser, J. Am. Chem. Soc., 1948, 70, 3619.
16. G.F. Hennion, J.J. Sheehan and D.E. Maloney, J. Am. Chem Soc., 1950, 72, 3544.
17. The procedure described is similar to that used for perhydrolysis of certain other acid chlorides (Y. Ogata and Y. Swak, Tetrahedron, 1967, 23, 3327.
18. M. E. Hanke, J. Am. Chem. Soc., 1923, 45, 1321.
19. G. A. Wiley, R. L. Hershkowitz, B. M. Rein and B. C. Chung, J. Am. Chem. Soc., 1964, 86, 964.
20. This is a general method for converting alcohols to alkyl halides. J.P. Schaefer and D.S. Weinberg, J. Org. Chem., 1965, 30, 2635.

21. J.V. Braun et al., Ber., 1918, 51, 82.
22. Dictionary of organic compounds by Chapman & Hall, 5th Edition, B-03209.
23. De Boer and Backer, Rec. Trav. Chim., 1954, 73, 229, 582.
24. A.M. Creighton and L.M. Jackman, J. Chem. Soc., 1960, 3138; Dictionary of organic compounds by Chapman & Hall, 5th Edition, D-04238.
24a. D. Walker and T.D. Waugh, J. Org. Chem., 1965, 30, 3240.
25a. Knotz, Osterr. Chem. Ztg., 1949, 50, 128; Inorganic Synthesis, 1953, 4, 71.
25b. J. F. Fletcher, J.C. Hamilton, I. Hechenbleikner, E.I. Hoigberg, B.J. Sertz and J. T. Cassaday, J. Am. Chem. Soc., 1948, 70, 3943.
25c. J. Omelanczuk, P. Kielbasinski, J. Michalski, J. Mikovajczyk, M. Mikovajczyk and A. Skowronska, Tetrahedron, 1975, 31. 2309.
25d. F. Gagiu and Al. Mavro, Bull. Soc. Chim Fr., 1967, 3, 1010; Chem. Abstr., 1967, 67, 82164b.
26. W.L. Truett and W. N. Moulton, J. Am. Chem. Soc., 1951, 73, 5913.
27. D.B. Denney and S. T. Ross, J. Org. Chem., 1962, 27, 998.
28. Heintz, Ann., Physik, 1860, 109, 331; 860, 111, 555; Sommelit, Ann. Chim. Phys., 1906, 9, 489.
29. D.D. Pratt and R. Robinson, J. Chem. Soc., 1923, 123, 752.
30. J. W. Farren, H. R. Fife, F. E. Clark and C. E. Garland, J. Am. Chem. Soc., 1925, 47, 2419.
31. Stephens, J. Soc. Chem. Ind., 1924, 43, 313T, 327T; Phelps and Tillotson, Am. J. Sci., 1908, 26, 264, 275.
32. Ladenburg and Wuchelhaus, Ann., 1869, 164, 152; A. Deutsch Ber., 1879, 12, 116.
33. P.L. Yadav, N.K. Jha and V. Ramakrishan, Ind. J. Chem., 1975, 13, 1095.
33a. A. Bowers, T. G. Halsall, E. R. H. Jones and A. J. Lenin, J. Chem. Soc., 1953, 2548.
34. J. Attenburrow, A.F.B. Cameron, J.H. Chapman, R. M. Evans, B.A. Hems, A.B.A. Jansen, and T. Walker, J. Chem. Soc., 1952,1094.
35. Weinland and Schmidit, Ber., 1905, 38, 2327.
36. E. Larson, Ber., 1930, 63, 1349.
37. Schmidt, Ber., 1919, 52, 409.
38. K.W. Rosenmund and G. Langer, Ber., 1923, 56, 2262.
39. Mannich and Thiele, Ber. Deut. Pharm. Ges., 1916, 26, 36.
40. Hartung, J. Am. Chem. Soc; 1928, 50, 3373; E.R. Alexander and A.C.Cope, J. Am. Chem. Soc., 1944, 66, 886.
41. Braun, J. Am. Chem. Soc., 1929, 51, 239; Dictionary of organic compounds by Chapmann & Hall, 5th Edition, P-00607

42. Fischer, Ann., 1878, 190, 79 ; Reychler, Ber., 1887, 20, 2463.
43. Milton W. Bullock and John J. Hand, J. Am. Chem. Soc., 1956, 78, 5854.
44. E. Bamberger et al., Ber., 1896, 29, 1834.
45. A. Ault, J. Chem. Educ., 1965, 42, 267.
46. Vogel, Text book of pratical organic chemistry, 4th Edition, Longman, p.727.
47. Dictionary of organic compounds by Chapman & Hall, 5th Edition, M-01019.
48. Dictionary of organic compounds by Chapman & Hall, 5th Edition, C-00734.
49. V.K.Ahluwalia et al., unpublished results.
50. A. N. Nesmejanow, Ber., 1929, 62, 1010.
51. Louis F. Fieser, Jour. Chem. Education, 1954, 31, 296.

51a. E.J. Corey and J.W. Suggs, Tetrahedron Lett, 1975, 2647

51b. Y.S. Cheng, W.L. Lur and S.H. Chen, Synthesis, 1980, 223.

52. A.A. Pavlic and H. Adkins, J. Am. Chem. Soc., 1946, 68, 1471.
53. H. Adkins and H.R. Billica, J. Am. Chem. Soc., 1948,70, 695 ; Murray and Raney, U.S Pat., 1, 628, 190 (Chem. Abstr., 1927, 21, 2116).

53a. S. Cacchi, synthesis, 1979, 356.

54. A. Albert and R. Royer, J, Chem. Soc., 1949, 1148.

54a. L. Ghosez, R. Montaigne, A. Roussel, H. Vanlierde and P. Mollet, Tetrahedron, 1971, 27, 615.

55. T.R. Seshadri and S. Varadarajan, Proc. Indian Acad. Sci., 1953, 37A, 145.
56. Dictionary of organic compounds by Chapman & Hall, 5th Edition, M-00665.
57. F.E. King and K.G. Neill, J. Chem. Soc., 1952, 4752.
58. J.M. Sehgal and T.R. Seshadri, Proc. Indian Acad. Sci., 1955, 42A, 36.
59. W. Cocker, T.B.H. McMurry and P.A. Staniland, J. Chem. Soc. 1965, 1034.
60. Mitter and Maitra, J. Indian Chem. Soc., 1936,13, 236.
61. Walter Kropp and H. Decker, Ber., 1909, 42, 1188.
62. G.A. Edwards, J. Chem. Soc., 1926, 744.
63. S.Takei, S. Miyajima and M. Ono, Ber., 1932, 65, 1047.
64. T.A. Govindachari, K. Nagarajan and P.C. Parthasarathy, J. Chem. Soc., 1957, 548.
65. S. Bhanu, Ph.D Thesis, Delhi University, 1972, p.108.

65a. R.L. Shriner and A.G. Schmidt, J. Am. Chem. Soc., 1929, 51, 3636.

66. S.K. Mukerjee, T. Saroja amd T.R. Seshadri, Indian J. Chemistry, 1969, 7, 671.

67. V.K. Ahluwalia and Sunita, Indian J. Chem., 1973, 11, 1241.
68. V.K. Ahluwalia and Sunita, Indian J. Chem., 1976, 14B, 682.
69. Kostanecki and Rossbach, Ber., 1896, 29, 1492.
70. Dictionary of organic compounds by Chapman & Hall, 5th Edition, H-03210.
71. Claisen and Ponder, Ann., 1884, 223, 138 ; L. Clasien, Ber., 1881, 14, 2468.
72. I.R. Dickson, I.M. Heilbron and F. Irring, J. Chem. Soc., 1927, 1888.
73. L.F. Fieser nad M. Fieser, J. Am . Chem. Soc., 1935, 57, 491.
74. C.E. Groves, J. Chem. Soc., 1873, 26, 209; 1884, 45, 294; Dictionary of organic compounds by Chapman & Hall , 5th Edition, N-00217.
75. Milller, J. Russ., Phys. Chem. Soc., 1884, 16, 414; F.R. Japp and N.H.J. Miller, J. Chem. Soc., 1881, 39, 220; Dictionary of organic compounds by Chapman & Hall, 5th Edition, N-00218.
76. L.F. Fieser, J. Am. Chem. Soc., 1948, 70, 3165.
77. A. Bernthsen and A. Semper, Ber., 1887, 20, 934.
78. R.N. Khanna et al, unpublished work.
79. L.F. Fieser, J. Biol. Chem., 1940, 133, 391.
80. Th. Zincke and M. Schmidt, Ber., 1894, 27, 2758.
81. P. Kr. Singh and R.N. Khanna, Synthetic Communications, 1993, 23, 2083.
82. S.W. Heinzman and J.R. Grunwell, Tet. Lett., 1980, 4305.
83. R.H. Thommson, J. Org. Chem., 1948, 13, 377.
84. Prem pal Singh, Ph.D. Thesis, Delhi University, 1990, p.140.
85. E. Nolting and T. Baumann, Ber., 1885, 18, 1150.
86. L.F. Fieser, J. Chem. Educ., 1954, 31, 291.
87. R.E. Buckles & E.A. Hausman, J. Am. Chem. Soc., 1948, 70, 415.
88. R.E. Buckles, J. Chem. Edn., 1950, 27, 210.
89. P.L. Sawhney, Ph.D. Thesis, Punjab University, 1959, p. 98.
90. Hewitt and Lewcock and Pope, J. Chem. Soc., 1912, 101, 605.
91. S. Natelson and S.P. Gottfried, J. Am. Chem. Soc., 1936, 58, 1432.